T0293841

Continuum Mechanics Modeling of Material Behavior

Continuum Mechanics Modeling of Material Behavior

Martin H. Sadd

Academic Press is an imprint of Elsevier
125 London Wall, London EC2Y 5AS, United Kingdom
525 B Street, Suite 1800, San Diego, CA 92101-4495, United States
50 Hampshire Street, 5th Floor, Cambridge, MA 02139, United States
The Boulevard, Langford Lane, Kidlington, Oxford OX5 1GB, United Kingdom

Library of Congress Cataloging-in-Publication Data
A catalog record for this book is available from the Library of Congress

British Library Cataloguing-in-Publication Data
A catalogue record for this book is available from the British Library

ISBN: 978-0-12-811474-2

For information on all Academic Press publications visit our website at https://www.elsevier.com/books-and-journals

Publisher: Katey Birtcher
Senior Acquisition Editor: Steven Merken
Content Development Specialist: Nate McFadden
Production Project Manager: Mohana Natarajan
Designer: Vicky Pearson Esser

Typeset by Thomson Digital

Contents

Preface

Continuum mechanics is a broad advanced study of material behavior based on the assumption that matter is continuously distributed in space. Initially such studies were limited to common solids and fluids that were present in engineering and scientific applications of the day. These classical studies went under the names *elasticity, plasticity, viscoelasticity, fluid mechanics, rheology, etc*. After a rebirth in the 1960's-70's, continuum mechanics began focusing on much more general and complicated nonlinear material behavior employing higher level mathematics. Advances in the field have continued into such areas as *multi-phase composites, graded materials, granular substances, cellular and porous solids, materials with microstructure, coupled multi-field problems, biological materials, etc.*

The mechanics of continuous media is often thought of as a somewhat *special* course which is more general and theoretical than others found in engineering curricula. Normally the study is broad in nature, and seeks to capture far-reaching principles that set the foundations for more specific applications in the mechanics of solids and fluids. Such an approach requires the use of tensor notation and other more advanced mathematical tools. As reflected in the literature, the community of scholars has generally agreed that continuum mechanics texts should include:

1. Tensor algebra and calculus, field theory, and related mathematical principles
2. Kinematics, deformation and rate of deformation measures
3. Force and stress descriptions
4. Conservation or balance principles
5. Constitutive equation development
6. Formulation and example solutions of particular theories

Items 1-4 are classical topics, and many existing continuum mechanics texts do an adequate job with these. However, it is with items 5 and 6 that the author feels the need for improved presentation and hence is the primary motivation for creating the current text. In this regard extensive coverage is provided in the development of a very broad class of constitutive relations in chapters 6–9. For a single semester course, coverage of all of these different constitutive theories would likely be overly ambitious. However, particular theories within these chapters can be selected for class presentation/discussion as per instructor or student interest, while other constitutive theories can be assigned for student term paper projects. The author has used this scheme many times with good success. In this fashion students gain a broad perspective of the vast range of continuum mechanics applications especially in contemporary areas of micromechanical modeling. Since much of graduate engineering classroom education is designed to help prepare students to conduct their program research, it is hoped that this broad coverage will aid in this effort.

Much of this material is an outgrowth of the author's class notes on the subject, coming from teaching numerous courses in continuum mechanics over the past several decades. The material represents a textbook for use in a beginning graduate level course on the subject. As such, emphasis is placed on organization, clarity and breadth, rather than mathematical rigor and highly specialized theoretical aspects. The text is certainly mathematical, but certain theoretical concepts are often avoided so as to not lose student interest. Examples are generously distributed in all chapters to provide extensive applications of the theoretical concepts and to illustrate how various theories are applied within the study.

Following the author's previous elasticity text, this book integrates the use of numerics by employing MATLAB software. This greatly aids in presenting and illustrating applications of particular theory through calculation and graphical display. Although other options (e.g. Mathematica) could also be used, the author has found MATLAB to be well suited for such application and is a very popular engineering software package. This software is used in many places for applications such as: tensor transformation operations, calculation of invariants and principal values and directions, two-dimensional deformation plotting, and calculation and plotting various specific results from example application problems. With numerical and graphical evaluations, application problems become more interesting and useful for student learning. Many of the MATLAB codes are listed in Appendix C and can also be accessed through the text's companion web site. This allows both instructors and students to easily integrate the numerics into homework exercises or to pursue further application cases of their own interest.

Contents Summary

Chapter 1 provides several introductory topics that the author finds necessary to discuss before delving into the heart of the subject matter. First, an introduction to materials, the continuum hypothesis, and length scale concepts are given. The need for tensor representation is then provided along with a brief introduction to the objectivity concept. A summary of the structure of continuum mechanics is given to provide a general overall picture of the subject, and the chapter ends with a brief historical summary to illustrate how we got to now.

Chapter 2 presents much of the mathematical tools and notation necessary for a modern treatment of continuum mechanics. Although some students may have seen some of these topics, the author feels that a reasonably comprehensive presentation will best serve the general student audience. Tensor analysis is one of the key features of this chapter, and both Cartesian and general tensors are presented in detail. Both index and direct notation are employed. Related topics of Cayley-Hamilton theorem, matrix polynomials and representation theorems for isotropic functions are also covered. The principle of objectivity or frame invariance is initially introduced in this chapter. Most all of these topics are to be used in later sections of the text.

Some use of MATLAB is incorporated to evaluate particular tensor algebra and calculus applications. The author feels that this material is best done in the beginning of the study; however, depending on the mathematical background of the students, some of this material could be skipped over and introduced later in the text at points of application.

Chapter 3 discusses motion, kinematics and deformation of continuous media. Lagrangian and Eulerian descriptions are presented, and many different strain and strain rate tensors are established for both large and small deformation. Strain compatibility is explored, and the concept of objectivity is investigated for various strain and rate of strain tensors. Formulation using the current configuration as reference and Rivlin-Ericksen tensors are also developed for later use. Finally various deformation measures using curvilinear cylindrical and spherical coordinates are presented. MATLAB is used to illustrate some particular two-dimensional deformation examples.

Chapter 4 addresses external and internal forces and various stress measures. The stress or traction vector is introduced and this leads to the definition of the Cauchy stress tensor. Definitions of principal, spherical, deviatoric, octahedral and vonMises stress are made. Piola-Kirchhoff stress tensors are defined and compared to Cauchy stress. Evaluation of objectivity of the various stress tensors is presented. Finally stress tensor components using curvilinear cylindrical and spherical coordinates are established. Use of MATLAB is again incorporated to evaluate and plot various stress tensor components to illustrate the field nature of such variables.

Chapter 5 presents the development of conservation or balance principles of mass, momentum and energy that are used in continuum mechanics. Both integral and differential forms are developed. In regard to energy, new thermodynamic field variables are introduced, and both the first and second laws of thermodynamics are discussed. The chapter ends with a summary of previously developed general relations and a listing of the associated unknown field variables.

Chapter 6 begins the presentation of constitutive equation development but limits the discussion to only the classical linear theories of elasticity, fluid mechanics, viscoelasticity and plasticity. The aim of this chapter is to explore each of the specific constitutive relations, develop the complete field equations and boundary conditions, and finally to present several closed-form analytical solutions to several problems of interest. This chapter sets the stage for more complex constitutive models in the following three chapters.

Chapter 7 explores constitutive theories and formulations for continuum problems that contain more than a single independent field behavior. Examples presented combine linear elastic deformation with thermal, fluid saturated and electro behaviors; leading to the formulation of thermoelasticity, poroelasticity and electroelasticity. Within each theory new constitutive relations are coupled with linear elasticity to develop the combined material model. The solution to many example problems are provided to demonstrate how the multiple fields produce coupled response.

Chapter 8 expands constitutive modeling into nonlinear behavior for both solids and fluids. Finite deformation theory is now used along with much more general constitutive principles including material frame indifference. Noll's general theory of simple materials forms the starting point in the presentation. Non-linear elasticity, non-linear viscous fluids and non-linear viscoelastic material models are discussed in detail. Numerous examples of these cases are provided.

Chapter 9 presents several different constitutive models that incorporate material microstructure. The discussion begins with concepts of the representative volume element, homogenization and length scales. The specific chosen material models include: micropolar elasticity, elasticity with voids, doublet mechanics, higher gradient elasticity, fabric tensor theories and damage mechanics. This collection is based on the elastic response, and this allows a relatively easy introduction to this type of important and contemporary modeling. Several example problems using these extended theories provide comparisons with the classical predictions to illustrate particular aspects of the newer models.

In general the text includes many worked-out examples to demonstrate the theory, and numerous exercises are given at the end of each chapter for student engagement and may also be used for class presentations. Appendices A and B offer a convenient summary of basic field equations and transformation relations between Cartesian, cylindrical and spherical coordinate systems. Appendix C provides a listing of many MATLAB codes used throughout the text. A humorous poem written by the author's doctoral advisor ends the text in Appendix D.

Web Support

The companion and instructor sites for this text can be accessed at https://www.elsevier.com/books/continuum-mechanics-modeling-of-material-behavior/sadd/978-0-12-811474-2. The companion site includes downloadable MATLAB codes listed in Appendix C. These codes will aid both students and instructors in developing codes for their own particular use and thus allow easy integration of the numerics. Instructors who register will be able to access PowerPoint lecture slides and solutions to the exercises featured in the text. Errata (when available) will be available at the companion site.

Feedback

The author is committed to continual improvement of engineering education and welcomes feedback from users of this book. Please feel free to send comments concerning suggested improvements or corrections via email (msadd@cox.net). It is likely that such feedback will be shared with the text's user community via the publisher's website.

Acknowledgements

Several individuals deserve acknowledgement for aiding in the development of this textbook. I would first recognize the many graduate students who have sat my continuum mechanics classes. They have been a repeated source of challenge and inspiration, and certainly influenced my efforts to find more effective ways to present this material.

I would also like to acknowledge the support of my institution, the University of Rhode Island, for providing time, resources and the intellectual climate that assisted my pursuit of this writing project. A special thank you to Prathmesh Naik Parrikar for helping with development of the Solutions Manual. Support from various members of the Elsevier editorial and production staff is greatly appreciated. An additional thank you goes to my wife Eve for her patience during my many months of disappearance and writing.

This book is dedicated to the professors who taught me this beautiful and challenging subject: Don Carlson, University of Illinois; Roger Fosdick, Raja Huilgol and Barry Bernstein, Illinois Institute of Technology. They all greatly stimulated my interest in continuum mechanics and the fascinating mathematics associated with the subject.

Martin H. Sadd

About the Author

Martin H. Sadd is Professor Emeritus of Mechanical Engineering at the University of Rhode Island. He received his Ph.D. in mechanics from the Illinois Institute of Technology in 1971 and then began his academic career at Mississippi State University. In 1979 he joined the faculty at Rhode Island and served as department chair from 1991–2000. He is a member of Phi Kappa Phi, Pi Tau Sigma, Tau Beta Pi, Sigma Xi, and is a Fellow of ASME. Professor Sadd's teaching background is in the area of solid mechanics with emphasis in elasticity, continuum mechanics, wave propagation, and computational methods. He has taught numerous courses in these fields at three academic institutions, several industries, and at a government laboratory.

Dr. Sadd's research has been in the area of analytical and computational modeling of materials under static and dynamic loading conditions. His recent work has involved micromechanical modeling of geomaterials including granular soil, rock, and concretes. He has authored over 75 publications and has given numerous presentations at national and international meetings. He is the author of *Elasticity: Theory, Applications and Numerics* (Third Ed.).

CHAPTER

Introduction

1

Continuum mechanics or mechanics of continuous media seeks to develop predictive mathematical models of material behavior relating the applied forces (mechanical and other types) to the material deformation and motion. In this chapter, we begin our study with a brief discussion on a few fundamental issues related to the theories pursued. Some basic material issues along with the continuum hypothesis will be discussed. An explanation for the need of tensors is given in an effort to justify our study and use of this important mathematical formalism. A brief overview of the general structure of continuum mechanics is given to provide an overall perspective of the field that we will explore in detail. The chapter ends with a little history outlining the major developments of this broad and encompassing field of study.

1.1 MATERIALS AND THE CONTINUUM HYPOTHESIS

Continuum mechanics is concerned with general modeling of material behavior that has been commonly observed by the scientific and engineering community. The study will be conducted using *Euclidian three-dimensional space*, and all phenomena will be governed by *Newtonian mechanics*. In pursuing such a study, we should first consider a few fundamental questions such as:

- How do we apply a continuum theory to real materials which are fundamentally not continuous?
- What types of materials are we interested in studying?
- What loading conditions are we concerned in applying to materials?
- What behaviors are we most interested in modeling?
- What are the length and time scales of the problem?

Let us now explore some of these questions and discuss possible answers and their consequences.

First, consider the issues of length and time scales of the problems we might be interested in modeling. In some cases, we may wish to study problems with very small length and time scales, whereas for other studies these scales may be larger by many orders. Terminology for various length scales are commonly referred to as nano (10^{-9} m), micro (10^{-6} m), and macro (10^{-3} to 10^{3} m, visible to the eye). For example, we might be interested in studying the behavior of small structural parts with millimeter length scales or in another problem trying to simulate the geomechanical response of earth layers with kilometer scales. Both length and time scales can have significant impact on

Continuum Mechanics Modeling of Material Behavior. http://dx.doi.org/10.1016/B978-0-12-811474-2.00001-0

the choice of the modeling irrespective of whether it be analytical or numerical. As we shall see, many *classical* continuum models of solids and fluids have no built-in length or time scales and thus can mathematically be used for problems of any size. However, more recent and sophisticated continuum models have been developed with one or more internal length scales in an effort to more accurately model micromechanical behaviors of heterogeneous materials. These issues will be further explored in Chapter 9.

Next, consider the fact that all real materials are not spatially continuously distributed. Actual materials are composed of extremely large numbers of atoms and molecules discretely distributed in space. For example, it has been estimated that there are about 10^{22} atoms in a cubic centimeter of many common solids, and a similar number would also estimate the number of molecules per cubic centimeter of water. Typically atoms and molecules are separated by very small distances on the order of 10^{-10} m. So, if we are interested in very small length scales on the order of 10^{-10} m, we may wish to use a modeling scheme that takes into account atomic quantum mechanics and/or discrete molecular theories. Such modeling has been done (Massobrio et al., 2012; Jensen, 2010), but it does not use traditional continuum mechanics, and the results are normally limited to nano length and time scales. In contrast, continuum mechanics is commonly used for predicting *macroscopic* material behavior over length scales of 10^{-6} to 10^{3} m and times scales of 10^{-6} to 10^{6} s. This region is where most engineering applications occur. There are also heterogeneous and composite materials that have considerable microstructures within our length scale of interest (10^{-6} to10^{3} m), thus requiring a modified continuum mechanics modeling approach. We will eventually look at such cases later in the text.

In order to explore some of these ideas further, consider the following two-dimensional example of a particulate reinforced composite material shown in Fig. 1.1. We simplify the problem by assuming that the particles are all square (dimensions $a \times a$) and equally spaced by the common size dimension a. The particles are embedded in a continuum matrix that holds everything together. The figure sequence (A)–(C) illustrates the regions of increasing particle numbers which also translates to decreasing relative length scale. It should be visually apparent that as we increase the number of

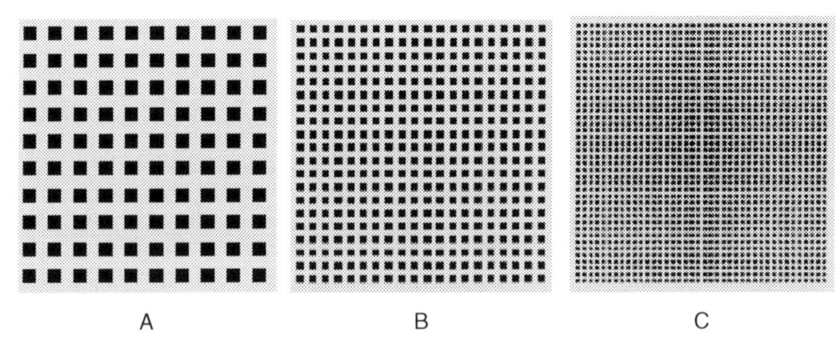

FIGURE 1.1

Example particulate composite samples: (A) 10×10 distribution; (B) 20×20 distribution; (C) 40×40 distribution.

particles in the sample, the overall material seems to become more of a continuum. Thus, while the 10×10 sample shown in Fig. 1.1A might warrant a detailed micromechanical particulate model, the 40×40 sample shown in Fig. 1.1C could be modeled as a continuum with appropriate averaged properties. This idea can be further explored by considering the mass density of the particulate sample as a function of the sample size. To make calculations easier, consider the mass density calculation based on an imaginary square area of increasing size. Starting on a single particle with an $a \times a$ dimension, the area is incrementally increased in each coordinate direction by a. Neglecting the density of the matrix material, the sequential density calculation (mass/area) can easily be made. A plot of the relative density (normalized by the particle density) versus the area dimension (normalized by a) is shown in Fig. 1.2. We observe significant fluctuations in the density for smaller sample areas, but as the area increases, the density variation diminishes and the density itself approaches a constant value (0.25 for this idealized case). Thus, we make the conclusion that as the relative sample size increases (relative length scale decreases), the local variation of density disappears and the material may be thought of as a continuum. The region where this variation becomes negligible is sometimes called the *minimum homogenization volume* or *representative volume element* or *representative elementary volume*. This concept can be extended to atomic, molecular, and other micromechanical mass distributions. We will have more to say about these topics later in the text.

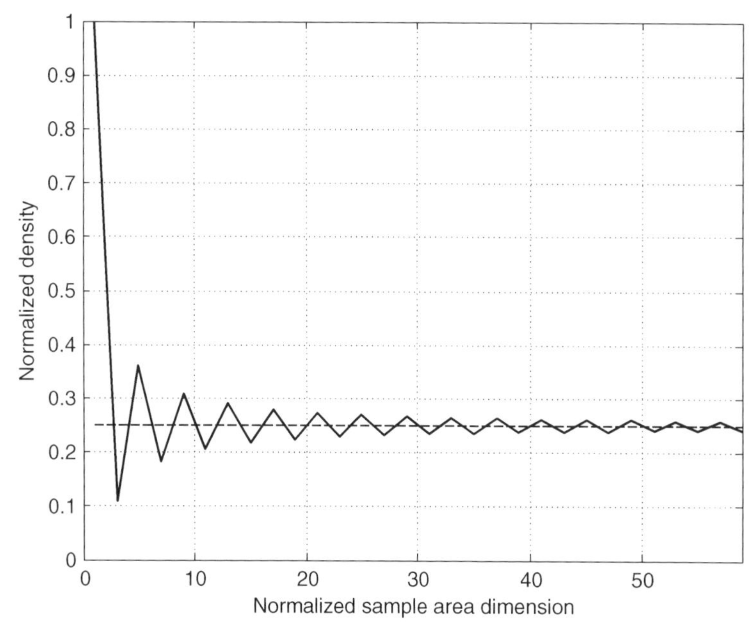

FIGURE 1.2 Relative mass density variation with sample size for particulate composite sample shown in Fig. 1.1.

With regard to the question on material types and behaviors, we are interested in modeling solids, fluids, and other materials that have both solid- and fluid-like behaviors together. Studies on solid materials will include both elastic and inelastic responses, whereas fluids will comprise Newtonian and non-Newtonian behaviors. Materials exhibiting both solid- and fluid-like responses will be included in various viscoelastic theories. We wish to establish models that can predict both mechanical and thermomechanical behaviors and this will require the use of the energy equation and second law of thermodynamics. Electromechanical response of linear elastic solids will also be briefly discussed. Loading conditions will then include mechanical, thermal, and electro types. We will also be exploring a few newer continuum theories that have applications to materials with internal microstructure. Our general study will normally include static and dynamic formulations, but detailed applications will generally emphasize static and quasi-static problems. Several of the terms just used have not been properly defined and this will be handled in subsequent chapters.

To complete our discussion in this section, let us refine our concept of the continuum hypothesis. Although most of us have an intuitive idea of a continuum, our scientific study requires we formulate a mathematical basis of this concept. We say that the real number system is a continuum because between any two distinct real numbers, there are an infinite number of other real numbers. Intuitively, we feel that time and space can be represented by real numbers and thus we identify time and space as a multidimensional continuum.

We can extend this concept of continuum to matter and speak of a continuous distribution of matter in space. This idea is illustrated again by considering the concept of *mass density*. If we let the amount of matter be measured by its mass, then consider a certain matter which permeates a particular region of space $\mathcal{V}_0$. Let us consider a point P in $\mathcal{V}_0$ and a sequence of subspaces $\mathcal{V}_0, \mathcal{V}_1, \mathcal{V}_2, \ldots$ converging on P as shown in Fig. 1.3. Let V_n be the volume of $\mathcal{V}_n$ and let M_n be the mass of $\mathcal{V}_n$. Then we define the mass density at the point P by

$$\rho(P) = \lim_{\substack{n\to\infty \\ V_n\to\infty}} \frac{M_n}{V_n} \tag{1.1.1}$$

If the mass density is well defined everywhere (at all points P) in $\mathcal{V}_0$, the mass is said to be continuously distributed in $\mathcal{V}_0$.

A similar scheme can be used to define densities of *force*, *linear and angular momentum*, *energy*, and other variables necessary to model the material. We then can say that a material continuum is one for which all density functions and modeling variables continuously exist in the mathematical sense. The *mechanics* of such materials is *continuum mechanics*. As we shall see, this type of theory is very useful in describing the gross or macroscopic behavior of a large variety of real materials.

Continuum mechanics generally ignores the fine details of the material's microstructure and replaces the discontinuous (likely nano- or microscopic) medium with a hypothetical or model continuum. The resulting model then describes the material behavior under the study using *field quantities* such as *displacements*, *velocities*,

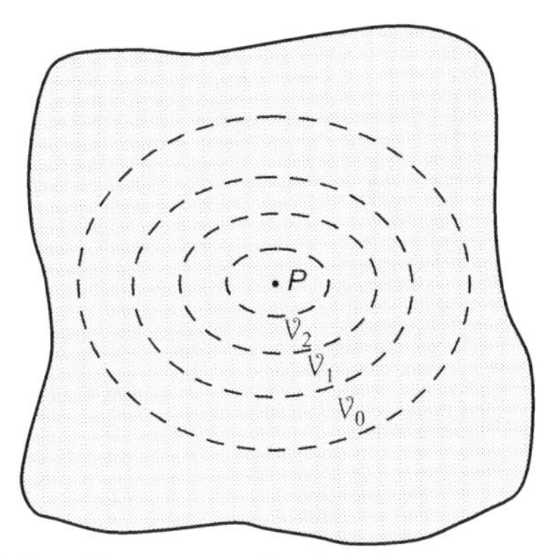

FIGURE 1.3 The concept of continuum mass density.

stresses, etc. which are *piecewise continuous*. The usefulness of continuum mechanics, however, can also be extended to cases which include inhomogeneity and discontinuity such as those used in micromechanical modeling.

1.2 NEED FOR TENSORS

Continuum mechanics is formulated in terms of many different types of variables: *scalars*, represented by a single value at each point in space (e.g. material density); *vectors*, expressible in terms of three components in a three-dimensional space (e.g. material displacement or velocity); *matrix variables*, which commonly require nine components to quantify (e.g. stress or strain). Other applications incorporate additional quantities that require even more components to characterize. Because of this complexity, continuum mechanics makes use of a *tensor formalism* which enables efficient representation of all variables and governing equations using a single standardized scheme. The tensor concept will be defined more precisely in Chapter 2, but for now we can simply say that scalars, vectors, matrices, and other higher-order variables can all be represented by tensors of various orders.

Another important point is that in order to develop a set of general laws and principles, the fundamental relations in continuum mechanics must be formulated in terms of quantities *that are independent of the coordinate frame used to describe the problem*. Thus, if two individuals using different coordinate frames observe a common physical event (see Fig. 1.4), it should always be possible to state a physical law governing the event, such that if the law is true for one observer, it will also be true for the other (once adjusted for the difference in coordinate frame). This concept is normally referred to as the principle of *objectivity*, *frame indifference*, or *isotropy of space*. This situation will require all relations to be written in an appropriate *tensor format* so as to guarantee the proper invariance. Thus, the form of any field equation

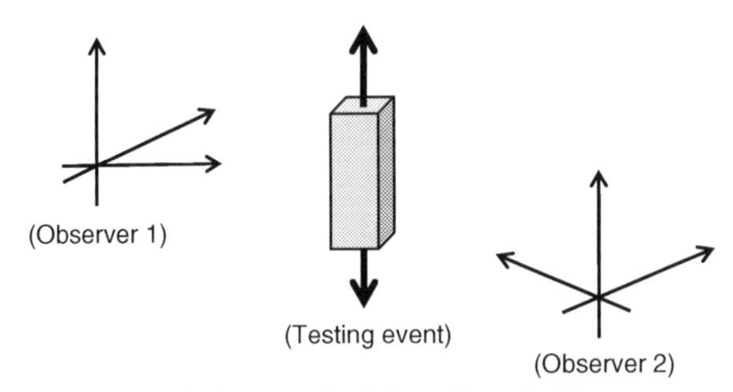

FIGURE 1.4 The concept of principle of objectivity or frame indifference.

can have general validity in any reference frame only if every term in the equation has the same tensorial characteristics. If this condition is not satisfied, a simple change in the coordinates will destroy the form of the relationship, thus indicating that the original form was only fortuitous and/or accidental.

For these reasons, use of tensors is essential for the formulation and solution of field equations in continuum mechanics. Consequently, the knowledge of tensor theory is required for this study, and we will devote appropriate detailed coverage of this topic in Chapter 2. In his text, Fung (1994) makes the insightful and poignant statement with regard to tensors: "A beautiful story needs a beautiful language to tell. Tensor is the language of mechanics."

1.3 STRUCTURE OF THE STUDY

Continuum mechanics may be generally considered as a four-part structure leading to a final model of particular materials as shown in Fig. 1.5. Kinematics deals with the geometrical relations between material motion, strain, displacement, and various rate-dependent variables. These topics will be covered in detail in Chapter 3. Additional fundamental relations include the concepts of force, traction, and stress at continuum points, and these are presented in Chapter 4. General balance principles involve the continuum interpretation of conservation of mass, momentum, and energy along with some fundamental thermodynamic laws, and these are given in Chapter 5. All of these are basic relations common to all continua irrespective of their material properties.

Constitutive relations characterize a particular material's macroscopic response to applied mechanical, thermal, or other types of loadings. Such relations are based on the material's internal constitution and commonly result in idealized material models such as *linear elastic solids* or *linear viscous fluids*. These models do not come directly from general principles but rather are often developed from observed experimental data. Combining constitutive laws with the general principle relations creates a closed system of field equations that contains sufficient number of equations

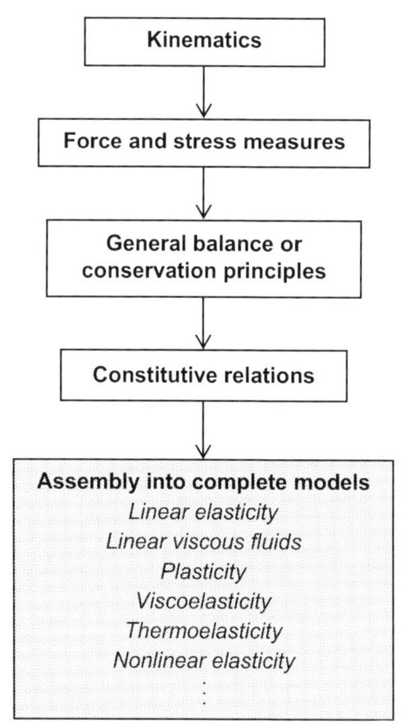

FIGURE 1.5 Structure of continuum mechanics.

to solve for all of the model unknowns.. It should be noted that solutions to particular problems within a given model require appropriate boundary and/or initial conditions.

This text will place emphasis on exploring a very wide collection of constitutive relations and developing the corresponding material model formulations. Such material behavior models will include classical linear theories of elasticity, fluid mechanics, viscoelasticity, and plasticity. Additional linear theories including multiple constitutive fields such as poro, thermo, and electro will also be developed. Nonlinear theories of solids and fluids including finite elasticity, nonlinear viscous fluids, and nonlinear viscoelastic materials will be presented. Finally, several relatively new continuum theories based on incorporation of material microstructure will be presented including: micropolar elasticity, elasticity with voids, nonlocal higher gradient elasticity, fabric tensor theories for granular materials, and damage mechanics.

1.4 A LITTLE HISTORY

Before embarking on the theoretical aspects of continuum mechanics, a very brief historical presentation of the subject will be given. This account is intended to provide a short overview of some of the major developments over the last few centuries, with emphasis on the twentieth century. Much more detailed historical information may be found in Maugin (2013), Truesdell and Toupin (1960), Truesdell and Noll (1965), Truesdell (1966), Soutas-Little (2011), Doraiswamy (2017), Walters (2017), and Tanner and Walters (1998).

Solid mechanics is perhaps one of the oldest branches of physical sciences, and many would say that it can be traced back to Archimedes (250 BC) who claimed that he could move the earth if he were given a proper place to stand and a lever long enough. Many prominent and brilliant scientists and mathematicians have made numerous contributions to continuum mechanics. Studies began with *rigid solids* which were Euclidean bodies that do not undergo deformation. Through human curiosity and societal needs to successfully build things, the study of material behavior has been explored for countless centuries. Early development of quantitative material behavior relations began in the 17th century with Galileo Galilei (nature of the resistance of solids), Robert Hooke (the linear elastic response), and of course Isaac Newton's *Principia*. The following century brought forward considerable important work including that of Leonhard Euler and Joseph-Louis Lagrange on deformation and strain. In the 19th century, Claude-Louis Navier presented results on the general equations of equilibrium and Augustin-Louis Cauchy published his work on the concept of stress at a point. During this time period, considerable work was developed on linear elasticity by Siméon Denis Poisson, Gabriel Léon Jean Baptiste Lamé, Albert Green, Adhémar Jean Claude Barré de Saint-Venant, George Airy, and others. Toward the latter half of the 1800s, Gabrio Piola, Gustav Kirchhoff, and Joseph Valentin Boussinesq initiated studies on the theory of finite deformations, and James Clerk Maxwell, William Lord Kelvin, and Ludwig Boltzmann proposed early viscoelastic constitutive material relations.

In regard to fluid mechanics, early work can be traced back to Leonardo da Vinci and others in the 16th century. Later in the 18th century, Leonhard Euler developed the concept of continuity of flow, and Daniel Bernoulli introduced the term *hydrodynamics* and formulated *Bernoulli's Principle*. Inviscid fluid mechanics was established, and work dealing with fluid friction was beginning. The next century brought the work of Navier and George Stokes together to formulate the famous governing relations for linearly viscous fluids—the *Navier–Stokes equations*. Even though these governing relations used a linear relation between stress and deformation rate, the equations are nonlinear and thus they proved to be very challenging to solve at this point in history. In the early part of the 20th century, Ludwig Prandtl provided important work to relate inviscid and viscous flows, and later developed important *boundary layer theory*.

The 19th century led to the development of the science of thermodynamics, which had its origins previously connected with heat engines. During the 1800s, scientists like Nicolas Léonard Carnot, James Prescott Joule, William John Rankine, and Rudof Clausius (initial definition of entropy) made fundamental contributions in this field, and in 1850 Lord Kelvin first labeled the study thermodynamics. Later in this century, Maxwell made studies of statistical thermodynamics and Josiah Willard Gibbs first defined *free energy* and *enthalpy*. During this time period, Jean-Baptiste Joseph Fourier made fundamental contributions in heat transfer.

In 1929, Eugene Bingham helped to create a branch of physics called *rheology* that specifically dealt with problems pertaining to the behavior of unusual liquids and solids. After consulting a language professor, Bingham took "rheo" from Greek, meaning "everything flows" and "-ology" meaning "the study of". That same year the *Society of Rheology* was formed eventually leading to a specialized journal presenting research on the subject matter. Rheology is classically defined as the study of flow of matter, primarily in a liquid state, but also as soft solids under conditions in which they respond with some combination of elastic, plastic, and viscous deformation behaviors. Rheology is obviously closely connected with continuum mechanics and many would categorize it as a subarea of study within the overall structure.

In general, most work before the 20th century was concerned with linear continuum theories of solids and fluids. By the turn of the 20th century, there was general acknowledgement of the existence of materials which could not be classified as Hookean solids or Newtonian fluids, and research began to come forward to describe behavior of such materials. Starting in the 1930s, rheologists Bingham and Markus Reiner began to explore some of these more complicated behaviors. In the 1940s and 50s, a great resurgence began that led to the development of much more general and mathematical material behavior formulations. Major contributions were made by Ronald Rivlin, James Gardner Oldroyd, and others.

During the 1950s, Clifford Truesdell established himself as the godfather of modern continuum mechanics or as he coined it *rational mechanics*. He began his brilliant work with a 1952 review article (later reprinted in 1966) that set down the modern fundamentals of elasticity and fluid mechanics (Truesdell, 1966). Among his numerous contributions were two monumental published works *Classical Field Theories* (1960) and *Nonlinear Field Theories* (1965) which developed a rigorous mathematical approach to continuum mechanics. He also initiated a new journal devoted to this field. Through this work, he stimulated many others to follow this mathematical path. Jerald Ericksen, Walter Noll, Richard Toupin, Bernard Coleman, and others made important contributions to the subject during the 1960s and 70s. Extensive research and publications on nonlinear theories of elasticity, viscoelasticity, plasticity, and non-Newtonian fluids occurred during this time period. Also, this era ushered in computational finite and boundary element methods that greatly aided solution capabilities for the application of continuum theories to problems of engineering interest.

The latter half of the 20th century produced a wide variety of work looking to extend traditional continuum theories to model behavior of multiphase composites, graded materials, granular substances, cellular and porous solids, materials with microstructure, coupled multifield problems, etc. For example, Ahmed Cemal Eringen developed theoretical foundations of micropolar and nonlocal theories of continuous media. Raymond Mindlin and Elias Aifantis also explored such developments by looking at continuously distributed microstructure and strain gradient theories of material behavior. Bernard Budiansky and Mark Kachanov studied elastic materials with distributed cracks, and Stephen Cowin and Jace Nuziato looked at materials with distributed voids. Following the previous fundamental work on soil mechanics by Karl von Terzaghi, Maurice Biot developed the theory of poroelasticity that incorporates the combined response of a fluid-saturated porous elastic material. Many other interesting new continuum theories of this type have been developed over the last few decades, and such work continues to the present day.

For well over a century, the scientific engineering community has successfully employed continuum mechanics to simulate the mechanical, thermal, and electrical/magnetic behavior of a broad class of materials under a variety of loading conditions. These theories have also provided necessary tools for the construction of *computational schemes* based on *finite differences* and *finite and boundary element methods*. By any measure, continuum mechanics has made significant contributions to our formulation and solution of engineering problems and has provided significant help in the design of safe and efficient structures and systems.

REFERENCES

Doraiswamy, D., 2017. n.d. The origins of rheology: a short historical excursion. Encyclopedia of Life Support Systems. http://www.eolss.net/.

Fung, Y.C., 1994. A First Course in Continuum Mechanics, Third ed Prentice-Hall, Englewood Cliffs, NJ.

Jensen, J.H., 2010. Molecular Modeling Basics. CRC Press, Boca Raton, FL.

Massobrio, C., Bulou, H., Goyhenex, C. (Eds.), 2012. Atomic-Scale Modeling of Nanosystems and Nanostructured Materials. Springer, Berlin.

Maugin, G.A., 2013. Continuum Mechanics through the Twentieth Century: A Concise Historical Perspective. Springer, Berlin.

Soutas-Little, R.W., 2011. History of continuum mechanics. Encyclopedia of Life Support Systems. http://www.eolss.net/sample-chapters/c05/e6-161-02-00.pdf.

Tanner, R.I., Walters, K., 1998. Rheology: An Historical Perspective. Elsevier.

Truesdell, C.A., 1966. Continuum Mechanics. I. The Mechanical Foundations of Elasticity and Fluid Dynamics. Gordon & Breach, New York.

Truesdell, C.A., Noll, W., 1965. Nonlinear field theories. In: Flugge, S. (Ed.), Encyclopedia of Physics. Springer, v. III/3.

Truesdell, C.A., Toupin, R.A., 1960. Classical field theories. In: Flugge, S. (Ed.), Encyclopedia of Physics. Springer, v. III/1.

Walters, K., 2017. History of Rheology, Encyclopedia of Life Support Systems. http://www.eolss.net/.

CHAPTER

Mathematical Preliminaries

2

As mentioned in Section 1.2, continuum mechanics must be formulated in terms of quantities that are independent of the coordinate system used to describe the problem. This requires the use of a tensor format to express all variables and relations. The main purpose of this chapter is to present the tensor language and theory necessary for the development of our study. The focus is mainly on Cartesian tensors, but will explore non-Cartesian curvilinear coordinates in the last two sections. Both tensor algebra and tensor calculus for field variables will be presented. Considerable time will be devoted to definitions, proper language, and notational issues that will appear many times in the later chapters. Additional related matrix and tensor mathematical topics will also be discussed. Further reading on the general topics of tensors can be found in Goodbody (1982), Simmons (1994), and Itskov (2015).

2.1 INDEX AND DIRECT NOTATION

Continuum mechanics is formulated in terms of many different types of variables: *scalars*, represented by a single value at each point in space (e.g. material density); *vectors*, expressible in terms of three components in a three-dimensional space (e.g. material displacement or velocity); *matrix variables*, which commonly require nine components to quantify (e.g. stress or strain). Other applications incorporate additional quantities that need even more components to characterize. Related operations between these variables should not depend on the coordinate system. Tensor notation and tensor operations offer a unified scheme to formulate all of the required mathematical relations and theories found in continuum mechanics. A scalar method of simply listing every single variable would prove to be totally inefficient for use in this study.

Generally two different tensor notational schemes have been constructed that enable us to list, develop, and manipulate the variables of interest. One method uses *index notation* to identify tensor variables and operations, whereas the second employs a *direct notation* without indices. Index notation generally provides more explicit information on the tensor operations with respect to a given coordinate system used to describe the problem. However, direct notation is a symbolic representational scheme that offers a more shorthand method with somewhat less details on operational information. These concepts will become clearer as we move forward in our study as we employ both notational methods in various places in the text.

Continuum Mechanics Modeling of Material Behavior. http://dx.doi.org/10.1016/B978-0-12-811474-2.00002-2

Index notation is a shorthand scheme that allows the representation of a whole set of elements or components by a single symbol with subscripts. For example, the three numbers a_1, a_2, and a_3 are denoted by the symbol a_i, where index i will normally have the range 1–3. In a similar fashion, A_{ij} represents the nine numbers A_{11}, A_{12}, A_{13}, A_{21}, A_{22}, A_{23}, A_{31}, A_{32}, and A_{33}. Although these representations can be written in any manner, it is common to use a scheme related to vector and matrix formats such that

$$a_i = \begin{bmatrix} a_1 \\ a_2 \\ a_3 \end{bmatrix}, \quad A_{ij} = \begin{bmatrix} A_{11} & A_{12} & A_{13} \\ A_{21} & A_{22} & A_{23} \\ A_{31} & A_{32} & A_{33} \end{bmatrix} \tag{2.1.1}$$

In the matrix format, A_{1j} represents the first row, whereas A_{i1} indicates the first column. Other columns and rows are indicated in similar fashion; thus, the first index represents the row, whereas the second denotes the column. Direct notation would simply use a bold character for the vector and matrix definitions in (2.1.1) by writing

$$\boldsymbol{a} = \begin{bmatrix} a_1 \\ a_2 \\ a_3 \end{bmatrix}, \quad \boldsymbol{A} = \begin{bmatrix} A_{11} & A_{12} & A_{13} \\ A_{21} & A_{22} & A_{23} \\ A_{31} & A_{32} & A_{33} \end{bmatrix} \tag{2.1.2}$$

Since the indices are not included, direct notation provides a less cluttered scheme but it also does not explicitly indicate the specific nature of the representation.

In general, a symbol $A_{ij...k}$ with N distinct indices represents 3^N distinct numbers or components. It should be apparent that a_i and a_j represent the same three numbers, and likewise a_{ij} and a_{mn} signify the same matrix. Addition, subtraction, multiplication, and equality of index symbols are defined in the normal fashion. For example, addition and subtraction are given by

$$a_i \pm b_i = \begin{bmatrix} a_1 \pm b_1 \\ a_2 \pm b_2 \\ a_3 \pm b_3 \end{bmatrix}, \quad A_{ij} \pm B_{ij} = \begin{bmatrix} A_{11} \pm B_{11} & A_{12} \pm B_{12} & A_{13} \pm B_{13} \\ A_{21} \pm B_{21} & A_{22} \pm B_{22} & A_{23} \pm B_{23} \\ A_{31} \pm B_{31} & A_{32} \pm B_{32} & A_{33} \pm B_{33} \end{bmatrix} \tag{2.1.3}$$

and scalar multiplication is specified as

$$\lambda a_i = \begin{bmatrix} \lambda a_1 \\ \lambda a_2 \\ \lambda a_3 \end{bmatrix}, \quad \lambda A_{ij} = \begin{bmatrix} \lambda A_{11} & \lambda A_{12} & \lambda A_{13} \\ \lambda A_{21} & \lambda A_{22} & \lambda A_{23} \\ \lambda A_{31} & \lambda A_{32} & \lambda A_{33} \end{bmatrix} \tag{2.1.4}$$

The multiplication of two symbols with different indices is called *outer multiplication*, and a simple example is given by

$$a_i b_j = \begin{bmatrix} a_1 b_1 & a_1 b_2 & a_1 b_3 \\ a_2 b_1 & a_2 b_2 & a_2 b_3 \\ a_3 b_1 & a_3 b_2 & a_3 b_3 \end{bmatrix} \tag{2.1.5}$$

The previous operations obey usual commutative, associative, and distributive laws, for example,

$$\begin{aligned}
&a_i + b_i = b_i + a_i \\
&A_{ij} b_k = b_k A_{ij} \\
&a_i + (b_i + c_i) = (a_i + b_i) + c_i \\
&a_i (B_{jk} c_l) = (a_i B_{jk}) c_l \\
&A_{ij} (b_k + c_k) = A_{ij} b_k + A_{ij} c_k
\end{aligned} \tag{2.1.6}$$

Note that the simple relations $a_i = b_i$ or $A_{ij} = B_{ij}$ imply that $a_1 = b_1$, $a_2 = b_2$, ... and $A_{11} = B_{11}$, $A_{12} = B_{12}$, ... However, relations of the form $a_i = b_j$ or $A_{ij} = B_{kl}$ have ambiguous meaning since the distinct indices on each term are not the same, and thus these types of expressions are to be avoided in this notational scheme. In general, the distinct subscripts on all individual terms in a given equation should match.

2.2 SUMMATION CONVENTION

It will be convenient to adopt the convention that if a subscript appears twice in the same term, then *summation* over that subscript from one to three is implied, for example,

$$\begin{aligned}
&A_{ii} = \sum_{i=1}^{3} A_{ii} = A_{11} + A_{22} + A_{33} \\
&A_{ij} b_j = \sum_{j=1}^{3} A_{ij} b_j = A_{i1} b_1 + A_{i2} b_2 + A_{i3} b_3
\end{aligned} \tag{2.2.1}$$

It should be apparent that $A_{ii} = A_{jj} = A_{kk} = \ldots$, and therefore the *repeated* subscripts or indices are sometimes called *dummy* subscripts. Unspecified indices that are not repeated are called *free* or *distinct* subscripts. The summation convention may be suspended by underlining one of the repeated indices or by writing *no sum*. The use of three or more repeated indices in the same term (e.g. A_{iii} or $A_{iij}B_{ij}$) has ambiguous meaning and is to be avoided. On a given symbol, the process of setting two free indices equal is called *contraction*. For example, A_{ii} is obtained from A_{ij} by contraction on i and j. The operation of outer multiplication of two indexed symbols followed by contraction with respect to one index from each symbol generates an *inner multiplication*, for example, $A_{ij}B_{jk}$ is an inner product obtained from the outer product $A_{ij}B_{mk}$ by contraction on indices j and m.

EXAMPLE 2.2.1 INDEX NOTATION EXAMPLES

Consider the following vector and matrix:

$$a_i = \begin{bmatrix} 1 \\ 4 \\ 2 \end{bmatrix}, \quad A_{ij} = \begin{bmatrix} 1 & 0 & 2 \\ 2 & 3 & 0 \\ 1 & 4 & 2 \end{bmatrix}$$

Evaluate the following expressions: $a_i a_i$, $a_i a_j$, A_{ii}, and $A_{ij} a_j$ and specify whether they are scalars, vectors, matrices, etc.

Solution: Using the standard definitions in Sections 2.1 and 2.2 ⇒

$$a_i a_i = 1^2 + 4^2 + 2^2 = 21 \text{ (scalar)}$$

$$a_i a_j = \begin{bmatrix} 1\times1 & 1\times4 & 1\times2 \\ 4\times1 & 4\times4 & 4\times2 \\ 2\times1 & 2\times4 & 2\times2 \end{bmatrix} = \begin{bmatrix} 1 & 4 & 2 \\ 4 & 16 & 8 \\ 2 & 8 & 4 \end{bmatrix} \text{(matrix)}$$

$$A_{ii} = 1 + 4 + 2 = 7 \text{ (scalar)}$$

$$A_{ij} a_j = \begin{bmatrix} 1\times1+0\times4+2\times2 \\ 2\times1+3\times4+0\times2 \\ 1\times1+4\times4+2\times2 \end{bmatrix} = \begin{bmatrix} 5 \\ 14 \\ 21 \end{bmatrix} \text{(vector)}$$

It is always a good idea to first review the distinct and summed indices in such expression to first determine the general nature of the expected final evaluation (i.e. scalar, vector, matrix, . . .).

2.3 SYMMETRIC AND ANTISYMMETRIC SYMBOLS

A symbol $A_{ij...m...n...k}$ is said to be *symmetric* with respect to index pair mn if

$$A_{ij...m...n...k} = A_{ij...n...m...k} \tag{2.3.1}$$

whereas it is *antisymmetric* or *skewsymmetric* if

$$A_{ij...m...n...k} = -A_{ij...n...m...k} \tag{2.3.2}$$

Note that if $A_{ij...m...n...k}$ is symmetric in mn, and $B_{pq...m...n...r}$ is antisymmetric in mn, then the product is zero

$$A_{ij...m...n...k} B_{pq...m...n...r} = 0 \tag{2.3.3}$$

A useful identity or decomposition for matrices may be written as

$$A_{ij} = \frac{1}{2}(A_{ij} + A_{ji}) + \frac{1}{2}(A_{ij} - A_{ji}) = A_{(ij)} + A_{[ij]} \tag{2.3.4}$$

The first term $A_{(ij)} = \frac{1}{2}(A_{ij} + A_{ji})$ is symmetric, whereas the second term $A_{[ij]} = \frac{1}{2}(A_{ij} - A_{ji})$ is antisymmetric, and thus an arbitrary symbol A_{ij} can always be expressed as the sum of symmetric and antisymmetric pieces. Note that if A_{ij} is symmetric, then it will have only six independent components. On the other hand, if A_{ij} is antisymmetric, then its diagonal terms $A_{\underline{ii}}$ (no sum on i) must be zero and it will only have three independent components. Since $A_{[ij]}$ has only three independent components, it can be related to a vector quantity with a single index (see Exercise 2.13).

EXAMPLE 2.3.1 DECOMPOSITION EXAMPLE

For the following matrix

$$A_{ij} = \begin{bmatrix} 1 & 1 & 2 \\ 2 & 1 & 3 \\ 1 & 2 & 4 \end{bmatrix}$$

determine the symmetric and antisymmetric decomposition matrices $A_{(ij)}$ and $A_{[ij]}$ specified by relation (2.3.4). Then justify the fact that the product $A_{(ij)}A_{[ij]}$ is zero.

Solution: Using the definitions in (2.3.4) ⇒

$$A_{(ij)} = \frac{1}{2}(A_{ij} + A_{ji}) = \frac{1}{2}\begin{bmatrix} 1 & 1 & 2 \\ 2 & 1 & 3 \\ 1 & 2 & 4 \end{bmatrix} + \frac{1}{2}\begin{bmatrix} 1 & 2 & 1 \\ 1 & 1 & 2 \\ 2 & 3 & 4 \end{bmatrix} = \begin{bmatrix} 1 & 3/2 & 3/2 \\ 3/2 & 1 & 5/2 \\ 3/2 & 5/2 & 4 \end{bmatrix}$$

$$A_{[ij]} = \frac{1}{2}(A_{ij} - A_{ji}) = \frac{1}{2}\begin{bmatrix} 1 & 1 & 2 \\ 2 & 1 & 3 \\ 1 & 2 & 4 \end{bmatrix} - \frac{1}{2}\begin{bmatrix} 1 & 2 & 1 \\ 1 & 1 & 2 \\ 2 & 3 & 4 \end{bmatrix} = \begin{bmatrix} 0 & -1/2 & 1/2 \\ 1/2 & 0 & 1/2 \\ -1/2 & -1/2 & 0 \end{bmatrix}$$

$$A_{(ij)}A_{[ij]} = A_{(11)}A_{[11]} + A_{(12)}A_{[12]} + A_{(13)}A_{[13]} + A_{(21)}A_{[21]} + A_{(22)}A_{[22]} + A_{(23)}A_{[23]} + \cdots = 0$$

Notice that $A_{(ij)}$ has only six independent components, whereas $A_{[ij]}$ has only three.

2.4 KRONECKER DELTA AND ALTERNATING SYMBOL

A useful special symbol commonly used in index notational schemes is the *Kronecker delta* defined by

$$\delta_{ij} = \begin{cases} 1 & \text{if } i = j \text{ (no sum)} \\ 0 & \text{if } i \neq j \end{cases} = \begin{bmatrix} 1 & 0 & 0 \\ 0 & 1 & 0 \\ 0 & 0 & 1 \end{bmatrix} \tag{2.4.1}$$

within usual matrix theory, it is observed that this symbol is simply the unit matrix $\boldsymbol{I}$, and we could use direct notation to write

$$\boldsymbol{I} = \begin{bmatrix} 1 & 0 & 0 \\ 0 & 1 & 0 \\ 0 & 0 & 1 \end{bmatrix} \tag{2.4.2}$$

Note that the Kronecker delta is a symmetric symbol with respect to its two indices. Particular useful properties of the Kronecker delta include

$$\begin{aligned}
&\delta_{ij} = \delta_{ji} \\
&\delta_{ii} = 3, \quad \delta_{\underline{ii}} = 1 \\
&\delta_{ij}a_j = a_i, \quad \delta_{ij}a_i = a_j \\
&\delta_{ij}a_{jk} = a_{ik}, \quad \delta_{jk}a_{ik} = a_{ij} \\
&\delta_{ij}a_{ij} = a_{ii}, \quad \delta_{ij}\delta_{ij} = 3
\end{aligned} \tag{2.4.3}$$

EXAMPLE 2.4.1 KRONECKER DELTA PROPERTY VERIFICATION

Explicitly justify the Kronecker delta property $\delta_{ij}a_j = a_i$.

Solution: Expand the basic relation using summation convention ⇒

$$\begin{aligned}\delta_{ij}a_j &= \delta_{i1}a_1 + \delta_{i2}a_2 + \delta_{i3}a_3 \\ &= \begin{cases} \delta_{11}a_1 + \delta_{12}a_2 + \delta_{13}a_3 = a_1, & i=1 \\ \delta_{21}a_1 + \delta_{22}a_2 + \delta_{23}a_3 = a_2, & i=2 \\ \delta_{31}a_1 + \delta_{32}a_2 + \delta_{33}a_3 = a_3, & i=3 \end{cases} \\ &= a_i\end{aligned}$$

Note that this relation is often called the *substitution property* associated with the Kronecker delta, and relations $(2.4.3)_{3,4,5}$ further expand on this property. We will make repeated use of such relations in future applications.

Another useful special symbol is the *alternating* or *permutation symbol* defined by

$$\varepsilon_{ijk} = \begin{cases} +1 & \text{if } ijk \text{ is an even permutation of } 1,2,3 \\ -1 & \text{if } ijk \text{ is an odd permutation of } 1,2,3 \\ 0 & \text{otherwise} \end{cases} \tag{2.4.4}$$

Consequently, $\varepsilon_{123} = \varepsilon_{231} = \varepsilon_{312} = 1$, $\varepsilon_{321} = \varepsilon_{132} = \varepsilon_{213} = -1$, $\varepsilon_{112} = \varepsilon_{131} = \varepsilon_{222} = \cdots = 0$. Therefore, of the 27 possible terms for the alternating symbol, three are equal to 1, three are equal to -1, and all others are zero. The alternating symbol is antisymmetric with respect to any pair of its indices.

2.5 DETERMINANTS

The alternating symbol proves to be useful in evaluating determinants and vector cross-products. The determinant a_{ij} of an array can be written in two equivalent forms as

$$\det[A_{ij}] = |A_{ij}| = \begin{vmatrix} A_{11} & A_{12} & A_{13} \\ A_{21} & A_{22} & A_{23} \\ A_{31} & A_{32} & A_{33} \end{vmatrix} = \varepsilon_{ijk}A_{1i}A_{2j}A_{3k} = \varepsilon_{ijk}A_{i1}A_{j2}A_{k3} \tag{2.5.1}$$

where the first index expression represents the row expansion, whereas the second is the column expansion. Note that indices i and j in the expression $\det[A_{ij}]$ are not subscripts in the usual sense defined in Section 2.1, because $\det[A_{ij}]$ denotes a single number through the operations specified in relation (2.5.1).

Note the following properties associated with determinants and the alternating symbol:

$$\begin{aligned}&\text{If} \quad C_{ij} = A_{ik}B_{kj}, \text{or } C_{ij} = A_{ik}B_{jk}, \text{or } C_{ij} = A_{ki}B_{kj} \\ &\text{then} \quad \det[C_{ij}] = \det[A_{ij}]\det[B_{ij}]\end{aligned} \tag{2.5.2}$$

$$\text{If} \quad A_{ik}A_{jk} = \delta_{ij}, \text{or } A_{ki}A_{kj} = \delta_{ij}, \text{then} \quad \det[A_{ij}] = \pm 1 \tag{2.5.3}$$

$$\varepsilon_{ijk}\varepsilon_{pqr}\det[A_{mn}] = \begin{vmatrix} A_{ip} & A_{iq} & A_{ir} \\ A_{jp} & A_{jq} & A_{jr} \\ A_{kp} & A_{kq} & A_{kr} \end{vmatrix} \tag{2.5.4}$$

$$\varepsilon_{ijk}\varepsilon_{pqr} = \begin{vmatrix} \delta_{ip} & \delta_{iq} & \delta_{ir} \\ \delta_{jp} & \delta_{jq} & \delta_{jr} \\ \delta_{kp} & \delta_{kq} & \delta_{kr} \end{vmatrix} \tag{2.5.5}$$

$$\varepsilon_{ijk}\varepsilon_{iqr} = \delta_{jq}\delta_{kr} - \delta_{jr}\delta_{kq} \tag{2.5.6}$$

$$\varepsilon_{ijk}\varepsilon_{ijr} = 2\delta_{kr} \tag{2.5.7}$$

$$\varepsilon_{ijk}\varepsilon_{ijk} = 6 \tag{2.5.8}$$

$$\det[A_{ij}] = \frac{1}{6}\varepsilon_{ijk}\varepsilon_{pqr}A_{ip}A_{jq}A_{kr} \tag{2.5.9}$$

$$\varepsilon_{ijk}A_{ip}A_{jq}A_{kr} = \varepsilon_{ijk}A_{pi}A_{qj}A_{rk} = \det[A_{mn}]\varepsilon_{pqr} \tag{2.5.10}$$

$$\frac{d}{dt}(\det[A_{ij}]) = \det[A_{ij}]A_{ij}^{-1}\frac{d}{dt}A_{ji} \tag{2.5.11}$$

$$\varepsilon_{ijk}a_i b_j c_k = \begin{vmatrix} a_1 & a_2 & a_3 \\ b_1 & b_2 & b_3 \\ c_1 & c_2 & c_3 \end{vmatrix} \tag{2.5.12}$$

2.6 VECTORS AND COORDINATE FRAMES

Let us first review a few fundamental properties of vectors. Traditionally, a vector is described as a quantity that has both a magnitude and direction, and with respect to a Cartesian coordinate system (frame), these characteristics can be illustrated geometrically by an arrow or directed line segment of particular magnitude and direction as shown in Fig. 2.1.

Let $\boldsymbol{a}$ denote an arbitrary vector. The rectangular *projections* of $\boldsymbol{a}$ on the x_1, x_2, and x_3 axes are denoted by a_1, a_2, and a_3, respectively. The *magnitude* or *length* of the vector $\boldsymbol{a}$ is given by

$$|\boldsymbol{a}| = (a_1^2 + a_2^2 + a_3^2)^{1/2} = (a_i a_i)^{1/2} \tag{2.6.1}$$

The cosine of the angle between the vector $\boldsymbol{a}$ and the x_i-axis may be expressed as

$$\cos(\boldsymbol{a}, x_i) = \frac{a_i}{|\boldsymbol{a}|} \tag{2.6.2}$$

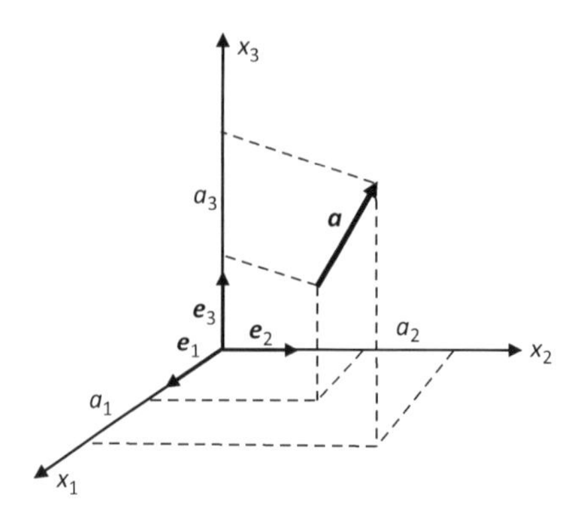

FIGURE 2.1

Vector representation.

Hence if we know the projections of $\boldsymbol{a}$, then we know both the magnitude and direction of the vector.

Eqs. (2.6.1) and (2.6.2) imply that two vectors $\boldsymbol{a}$ and $\boldsymbol{b}$ are equal (have the same magnitude and direction) if their corresponding projections are equal, $a_i = b_i$. In addition, the *parallelogram rule* for vector addition, $\boldsymbol{c} = \boldsymbol{a} + \boldsymbol{b}$, follows the indicial notation scheme introduced earlier, that is, $c_i = a_i + b_i$.

We define the three unit vectors along the x_1, x_2, and x_3 axes by $\boldsymbol{e}_1$, $\boldsymbol{e}_2$, and $\boldsymbol{e}_3$, respectively. This *vector triad* $\{\boldsymbol{e}_i\}$ forms what is called an *orthonormal basis* for the coordinate system. Any and every vector within this system may be expressed in terms of this orthonormal basis by the usual expression

$$\boldsymbol{a} = a_1\boldsymbol{e}_1 + a_2\boldsymbol{e}_2 + a_3\boldsymbol{e}_3 = a_i\boldsymbol{e}_i \tag{2.6.3}$$

The numbers a_i are normally called the *Cartesian components* of $\boldsymbol{a}$ in the particular system.

If $\{\boldsymbol{e}_i\}$ forms a *right-handed triad*, then

$$\begin{aligned} \boldsymbol{e}_1 \times \boldsymbol{e}_2 &= \boldsymbol{e}_3 \\ \boldsymbol{e}_2 \times \boldsymbol{e}_3 &= \boldsymbol{e}_1 \\ \boldsymbol{e}_3 \times \boldsymbol{e}_1 &= \boldsymbol{e}_2 \end{aligned} \tag{2.6.4}$$

where the cross operation, $\times$, means the *vector cross product*, which will be defined later. Furthermore, from the definition of $\boldsymbol{e}_i$ and relations (2.6.4), it follows that

$$\begin{aligned} \boldsymbol{e}_i \cdot \boldsymbol{e}_j &= \delta_{ij} \\ \boldsymbol{e}_i \times \boldsymbol{e}_j &= \varepsilon_{ijk}\boldsymbol{e}_k \end{aligned} \tag{2.6.5}$$

where the dot operation, $\cdot$,indicates the common *dot product* to be defined later.

2.7 CHANGES IN COORDINATE FRAMES: ORTHOGONAL TRANSFORMATIONS

It will be necessary to express continuum field variables and equations in several different coordinate systems. This situation requires the development of particular transformation rules for scalar, vector, matrix, and higher-order variables. This concept is fundamentally connected with the basic definitions of tensor variables and their related tensor transformation laws. At the moment, we will restrict our discussion to transformations only between Cartesian coordinate systems, and thus consider the two systems shown in Fig. 2.2. The two Cartesian frames (x_1, x_2, x_3) and (x'_1, x'_2, x'_3) differ only by orientation, and the unit basis vectors for each frame are $\{\boldsymbol{e}_i\} = \{\boldsymbol{e}_1, \boldsymbol{e}_2, \boldsymbol{e}_3\}$ and $\{\boldsymbol{e}'_i\} = \{\boldsymbol{e}'_1, \boldsymbol{e}'_2, \boldsymbol{e}'_3\}$.

Let Q_{ij} denote the cosine of the angle between the x'_i-axis and the x_j-axis:

$$Q_{ij} = \cos(x'_i, x_j) \tag{2.7.1}$$

Using this definition, the basis vectors in the primed coordinate frame can be easily expressed in terms of those in the unprimed frame by the relations

$$\begin{aligned} \boldsymbol{e}'_1 &= Q_{11}\boldsymbol{e}_1 + Q_{12}\boldsymbol{e}_2 + Q_{13}\boldsymbol{e}_3 \\ \boldsymbol{e}'_2 &= Q_{21}\boldsymbol{e}_1 + Q_{22}\boldsymbol{e}_2 + Q_{23}\boldsymbol{e}_3 \\ \boldsymbol{e}'_3 &= Q_{31}\boldsymbol{e}_1 + Q_{32}\boldsymbol{e}_2 + Q_{33}\boldsymbol{e}_3 \end{aligned} \tag{2.7.2}$$

or in index notation

$$\boldsymbol{e}'_i = Q_{ij}\boldsymbol{e}_j \tag{2.7.3}$$

Likewise, the opposite transformation can be written using the same format as

$$\boldsymbol{e}_i = Q_{ji}\boldsymbol{e}'_j \tag{2.7.4}$$

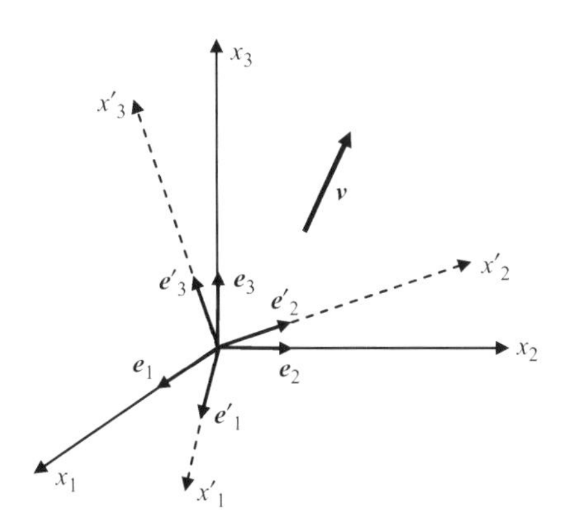

FIGURE 2.2

Change of Cartesian coordinate frames.

Now an arbitrary vector $\boldsymbol{v}$ (see Fig. 2.2) can be written in either of the two coordinate systems as

$$\begin{aligned} \boldsymbol{v} &= v_1\boldsymbol{e}_1 + v_2\boldsymbol{e}_2 + v_3\boldsymbol{e}_3 = v_i\boldsymbol{e}_i \\ &= v_1'\boldsymbol{e}_1' + v_2'\boldsymbol{e}_2' + v_3'\boldsymbol{e}_3' = v_i'\boldsymbol{e}_i' \end{aligned} \tag{2.7.5}$$

Substituting form (2.7.4) into $(2.7.5)_1$ gives

$$\boldsymbol{v} = v_i Q_{ji}\boldsymbol{e}_j'$$

but from $(2.7.5)_2$, $\boldsymbol{v} = v_j'\boldsymbol{e}_j'$, and so after renaming some indices we find

$$v_i' = Q_{ij}v_j \tag{2.7.6}$$

In a similar fashion using (2.7.3) in $(2.7.5)_2$ gives

$$v_i = Q_{ji}v_j' \tag{2.7.7}$$

Relations (2.7.6) and (2.7.7) constitute the transformation laws for the Cartesian components of a vector under a change in rectangular Cartesian coordinate frames. It should be understood that under such transformations, the vector itself is unaltered (retaining original length and orientation) and only its components are changed. Consequently, knowing the components of a vector in one frame, relations (2.7.6) or (2.7.7) can be used to calculate components in any other frame.

The fact that transformations are being made only between orthogonal coordinate systems places some particular restrictions on the transformation or direction cosine matrix Q_{ij}. These can be determined by using (2.7.6) and (2.7.7) together to get

$$v_i = Q_{ji}v_j' = Q_{ji}Q_{jk}v_k \tag{2.7.8}$$

From the properties of the Kronecker delta, this expression can be written as

$$\delta_{ik}v_k = Q_{ji}Q_{jk}v_k \quad \text{or} \quad (Q_{ji}Q_{jk} - \delta_{ik})v_k = 0$$

and since this relation is true for all vectors v_k, the expression in parentheses must be zero giving the result

$$Q_{ji}Q_{jk} = \delta_{ik} \tag{2.7.9}$$

In a similar fashion, relations (2.7.6) and (2.7.7) can be used to eliminate v_i (instead of v'_i) to get

$$Q_{ij}Q_{kj} = \delta_{ik} \tag{2.7.10}$$

Relations (2.7.9) and (2.7.10) comprise the *orthogonality conditions* that Q_{ij} must satisfy. Taking the determinant of either relation and using previous relation (2.5.2) gives another related result

$$\det[Q_{ij}] = \pm 1 \tag{2.7.11}$$

Matrices that satisfy relations (2.7.9)–(2.7.11) are called *orthogonal*, and the transformations given by (2.7.6) and (2.7.7) are therefore referred to as *orthogonal transformations*. An orthogonal matrix (transformation) is said to be *proper* or according to whether $\det[Q_{ij}] = +1$ or -1. We will normally choose proper orthogonal transformations which will maintain right-handed coordinate systems and ensure a few other convenient properties. For the case $\det[Q_{ij}] = +1$, Q_{ij} corresponds to a *rotation*, whereas for $\det[Q_{ij}] = -1$, Q_{ij} corresponds to a *reflection*. The set of all orthogonal transformations constitutes a *group* in the mathematical sense with respect to composition (multiplication).

2.8 CARTESIAN TENSORS AND TRANSFORMATION LAWS

We now proceed to define what is a tensor. As will become evident, we actually cannot precisely define a tensor but rather we describe how a tensor transforms and present some additional general properties. Scalars, vectors, matrices, and higher-order quantities can be represented by a general index notational scheme. Using this approach, all quantities may then be referred to as tensors of different orders. The previously presented transformation properties of a vector can then be used to establish the general transformation properties of these tensors. Restricting the transformations to those only between Cartesian coordinate systems, the general set of transformation relations for various orders can thus be written as

$$
\begin{aligned}
a' &= a, && \text{zero order (scalar)} \\
a'_i &= Q_{ip}a_p, && \text{first order (vector)} \\
A'_{ij} &= Q_{ip}Q_{jq}A_{pq}, && \text{second order (}matrix\text{)} \\
A'_{ijk} &= Q_{ip}Q_{jq}Q_{kr}A_{pqr}, && \text{third order} \\
A'_{ijkl} &= Q_{ip}Q_{jq}Q_{kr}Q_{ls}A_{pqrs}, && \text{fourth order} \\
&\vdots && \vdots \\
A'_{ijk\ldots m} &= Q_{ip}Q_{jq}Q_{kr}\cdots Q_{mt}A_{pqr\ldots t} && \text{general order}
\end{aligned}
\tag{2.8.1}
$$

Note that according to these definitions, a scalar is a zeroth-order tensor, a vector is a tensor of order 1, and a matrix is a tensor of order 2. Relations (2.8.1) then specify the transformation rules for the components of Cartesian tensors of any order under the rotation Q_{ij}. This transformation theory will prove to be very valuable in determining the displacement, stress, and strain in different coordinate directions. Appendix B contains several different such transformation applications between Cartesian and other curvilinear coordinate systems. It will often be more convenient to write the first- and second-order transformation laws (for vector $\boldsymbol{a}$ and square matrix $\boldsymbol{A}$) in direct notation as

$$
\begin{aligned}
\boldsymbol{a}' &= \boldsymbol{Qa} && \text{first order (vector)} \\
\boldsymbol{A}' &= \boldsymbol{QAQ}^T && \text{second order (matrix)}
\end{aligned}
\tag{2.8.2}
$$

where $\boldsymbol{Q}^T$ is the transpose of $\boldsymbol{Q}$ defined in Section 2.10.

The distinction between the components and the tensor itself should be understood. Recall that a vector $\boldsymbol{v}$ can be expressed as

$$\begin{aligned} \boldsymbol{v} &= v_1\boldsymbol{e}_1 + v_2\boldsymbol{e}_2 + v_3\boldsymbol{e}_3 = v_i\boldsymbol{e}_i \\ &= v_1'\boldsymbol{e}_1' + v_2'\boldsymbol{e}_2' + v_3'\boldsymbol{e}_3' = v_i'\boldsymbol{e}_i' \end{aligned} \tag{2.8.3}$$

In a similar fashion, a second-order tensor $\boldsymbol{A}$ can be written as

$$\begin{aligned} \boldsymbol{A} = {} & A_{11}\boldsymbol{e}_1\boldsymbol{e}_1 + A_{12}\boldsymbol{e}_1\boldsymbol{e}_2 + A_{13}\boldsymbol{e}_1\boldsymbol{e}_3 \\ & + A_{21}\boldsymbol{e}_2\boldsymbol{e}_1 + A_{22}\boldsymbol{e}_2\boldsymbol{e}_2 + A_{23}\boldsymbol{e}_2\boldsymbol{e}_3 \\ & + A_{31}\boldsymbol{e}_3\boldsymbol{e}_1 + A_{32}\boldsymbol{e}_3\boldsymbol{e}_2 + A_{33}\boldsymbol{e}_3\boldsymbol{e}_3 \\ = {} & A_{ij}\boldsymbol{e}_i\boldsymbol{e}_j = A_{ij}'\boldsymbol{e}_i'\boldsymbol{e}_j' \end{aligned} \tag{2.8.4}$$

and similar schemes can be used to represent tensors of higher order. The representation used in Eq. (2.8.4) is commonly called *dyadic notation* using dyadic products $\boldsymbol{e}_i\boldsymbol{e}_j$. Another common notation for such products is to use a *tensor product* notation $\boldsymbol{e}_i \otimes \boldsymbol{e}_j$. Additional information on dyadic notation can be found in Weatherburn (1948) and Chou and Pagano (1967).

Second-order tensors can also be defined as *linear vector transformations*. For example, we define the second-order tensor $\boldsymbol{T}$ to have the property that transforms vectors $\boldsymbol{a}$ into other vectors $\boldsymbol{b}$ through the simple inner product operation

$$\boldsymbol{Ta} = \boldsymbol{b} \tag{2.8.5}$$

or in index notation

$$T_{ij}a_j = b_i \tag{2.8.6}$$

Relations (2.8.3) and (2.8.4) indicate that any tensor can be expressed in terms of components in any coordinate system, and it is only the components that change under coordinate transformation. For example, the state of stress at a point in a continuum solid will depend on the problem geometry and applied loadings. As will be shown later, these stress components are those of a second-order tensor and will therefore obey transformation law $(2.8.1)_3$. However, although the components of the stress tensor will change with the choice of coordinates, the stress tensor itself will not.

An important property of a tensor is that if we know its components in one coordinate system, then we can find them in any other coordinate frame using the appropriate transformation law. Since the components of Cartesian tensors are representable by indexed symbols, the operations of equality, addition, subtraction, multiplication, etc. are defined in a manner consistent with the indicial notation procedures previously discussed. The terminology tensor without the adjective Cartesian, usually refers to a more general scheme, where the coordinates are not necessarily rectangular Cartesian and the transformations between coordinates are not always orthogonal. Such general tensor theory will be briefly discussed in Section 2.19.

EXAMPLE 2.8.1 TRANSFORMATION EXAMPLES

The components of a first- and second-order tensor in a particular coordinate frame are given by

$$a_i = \begin{bmatrix} 1 \\ 2 \\ 4 \end{bmatrix}, \quad A_{ij} = \begin{bmatrix} 1 & 0 & 2 \\ 0 & 2 & 1 \\ 1 & 2 & 4 \end{bmatrix}$$

Determine the components of each tensor in a new coordinate system found through a rotation of 60° ($\pi/6$ radians) about the x_3-axis. Choose a counterclockwise rotation when viewing down the negative x_3-axis (see Fig. 2.3).

Solution: The original and primed coordinate systems are shown in Fig. 2.3. The solution starts by determining the rotation matrix for this case

$$Q_{ij} = \begin{bmatrix} \cos 60^\circ & \cos 30^\circ & \cos 90^\circ \\ \cos 150^\circ & \cos 60^\circ & \cos 90^\circ \\ \cos 90^\circ & \cos 90^\circ & \cos 0^\circ \end{bmatrix} = \begin{bmatrix} 1/2 & \sqrt{3}/2 & 0 \\ -\sqrt{3}/2 & 1/2 & 0 \\ 0 & 0 & 1 \end{bmatrix}$$

The transformation for the vector quantity follows from Eq. $(2.8.1)_2$, and using matrix products from the direct notation $(2.8.2)_1$

$$a'_i = Q_{ij}a_j = \begin{bmatrix} 1/2 & \sqrt{3}/2 & 0 \\ -\sqrt{3}/2 & 1/2 & 0 \\ 0 & 0 & 1 \end{bmatrix} \begin{bmatrix} 1 \\ 2 \\ 4 \end{bmatrix} = \begin{bmatrix} 1/2+\sqrt{3} \\ 1-\sqrt{3}/2 \\ 4 \end{bmatrix}$$

and the second-order tensor (matrix) transforms according to $(2.8.1)_3$, and using matrix products from the direct notation $(2.8.2)_2$

$$A'_{ij} = Q_{ip}Q_{jq}A_{pq} = \begin{bmatrix} 1/2 & \sqrt{3}/2 & 0 \\ -\sqrt{3}/2 & 1/2 & 0 \\ 0 & 0 & 1 \end{bmatrix} \begin{bmatrix} 1 & 0 & 2 \\ 0 & 2 & 1 \\ 1 & 2 & 4 \end{bmatrix} \begin{bmatrix} 1/2 & \sqrt{3}/2 & 0 \\ -\sqrt{3}/2 & 1/2 & 0 \\ 0 & 0 & 1 \end{bmatrix}^T$$

$$= \begin{bmatrix} 7/4 & \sqrt{3}/4 & 1+\sqrt{3}/2 \\ \sqrt{3}/4 & 5/4 & 1/2-\sqrt{3} \\ 1/2+\sqrt{3} & 1-\sqrt{3}/2 & 4 \end{bmatrix}$$

While simple transformation cases can be worked out by hand, for more general cases it is much more convenient to use a computational scheme to evaluate the necessary matrix multiplications required in the transformation laws (2.8.1). MATLAB or Mathematica software are ideally suited to carry out such calculations, and an example MATLAB program to evaluate the transformation of second-order tensors is given by the code C-1 in Appendix C.

Using the general transformation relations, several properties can be established as listed in Table 2.1.

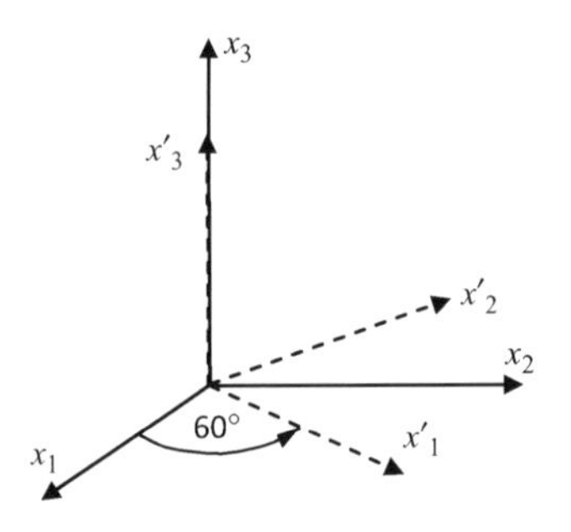

FIGURE 2.3

Coordinate transformation Example 2.8.1.

Table 2.1 Tensor properties

1. The operations addition, scalar multiplication, and subtraction yield Cartesian tensors of the same order
2. If the components of a Cartesian tensor are all zero in one frame, then they are all zero in any frame
3. Two tensors are equal if their components in any one frame are equal
4. If a tensor is symmetric (antisymmetric) in one frame, then it is symmetric (antisymmetric) in any other frame
5. If $A_{ij...k}$ is a tensor of order M and $B_{pq...r}$ is a tensor of order N, then the outer product $A_{ij...k}B_{pq...r}$ is a tensor of order $M + N$
6. If $A_{ij...m...n...k}$ is a tensor of order N, then $A_{ij...m...m...k}$ is a tensor of order $N - 2$
7. If $A_{ij}B_{ij} = c$ = scalar, and if B_{ij} is a second-order tensor, then A_{ij} are the components of a second-order tensor
8. If $A_{ij}b_j = c_i$, and if b_j and c_i are both vectors, then A_{ij} are the components of a second-order tensor

EXAMPLE 2.8.2 TENSOR PROPERTY JUSTIFICATION

Justify tensor property 5 stated in Table 2.1.

Solution: To justify this property, consider the specific outer product case of two second-order tensors $A_{ij}B_{kl}$. Since both tensors must satisfy their own transformation law, we can write

$$A'_{ij}B'_{kl} = Q_{ip}Q_{jq}A_{pq}Q_{kr}Q_{ls}B_{rs}$$

and then by simple rearrangement, we have

$$A'_{ij}B'_{kl} = Q_{ip}Q_{jq}Q_{kr}Q_{ls}A_{pq}B_{rs}$$

which for the product term is the standard form for the transformation of fourth-order tensors given by $(2.8.1)_5$.

Next, consider the transformation of the Kronecker delta. Starting with the standard transformation law $(2.8.1)_3$ and then incorporating the substitution property of the Kronecker delta along with the orthogonality relation (2.7.10), we find

$$\delta'_{ij} = Q_{ip}Q_{jq}\delta_{pq} = Q_{ip}Q_{jp} = \delta_{ij} \tag{2.8.7}$$

Therefore, the components of the Kronecker delta are the same in every frame. Tensors with this property are called *isotropic*. We will investigate these types of tensors in more detail in Section 2.15.

Consider next the transformation behavior of the alternating symbol. Again from transformation law $(2.8.1)_4$, we find

$$\varepsilon'_{ijk} = Q_{ip}Q_{jq}Q_{kr}\varepsilon_{pqr}$$

From property (2.5.10) of the alternating symbol, $Q_{ip}Q_{jq}Q_{kr}\varepsilon_{pqr} = \det[Q_{mn}]\varepsilon_{ijk}$, and since from the orthogonality conditions, $\det[Q_{mn}] = \pm 1$, we find $\varepsilon'_{ijk} = \pm\varepsilon_{ijk}$. Choosing only proper orthogonal transformations eliminates the plus/minus ambiguity and then implies that

$$\varepsilon'_{ijk} = \varepsilon_{ijk} \tag{2.8.8}$$

Thus, the alternating symbol also remains the same in all coordinate systems.

2.9 OBJECTIVITY BETWEEN DIFFERENT REFERENCE FRAMES

As previously discussed in Section 1.2, the fundamental relations in continuum mechanics must be formulated in terms of quantities that are independent of the coordinate frame used to describe the problem. So, if two individuals using different coordinate frames observe a common physical event, it should always be possible to state a physical law governing the event so that if the law is true for one observer, it will also be true for the other (once adjusted for the difference in coordinate frame). In this section, we will explore the basic concept of changes in reference frames and how this affects a few specific variables like distance, velocity, and acceleration. Later, we shall explore how additional continuum mechanics tensor variables transform under such reference frame changes, and if these variables transform in a particular way they will be labeled as *objective*. As we develop constitutive equations for material models, we will normally require these relations to contain only objective variables, and this will then place restrictions on general constitutive forms.

Reflecting back on the simple coordinate frame changes in Fig. 2.2, we now consider the more general transformation between two reference frames as shown in Fig. 2.4. Note that the two frames shown now include both time-dependent translational and rotational differences relative to each other. It should be kept in mind that frames of reference are not the same as coordinate systems, as a given observer can choose any coordinate system to observe and analyze events. Using

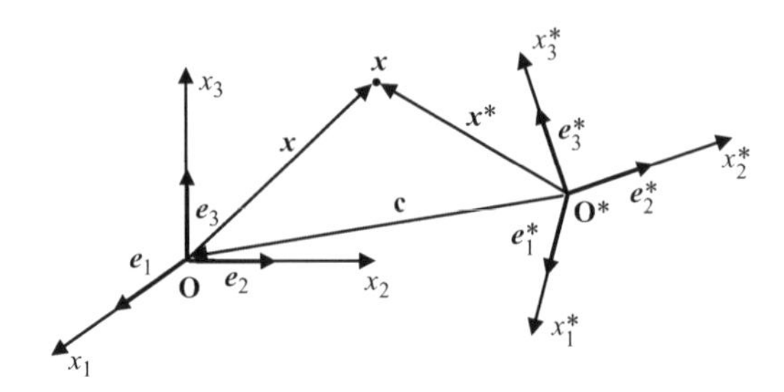

FIGURE 2.4

Objectivity between two reference frames.

Cartesian coordinates, we designate each frame with observers **O** and **O***, and in a Euclidean three-dimensional space and time, each observer can measure relative positions of points in space and intervals of time. With this is mind, we consider an event in **O** as the position/time pair $\{\boldsymbol{x}, t\}$ and in **O*** the event is $\{\boldsymbol{x}^*, t^*\}$. We can assume a simple time shift $t^* - a$, where a is a constant that can be set to zero without loss in generality. Note that the time clocks run at the same rate in each reference frame.

It can be shown that the most general change of frame is given by

$$\boldsymbol{x}^* = \boldsymbol{c}(t) + \boldsymbol{Q}(t)\boldsymbol{x}, \quad x_i^* = c_i(t) + Q_{ij}(t)x_j \tag{2.9.1}$$

where $\boldsymbol{c}(t)$ is the relative displacement between the two reference frames, and $\boldsymbol{Q}(t)$ is a second-order orthogonal tensor accounting for the relative rotational differences. Section 2.7 defined and listed various properties for the rotation tensor $\boldsymbol{Q}(t)$.

Under the frame change given by (2.9.1), let us first consider the displacement between the neighboring points $\boldsymbol{x}$ and $\boldsymbol{y}$. We start with the relations

$$\begin{aligned}\boldsymbol{x}^* &= \boldsymbol{c}(t) + \boldsymbol{Q}(t)\boldsymbol{x} \\ \boldsymbol{y}^* &= \boldsymbol{c}(t) + \boldsymbol{Q}(t)\boldsymbol{y}\end{aligned}$$

Subtracting these two defines the displacements $\boldsymbol{u}^* = \boldsymbol{y}^* - \boldsymbol{x}^*$ and $\boldsymbol{u} = \boldsymbol{y} - \boldsymbol{x}$ and establishes the transformation relation

$$\boldsymbol{u}^* = \boldsymbol{Q}(t)\boldsymbol{u} \tag{2.9.2}$$

which is exactly the same as the general tensor transformation law for vectors given in Sections 2.7 and 2.8. This is the required transformation behavior for first-order tensors in order to be called objective. Note also that it follows from (2.9.2) that $\boldsymbol{u}^*| = |\boldsymbol{u}|$ and thus displacement distance or magnitude is invariant as expected for this and all other scalars.

It was pointed out in Section 2.8 that second-order tensors can be defined as linear vector transformations (see Eq. (2.8.5)). Using this definition, we can determine the objective definition that is applied to second-order tensors. Starting with linear vector transformation relation (2.8.5), the second-order tensor $\boldsymbol{T}$ transforms vector $\boldsymbol{a}$ into another vector $\boldsymbol{b}$, through the relation $\boldsymbol{Ta} = \boldsymbol{b}$. Now we wish to have this relation be objective so that $\boldsymbol{T}^*\boldsymbol{a}^* = \boldsymbol{b}^*$. Also each vector should satisfy the objectivity relation (2.9.2), and so $\boldsymbol{a}^* = \boldsymbol{Q}\boldsymbol{a}$, $\boldsymbol{b}^* = \boldsymbol{Q}\boldsymbol{b}$. Combing these results together gives

$$\boldsymbol{T}^*\boldsymbol{a}^* = \boldsymbol{b}^* \Rightarrow \boldsymbol{T}^*\boldsymbol{Q}\boldsymbol{a} = \boldsymbol{Q}\boldsymbol{b} \Rightarrow \boldsymbol{T}^*\boldsymbol{Q}\boldsymbol{a} = \boldsymbol{Q}\boldsymbol{T}\boldsymbol{a}$$
$$\therefore \boldsymbol{T}^*\boldsymbol{Q} = \boldsymbol{Q}\boldsymbol{T}$$

post multiplying by $\boldsymbol{Q}^T$ then gives the expected transformation relation for objective second-order tensors

$$\boldsymbol{T}^* = \boldsymbol{Q}(t)\boldsymbol{T}\boldsymbol{Q}^T(t) \tag{2.9.3}$$

Therefore, first- and second-order tensors are called objective or frame indifferent if they satisfy relations (2.9.2) and (2.9.3).

Next, consider the velocity vector transformation behavior. Differentiating (2.9.1) with respect to time gives

$$\begin{aligned} \boldsymbol{v}^* &= \frac{d\boldsymbol{x}^*}{dt} = \dot{\boldsymbol{c}}(t) + \dot{\boldsymbol{Q}}(t)\boldsymbol{x} + \boldsymbol{Q}(t)\dot{\boldsymbol{x}} \Rightarrow \\ \boldsymbol{v}^* &= \dot{\boldsymbol{c}}(t) + \dot{\boldsymbol{Q}}(t)\boldsymbol{x} + \boldsymbol{Q}(t)\boldsymbol{v} = \dot{\boldsymbol{c}}(t) + \boldsymbol{Q}(t)\boldsymbol{v} + \dot{\boldsymbol{Q}}(t)\boldsymbol{Q}^T(t)[\boldsymbol{x}^* - \boldsymbol{c}] \end{aligned} \tag{2.9.4}$$

Note that the quantity $\dot{\boldsymbol{Q}}(t)$ would represent a relative angular velocity between the two reference frames $\mathbf{O}^*$ and $\mathbf{O}$. We thus conclude that the velocity is not in general objective unless $\dot{\boldsymbol{c}}(t) + \dot{\boldsymbol{Q}}(t)\boldsymbol{x}$ were to vanish which could be accomplished by the special case of time-independent changes of observer.

Finally, let us explore the next time derivative to calculate the acceleration

$$\begin{aligned} \boldsymbol{a}^* &= \frac{d\boldsymbol{v}^*}{dt} = \ddot{\boldsymbol{c}}(t) + \ddot{\boldsymbol{Q}}(t)\boldsymbol{x} + \dot{\boldsymbol{Q}}(t)\dot{\boldsymbol{x}} + \dot{\boldsymbol{Q}}(t)\dot{\boldsymbol{x}} + \boldsymbol{Q}(t)\ddot{\boldsymbol{x}} \Rightarrow \\ \boldsymbol{a}^* &= \ddot{\boldsymbol{c}}(t) + \ddot{\boldsymbol{Q}}(t)\boldsymbol{x} + 2\dot{\boldsymbol{Q}}(t)\boldsymbol{v} + \boldsymbol{Q}(t)\boldsymbol{a} \end{aligned} \tag{2.9.5}$$

and like the velocity, the acceleration is also not in general objective.

In summary, we have seen that under the general change in reference frame given by relation (2.9.1), objective scalars remain the same; objective vectors must satisfy (2.9.2); and objective second-order tensors must obey relation (2.9.3). We have set the stage for more detailed use of the principle of objectivity or frame indifference for many other variables to come in our study. As we have seen already, not all variables in continuum mechanics will be objective.

2.10 VECTOR AND MATRIX ALGEBRA

Continuum mechanics will require the use of many standard algebraic operations among vector, matrix, and tensor variables. These operations include dot and cross products of vectors and numerous matrix/tensor products. All of these operations can be expressed efficiently using compact tensor index or direct notation. We now present many of these operations.

First, consider some particular vector products. Given two vectors $\boldsymbol{a}$ and $\boldsymbol{b}$, with Cartesian components a_i and b_i, the *scalar* or *dot product* is defined by

$$\boldsymbol{a}\cdot\boldsymbol{b}=a_1b_1+a_2b_2+a_3b_3=a_ib_i \tag{2.10.1}$$

Since all indices in this expression are repeated, the quantity must be a scalar, that is, a tensor of order zero. The magnitude of a vector can then be expressed as

$$|\boldsymbol{a}|=(\boldsymbol{a}\cdot\boldsymbol{a})^{1/2}=(a_ia_i)^{1/2} \tag{2.10.2}$$

The *vector* or *cross product* between two vectors $\boldsymbol{a}$ and $\boldsymbol{b}$ can be written as

$$\boldsymbol{a}\times\boldsymbol{b}=\begin{vmatrix}\boldsymbol{e}_1 & \boldsymbol{e}_2 & \boldsymbol{e}_3\\ a_1 & a_2 & a_3\\ b_1 & b_2 & b_3\end{vmatrix}=\varepsilon_{ijk}a_jb_k\boldsymbol{e}_i \tag{2.10.3}$$

where $\boldsymbol{e}_i$ are the unit basis vectors for the coordinate system. Note that the cross product gives a vector resultant whose components are $\varepsilon_{ijk}a_jb_k$. Another common vector product is the *scalar triple product* defined by

$$\boldsymbol{a}\cdot(\boldsymbol{b}\times\boldsymbol{c})=\begin{vmatrix}a_1 & a_2 & a_3\\ b_1 & b_2 & b_3\\ c_1 & c_2 & c_3\end{vmatrix}=\varepsilon_{ijk}a_ib_jc_k \tag{2.10.4}$$

For this triple product, the cross operation must be done first, and the end result is then a scalar which is easily verified by observing the index notation form that has no free indices.

Next, consider some common matrix products. Using the usual direct notation for matrices and vectors, common products between a matrix $\boldsymbol{A}=[\boldsymbol{A}]$ with a vector $\boldsymbol{a}$ can be written as

$$\begin{aligned}\boldsymbol{A}\boldsymbol{a}&=[A]\{\boldsymbol{a}\}=A_{ij}a_j=a_jA_{ij}\\ \boldsymbol{a}^TA&=\{\boldsymbol{a}\}^T[A]=a_iA_{ij}=A_{ij}a_i\end{aligned} \tag{2.10.5}$$

where $\boldsymbol{a}^T$ denotes the *transpose* and for a vector quantity this simply changes the (3×1) column matrix into a (1×3) row matrix. Note that each of these products results in a vector resultant. These types of expressions generally involve various inner products within the index notational scheme, and as previously noted once the summation index is properly specified, the order of listing the product terms will not change the result. Considering next the various products among two matrices $\boldsymbol{A}$ and $\boldsymbol{B}$, several different combinations are commonly encountered:

$$\begin{aligned}\boldsymbol{AB}&=A_{ij}B_{jk}\\ \boldsymbol{AB}^T&=A_{ij}B_{kj}\\ \boldsymbol{A}^T\boldsymbol{B}&=A_{ji}B_{jk}\\ (\boldsymbol{AB})^T&=\boldsymbol{B}^T\boldsymbol{A}^T=(A_{ij}B_{jk})^T=A_{kj}B_{ji}\\ tr(\boldsymbol{AB})&=A_{ij}B_{ji}\\ tr(\boldsymbol{AB}^T)&=tr(\boldsymbol{A}^T\boldsymbol{B})=A_{ij}B_{ij}\\ \boldsymbol{A}^2&=\boldsymbol{AA}=A_{ij}A_{jk}\\ \boldsymbol{A}^3&=\boldsymbol{AAA}=A_{ij}A_{jk}A_{kl}\end{aligned} \tag{2.10.6}$$

where $\boldsymbol{A}^T$ indicates the *transpose* and $tr\,\boldsymbol{A}$ is the *trace* of the matrix defined by

$$A_{ij}^T = A_{ji}, \quad \boldsymbol{A}^T = A_{ji}\boldsymbol{e}_i \otimes \boldsymbol{e}_j$$
$$tr\,\boldsymbol{A} = A_{ii} = A_{11} + A_{22} + A_{33} \tag{2.10.7}$$

Similar to vector products, once the summation index is properly specified, the results in (2.10.6) will not depend on the order of listing the product terms. Note that this does not imply that $\boldsymbol{AB} = \boldsymbol{BA}$, which is certainly not true.

The *inverse* of a second-order tensor (square matrix) $\boldsymbol{A}$ is written as $\boldsymbol{A}^{-1}$ with the following definition properties written in both index and direct notation:

$$A_{ij}^{-1}A_{jk} = A_{ij}A_{jk}^{-1} = \delta_{ik}$$
$$\boldsymbol{A}^{-1}\boldsymbol{A} = \boldsymbol{A}\boldsymbol{A}^{-1} = \boldsymbol{I} \tag{2.10.8}$$

The matrix $\boldsymbol{A}$ is said to be *nonsingular* if it has a nonzero determinant, and if this is the case the matrix inverse is given by the following index notation relation:

$$A_{ij}^{-1} = \frac{1}{2\det[\boldsymbol{A}]}\varepsilon_{ikl}\varepsilon_{jqr}A_{qk}A_{rl} \tag{2.10.9}$$

Note a couple of other useful properties:

$$(\boldsymbol{AB})^{-1} = \boldsymbol{B}^{-1}\boldsymbol{A}^{-1}$$
$$(\boldsymbol{A}^T)^{-1} = (\boldsymbol{A}^{-1})^T \tag{2.10.10}$$

2.11 PRINCIPAL VALUES, DIRECTIONS, AND INVARIANTS OF SYMMETRIC SECOND-ORDER TENSORS

It should be apparent from the tensor transformation concepts previously discussed that there might exist particular coordinate systems where the components of a tensor will take on maximum, minimum, or other special values. This concept can be easily visualized by considering the components of a vector shown in Fig. 2.1. If we choose a particular coordinate system that has been rotated so that the x_3-axis lies along the direction of the vector itself, then the vector will have components $\boldsymbol{u} = [0, 0, |\boldsymbol{u}|]$. For this case, two of the components have been reduced to zero while the remaining component becomes the largest possible (the total magnitude).

This situation is most useful for symmetric second-order tensors that will eventually represent the stress and strain at a point in a continuum material. The direction determined by the unit vector $\boldsymbol{n}$ is said to be a *principal direction* or *eigenvector* of the symmetric second-order tensor A_{ij} if there exists a parameter λ such that

$$A_{ij}n_j = \lambda n_i \tag{2.11.1}$$

The parameter λ is called the *principal value* or *eigenvalue* of the tensor. Relation (2.11.1) can be rewritten as

$$(A_{ij} - \lambda\delta_{ij})n_j = 0$$

and this expression is recognized as a homogeneous system of three linear algebraic equations in the unknowns n_1, n_2, and n_3. The system possesses a nontrivial solution if and only if the determinant of its coefficient matrix vanishes, that is,

$$\det[A_{ij} - \lambda\delta_{ij}] = 0$$

Expanding the determinant produces a cubic equation in terms of λ:

$$\det[A_{ij} - \lambda\delta_{ij}] = -\lambda^3 + I_A\lambda^2 - II_A\lambda + III_A = 0 \tag{2.11.2}$$

where

$$\begin{aligned} I_A &= A_{ii} = A_{11} + A_{22} + A_{33} \\ II_A &= \frac{1}{2}(A_{ii}A_{jj} - A_{ij}A_{ij}) = \begin{vmatrix} A_{11} & A_{12} \\ A_{21} & A_{22} \end{vmatrix} + \begin{vmatrix} A_{22} & A_{23} \\ A_{32} & A_{33} \end{vmatrix} + \begin{vmatrix} A_{11} & A_{13} \\ A_{31} & A_{33} \end{vmatrix} \\ III_A &= \det[A_{ij}] \end{aligned} \tag{2.11.3}$$

The scalars I_A, II_A, and III_A are called the *fundamental invariants* of the tensor A_{ij} and the relation (2.11.2) is known as the *characteristic equation*. As expected by their name, the three invariants do not change value under coordinate transformation.

Another alternate set of invariants of a tensor that is sometimes used is given by

$$I_1 = tr\,\boldsymbol{A}, \quad I_2 = tr\,\boldsymbol{A}^2, \quad I_3 = tr\,\boldsymbol{A}^3 \tag{2.11.4}$$

These two sets of invariants are not independent and are related by the relations

$$I_1 = I_A, \quad I_2 = I_A^2 - 2II_A, \quad I_3 = I_A^3 - 3II_A I_A + 3III_A \tag{2.11.5}$$

The roots of the characteristic equation determine the allowable values for λ, and each of these may be back-substituted into relation (2.11.1) to solve for the associated principal direction $\boldsymbol{n}$. Normally, the principal directions are normalized so that $n_1^2 + n_2^2 + n_3^2 = 1$, thus adding another needed equation to the system to be solved (see Example 2.11.1).

Under the condition that the components A_{ij} are real, it can be shown that all three roots λ_1, λ_2, and λ_3 of the cubic equation (2.11.2) must be real. Furthermore, if these roots are distinct, the principal directions associated with each principal value will be orthogonal. Thus, we can conclude that every symmetric second-order tensor has at least three mutually perpendicular principal directions and at most three distinct principal values which are the roots of the characteristic equation. Denoting the principal directions $\boldsymbol{n}^{(1)}$, $\boldsymbol{n}^{(2)}$, and $\boldsymbol{n}^{(3)}$ corresponding to the principal values λ_1, λ_2, and λ_3, three possibilities arise:

1. All three principal values distinct; thus, the three corresponding principal directions are unique (except for sense).

2. Two principal values equal ($\lambda_1 \neq \lambda_2 = \lambda_3$); the principal direction $\boldsymbol{n}^{(1)}$ is unique (except for sense) and every direction perpendicular to $\boldsymbol{n}^{(1)}$ is a principal direction associated with λ_2 and λ_3.
3. All three principal values equal; every direction is principal and the tensor is isotropic, as per discussion in the previous section.

Therefore, according to what we have presented, it is always possible to identify a right-handed Cartesian coordinate system with axes that lie along the principal directions of any given symmetric second-order tensor. Such axes are called the *principal axes* of the tensor. For this case, the basis vectors are actually the unit principal directions $[\boldsymbol{n}^{(1)}, \boldsymbol{n}^{(2)}, \boldsymbol{n}^{(3)}]$, and it can be shown from (2.8.3) that with respect to principal axes the tensor components can be calculated using the relation

$$A_{ij} = \boldsymbol{e}_i \cdot \boldsymbol{A}\boldsymbol{e}_j = \boldsymbol{n}^{(i)} \cdot \boldsymbol{A}\boldsymbol{n}^{(j)} \tag{2.11.6}$$

which gives

$$\begin{aligned}
A_{11} &= \boldsymbol{n}^{(1)} \cdot \boldsymbol{A}\boldsymbol{n}^{(1)} = \boldsymbol{n}^{(1)} \cdot (\lambda_1 \boldsymbol{n}^{(1)}) = \lambda_1 \\
A_{22} &= \boldsymbol{n}^{(2)} \cdot \boldsymbol{A}\boldsymbol{n}^{(2)} = \boldsymbol{n}^{(2)} \cdot (\lambda_2 \boldsymbol{n}^{(2)}) = \lambda_2 \\
A_{33} &= \boldsymbol{n}^{(3)} \cdot \boldsymbol{A}\boldsymbol{n}^{(3)} = \boldsymbol{n}^{(3)} \cdot (\lambda_3 \boldsymbol{n}^{(3)}) = \lambda_3 \\
A_{12} &= \boldsymbol{n}^{(1)} \cdot \boldsymbol{A}\boldsymbol{n}^{(2)} = \boldsymbol{n}^{(1)} \cdot (\lambda_2 \boldsymbol{n}^{(2)}) = 0 = A_{21} \\
A_{23} &= \boldsymbol{n}^{(2)} \cdot \boldsymbol{A}\boldsymbol{n}^{(3)} = \boldsymbol{n}^{(2)} \cdot (\lambda_3 \boldsymbol{n}^{(3)}) = 0 = A_{32} \\
A_{31} &= \boldsymbol{n}^{(3)} \cdot \boldsymbol{A}\boldsymbol{n}^{(1)} = \boldsymbol{n}^{(3)} \cdot (\lambda_1 \boldsymbol{n}^{(1)}) = 0 = A_{13}
\end{aligned}$$

Collecting these results indicates that with respect to the principal axes, the tensor reduces to the diagonal form:

$$A_{ij} = \begin{bmatrix} \lambda_1 & 0 & 0 \\ 0 & \lambda_2 & 0 \\ 0 & 0 & \lambda_3 \end{bmatrix} \tag{2.11.7}$$

Note that the fundamental invariants defined by relations (2.11.3) can be expressed in terms of the principal values as

$$\begin{aligned}
I_A &= \lambda_1 + \lambda_2 + \lambda_3 \\
II_A &= \lambda_1\lambda_2 + \lambda_2\lambda_3 + \lambda_3\lambda_1 \\
III_A &= \lambda_1\lambda_2\lambda_3
\end{aligned} \tag{2.11.8}$$

The eigenvalues of a symmetric second-order tensor have important *extremal properties*. If we consider an arbitrary unit vector $\boldsymbol{e}_1 = \alpha\boldsymbol{n}^{(1)} + \beta\boldsymbol{n}^{(2)} + \gamma\boldsymbol{n}^{(3)}$, then $\alpha^2 + \beta^2 + \gamma^2 = 1$. Using relation (2.11.6) again

$$A_{11} = \boldsymbol{e}_1 \cdot \boldsymbol{A}\boldsymbol{e}_1 = \lambda_1\alpha^2 + \lambda_2\beta^2 + \lambda_3\gamma^2$$

Without loss in generality, let $\lambda_1 \geq \lambda_2 \geq \lambda_3$, and then since $\alpha^2 + \beta^2 + \gamma^2 = 1$, we can write

$$\lambda_1 = \lambda_1(\alpha^2 + \beta^2 + \gamma^2) \geq \lambda_1\alpha^2 + \lambda_2\beta^2 + \lambda_3\gamma^2$$

thus proving that $\lambda_1 \geq A_{11}$. Likewise

$$\lambda_3 = \lambda_3(\alpha^2 + \beta^2 + \gamma^2) \leq \lambda_1\alpha^2 + \lambda_2\beta^2 + \lambda_3\gamma^2$$

and so $\lambda_3 \leq A_{11}$.

Thus, we can conclude that the maximum (minimum) principal value of $\boldsymbol{A}$ is the maximum (minimum) value of any diagonal element of the second-order tensor in any coordinate frame. Or stated in another way: if we arbitrarily rank the principal values such that $\lambda_1 > \lambda_2 > \lambda_3$, then λ_1 will be the largest of all possible diagonal elements, whereas λ_3 will be the smallest diagonal element possible. Several of these concepts will be applied in our subsequent studies as we seek the largest stress or strain components in a continuum material.

EXAMPLE 2.11.1 PRINCIPAL VALUE PROBLEM

Determine the invariants, and principal values and directions of the following symmetric second-order tensor:

$$A_{ij} = \begin{bmatrix} -1 & 1 & 0 \\ 1 & -1 & 0 \\ 0 & 0 & 1 \end{bmatrix}$$

Solution: The invariants follow from relations (2.11.3)

$$I_A = a_{ii} = -1 - 1 + 1 = -1$$
$$II_A = \begin{vmatrix} -1 & 1 \\ 1 & -1 \end{vmatrix} + \begin{vmatrix} -1 & 0 \\ 0 & 1 \end{vmatrix} + \begin{vmatrix} -1 & 0 \\ 0 & 1 \end{vmatrix} = 0 - 1 - 1 = -2$$
$$III_A = \begin{vmatrix} -1 & 1 & 0 \\ 1 & -1 & 0 \\ 0 & 0 & 1 \end{vmatrix} = 0$$

The characteristic equation then becomes

$$\det[A_{ij} - \lambda\delta_{ij}] = -\lambda^3 - \lambda^2 + 2\lambda = 0$$
$$\Rightarrow \lambda(\lambda^2 + \lambda - 2) = 0 \Rightarrow \lambda(\lambda - 1)(\lambda + 2) = 0$$
$$\therefore \lambda_1 = 1, \quad \lambda_2 = 0, \quad \lambda_3 = -2$$

Thus, for this case, all principal values are distinct.

For the $\lambda_1 = 1$ root, Eq. (2.11.1) gives the system

$$-2n_1^{(1)} + n_2^{(1)} = 0$$
$$n_1^{(1)} - 2n_2^{(1)} = 0$$
$$0 = 0$$

which gives a solution $n_1^{(1)} = n_2^{(1)} = 0$ and leaves $n_3^{(1)}$ undefined. Adding the normalization equation $n_1^{(1)^2} + n_2^{(1)^2} + n_3^{(1)^2} = 1$ implies that $n_3^{(1)} = 1$, and thus $\boldsymbol{n}^{(1)} = \pm\boldsymbol{e}_3$. In a similar fashion, the other two principal directions are found to be $\boldsymbol{n}^{(2)} = \pm\frac{1}{\sqrt{2}}(\boldsymbol{e}_1 + \boldsymbol{e}_2), \boldsymbol{n}^{(3)} = \pm\frac{1}{\sqrt{2}}(\boldsymbol{e}_1 - \boldsymbol{e}_2)$. It is easily verified that these directions are mutually orthogonal. Fig. 2.5 illustrates their directions with respect to the given coordinate system and this establishes the right-handed principal coordinate axes (x_1', x_2', x_3'). For this case, the transformation matrix Q_{ij} defined by (2.7.1) becomes

$$Q_{ij} = \begin{bmatrix} 0 & 0 & 1 \\ 1/\sqrt{2} & 1/\sqrt{2} & 0 \\ 1/\sqrt{2} & -1/\sqrt{2} & 0 \end{bmatrix}$$

Notice the eigenvectors actually form the rows of the Q-matrix.

Using this in the transformation law $(2.8.1)_3$, the components of the given second-order tensor become

$$A_{ij}' = \begin{bmatrix} 1 & 0 & 0 \\ 0 & 0 & 0 \\ 0 & 0 & -2 \end{bmatrix}$$

This result then validates the general theory given by relation (2.11.7) indicating that the tensor should take on diagonal form with the principal values as the elements.

Only simple second-order tensors will lead to a characteristic equation that is factorable, thus allowing an easy solution via hand calculation. Most of the other cases will normally develop a general cubic equation and thus a more complicated system to solve for the principal directions. Again particular routines within the MATLAB package offer convenient tools to solve these more general problems. Code C2 in Appendix C provides a simple code to determine the principal values and directions for symmetric second-order tensors.

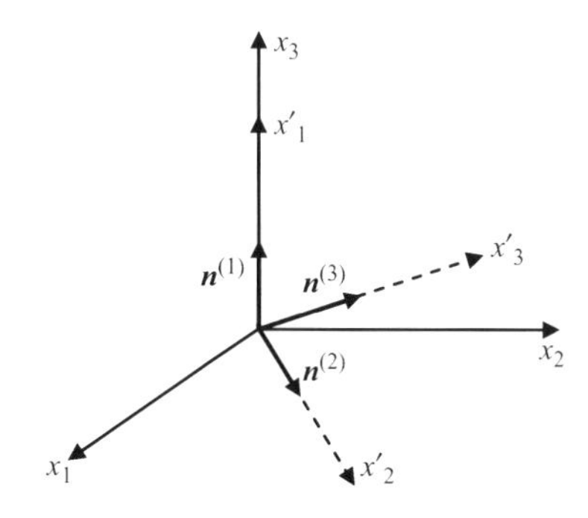

FIGURE 2.5

Principal axes for Example 2.11.1.

2.12 SPHERICAL AND DEVIATORIC SECOND-ORDER TENSORS

In particular applications, it is often convenient to decompose second-order tensors into two parts called *spherical* and *deviatoric tensors*. Considering a general second-order tensor A_{ij}, the spherical part is defined by

$$\tilde{A}_{ij} = \frac{1}{3} A_{kk} \delta_{ij} \tag{2.12.1}$$

while the deviatoric part is specified as

$$\hat{A}_{ij} = A_{ij} - \frac{1}{3} A_{kk} \delta_{ij} \tag{2.12.2}$$

Note that the sum of the two parts then gives the original tensor A_{ij}

$$A_{ij} = \tilde{A}_{ij} + \hat{A}_{ij} \tag{2.12.3}$$

The spherical part is an isotropic tensor, being the same in all coordinate systems (as per discussion in Section 2.8). It can be shown that the principal directions of the deviatoric tensor are the same as those of the original tensor itself and that $tr\hat{\mathbf{A}} = \hat{A}_{kk} = 0$. These definitions will prove to be useful in later sections of the text dealing with strain, stress, and constitutive relations.

EXAMPLE 2.12.1 SPHERICAL AND DEVIATORIC DECOMPOSITION

Determine the spherical and deviatoric parts of the given second-order tensor:

$$A_{ij} = \begin{bmatrix} 2 & 2 & 1 \\ 0 & 2 & 4 \\ 2 & 0 & 2 \end{bmatrix}$$

Solution: From definitions (2.12.1) and (2.12.2)

$$\tilde{A}_{ij} = \frac{1}{3} A_{kk} \delta_{ij} = \frac{1}{3}(6)\delta_{ij} = 2\delta_{ij} = 2\begin{bmatrix} 1 & 0 & 0 \\ 0 & 1 & 0 \\ 0 & 0 & 1 \end{bmatrix}$$

$$\hat{A}_{ij} = A_{ij} - \frac{1}{3} A_{kk} \delta_{ij} = \begin{bmatrix} 2 & 2 & 1 \\ 0 & 2 & 4 \\ 2 & 0 & 2 \end{bmatrix} - 2\begin{bmatrix} 1 & 0 & 0 \\ 0 & 1 & 0 \\ 0 & 0 & 1 \end{bmatrix} = \begin{bmatrix} 0 & 2 & 1 \\ 0 & 0 & 4 \\ 2 & 0 & 0 \end{bmatrix}$$

A simple check will verify that $A_{ij} = \tilde{A}_{ij} + \hat{A}_{ij}$ and $tr\,\hat{A} = 0$.

2.13 CAYLEY–HAMILTON THEOREM AND MATRIX POLYNOMIALS

The Cayley–Hamilton theorem states that a square matrix or second-order tensor satisfies its own characteristic equation. So, for application to a symmetric second-order tensor $\boldsymbol{A}$, we can write

$$\boldsymbol{A}^3 - I_A \boldsymbol{A}^2 + II_A \boldsymbol{A} - III_A \boldsymbol{I} = 0 \tag{2.13.1}$$

where scalars I_A, II_A, and III_A are the fundamental invariants of tensor $\boldsymbol{A}$ previously defined in (2.11.3).

Clearly result (2.13.1) then allows one to express $\boldsymbol{A}^3$ in terms of $\boldsymbol{A}^2$ and $\boldsymbol{A}$:

$$\boldsymbol{A}^3 = I_A \boldsymbol{A}^2 - II_A \boldsymbol{A} + III_A \boldsymbol{I} \tag{2.13.2}$$

Multiplying (2.13.2) by $\boldsymbol{A}$ gives an expression for $\boldsymbol{A}^4$, and back-substituting this with (2.13.2) to eliminate $\boldsymbol{A}^3$ gives a relationship for $\boldsymbol{A}^4$ in terms of $\boldsymbol{A}$ and $\boldsymbol{A}^2$:

$$\boldsymbol{A}^4 = (I_A^2 - II_A)\boldsymbol{A}^2 + (III_A - I_A II_A)\boldsymbol{A} + I_A III_A \boldsymbol{I} \tag{2.13.3}$$

We can then continue this process to get

$$\begin{aligned}\boldsymbol{A}^5 = {} & (I_A^3 - 2I_A II_A + III_A)\boldsymbol{A}^2 + (I_A III_A - I_A^2 II_A + II_A^2)\boldsymbol{A} \\ & + (I_A^2 III_A - II_A III_A)\boldsymbol{I}\end{aligned} \tag{2.13.4}$$

It then follows that by using this procedure, one can express any power of $\boldsymbol{A}$ greater than or equal to 3 in terms of $\boldsymbol{A}$ and $\boldsymbol{A}^2$. Consequently, any *polynomial representation* of $\boldsymbol{A}$

$$f(\boldsymbol{A}) = c_0 \boldsymbol{I} + c_1 \boldsymbol{A} + c_2 \boldsymbol{A}^2 + c_3 \boldsymbol{A}^3 + \cdots + c_n \boldsymbol{A}^n \tag{2.13.5}$$

where c_i are *constants*, can be expressed as

$$f(\boldsymbol{A}) = k_0 \boldsymbol{I} + k_1 \boldsymbol{A} + k_2 \boldsymbol{A}^2 \tag{2.13.6}$$

where k_i are now polynomial functions of the invariants I_A, II_A, and III_A. Result (2.13.6) is useful in developing nonlinear constitutive equations in continuum mechanics, and we will put this into play in later chapters. On a related topic, the Cayley–Hamilton theorem is also useful to express $\boldsymbol{A}^2$ in terms of $\boldsymbol{A}$ its inverse $\boldsymbol{A}^{-1}$ through the relation

$$\boldsymbol{A}^2 = -II_A \boldsymbol{I} + I_A \boldsymbol{A} + III_A \boldsymbol{A}^{-1} \tag{2.13.7}$$

Using this result, representation relation (2.13.6) can be expressed in the alternative form

$$f(\boldsymbol{A}) = K_0 \boldsymbol{I} + K_1 \boldsymbol{A} + K_2 \boldsymbol{A}^{-1} \tag{2.13.8}$$

2.14 REPRESENTATION THEOREMS

Many applications in continuum mechanics associated with constitutive equation development require representation of *scalar- and tensor-valued functions.* These particular functions involve tensor arguments. In general, these relations should be *form invariant*, that is, if the coordinate system is changed, then with the appropriate tensorial changes of each tensor in the equation, the basic form of the relation should remain invariant. Several useful representation theorems are hereby listed.

Scalar-Valued Theorem

A scalar-valued function $f(\mathbf{A})$ of a symmetric second-order tensor ($\mathbf{A} = \mathbf{A}^T$), is form invariant, that is, $f(\mathbf{A}) = f(\mathbf{QAQ}^T)$, if

$$f(\mathbf{A}) = f(I_A, II_A, III_A) \tag{2.14.1}$$

Note that if f is also to be a *linear function*, then form (2.14.1) must reduce to

$$f(\mathbf{A}) = c_0 I_A = c_0(\mathrm{tr}\,\mathbf{A}) = c_0 A_{ii} \tag{2.14.2}$$

with c_0 being an arbitrary constant.

Tensor-Valued Theorem

A symmetric tensor-valued function $\boldsymbol{f}(\mathbf{A})$ of a symmetric second-order tensor ($\mathbf{A} = \mathbf{A}^T$) is form invariant, that is, $\mathbf{Q}\boldsymbol{f}(\mathbf{A})\mathbf{Q}^T = \boldsymbol{f}(\mathbf{QAQ}^T)$, if

$$\boldsymbol{f}(\mathbf{A}) = c_0\mathbf{I} + c_1\mathbf{A} + c_2\mathbf{A}^2 \tag{2.14.3}$$

where $c_i = c_i(I_A, II_A, III_A)$. Note that if $\boldsymbol{f}$ is *linear* in $\mathbf{A}$, then form (2.14.3) must reduce to

$$\boldsymbol{f}(\mathbf{A}) = c_0 I_A \mathbf{I} + c_1\mathbf{A} \tag{2.14.4}$$

Relations (2.14.3) and (2.14.4) are useful in elasticity theories where the stress tensor is a function of the current value of the strain tensor.

Tensor-Valued Theorem with Two Arguments

A symmetric tensor-valued function $\boldsymbol{f}(\mathbf{A},\mathbf{B})$ of two symmetric second-order tensors ($\mathbf{A} = \mathbf{A}^T$, $\mathbf{B} = \mathbf{B}^T$) is form invariant, that is, $\mathbf{Q}\boldsymbol{f}(\mathbf{A},\mathbf{B})\mathbf{Q}^T = \boldsymbol{f}(\mathbf{QAQ}^T, \mathbf{QBQ}^T)$, if

$$\begin{aligned}\boldsymbol{f}(\mathbf{A},\mathbf{B}) = c_0\mathbf{I} &+ c_1\mathbf{A} + c_2\mathbf{A}^2 + c_3\mathbf{B} + c_4\mathbf{B}^2 \\ &+ c_5(\mathbf{AB} + \mathbf{BA}) + c_6(\mathbf{A}^2\mathbf{B} + \mathbf{BA}^2) \\ &+ c_7(\mathbf{AB}^2 + \mathbf{B}^2\mathbf{A}) + c_8(\mathbf{A}^2\mathbf{B}^2 + \mathbf{B}^2\mathbf{A}^2)\end{aligned} \tag{2.14.5}$$

where the coefficients c_i are functions of the invariants: $tr\mathbf{A}$, $tr\mathbf{A}^2$, $tr\mathbf{A}^3$, $tr\mathbf{B}$, $tr\mathbf{B}^2$, $tr\mathbf{B}^3$, $tr(\mathbf{AB})$, $tr(\mathbf{A}^2\mathbf{B})$, $tr(\mathbf{AB}^2)$, $tr(\mathbf{A}^2\mathbf{B}^2)$.

2.15 ISOTROPIC TENSORS

As pointed out in Section 2.8, a tensor is called *isotropic if its components are the same in all coordinate frames.* In that section, we showed that the Kronecker delta and the alternating symbol had such properties. We now wish to explore a more complete collection of general forms of isotropic tensors of various orders. For a general tensor $\boldsymbol{A}$ to be isotropic, it must satisfy the transformation relation

$$A_{ijk\dots m} = Q_{ip}Q_{jq}Q_{kr}\cdots Q_{mt}A_{pqr\dots t} \tag{2.15.1}$$

Using this result, we can construct general three-dimensional forms for orders up to 4 as listed in Table 2.2. Notice that the fourth-order case involves products of the Kronecker delta which is an expected result.

Table 2.2 General forms of isotropic tensors

Zeroth order	All tensors (scalars)
First order	Zero or null vector
Second order	$C\delta_{ij}$, for any constant C
Third order	$C\varepsilon_{ijk}$, for any constant C
Fourth order	$\alpha\delta_{ij}\delta_{kl} + \beta\delta_{ik}\delta_{jl} + \gamma\delta_{il}\delta_{jk}$, for any constants α, β, and γ

2.16 POLAR DECOMPOSITION THEOREM

A useful multiplicative decomposition in tensor theory is known as the *Polar Decomposition Theorem*, which states that any nonsingular second-order tensor $\boldsymbol{A}$ may be uniquely written as the product

$$\boldsymbol{A} = \boldsymbol{RU} = \boldsymbol{VR}, \quad A_{ij} = R_{ik}U_{kj} = V_{ik}R_{kj} \tag{2.16.1}$$

where $\boldsymbol{R}$ is a proper orthogonal tensor, and $\boldsymbol{U}$ and $\boldsymbol{V}$ are symmetric positive definite tensors (see definition in Exercise 2.20). It can be shown that

$$\begin{aligned} \boldsymbol{U}^2 &= \boldsymbol{A}^T\boldsymbol{A}, \quad U_{ik}U_{kj} = A_{ki}A_{kj} \\ \boldsymbol{V}^2 &= \boldsymbol{AA}^T, \quad V_{ik}V_{kj} = A_{ik}A_{jk} \end{aligned} \tag{2.16.2}$$

This result will be useful to decompose particular deformation tensors.

2.17 CALCULUS OF CARTESIAN FIELD TENSORS

Most variables used in continuum mechanics are field variables that depend on the spatial coordinates used to formulate the problem under study. In addition, for time-dependent problems, these variables could also have temporal variation; however, at the moment, we will hold off on including time dependency. Thus, with respect to a Cartesian coordinate system, our scalar, vector, matrix, and other general tensor variables will be functions of the spatial coordinates (x_1, x_2, x_3). Since many of our equations will involve differential and integral operations, it is necessary to have an understanding of the calculus of Cartesian tensor fields. The field concept for tensor components can be expressed as

$$\begin{aligned} a &= a(x_1,x_2,x_3) = a(x_m) = a(\mathbf{x}) \\ a_i &= a_i(x_1,x_2,x_3) = a_i(x_m) = \boldsymbol{a}(\mathbf{x}) \\ A_{ij} &= A_{ij}(x_1,x_2,x_3) = A_{ij}(x_m) = \boldsymbol{A}(\mathbf{x}) \\ &\vdots \end{aligned} \tag{2.17.1}$$

The transformation law for the components of a Cartesian tensor field is then

$$A'_{ij\cdots k}(x'_m) = Q_{ip}Q_{jq}\cdots Q_{ks}A_{ij\cdots k}(x_m) \tag{2.17.2}$$

where $x'_m = Q_{mn}x_n$.

It is convenient to introduce the *comma notation* for partial differentiation

$$a_{,i} = \frac{\partial}{\partial x_i}a, \quad a_{i,j} = \frac{\partial}{\partial x_j}a_i, \quad A_{ij,k} = \frac{\partial}{\partial x_k}A_{ij}, \ldots \tag{2.17.3}$$

If the differentiation index is distinct as shown in the examples in relation (2.17.3), the order of the tensor will be increased by 1. For example, the derivative operation on a vector $a_{i,j}$ produces a second-order tensor or matrix given by

$$a_{i,j} = \begin{bmatrix} \dfrac{\partial a_1}{\partial x_1} & \dfrac{\partial a_1}{\partial x_2} & \dfrac{\partial a_1}{\partial x_3} \\ \dfrac{\partial a_2}{\partial x_1} & \dfrac{\partial a_2}{\partial x_2} & \dfrac{\partial a_2}{\partial x_3} \\ \dfrac{\partial a_3}{\partial x_1} & \dfrac{\partial a_3}{\partial x_2} & \dfrac{\partial a_3}{\partial x_3} \end{bmatrix} \tag{2.17.4}$$

We can formally prove this general property for the case of differentiating a second-order tensor. Starting with the standard transformation relation

$$A'_{ij} = Q_{ip}Q_{jq}A_{pq}$$

differentiate with respect to x'_k:

$$A'_{ij,k} = \frac{\partial}{\partial x'_k}(Q_{ip}Q_{jq}A_{pq}) = Q_{ip}Q_{jq}\frac{\partial A_{pq}}{\partial x'_k} = Q_{ip}Q_{jq}\frac{\partial A_{pq}}{\partial x_r}\frac{\partial x_r}{\partial x'_k}$$

Note we used the fact that the transformation matrices are constants and employed the chain rule. Now since $x_r = Q_{sr}x'_s$, $\dfrac{\partial x_r}{\partial x'_k} = Q_{sr}\dfrac{\partial x'_s}{\partial x'_k} = Q_{sr}\delta_{sk} = Q_{kr}$. Putting all of this together gives

$$A'_{ij,k} = Q_{ip}Q_{jq}Q_{kr}A_{pq,r}$$

and thus $A_{ij,k}$ must be a third-order tensor since it satisfies the standard transformation law.

Consider next the *directional derivative* of a scalar field function f with respect to direction s:

$$\frac{df}{ds} = \frac{\partial f}{\partial x_1}\frac{dx_1}{ds} + \frac{\partial f}{\partial x_2}\frac{dx_2}{ds} + \frac{\partial f}{\partial x_3}\frac{dx_3}{ds} \tag{2.17.5}$$

Noting that the unit vector in direction s can be written as

$$\boldsymbol{n} = \frac{dx_1}{ds}\boldsymbol{e}_1 + \frac{dx_2}{ds}\boldsymbol{e}_2 + \frac{dx_3}{ds}\boldsymbol{e}_3 \tag{2.17.6}$$

and so the directional derivative can be expressed as the following scalar product:

$$\frac{df}{ds} = \boldsymbol{n}\cdot\nabla f \tag{2.17.7}$$

where ∇f is called the *gradient* of the scalar function f and is defined by

$$\nabla f = \text{grad}\, f = \boldsymbol{e}_1\frac{\partial f}{\partial x_1} + \boldsymbol{e}_2\frac{\partial f}{\partial x_2} + \boldsymbol{e}_3\frac{\partial f}{\partial x_3} \tag{2.17.8}$$

and the symbolic vector differential operator ∇ called the *del operator* is

$$\nabla = \boldsymbol{e}_1\frac{\partial}{\partial x_1} + \boldsymbol{e}_2\frac{\partial}{\partial x_2} + \boldsymbol{e}_3\frac{\partial}{\partial x_3} \tag{2.17.9}$$

These and other useful operations can be expressed in Cartesian tensor notation. Given the scalar field ϕ and vector field $\boldsymbol{u}$, the following common differential operations can then be written in index notation:

$$\begin{aligned}
&\text{Gradient of a scalar:} && \nabla\phi = \text{grad}\,\phi = \phi_{,i}\boldsymbol{e}_i\\
&\text{Gradient of a vector:} && \nabla\boldsymbol{u} = \text{grad}\,\boldsymbol{u} = u_{i,j}\boldsymbol{e}_i\boldsymbol{e}_j\\
&\text{Laplacian of a scalar:} && \nabla^2\phi = \nabla\cdot\nabla\phi = \phi_{,ii}\\
&\text{Divergence of a vector:} && \nabla\cdot\boldsymbol{u} = \text{div}\,\boldsymbol{u} = u_{i,i}\\
&\text{Curl of a vector:} && \nabla\times\boldsymbol{u} = \text{curl}\,\boldsymbol{u} = \varepsilon_{ijk}u_{k,j}\boldsymbol{e}_i\\
&\text{Laplacian of a vector:} && \nabla^2\boldsymbol{u} = u_{i,kk}\boldsymbol{e}_i
\end{aligned} \tag{2.17.10}$$

These operations will appear in many places in continuum mechanics.

If ϕ and ψ are scalar fields and $\boldsymbol{u}$ and $\boldsymbol{v}$ are vector fields, several useful identities exist:

$$\begin{gathered}
\nabla(\phi\psi) = (\nabla\phi)\psi + \phi(\nabla\psi)\\
\nabla^2(\phi\psi) = (\nabla^2\phi)\psi + \phi(\nabla^2\psi) + 2\nabla\phi\cdot\nabla\psi\\
\nabla\cdot(\phi\boldsymbol{u}) = \nabla\phi\cdot\boldsymbol{u} + \phi(\nabla\cdot\boldsymbol{u})\\
\nabla\times(\phi\boldsymbol{u}) = \nabla\phi\times\boldsymbol{u} + \phi(\nabla\times\boldsymbol{u})\\
\nabla\cdot(\boldsymbol{u}\times\boldsymbol{v}) = \boldsymbol{v}\cdot(\nabla\times\boldsymbol{u}) - \boldsymbol{u}\cdot(\nabla\times\boldsymbol{v})\\
\nabla\times\nabla\phi = 0\\
\nabla\cdot\nabla\phi = \nabla^2\phi\\
\nabla\cdot\nabla\times\boldsymbol{u} = 0\\
\nabla\times(\nabla\times\boldsymbol{u}) = \nabla(\nabla\cdot\boldsymbol{u}) - \nabla^2\boldsymbol{u}\\
\boldsymbol{u}\times(\nabla\times\boldsymbol{u}) = \frac{1}{2}\nabla(\boldsymbol{u}\cdot\boldsymbol{u}) - \boldsymbol{u}\cdot\nabla\boldsymbol{u}
\end{gathered} \tag{2.17.11}$$

Each of these identities can be easily justified using index notation from definition relations (2.17.10).

Next consider some results from vector/tensor integral calculus. We will simply list some theorems that will have later use in the development of continuum mechanics theory.

Divergence or Gauss Theorem

Let S be a piecewise continuous surface bounding the region of space V. If a vector field $\boldsymbol{u}$ is continuous and has continuous first derivatives in V, then

$$\iint_S \boldsymbol{u}\cdot\boldsymbol{n}\,dS = \iiint_V \nabla\cdot\boldsymbol{u}\,dV \tag{2.17.12}$$

where $\boldsymbol{n}$ is the outer unit normal vector to surface S. This result is also true for tensors of any order, that is,

$$\iint_S A_{ij...k}n_k\,dS = \iiint_V A_{ij...k,k}\,dV \tag{2.17.13}$$

Stokes Theorem

Let S be an open two-sided surface bounded by a piecewise continuous, simple closed curve C. If $\boldsymbol{u}$ is continuous and has continuous first derivatives on S, then

$$\oint_C \boldsymbol{u}\cdot d\boldsymbol{r} = \iint_S (\nabla\times\boldsymbol{u})\cdot\boldsymbol{n}\,dS \tag{2.17.14}$$

where the positive sense for the line integral is for the region S to lie to the left as one traverses curve C, and $\boldsymbol{n}$ is the unit normal vector to S. Again, this result is also valid for tensors of arbitrary order, and so

$$\oint_C A_{ij...k}\,dx_k = \iint_S \varepsilon_{rsk}A_{ij...k,s}n_r\,dS \tag{2.17.15}$$

It can be shown that both Divergence and Stokes Theorems can be generalized so that the dot product in (2.17.12) and (2.17.14) can be replaced with a cross product.

Localization Theorem

Let $F_{ij...k}$ be a continuous tensor field of any order defined in an arbitrary region V. If the integral of $F_{ij...k}$ over V vanishes, then F_{ijk} must vanish in V, that is,

$$\iiint_V F_{ij...k}\,dV = 0 \Rightarrow F_{ij...k} = 0 \in V \tag{2.17.16}$$

This result will be repeatedly used in Chapter 5 in going through the conversion process from integral to differential forms related to conservation laws.

Further information on vector differential and integral calculus can be found in the text by Hildebrand (1976) and Kreyszig (2011).

2.18 ORTHOGONAL CURVILINEAR COORDINATE SYSTEMS

In many applications, continuum mechanics is applied to problems that have curved domain and boundary geometry commonly including circular, cylindrical, and spherical surfaces. For such applications, it is highly desirable to use a curvilinear coordinate system to formulate and solve such problems. This requires redevelopment of some previous results in orthogonal curvilinear coordinates. Before pursuing these general steps, we will review the two most common curvilinear systems, *cylindrical* and *spherical coordinates*. The cylindrical coordinate system shown in Fig. 2.6 uses (r, θ, z) coordinates to describe spatial geometry. Relations between the Cartesian and cylindrical systems are given by

$$\begin{aligned} &x_1 = r\cos\theta, \quad x_2 = r\sin\theta, \quad x_3 = z \\ &r = \sqrt{x_1^2 + x_2^2}, \quad \theta = \tan^{-1}\frac{x_2}{x_1}, \quad z = x_3 \end{aligned} \tag{2.18.1}$$

The spherical coordinate system shown in Fig. 2.7 uses (R, ϕ, θ) coordinates to describe geometry, and the relations between Cartesian and spherical coordinates are

$$\begin{aligned} &x_1 = R\cos\theta\sin\phi, \quad x_2 = R\sin\theta\sin\phi, \quad x_3 = R\cos\phi \\ &R = \sqrt{x_1^2 + x_2^2 + x_3^2}, \quad \phi = \cos^{-1}\frac{x_3}{\sqrt{x_1^2 + x_2^2 + x_3^2}}, \quad \theta = \tan^{-1}\frac{x_2}{x_1} \end{aligned} \tag{2.18.2}$$

The unit basis vectors for each of these curvilinear systems are illustrated in Figs. 2.6 and 2.7, and these represent unit tangent vectors along each of the three orthogonal coordinate curves.

Although primary use of curvilinear systems will employ cylindrical and spherical coordinates, we briefly present a general discussion valid for arbitrary coordinate systems. Consider the general case where three orthogonal curvilinear coordinates

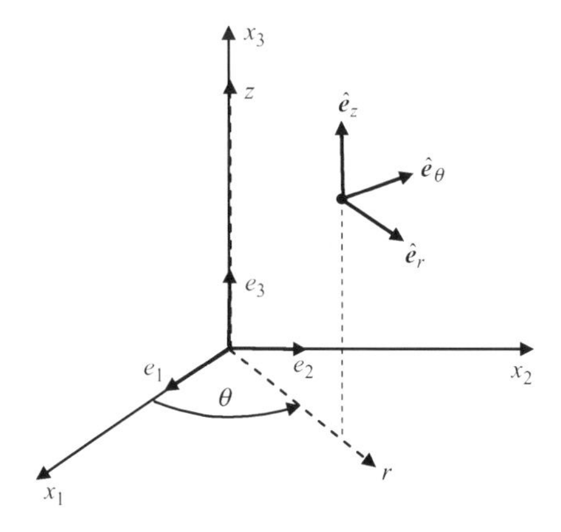

FIGURE 2.6

Cylindrical coordinate system.

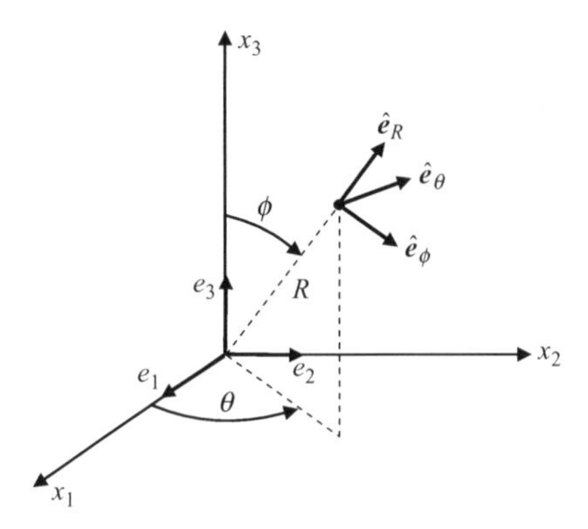

FIGURE 2.7

Spherical coordinate system.

are denoted by ξ^1, ξ^2, and ξ^3 whereas the Cartesian coordinates are defined by x^1, x^2, and x^3 (see Fig. 2.8). We assume there exists invertible coordinate transformations between these systems specified by

$$\xi^m = \xi^m(x^1, x^2, x^3), \quad x^m = x^m(\xi^1, \xi^2, \xi^3) \tag{2.18.3}$$

In the curvilinear system, an arbitrary differential length in space can be expressed by

$$(ds)^2 = (h_1\, d\xi^1)^2 + (h_2\, d\xi^2)^2 + (h_3\, d\xi^3)^2 \tag{2.18.4}$$

where h_1, h_2, and h_3 are called *scale factors* that are in general nonnegative functions of position. Let $\boldsymbol{e}_k$ be the fixed Cartesian basis vectors and $\hat{\boldsymbol{e}}_k$ the curvilinear basis (see Fig. 2.8). Using similar concepts from the transformations discussed in Section 2.8, the curvilinear basis can be expressed in terms of the Cartesian basis as

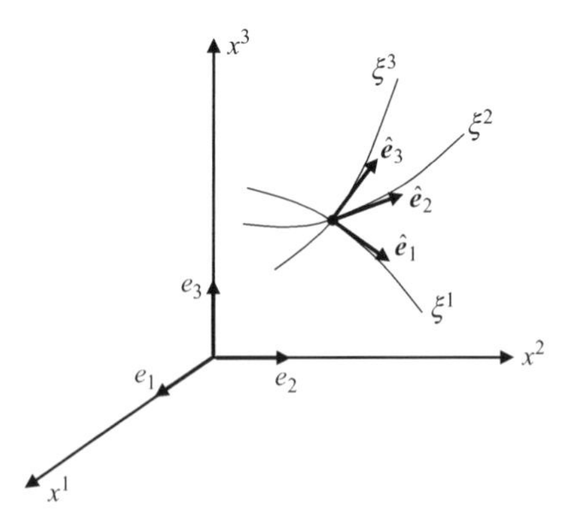

FIGURE 2.8

General curvilinear coordinates.

$$\begin{aligned}
\hat{\boldsymbol{e}}_1 &= \frac{dx^k}{ds_1}\boldsymbol{e}_k = \frac{1}{h_1}\frac{\partial x^k}{\partial \xi^1}\boldsymbol{e}_k \\
\hat{\boldsymbol{e}}_2 &= \frac{dx^k}{ds_2}\boldsymbol{e}_k = \frac{1}{h_2}\frac{\partial x^k}{\partial \xi^2}\boldsymbol{e}_k \\
\hat{\boldsymbol{e}}_3 &= \frac{dx^k}{ds_3}\boldsymbol{e}_k = \frac{1}{h_3}\frac{\partial x^k}{\partial \xi^3}\boldsymbol{e}_k
\end{aligned} \tag{2.18.5}$$

where we have used (2.18.4). Using the fact that $\hat{\boldsymbol{e}}_i \cdot \hat{\boldsymbol{e}}_j = \delta_{ij}$, relation (2.18.5) gives

$$\begin{aligned}
(h_1)^2 &= \frac{\partial x^k}{\partial \xi^1}\frac{\partial x^k}{\partial \xi^1} \\
(h_2)^2 &= \frac{\partial x^k}{\partial \xi^2}\frac{\partial x^k}{\partial \xi^2} \\
(h_3)^2 &= \frac{\partial x^k}{\partial \xi^3}\frac{\partial x^k}{\partial \xi^3}
\end{aligned} \tag{2.18.6}$$

It follows from (2.18.5) that the quantity

$$Q_r^k = \frac{1}{h_r}\frac{\partial x^k}{\partial \xi^r} \text{(no sum on } r\text{)} \tag{2.18.7}$$

represents the transformation tensor giving the curvilinear basis in terms of the Cartesian basis. This concept is similar to the transformation tensor Q_{ij} defined by (2.7.1) that was used between Cartesian systems.

The *physical components* of a vector or tensor are simply the components in a local set of Cartesian axes tangent to the curvilinear coordinate curves at any point in space. Thus, using transformation relation (2.18.7), the physical components of a tensor $\boldsymbol{A}$ in a general curvilinear system are given by

$$A_{\langle ij\ldots k\rangle} = Q_i^p Q_j^q \cdots Q_k^s A_{pq\ldots s} \tag{2.18.8}$$

where $A_{pq\ldots s}$ are the components in a fixed Cartesian frame. Note that the tensor can be expressed in either system as

$$\begin{aligned}
\boldsymbol{A} &= A_{ij\ldots k}\boldsymbol{e}_i\boldsymbol{e}_j\cdots\boldsymbol{e}_k \\
&= A_{\langle ij\ldots k\rangle}\hat{\boldsymbol{e}}_i\hat{\boldsymbol{e}}_j\cdots\hat{\boldsymbol{e}}_k
\end{aligned} \tag{2.18.9}$$

Since many applications involve differentiation of tensors, we must consider the differentiation of the curvilinear basis vectors. The Cartesian basis system $\boldsymbol{e}_k$ is fixed in orientation and therefore $\dfrac{\partial \boldsymbol{e}_k}{\partial x^j} = \dfrac{\partial \boldsymbol{e}_k}{\partial \xi^j} = 0$. However, derivatives of the curvilinear basis will not in general vanish, and differentiation of relations (2.18.5) gives the following results

$$\begin{aligned}
\frac{\partial \hat{\boldsymbol{e}}_m}{\partial \xi^m} &= -\frac{1}{h_n}\frac{\partial h_m}{\partial \xi^n}\hat{\boldsymbol{e}}_n - \frac{1}{h_r}\frac{\partial h_m}{\partial \xi^r}\hat{\boldsymbol{e}}_r, \quad m \neq n \neq r \\
\frac{\partial \hat{\boldsymbol{e}}_m}{\partial \xi^n} &= \frac{1}{h_m}\frac{\partial h_n}{\partial \xi^m}\hat{\boldsymbol{e}}_n, \quad m \neq n, \quad \text{no sum on repeated indices}
\end{aligned} \tag{2.18.10}$$

Using these results, the derivative of any tensor can be evaluated. Consider the first-order case of the derivative of a vector $\boldsymbol{u}$:

$$\frac{\partial}{\partial \xi^n}\boldsymbol{u} = \frac{\partial}{\partial \xi^n}(u_{\langle m\rangle}\hat{\boldsymbol{e}}_m) = \frac{\partial u_{\langle m\rangle}}{\partial \xi^n}\hat{\boldsymbol{e}}_m + u_{\langle m\rangle}\frac{\partial \hat{\boldsymbol{e}}_m}{\partial \xi^n} \tag{2.18.11}$$

The last term can be evaluated using (2.18.10), and thus the derivative of $\boldsymbol{u}$ can be expressed in terms of curvilinear components. Following a similar scheme, the derivative of a second-order tensor $\boldsymbol{A}$ is given by

$$\frac{\partial}{\partial \xi^r}\boldsymbol{A} = \frac{\partial}{\partial \xi^r}(A_{\langle mn\rangle}\hat{\boldsymbol{e}}_m\hat{\boldsymbol{e}}_n) = \frac{\partial A_{\langle mn\rangle}}{\partial \xi^r}\hat{\boldsymbol{e}}_m\hat{\boldsymbol{e}}_n + A_{\langle mn\rangle}\frac{\partial \hat{\boldsymbol{e}}_m}{\partial \xi^r}\hat{\boldsymbol{e}}_n + A_{\langle mn\rangle}\hat{\boldsymbol{e}}_m\frac{\partial \hat{\boldsymbol{e}}_n}{\partial \xi^r} \tag{2.18.12}$$

Analogous patterns follow for derivatives of higher-order tensors.

All vector differential operators of gradient, divergence, curl, etc. can be expressed in any general curvilinear system using these techniques. For example, the vector differential operator previously defined in Cartesian coordinates in (2.17.8) is given by

$$\nabla = \hat{\boldsymbol{e}}_1\frac{1}{h_1}\frac{\partial}{\partial \xi^1} + \hat{\boldsymbol{e}}_2\frac{1}{h_2}\frac{\partial}{\partial \xi^2} + \hat{\boldsymbol{e}}_3\frac{1}{h_3}\frac{\partial}{\partial \xi^3} = \sum_i \hat{\boldsymbol{e}}_i\frac{1}{h_i}\frac{\partial}{\partial \xi^i} \tag{2.18.13}$$

This then leads to the construction of the other common forms

Gradient of a scalar: $$\nabla f = \hat{\boldsymbol{e}}_1\frac{1}{h_1}\frac{\partial f}{\partial \xi^1} + \hat{\boldsymbol{e}}_2\frac{1}{h_2}\frac{\partial f}{\partial \xi^2} + \hat{\boldsymbol{e}}_3\frac{1}{h_3}\frac{\partial f}{\partial \xi^3} = \sum_i \hat{\boldsymbol{e}}_i\frac{1}{h_i}\frac{\partial f}{\partial \xi^i} \tag{2.18.14}$$

Divergence of a vector: $$\nabla\cdot\boldsymbol{u} = \frac{1}{h_1h_2h_3}\sum_i\frac{\partial}{\partial \xi^i}\left(\frac{h_1h_2h_3}{h_i}u_{\langle i\rangle}\right) \tag{2.18.15}$$

Laplacian of a scalar: $$\nabla^2\phi = \frac{1}{h_1h_2h_3}\sum_i\frac{\partial}{\partial \xi^i}\left(\frac{h_1h_2h_3}{(h_i)^2}\frac{\partial \phi}{\partial \xi^i}\right) \tag{2.18.16}$$

Curl of a vector: $$\nabla\times\boldsymbol{u} = \sum_i\sum_j\sum_k\frac{\varepsilon_{ijk}}{h_jh_k}\frac{\partial}{\partial \xi^j}(u_{\langle k\rangle}h_k)\hat{\boldsymbol{e}}_i \tag{2.18.17}$$

Gradient of a vector: $$\nabla\boldsymbol{u} = \sum_i\sum_j\frac{\hat{\boldsymbol{e}}_i}{h_i}\left(\frac{\partial u_{\langle j\rangle}}{\partial \xi^i}\hat{\boldsymbol{e}}_j + u_{\langle j\rangle}\frac{\partial \hat{\boldsymbol{e}}_j}{\partial \xi^i}\right) \tag{2.18.18}$$

Laplacian of a vector: $$\nabla^2 u = \left(\sum_i\frac{\hat{\boldsymbol{e}}_i}{h_i}\frac{\partial}{\partial \xi^i}\right)\cdot\left(\sum_j\sum_k\frac{\hat{\boldsymbol{e}}_k}{h_k}\left[\frac{\partial u_{\langle j\rangle}}{\partial \xi^k}\hat{\boldsymbol{e}}_j + u_{\langle j\rangle}\frac{\partial \hat{\boldsymbol{e}}_j}{\partial \xi^k}\right]\right) \tag{2.18.19}$$

It should be noted that these curvilinear forms are significantly different from those previously given in relations (2.17.10) for Cartesian coordinates. Curvilinear systems will add additional terms not found in rectangular coordinates. Other operations on higher-order tensors can be developed in a similar fashion (see Malvern 1969,

Appendix II). Specific transformation relations and field equations in cylindrical and spherical coordinate systems are given in Appendices A and B. Further discussion of these results will be taken up in later chapters.

EXAMPLE 2.18.1 POLAR COORDINATES

Determine the various vector differential operators for a two-dimensional polar coordinate system as shown in Fig. 2.9.

Solution: The differential length relation (2.18.4) for this case can be written as

$$(ds)^2 = (dr)^2 + (r\,d\theta)^2$$

and thus $h_1 = 1$ and $h_2 = r$. From relations (2.18.5) or simply using the geometry shown in Fig. 2.9,

$$\begin{aligned}\hat{\boldsymbol{e}}_r &= \cos\theta\,\boldsymbol{e}_1 + \sin\theta\,\boldsymbol{e}_2\\ \hat{\boldsymbol{e}}_\theta &= -\sin\theta\,\boldsymbol{e}_1 + \cos\theta\,\boldsymbol{e}_2\end{aligned}$$

and so

$$\frac{\partial\hat{\boldsymbol{e}}_r}{\partial\theta} = \hat{\boldsymbol{e}}_\theta,\quad \frac{\partial\hat{\boldsymbol{e}}_\theta}{\partial\theta} = -\hat{\boldsymbol{e}}_r,\quad \frac{\partial\hat{\boldsymbol{e}}_r}{\partial r} = \frac{\partial\hat{\boldsymbol{e}}_\theta}{\partial r} = 0$$

The basic vector differential operations then follow from relations (2.18.13)–(2.18.19) to be

$$\begin{aligned}
\nabla &= \hat{\boldsymbol{e}}_r\frac{\partial}{\partial r} + \hat{\boldsymbol{e}}_\theta\frac{1}{r}\frac{\partial}{\partial\theta}\\
\nabla f &= \hat{\boldsymbol{e}}_r\frac{\partial f}{\partial r} + \hat{\boldsymbol{e}}_\theta\frac{1}{r}\frac{\partial f}{\partial\theta}\\
\nabla\cdot\boldsymbol{u} &= \frac{1}{r}\frac{\partial}{\partial r}(ru_r) + \frac{1}{r}\frac{\partial u_\theta}{\partial\theta}\\
\nabla^2 f &= \frac{1}{r}\frac{\partial}{\partial r}\left(r\frac{\partial f}{\partial r}\right) + \frac{1}{r^2}\frac{\partial^2 f}{\partial\theta^2}\\
\nabla\times\boldsymbol{u} &= \left(\frac{1}{r}\frac{\partial}{\partial r}(ru_\theta) - \frac{1}{r}\frac{\partial u_r}{\partial\theta}\right)\hat{\boldsymbol{e}}_z\\
\nabla\boldsymbol{u} &= \frac{\partial u_r}{\partial r}\hat{\boldsymbol{e}}_r\hat{\boldsymbol{e}}_r + \frac{1}{r}\left(\frac{\partial u_r}{\partial\theta} - u_\theta\right)\hat{\boldsymbol{e}}_r\hat{\boldsymbol{e}}_\theta + \frac{\partial u_\theta}{\partial r}\hat{\boldsymbol{e}}_\theta\hat{\boldsymbol{e}}_r + \frac{1}{r}\left(\frac{\partial u_\theta}{\partial\theta} + u_r\right)\hat{\boldsymbol{e}}_\theta\hat{\boldsymbol{e}}_\theta\\
\nabla^2\boldsymbol{u} &= \left(\nabla^2 u_r - \frac{2}{r^2}\frac{\partial u_\theta}{\partial\theta} - \frac{u_r}{r^2}\right)\hat{\boldsymbol{e}}_r + \left(\nabla^2 u_\theta + \frac{2}{r^2}\frac{\partial u_r}{\partial\theta} - \frac{u_\theta}{r^2}\right)\hat{\boldsymbol{e}}_\theta
\end{aligned} \tag{2.18.20}$$

where $\boldsymbol{u} = u_r\hat{\boldsymbol{e}}_r + u_\theta\hat{\boldsymbol{e}}_\theta$, $\hat{\boldsymbol{e}}_z = \hat{\boldsymbol{e}}_r\times\hat{\boldsymbol{e}}_\theta$. Notice that the Laplacian of a vector does not simply pass through and operate on each of the individual components as in the Cartesian case. Additional terms are generated due to the curvature of the polar coordinate system. Similar relations can be developed for cylindrical and spherical coordinate systems (see Exercises 2.26 and 2.27).

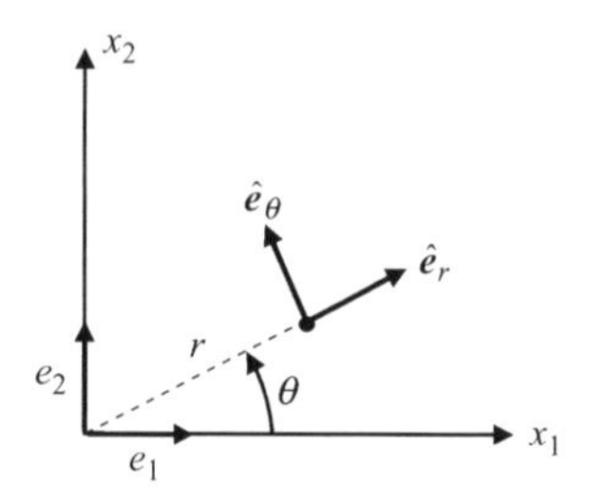

FIGURE 2.9

Polar coordinate system.

2.19 GENERAL TENSORS

Rather than using the concepts scale factors and physical components presented in the previous section, we now present a more general scheme to deal with curvilinear coordinate systems. With the exception of Section 2.18, our previous discussions were valid only for Cartesian coordinates. Transformation laws given in relations (2.8.1) were valid only between such reference frames. Another more fundamental scheme to handle continuum mechanics formulation in curvilinear coordinate systems is to employ *general tensor theory* that is not restricted to only Cartesian systems. A very brief introduction into general tensors and their application for equations in continuum mechanics will now be presented.

We start with the general transformation relations (2.18.3) between rectangular coordinates x^i and orthogonal curvilinear coordinates ξ^i:

$$\xi^m = \xi^m(x^1, x^2, x^3), \quad x^m = x^m(\xi^1, \xi^2, \xi^3) \tag{2.19.1}$$

where the use of both subscripts and superscripts will now be used. Fig. 2.10 illustrates the two coordinate systems, and the position vector $\boldsymbol{r}$ in the Cartesian coordinate system is given by

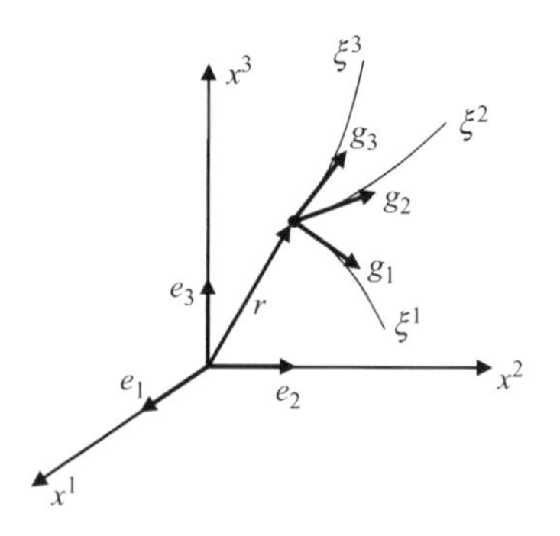

FIGURE 2.10

Cartesian and curvilinear coordinates.

$$\boldsymbol{r} = x^k \boldsymbol{e}_k \tag{2.19.2}$$

where now summations are carried from subscripts to superscripts.

The *covariant base vectors* $\boldsymbol{g}_k$, tangent to the curvilinear system, then follows to be

$$\boldsymbol{g}_k = \frac{\partial x^m}{\partial \xi^k} \boldsymbol{e}_m \tag{2.19.3}$$

It should be noted that these base vectors are *not unit vectors*, but they will be orthogonal for our case, since we are considering only orthogonal curvilinear coordinate systems.

From relation (2.19.3), it follows that

$$\boldsymbol{e}_m = \frac{\partial \xi^k}{\partial x^m} \boldsymbol{g}_k \tag{2.19.4}$$

and thus the differential position vector $d\boldsymbol{r}$ can be written as

$$d\boldsymbol{r} = \frac{\partial \boldsymbol{r}}{\partial \xi^k} d\xi^k = \boldsymbol{g}_k \, d\xi^k \tag{2.19.5}$$

This allows the calculation of the square of the arc length (see Eq. (2.18.4)):

$$(ds)^2 = d\boldsymbol{r} \cdot d\boldsymbol{r} = \boldsymbol{g}_k \cdot \boldsymbol{g}_m \, d\xi^k \, d\xi^m = g_{km} \, d\xi^k \, d\xi^m \tag{2.19.6}$$

where g_{ij} is called the *covariant metric tensor* defined by

$$g_{ij} = \boldsymbol{g}_i \cdot \boldsymbol{g}_j = \frac{\partial x^m}{\partial \xi^i} \frac{\partial x^n}{\partial \xi^j} \delta_{mn} \tag{2.19.7}$$

The terminology metric tensor follows from the fact that g_{ij} provides a means to calculate lengths and angles in space. Since the coordinates are orthogonal, $g_{ij} = 0$ for $i \neq j$, and in general the components of the metric tensor are functions of the spatial coordinates. We thus could express (2.19.6) as

$$(ds)^2 = g_{11}(d\xi^1)^2 + g_{22}(d\xi^2)^2 + g_{33}(d\xi^3)^2 \tag{2.19.8}$$

note the similarity with the previous relation (2.18.4).

Another set of base vectors $\boldsymbol{g}^k$ can also be defined called the *reciprocal base vectors* and may be found from the solution to the equations

$$\boldsymbol{g}^k \cdot \boldsymbol{g}_m = \delta^k_m \tag{2.19.9}$$

where δ^k_m is the Kronecker delta with upper and lower indices. It can be shown that the reciprocal base vectors can be written as

$$\boldsymbol{g}^k = g^{km} \boldsymbol{g}_m \tag{2.19.10}$$

where g^{km} is the contravariant metric tensor given by

$$g^{ij} = \boldsymbol{g}^i \cdot \boldsymbol{g}^j = \frac{\partial \xi^i}{\partial x^m} \frac{\partial \xi^j}{\partial x^n} \delta^{mn} \tag{2.19.11}$$

with properties

$$g^{ij} g_{jk} = \delta^i_k \tag{2.19.12}$$

Note that with orthogonal coordinates $g^{11} = 1/g_{11}, g^{22} = 1/g_{22}, g^{33} = 1/g_{33}$.

Now any vector $\boldsymbol{u}$ (or tensor) may be decomposed into its *covariant components* u_k or *contravariant components* u^k by the expressions

$$\boldsymbol{u} = u_k \boldsymbol{g}^k = u^k \boldsymbol{g}_k \tag{2.19.13}$$

The covariant and contravariant components are not identical, and since the base vectors are not unit vectors these components will not necessarily have the same physical dimensions (units) as the vector $\boldsymbol{u}$ itself. Using the metric tensors, one can raise and lower indices as follows:

$$u_k = g_{km} u^m, \quad u^k = g^{km} u_m \tag{2.19.14}$$

By decomposing a vector $\boldsymbol{u}$ into its components along *unit vectors* $\hat{\boldsymbol{e}}_k$ which lie along the coordinate curves, one obtains the *physical components*, $u^{\langle k \rangle}$ of the vector, that is,

$$\boldsymbol{u} = u^{\langle k \rangle} \hat{\boldsymbol{e}}_k \tag{2.19.15}$$

where the unit vectors $\hat{\boldsymbol{e}}_k$ are defined by

$$\hat{\boldsymbol{e}}_k = \frac{\boldsymbol{g}_k}{(g_{kk})^{1/2}} (\text{no sum on } k) \tag{2.19.16}$$

Next considering spatial derivatives of tensor fields within this more general scheme, we follow similar steps as done in the previous section. As before, both covariant and contravariant base vectors vary with position in space, so their spatial derivatives will not be zero. One can show that

$$\frac{\partial \boldsymbol{g}_k}{\partial \xi^j} = \Gamma^k_{ij} \boldsymbol{g}_k, \quad \frac{\partial \boldsymbol{g}^k}{\partial \xi^j} = -\Gamma^i_{kj} \boldsymbol{g}^k \tag{2.19.17}$$

where Γ^i_{kj} are known as the *Christoffel symbols of the second kind* given by

$$\Gamma^i_{kj} = \Gamma^i_{jk} = \frac{\partial^2 x^m}{\partial \xi^k \partial \xi^j} \frac{\partial \xi^i}{\partial x^m} = \frac{1}{2} (g_{km,j} + g_{jm,k} - g_{kj,m}) g^{mi} \tag{2.19.18}$$

We then can define the *covariant derivative* of a contravariant vector as

$$u^i_{;j} = u^i_{,j} + \Gamma^i_{kj} u^k, \quad \text{where } u^i_{,j} = \frac{\partial u^i}{\partial \xi^j} \tag{2.19.19}$$

while the covariant derivative of a covariant vector is given by

$$u_{i;j} - u_{i,j} - \Gamma^k_{ij} u_k \tag{2.19.20}$$

For the special case of *transformations only between Cartesian frames*, relation (2.19.1) becomes

$$\xi^i = Q^i_j x^j + a^i \tag{2.19.21}$$

where Q^i_j is the constant rotation matrix defined by (2.7.1) with orthogonal property $Q^i_k Q^j_k = \delta^{ij}$, and a^i is a constant translation vector. For this case,

$$\frac{\partial \xi^i}{\partial x^k} = Q^i_k \Rightarrow g^{ij} = Q^i_m Q^j_n \delta^{mn} = \delta^{ij}, \quad \Gamma^i_{jk} = 0 \tag{2.19.22}$$

thus, all Christoffel symbols vanish for the case of Cartesian coordinates, and the covariant derivatives reduce to the ordinary partial derivatives.

For higher-order tensors, the covariant derivative is defined in an analogous manner. For example, for second-order tensors A^{ij} and A_{ij}, we have

$$\begin{aligned} A^{ij}_{;k} &= A^{ij}_{,k} + \Gamma^i_{km} A^{mj} + \Gamma^j_{km} A^{im} \\ A_{ij;k} &= A_{ij,k} - \Gamma^m_{ik} A_{mj} - \Gamma^m_{jk} A_{im} \end{aligned} \tag{2.19.23}$$

Details on the derivation of these derivative expressions can be found in Malvern (1969). Using relations (2.19.19) and (2.19.20) allows development of the standard vector differential operators in curvilinear coordinate systems previously present in Section 2.18. Other differential relations involving second-order tensors follow from relations (2.19.23). The following example illustrates many of the basic applications for the specific case of a polar coordinate curvilinear system.

EXAMPLE 2.19.1 POLAR COORDINATES

For a two-dimensional polar coordinate system as shown in Fig. 2.9, determine

(a) covariant and reciprocal base vectors;
(b) covariant and contravariant metric tensors;
(c) relations between the physical, covariant, and contravariant components of a vector $\boldsymbol{u}$;
(d) Christoffel symbols;
(e) gradient of vector $\boldsymbol{u}$ and express it in physical components.

Solution: For polar coordinates with $\xi^1 = r$ and $\xi^2 = \theta$, the transformation relations (2.19.1) take the form

$$\begin{aligned} &\xi^1 = \sqrt{(x^1)^2 + (x^2)^2}, \quad \xi^2 = \tan^{-1}(x^2/x^1) \\ &x^1 = \xi^1 \cos\xi^2, \quad x^2 = \xi^1 \sin\xi^2 \end{aligned}$$

(a) From relation (2.19.3)

$$\boldsymbol{g}_1 = \frac{\partial x^m}{\partial \xi^1}\boldsymbol{e}_m = \cos\xi^2 \boldsymbol{e}_1 + \sin\xi^2 \boldsymbol{e}_2$$
$$\boldsymbol{g}_2 = \frac{\partial x^m}{\partial \xi^2}\boldsymbol{e}_m = -\xi^1 \sin\xi^2 \boldsymbol{e}_1 + \xi^1 \cos\xi^2 \boldsymbol{e}_2$$

Using (2.19.9) $\boldsymbol{g}^k \cdot \boldsymbol{g}_m = \delta^k_m \Rightarrow$

$$\boldsymbol{g}^1 \cdot \boldsymbol{g}_1 = 1 \Rightarrow \boldsymbol{g}^1 = \boldsymbol{g}_1 = \cos\xi^2 \boldsymbol{e}_1 + \sin\xi^2 \boldsymbol{e}_2$$
$$\boldsymbol{g}^2 \cdot \boldsymbol{g}_2 = 1 \Rightarrow \boldsymbol{g}^2 = \frac{1}{(\xi^1)^2}\boldsymbol{g}_2 = \frac{1}{(\xi^1)}\left(-\sin\xi^2 \boldsymbol{e}_1 + \cos\xi^2 \boldsymbol{e}_2\right)$$

(b) Eqs. (2.19.7) and (2.19.12), then give

$$g_{11} = \boldsymbol{g}_1 \cdot \boldsymbol{g}_1 = 1 = 1/g^{11}$$
$$g_{22} = \boldsymbol{g}_2 \cdot \boldsymbol{g}_2 = (\xi^1)^2 = 1/g^{22}$$
$$g_{12} = g_{21} = 0, \quad g^{12} = g^{21} = 0$$

(c) Using (2.19.16) $\hat{\boldsymbol{e}}_k = \dfrac{\boldsymbol{g}_k}{(g_{kk})^{1/2}}$ (no sum on k) $\Rightarrow$

$$\hat{\boldsymbol{e}}_1 = \frac{\boldsymbol{g}_1}{(g_{11})^{1/2}} = \boldsymbol{g}_1 = \boldsymbol{g}^1, \quad \hat{\boldsymbol{e}}_2 = \frac{\boldsymbol{g}_2}{(g_{22})^{1/2}} = \frac{1}{\xi^1}\boldsymbol{g}_2 = \xi^1 \boldsymbol{g}^2$$

(d) The Christoffel symbols follow from (2.19.18) $\Gamma^i_{kj} = \dfrac{1}{2}\left(g_{km,j} + g_{jm,k} - g_{kj,m}\right)$ $\Rightarrow$

$$\Gamma^1_{22} = \frac{1}{2}\left(g_{21,2} + g_{21,2} - g_{22,1}\right)g^{11} = -\xi^1$$
$$\Gamma^2_{12} = \Gamma^2_{21} = \frac{1}{2}\left(g_{22,1} + g_{12,2} - g_{21,2}\right)g^{22} = \frac{1}{\xi^1}$$

and all others are zero, for example

$$\Gamma^1_{11} = \frac{1}{2}\left(g_{11,1} + g_{11,1} - g_{11,1}\right)g^{11} = 0$$
$$\Gamma^2_{11} = \frac{1}{2}\left(g_{11,1} + g_{11,1} - g_{11,1}\right)g^{22} = 0$$

...

(e) Using the contravariant form from (2.19.13), a vector $\boldsymbol{u}$ can be expressed by $\boldsymbol{u}=u^k\boldsymbol{g}_k$, and so the gradient operation is expressed using (2.19.19)

$$\text{grad}\,\boldsymbol{u} = u^i_{;j}\boldsymbol{g}_i\boldsymbol{g}^j = \left(u^i_{,j} + \Gamma^i_{kj}u^k\right)\boldsymbol{g}_i\boldsymbol{g}^j = \left(\frac{\partial u^i}{\partial \xi^j} + \Gamma^i_{kj}u^k\right)\boldsymbol{g}_i\boldsymbol{g}^j$$
$$= \frac{\partial u^1}{\partial \xi^1}\boldsymbol{g}_1\boldsymbol{g}^1 + \left(\frac{\partial u^1}{\partial \xi^2} + \Gamma^1_{22}u^2\right)\boldsymbol{g}_1\boldsymbol{g}^2 + \left(\frac{\partial u^2}{\partial \xi^1} + \Gamma^2_{21}u^2\right)\boldsymbol{g}_2\boldsymbol{g}^1 + \left(\frac{\partial u^2}{\partial \xi^2} + \Gamma^2_{12}u^1\right)\boldsymbol{g}_2\boldsymbol{g}^2$$
$$= \frac{\partial u^1}{\partial \xi^1}\boldsymbol{g}_1\boldsymbol{g}^1 + \left(\frac{\partial u^1}{\partial \xi^2} - \xi^1 u^2\right)\boldsymbol{g}_1\boldsymbol{g}^2 + \left(\frac{\partial u^2}{\partial \xi^1} + \frac{1}{\xi^1}u^2\right)\boldsymbol{g}_2\boldsymbol{g}^1 + \left(\frac{\partial u^2}{\partial \xi^2} + \frac{1}{\xi^1}u^1\right)\boldsymbol{g}_2\boldsymbol{g}^2$$

Using part (c), we can easily translate these results in terms of the physical components as

$$\begin{aligned}
\operatorname{grad} \boldsymbol{u} &= \frac{\partial u^1}{\partial \xi^1}\hat{\boldsymbol{e}}_1\hat{\boldsymbol{e}}_1 + \left(\frac{1}{\xi^1}\frac{\partial u^1}{\partial \xi^2} - u^2\right)\hat{\boldsymbol{e}}_1\hat{\boldsymbol{e}}_2 + \left(\xi^1\frac{\partial u^2}{\partial \xi^1} + u^2\right)\hat{\boldsymbol{e}}_2\hat{\boldsymbol{e}}_1 + \left(\frac{\partial u^2}{\partial \xi^2} + \frac{1}{\xi^1}u^1\right)\hat{\boldsymbol{e}}_2\hat{\boldsymbol{e}}_2 \\
&= \frac{\partial u^{\langle 1\rangle}}{\partial \xi^1}\hat{\boldsymbol{e}}_1\hat{\boldsymbol{e}}_1 + \left(\frac{1}{\xi^1}\frac{\partial u^{\langle 1\rangle}}{\partial \xi^2} - \frac{u^{\langle 2\rangle}}{\xi^1}\right)\hat{\boldsymbol{e}}_1\hat{\boldsymbol{e}}_2 + \left(\xi^1\frac{\partial}{\partial \xi^1}\left(\frac{u^{\langle 2\rangle}}{\xi^1}\right) + \frac{u^{\langle 2\rangle}}{\xi^1}\right)\hat{\boldsymbol{e}}_2\hat{\boldsymbol{e}}_1 \\
&\quad + \left(\frac{\partial}{\partial \xi^2}\left(\frac{u^{\langle 2\rangle}}{\xi^1}\right) + \frac{1}{\xi^1}u^{\langle 1\rangle}\right)\hat{\boldsymbol{e}}_2\hat{\boldsymbol{e}}_2 \\
&= \frac{\partial u^{\langle 1\rangle}}{\partial \xi^1}\hat{\boldsymbol{e}}_1\hat{\boldsymbol{e}}_1 + \left(\frac{1}{\xi^1}\frac{\partial u^{\langle 1\rangle}}{\partial \xi^2} - \frac{u^{\langle 2\rangle}}{\xi^1}\right)\hat{\boldsymbol{e}}_1\hat{\boldsymbol{e}}_2 + \left(\frac{\partial u^{\langle 2\rangle}}{\partial \xi^1}\right)\hat{\boldsymbol{e}}_2\hat{\boldsymbol{e}}_1 + \left(\frac{1}{\xi^1}\frac{\partial u^{\langle 2\rangle}}{\partial \xi^2} + \frac{u^{\langle 1\rangle}}{\xi^1}\right)\hat{\boldsymbol{e}}_2\hat{\boldsymbol{e}}_2
\end{aligned}$$

which matches with the corresponding result from the previous Example 2.18.1.

REFERENCES

Chou, P.C., Pagano, N.J., 1967. Elasticity—Tensor, Dyadic and Engineering Approaches. D. Van Nostrand, Princeton, NJ.

Goodbody, A.M., 1982. Cartesian Tensors: With Applications to Mechanics, Fluid Mechanics and Elasticity. Ellis Horwood, New York.

Hildebrand, F.B., 1976. Advanced Calculus for Applications, Second ed Prentice-Hall, Englewood Cliffs, NJ.

Itskov, M., 2015. Tensor Algebra and Tensor Analysis for Engineers, Fourth ed Springer, New York.

Kreyszig, E., 2011. Advanced Engineering Mathematics, Tenth ed. John Wiley, New York.

Malvern, L.E., 1969. Introduction to the Mechanics of a Continuous Medium. Prentice-Hall, Englewood Cliffs, NJ.

Simmons, J.G., 1994. A Brief on Tensor Analysis. Springer, New York.

Weatherburn, C.E., 1948. Advanced Vector Analysis. Open Court, LaSalle, IL.

EXERCISES

2.1 For the following vector and matrix pairs, compute the expressions $a_i a_i$, $a_i a_j$, A_{ii}, $A_{ij} a_j$, and specify whether they are scalars, vectors, or matrices:

(a) $a_i = \begin{bmatrix} 1 \\ 2 \\ 1 \end{bmatrix}, \quad A_{ij} = \begin{bmatrix} 1 & 0 & 1 \\ 1 & 2 & 0 \\ 1 & 2 & 1 \end{bmatrix}$

(b) $a_i = \begin{bmatrix} 0 \\ 2 \\ 1 \end{bmatrix}, \quad A_{ij} = \begin{bmatrix} 1 & 2 & 1 \\ 1 & 0 & 0 \\ 1 & 2 & 1 \end{bmatrix}$

(c) $a_i = \begin{bmatrix} 2 \\ 1 \\ 1 \end{bmatrix}$, $A_{ij} = \begin{bmatrix} 0 & 2 & 1 \\ 3 & 2 & 0 \\ 1 & 0 & 4 \end{bmatrix}$

2.2 Decompose the matrices A_{ij} in Exercise 2.1 into the sum of symmetric $A_{(ij)} = \frac{1}{2}(A_{ij} + A_{ji})$ and antisymmetric $A_{[ij]} = \frac{1}{2}(A_{ij} - A_{ji})$ matrices. For each case, verify that $A_{(ij)}$ and $A_{[ij]}$ satisfy the conditions specified in the last paragraph of Section 2.3.

2.3 Verify in general that the product $A_{ij}B_{ij}$ will vanish if A_{ij} is symmetric and B_{ij} is antisymmetric.

2.4 Given the three matrices: A_{ij} (symmetric), B_{ij} (antisymmetric), and C_{ij} (general), show that

$$\begin{aligned} A_{ij}C_{ij} &= A_{ij}C_{ji} = A_{ij}C_{(ij)} \\ B_{ij}C_{ij} &= -B_{ij}C_{ji} = B_{ij}C_{[ij]} \\ \det(B_{ij}) &= 0 \end{aligned}$$

2.5 Explicitly verify the following properties of the Kronecker delta:

$$\delta_{ij}a_{jk} = a_{ik}, \quad \delta_{ij}a_{kj} = a_{ki}$$

2.6 Formally expand the determinant relation (2.5.1) and justify that either index notation form yields a result that matches the traditional expansion of $\det[A_{ij}]$.

2.7 Explicitly justify relation (2.5.2) for the 2×2 case, and then use this result to verify orthogonal properties (2.5.3).

2.8 Using property (2.5.4), verify relations (2.5.5)–(2.5.8).

2.9 Using the general definitions of the dot and cross product given in Section 2.10, explicitly justify properties (2.6.5).

2.10 Consider a new set of axes obtained by rotating the original system 90° about the x_2-axis. Positive rotation is the same as defined in Fig. 2.3. Determine the transformation matrix Q_{ij}, and find the components of A_{ij} and a_i given in Exercise 2.1 in the new system. Verify that the length of the vector $|\boldsymbol{a}| = (a_i a_i)^{1/2}$ and the quantity A_{ii} are the same in each coordinate system, that is, they are invariant with respect to the rotation.

2.11 Differentiate with respect to time the orthogonality condition $\boldsymbol{QQ}^T = \boldsymbol{I}$, and show that

$$\dot{\boldsymbol{Q}}\boldsymbol{Q}^T = -\boldsymbol{Q}\dot{\boldsymbol{Q}}^T = -(\dot{\boldsymbol{Q}}\boldsymbol{Q}^T)^T$$

2.12 Consider the two-dimensional coordinate transformation shown in Fig. 2.9. Through the counter-clockwise rotation θ, a new polar coordinate system is created. Show that the transformation matrix for this case is given by

$$Q_{ij} = \begin{bmatrix} \cos\theta & \sin\theta \\ -\sin\theta & \cos\theta \end{bmatrix}$$

If $b_i = \begin{bmatrix} b_1 \\ b_2 \end{bmatrix}, A_{ij} = \begin{bmatrix} A_{11} & A_{12} \\ A_{21} & A_{22} \end{bmatrix}$ are the components of a first- and second-order tensor in the x_1,x_2-system, calculate their components in a rotated Cartesian system aligned with the polar coordinates.

2.13 The *axial* or *dual vector* a_i, of a second-order tensor A_{ij} is defined by $a_i = -\frac{1}{2}\varepsilon_{ijk}A_{jk}$. Show that the dual vector of a symmetric tensor is zero, and thus one can write $a_i = -\frac{1}{2}\varepsilon_{ijk}A_{[jk]}$ or in direct notation $\boldsymbol{a} = -A_{23}\boldsymbol{e}_1 + A_{13}\boldsymbol{e}_2 - A_{12}\boldsymbol{e}_3$. Also verify that one can invert the previous expression to get $A_{[jk]} = -\varepsilon_{ijk}a_i = -\varepsilon_{jki}a_i$.

2.14 Verify properties 7 and 8 in Table 2.1.

2.15 Show that the relation (2.10.9) for the inverse of a matrix satisfies the defining statement (2.10.8).

2.16 Verify the relationships (2.11.5) that express the invariants I_1, I_2, and I_3 in terms of the fundamental invariants I_A, II_A, and III_A. Next, invert these relations and express the fundamental invariants in terms of I_1, I_2, and I_3.

2.17 Verify that the fundamental invariants can be expressed in terms of the principal values as given by relations (2.11.8).

2.18 Determine the invariants, and principal values and directions of the following symmetric second-order tensors (matrices). Use the determined principal directions to establish a principal coordinate system, and following the procedures in Example 2.11.1, formally transform (rotate) the given matrix into the principal system to arrive at the appropriate diagonal form.

(a) $\begin{bmatrix} -2 & 1 & 0 \\ 1 & -2 & 0 \\ 0 & 0 & 0 \end{bmatrix}$ (Answer: $\lambda_i = -3, -1, 0$)

(b) $\begin{bmatrix} -1 & 1 & 0 \\ 1 & -1 & 0 \\ 0 & 0 & 2 \end{bmatrix}$ (Answer: $\lambda_i = -2, 0, 2$)

(c) $\begin{bmatrix} 6 & -3 & 0 \\ -3 & 6 & 0 \\ 0 & 0 & 6 \end{bmatrix}$ (Answer: $\lambda_i = 3, 6, 9$)

You may check your work using the MATLAB code C-2.

2.19 Consider the principal value problem for matrix $\boldsymbol{A}$, $\boldsymbol{An} = \lambda_{(i)}\boldsymbol{n}$. First show that the principal values of $\boldsymbol{A}^{-1}$ are given by $1/\lambda_{(i)}$ and that the principal directions of $\boldsymbol{A}^{-1}$ are the same as $\boldsymbol{A}$. Then using principal coordinates, show that $I_{A^{-1}} = II_A / III_A$.

2.20 A symmetric tensor A_{ij} is said to be *positive definite*, if for all vectors b_i, $A_{ij}b_ib_j \geq 0$, with equality only if $b_i = 0$. Show that a symmetric tensor A_{ij} is positive definite if and only if (iff) all of its eigenvalues are positive.

2.21 Calculate the spherical and deviatoric parts of the following second-order tensors:

$$(a)\ A_{ij} = \begin{bmatrix} 3 & 0 & 1 \\ 1 & 2 & 0 \\ 1 & 2 & 4 \end{bmatrix} \quad (b)\ A_{ij} = \begin{bmatrix} 0 & 0 & 1 \\ 1 & 2 & 0 \\ 1 & 2 & 1 \end{bmatrix} \quad (c)\ A_{ij} = \begin{bmatrix} 0 & 2 & 1 \\ 3 & 2 & 0 \\ 1 & 0 & 4 \end{bmatrix}$$

2.22 Verify relations (2.13.3) and (2.13.4) to express $\boldsymbol{A}^4$ and $\boldsymbol{A}^5$ in terms of powers ≤ 2.

2.23 Using the Polar Decomposition Theorem (2.16.1), establish relations (2.16.2).

2.24 Calculate the quantities $\nabla \cdot \boldsymbol{u}, \nabla \times \boldsymbol{u}, \nabla^2 \boldsymbol{u}, \nabla \boldsymbol{u}, tr(\nabla \boldsymbol{u})$ for the following Cartesian vector fields:

(a) $\boldsymbol{u} = x_1\boldsymbol{e}_1 + x_1x_2\boldsymbol{e}_2 + 2x_1x_2x_3\boldsymbol{e}_3$

(b) $\boldsymbol{u} = x_1^2\boldsymbol{e}_1 + 2x_1x_2\boldsymbol{e}_2 + x_3^3\boldsymbol{e}_3$

(c) $\boldsymbol{u} = x_2^2\boldsymbol{e}_1 + 2x_2x_3\boldsymbol{e}_2 + 4x_1^2\boldsymbol{e}_3$

2.25 Using index notation, verify the following vector identities:

(a) $(2.17.11)_{1,2,3}$

(b) $(2.17.11)_{4,5,6,7}$

(c) $(2.17.11)_{8,9,10}$

2.26 Extend the polar coordinate results in Example 2.18.1 and determine the forms of $\nabla f, \nabla \cdot \boldsymbol{u}$, $\nabla^2 f$ and $\nabla \times \boldsymbol{u}$ for a three-dimensional cylindrical coordinate system as shown in Fig. 2.6.

2.27 For the spherical coordinate system (R, ϕ, θ) in Fig. 2.7, show that

$$h_1 = 1, h_2 = R, h_3 = R\sin\phi$$

and the standard vector operations are given by

$$\nabla f = \hat{\boldsymbol{e}}_R \frac{\partial f}{\partial R} + \hat{\boldsymbol{e}}_\phi \frac{1}{R}\frac{\partial f}{\partial \phi} + \hat{\boldsymbol{e}}_\theta \frac{1}{R\sin\phi}\frac{\partial f}{\partial \theta}$$

$$\nabla \cdot \boldsymbol{u} = \frac{1}{R^2}\frac{\partial}{\partial R}(R^2 u_R) + \frac{1}{R\sin\phi}\frac{\partial}{\partial \phi}(\sin\phi\, u_\phi) + \frac{1}{R\sin\phi}\frac{\partial u_\theta}{\partial \theta}$$

$$\nabla^2 f = \frac{1}{R^2}\frac{\partial}{\partial R}\left(R^2\frac{\partial f}{\partial R}\right) + \frac{1}{R^2\sin\phi}\frac{\partial}{\partial \phi}\left(\sin\phi\frac{\partial f}{\partial \phi}\right) + \frac{1}{R^2\sin^2\phi}\frac{\partial^2 f}{\partial \theta^2}$$

$$\nabla \times \boldsymbol{u} = \hat{\boldsymbol{e}}_R\left[\frac{1}{R\sin\phi}\left(\frac{\partial}{\partial \phi}(\sin\phi\, u_\theta) - \frac{\partial u_\phi}{\partial \theta}\right)\right] + \hat{\boldsymbol{e}}_\phi\left[\frac{1}{R\sin\phi}\frac{\partial u_R}{\partial \theta} - \frac{1}{R}\frac{\partial}{\partial R}(Ru_\theta)\right]$$
$$+\hat{\boldsymbol{e}}_\theta\left[\frac{1}{R}\left(\frac{\partial}{\partial R}(Ru_\phi) - \frac{\partial u_R}{\partial \phi}\right)\right]$$

2.28 For the case of *spherical coordinates*, determine: g_{ij}, g^{ij}, Γ^i_{jk} (calculate only those shown in answer).

Ans. Using $\xi^1 = R$, $\xi^2 = \phi$, $\xi^3 = \theta$

$$g_{ij} = \begin{bmatrix} 1 & 0 & 0 \\ 0 & R^2 & 0 \\ 0 & 0 & R^2 \sin^2 \phi \end{bmatrix}, \quad \Gamma^1_{11} = 0, \quad \Gamma^3_{13} = \frac{1}{R}, \quad \Gamma^2_{33} = -\sin\phi\cos\phi, \quad \Gamma^3_{23} = \cot\phi$$

CHAPTER

3 Kinematics of Motion and Deformation Measures

3.1 MATERIAL BODY AND MOTION

We wish to start with a general definition of a *material body* which can describe a broad class of continuum materials. For this purpose, we define a material body B as a continuum set of particles or material points X (see Fig. 3.1). Note that these particles are not discrete mass points as in Newtonian mechanics, but instead they are infinitesimally small portions of a continuous media with definable mass density (see discussions in Section 1.1). For each of these particles, we assign a one-to-one mapping to spatial points $\boldsymbol{X}$ in a three-dimensional Euclidean space that the particles occupy at a given instant of time t_o. This then establishes a particular *configuration* of body B, and we will call this the *reference configuration*. This could be selected as the initial configuration at $t = 0$, or for solids it might be taken as the undeformed configuration. However, the choice of the reference configuration is completely arbitrary.

As the continuum deforms at subsequent times particles move, and this motion can be expressed by the simple relation

$$\boldsymbol{x} = \chi(\boldsymbol{X}, t) \tag{3.1.1}$$

Thus, particle X at location $\boldsymbol{X}$ in the reference configuration moves to a new location $\boldsymbol{x}$ in the *current configuration* at time t as shown in Fig. 3.1. Note that when $t = t_o$, relation (3.1.1) gives $\boldsymbol{X} = \chi(\boldsymbol{X}, t_0)$. Certain requirements must be made on the allowable types of motion in continuum mechanics. First, it is required that the motion (mapping) be one to one, so that two particles do not end up at the same place at later times. This is often referred to as the *impenetrability of matter*. We also require that the inverse motion exists

$$\boldsymbol{X} = \chi^{-1}(\boldsymbol{x}, t) \tag{3.1.2}$$

and that the motion and its inverse are continuously differentiable functions. Under these conditions, it follows that the *Jacobian determinant* $J = \det(\partial \boldsymbol{x} / \partial \boldsymbol{X})$ cannot vanish, and in fact we will normally assume

$$0 < \det\left(\frac{\partial \boldsymbol{x}}{\partial \boldsymbol{X}}\right) < \infty \tag{3.1.3}$$

Continuum Mechanics Modeling of Material Behavior. http://dx.doi.org/10.1016/B978-0-12-811474-2.00003-4

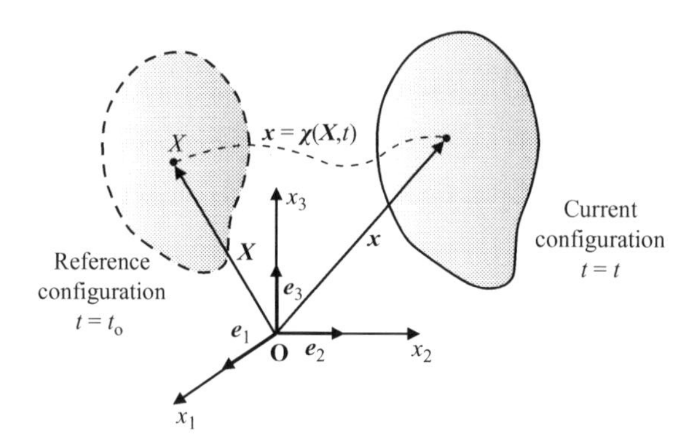

FIGURE 3.1

Continuum motion.

Later in our studies, we will develop some continuum theories for material models that will assume small infinitesimal deformations which will not require the rigor of the two configuration situation shown in Fig. 3.1.

EXAMPLE 3.1.1 SIMPLE SHEARING MOTION

Explore basic features of the following *simple shearing motion* specified in direct vector form as

$$\boldsymbol{x} = \boldsymbol{X} + \gamma X_2 \boldsymbol{e}_1 \tag{3.1.4}$$

where γ is the amount of shear and could be a function of time.

Solution: In component form, this motion is expressed as

$$\begin{aligned} x_1 &= X_1 + \gamma X_2 \\ x_2 &= X_2 \\ x_3 &= X_3 \end{aligned} \tag{3.1.5}$$

This motion is confined in the X_1,X_2-plane and is perhaps best illustrated if we consider the deformation of a square element that might be part of a larger body. The reference configuration is shown in Fig. 3.2 as the dotted square with corners *ABCD*. Under the given motion, the reference element shears into a rhombus shape shown by the solid line. Points *A* and *B* do not move because they lie on the line $X_2 = 0$. Points *C* and *D* move only in the X_1-direction in direct proportion to their X_2 coordinate value. We will often use this simple type of deformation field in later chapters to demonstrate particular features of material constitutive laws. Note that for this simple example we can easily determine the inverse relation (3.1.2):

$$\begin{aligned} X_1 &= x_1 - \gamma x_2 \\ X_2 &= x_2 \\ X_3 &= x_3 \end{aligned} \tag{3.1.6}$$

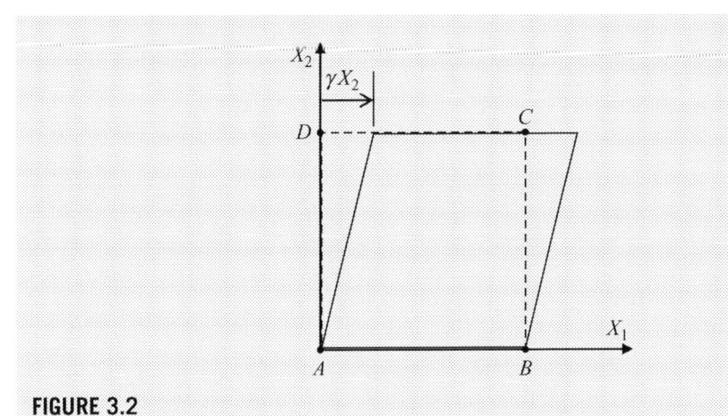

FIGURE 3.2

Simple shearing motion.

EXAMPLE 3.1.2 MATLAB COMPUTER PLOTS OF DEFORMATION/MOTION

Use MATLAB Code C-3 to make computer plots of the deformation of a unit square and unit circle for the following two-dimensional motion:

$$x_1 = 1.2X_1 + 0.2X_2 + 1.2$$
$$x_2 = 0.2X_1 + 1.2X_2 + 1.2$$
$$x_3 = X_3$$

Solution: Running the code with the given motion yields Fig. 3.3.

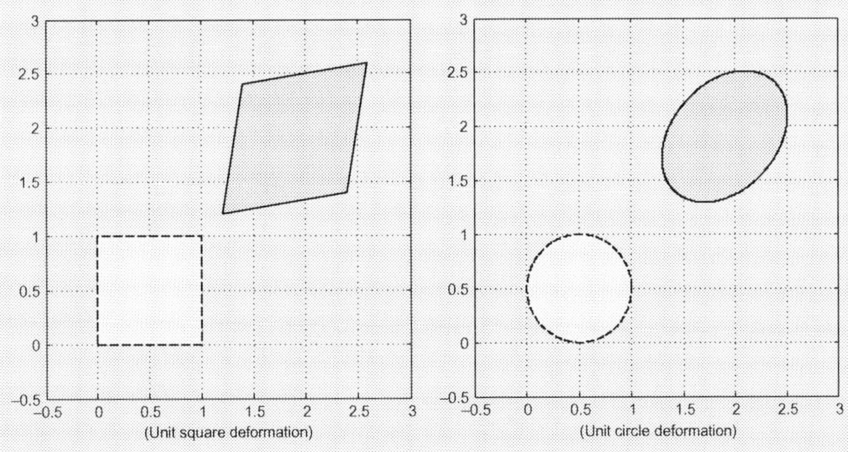

FIGURE 3.3

Computer deformation plots.

3.2 LAGRANGIAN AND EULERIAN DESCRIPTIONS

Motion equation (3.1.1) $x_i = \chi_i(X_1, X_2, X_3, t)$ can be thought of as a relationship between the *reference* or *material coordinates* X_i and the *current* or *spatial coordinates* x_i. Thus, if we know the material coordinates of a given particle, this relation will allow us to determine the particle's location in the current configuration. Likewise, the inverse equation (3.1.2). $X_i = \chi_i^{-1}(x_1, x_2, x_3, t)$ gives the opposite relationship, and consequently we can use this form to determine the material coordinates of a particular particle in the current configuration. Example 3.1.1 demonstrates these points for the simple shearing case.

All spatial field variables in continuum mechanics (density, temperature, displacement, strain, stress, etc.) can be described in terms of either the material coordinates X_i or the spatial coordinates x_i. A material coordinate description can be easily transformed into a spatial form or a spatial form transformed into a material description by using (3.1.1) or (3.1.2). Following particles, we can express tensor quantities as functions identified by the material coordinates (X_1, X_2, X_3). Such a description is known as the *Lagrangian*, *material*, or *reference description*. Using the other scheme, we can observe changes at *fixed locations* and thus express tensor quantities as functions of *position coordinates* (x_1, x_2, x_3). Such a description is known as an *Eulerian or spatial description*. Notice that as time progresses, different particles will occupy the same spatial position, and thus a spatial description will not provide specific information about particle properties during the motion. The following example illustrates some of the basic features of each description.

EXAMPLE 3.2.1 EULERIAN AND LAGRANGIAN TEMPERATURE DESCRIPTIONS UNDER SIMPLE SHEARING MOTION

Under simple shearing motion from Example 3.1.1 where $\gamma = \gamma_o t$, consider a continuum with spatial temperature field distribution $\theta = Ax_1 + Btx_2 + Cx_3$, where A, B, and C are constants. Find the material description of the temperature and its time rate of change for particular material particles.

Solution: The temperature is easily converted into a material description by using (3.1.5):

$$\begin{aligned}\theta &= Ax_1 + Btx_2 + Cx_3 \\ &= A(X_1 + \gamma_o tX_2) + BtX_2 + C_3X_3 \\ &= AX_1 + (A\gamma_o + B)tX_2 + C_3X_3\end{aligned}$$

The time rate of change in temperature for a given particle is

$$\left(\frac{\partial \theta}{\partial t}\right)_{X_i-\text{fixed}} = \frac{\partial}{\partial t}\left(AX_1 + (A\gamma_o + B)tX_2 + C_3X_3\right) = (A\gamma_o + B)X_2 = (A\gamma_o + B)x_2$$

3.3 MATERIAL TIME DERIVATIVE

The time rate of change of a tensor quantity following a material particle is known as the *material time derivative* and is commonly denoted by *D/Dt*. As we did in the previous example, when the material description of a given tensor field $\boldsymbol{T}$ is used, such a derivative is calculated in the straightforward manner

$$\frac{D\boldsymbol{T}}{Dt} = \frac{\partial}{\partial t}\boldsymbol{T}(X_1, X_2, X_3, t)\Big|_{X_i - fixed} \tag{3.3.1}$$

However, when the spatial description is used for tensor $\boldsymbol{T}$, the time derivative is a bit more complex since the spatial coordinates themselves are now functions of time. This requires use of the chain rule

$$\begin{aligned}\frac{D\boldsymbol{T}}{Dt} &= \frac{\partial}{\partial t}\boldsymbol{T}(x_1, x_2, x_3, t)\Big|_{X_i - fixed} \\ &= \frac{\partial \boldsymbol{T}}{\partial x_1}\frac{\partial x_1}{\partial t} + \frac{\partial \boldsymbol{T}}{\partial x_2}\frac{\partial x_2}{\partial t} + \frac{\partial \boldsymbol{T}}{\partial x_3}\frac{\partial x_3}{\partial t} + \left(\frac{\partial \boldsymbol{T}}{\partial t}\right)_{x_i - fixed} \\ &= \frac{\partial \boldsymbol{T}}{\partial x_i}\frac{\partial x_i}{\partial t} + \left(\frac{\partial \boldsymbol{T}}{\partial t}\right)_{x_i - fixed}\end{aligned} \tag{3.3.2}$$

Notice that we have used a little hybrid mix of direct and index notation. The quantities $\frac{\partial x_i}{\partial t}$ are taken with X_i fixed, and hence actually represent the velocity v_i of the continuum particle. Consequently, (3.3.2) can be expressed as

$$\frac{D\boldsymbol{T}}{Dt} = \frac{\partial \boldsymbol{T}}{\partial t} + \boldsymbol{T}_{,i}v_i = \frac{\partial \boldsymbol{T}}{\partial t} + \boldsymbol{v}\cdot\nabla\boldsymbol{T} \tag{3.3.3}$$

EXAMPLE 3.3.1 MATERIAL TIME DERIVATIVE OF EXAMPLE 3.2.1

Consider again the temperature distribution problem given in Example 3.2.1 under simple shearing motion. Using the spatial distribution form $\theta = Ax_1 + Btx_2 + Cx_3$, calculate the material time derivative of the temperature field.

Solution: We use relation (3.3.3) for the temperature (zeroth-order tensor) field

$$\frac{D\theta}{Dt} = \frac{\partial\theta}{\partial t} + \theta_{,i}v_i = \frac{\partial\theta}{\partial t} + \boldsymbol{v}\cdot\nabla\theta$$

First, we need the velocity field $v_i = \frac{\partial x_i}{\partial t}$. Using the shearing motion relations (3.1.5), we get

$$\begin{aligned} v_1 &= \gamma_o X_2 = \gamma_o X_2 \\ v_2 &= v_3 = 0 \end{aligned}$$

Next calculating the temperature gradient $\nabla\theta = \left\{ \frac{\partial\theta}{\partial x_1}, \frac{\partial\theta}{\partial x_2}, \frac{\partial\theta}{\partial x_3} \right\} = \{A, Bt, C\}$. Using these results,

$$\frac{D\theta}{Dt} = \frac{\partial\theta}{\partial t} + \boldsymbol{v} \cdot \nabla\theta = Bx_2 + A\gamma_o x_2 = (A\gamma_o + B)x_2$$

which is the same result we found in the previous example using a different method.

3.4 VELOCITY AND ACCELERATION

As discussed previously, the velocity of a continuum particle is given by its time rate of change of position

$$v_i(\boldsymbol{X}, t) = \frac{\partial}{\partial t} x_i(\boldsymbol{X}, t)\bigg|_{X_i - \text{fixed}} = \frac{Dx_i}{Dt} \tag{3.4.1}$$

Likewise, the acceleration is the time rate of change of the velocity

$$a_i(\boldsymbol{X}, t) = \frac{\partial}{\partial t} v_i(\boldsymbol{X}, t)\bigg|_{X_i - \text{fixed}} = \frac{Dv_i}{Dt} \tag{3.4.2}$$

If the velocity is expressed in spatial form $v_i(\boldsymbol{x}, t)$, then the acceleration follows from relation (3.3.3):

$$\boldsymbol{a}(\boldsymbol{x}, t) = \frac{\partial \boldsymbol{v}}{\partial t} + \boldsymbol{v}.\nabla\boldsymbol{v} \tag{3.4.3}$$

or in index notation

$$a_i(\boldsymbol{x}, t) = \frac{\partial v_i}{\partial t} + v_j v_{i,j} \tag{3.4.4}$$

Note that $v_{i,j}$ is the spatial *velocity gradient tensor* (see relation (2.17.4)). Thus, if the velocity is expressed in terms of material coordinates, the acceleration is found by simple partial differentiation with respect to time, relation (3.4.2). On the other hand, if the velocity is specified in spatial coordinates, the acceleration is determined by a more complicated procedure given by relations (3.3.3) or (3.4.4).

EXAMPLE 3.4.1 VELOCITY AND ACCELERATION FROM A MATERIAL MOTION

The motion of a continuum is given in material coordinates by

$$\begin{aligned} x_1 &= X_1 + AtX_2 \\ x_2 &= X_2 + Bt^2 X_1 \\ x_3 &= X_3 \end{aligned}$$

where A and B are constants. Determine

(a) the velocity and acceleration field in material coordinates;

(b) the path in the space of a particular particle that was located at $\boldsymbol{X} = \{1,2,4\}$.

Solution: Since the motion is given in material coordinates, we can simply use relations (3.4.1) and (3.4.2) to get

$$v_i(X,t) = \frac{\partial}{\partial t} x_i(X,t)|_{X_i\text{-fixed}} \Rightarrow \begin{array}{c} v_1 = AX_2 \\ v_2 = 2BtX_1 \\ v_3 = 0 \end{array}$$

$$a_i(X,t) = \frac{\partial}{\partial t} v_i(X,t)|_{X_i\text{-fixed}} \Rightarrow \begin{array}{c} a_1 = 0 \\ a_2 = 2BX_1 \\ a_3 = 0 \end{array}$$

For the particle located at $\boldsymbol{X} = \{1,2,4\}$, motion now specifies $x_1 = 1 + 2At$, $x_2 = 2 + Bt^2, x_3 = 4$. Since the motion is simply a time-parameterized expression, we can simply eliminate the time from these motion expressions to get

$$x_2 = \frac{B}{4A^2} x_1(x_1 - 2) + \frac{B}{4A^2} + 2, \quad x_3 = 4$$

and so the particle path is a parabolic form in the plane $x_3 = 4$.

EXAMPLE 3.4.2 ACCELERATION FROM SPATIAL VELOCITY FIELD FOR RIGID BODY ROTATION

Determine the acceleration field for a rigid body rotation about the x_3-aixs with a velocity field given by $\boldsymbol{v} = -\omega x_2 \boldsymbol{e}_1 + \omega x_1 \boldsymbol{e}_2 + 0\boldsymbol{e}_3$, where ω is the constant angular velocity.

Solution: First, calculate the velocity gradient tensor

$$v_{i,j} = \begin{bmatrix} 0 & -\omega & 0 \\ \omega & 0 & 0 \\ 0 & 0 & 0 \end{bmatrix}$$

$$v_{i,j}v_j = \begin{bmatrix} 0 & -\omega & 0 \\ \omega & 0 & 0 \\ 0 & 0 & 0 \end{bmatrix} \begin{bmatrix} -\omega x_2 \\ \omega x_1 \\ 0 \end{bmatrix} = \begin{bmatrix} -\omega^2 x_1 \\ -\omega^2 x_2 \\ 0 \end{bmatrix}, \quad \frac{\partial v_i}{\partial t} = 0$$

Using (3.4.4)

$$\Rightarrow a_i(\mathbf{x},t) = \frac{\partial v_i}{\partial t} v_j v_{i,j} = -\omega^2 \begin{bmatrix} x_1 \\ x_2 \\ 0 \end{bmatrix}$$

3.5 DISPLACEMENT AND DEFORMATION GRADIENT TENSORS

We now begin our exploration of a few fundamental kinematical variables related to continuum deformation. Fig. 3.4 illustrates again the basic concept of material motion between the reference configuration and the current configuration. Consider two neighboring points $\boldsymbol{P}$ and $\boldsymbol{Q}$ in the reference configuration. Through the general motion $\boldsymbol{x} = \chi(\boldsymbol{X},t)$, these two points get mapped into $\boldsymbol{P}'$ and $\boldsymbol{Q}'$ in the current configuration. Point $\boldsymbol{P}$ located at $\boldsymbol{X}$ then undergoes a *displacement* $\boldsymbol{u}$ so that it arrives at point $\boldsymbol{P}'$ located at $\boldsymbol{x}$.

We thus can write

$$\boldsymbol{x} = \boldsymbol{X} + \boldsymbol{u}(\boldsymbol{X},t) \tag{3.5.1}$$

the neighboring point $\boldsymbol{Q}$ then has a similar displacement relation

$$\boldsymbol{x} + d\boldsymbol{x} = \boldsymbol{X} + d\boldsymbol{X} + \boldsymbol{u}(\boldsymbol{X} + d\boldsymbol{X},\mathrm{t}) \tag{3.5.2}$$

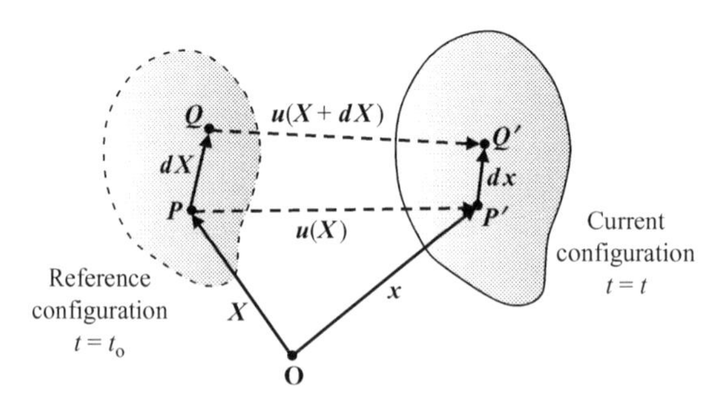

FIGURE 3.4

Displacements in continuum motion.

combining these relations gives

$$\begin{aligned} d\boldsymbol{x} &= d\boldsymbol{X} + \boldsymbol{u}(\boldsymbol{X} + d\boldsymbol{X}, t) - \boldsymbol{u}(\boldsymbol{X}, t) \\ &= d\boldsymbol{X} + \frac{\partial \boldsymbol{u}}{\partial \mathbf{X}} d\boldsymbol{X} \\ &= d\boldsymbol{X} + \nabla \boldsymbol{u}\, d\boldsymbol{X} \end{aligned} \tag{3.5.3}$$

The term $\frac{\partial \boldsymbol{u}}{\partial \boldsymbol{X}} = \nabla \boldsymbol{u}$ is the *displacement gradient tensor* with respect to material coordinates and it can be expressed in full matrix form as

$$\nabla \boldsymbol{u} = \frac{\partial \boldsymbol{u}}{\partial \boldsymbol{X}} = \begin{bmatrix} \frac{\partial u_1}{\partial X_1} & \frac{\partial u_1}{\partial X_2} & \frac{\partial u_1}{\partial X_3} \\ \frac{\partial u_2}{\partial X_1} & \frac{\partial u_2}{\partial X_2} & \frac{\partial u_2}{\partial X_3} \\ \frac{\partial u_3}{\partial X_1} & \frac{\partial u_3}{\partial X_2} & \frac{\partial u_3}{\partial X_3} \end{bmatrix} \tag{3.5.4}$$

it also follows that

$$d\boldsymbol{x} = \frac{\partial \boldsymbol{x}}{\partial \boldsymbol{X}} d\boldsymbol{x} = \boldsymbol{F}\, d\boldsymbol{X} \tag{3.5.5}$$

where

$$\boldsymbol{F} = \frac{\partial \boldsymbol{x}}{\partial \boldsymbol{X}}, \quad F_{ij} = \frac{\partial x_i}{\partial X_j} \tag{3.5.6}$$

is the *deformation gradient tensor.* Note that this tensor is also taken with respect to the material coordinates and can be written out in matrix form

$$\boldsymbol{F} = \frac{\partial \boldsymbol{x}}{\partial \boldsymbol{X}} = \begin{bmatrix} \frac{\partial x_1}{\partial X_1} & \frac{\partial x_1}{\partial X_2} & \frac{\partial x_1}{\partial X_3} \\ \frac{\partial x_2}{\partial X_1} & \frac{\partial x_2}{\partial X_2} & \frac{\partial x_2}{\partial X_3} \\ \frac{\partial x_3}{\partial X_1} & \frac{\partial x_3}{\partial X_2} & \frac{\partial x_3}{\partial X_3} \end{bmatrix} \tag{3.5.7}$$

Combining (3.5.3) and (3.5.5), a simple relationship can easily be derived between the deformation gradient and the displacement gradient

$$\boldsymbol{F} = \boldsymbol{I} + \nabla \boldsymbol{u} \tag{3.5.8}$$

EXAMPLE 3.5.1 DISPLACEMENT AND DEFORMATION GRADIENT TENSORS FOR SIMPLE SHEARING MOTION

For the simple shearing motion given in Example 3.1.1, determine the displacement field, displacement gradient tensor, and the deformation gradient tensor.

Solution: From relation (3.5.1), $\boldsymbol{u}(\boldsymbol{X},t)=\boldsymbol{x}-\boldsymbol{X}$, and so for the motion (3.1.5),

$$\begin{aligned} u_1 &= x_1 - X_1 = \gamma X_2 \\ u_2 &= x_2 - X_2 = 0 \\ u_3 &= x_3 - X_3 = 0 \end{aligned}$$

The displacement gradient tensor follows from $\dfrac{\partial \boldsymbol{u}}{\partial \boldsymbol{X}} = \nabla \boldsymbol{u} = \begin{bmatrix} 0 & \gamma & 0 \\ 0 & 0 & 0 \\ 0 & 0 & 0 \end{bmatrix}$

The deformation gradient is calculated from $\boldsymbol{F} = \dfrac{\partial \boldsymbol{x}}{\partial \boldsymbol{X}} = \begin{bmatrix} 1 & \gamma & 0 \\ 0 & 1 & 0 \\ 0 & 0 & 1 \end{bmatrix}$

Note that these results satisfy the relation $\boldsymbol{F}=\boldsymbol{I}+\nabla\boldsymbol{u}$.

EXAMPLE 3.5.2 DISPLACEMENTS AND DEFORMATION GRADIENT TENSOR FOR EXTENSIONAL MOTION

Three-dimensional *extensional motion* may be specified by

$$\begin{aligned} x_1 &= \lambda_1 X_1 \\ x_2 &= \lambda_2 X_2 \\ x_3 &= \lambda_3 X_3 \end{aligned} \tag{3.5.9}$$

where λ_i are constants often called the *stretch ratios*. Determine the displacement field, displacement gradient tensor, and the deformation gradient tensor for this motion.

Solution: From relation (3.5.1), $\boldsymbol{u}(\boldsymbol{X},t)=\boldsymbol{x}-\boldsymbol{X}$, and so for the motion (3.5.9),

$$\begin{aligned} u_1 &= x_1 - X_1 = (\lambda_1 - 1)X_1 \\ u_2 &= x_2 - X_2 = (\lambda_2 - 1)X_2 \\ u_3 &= x_3 - X_3 = (\lambda_3 - 1)X_3 \end{aligned}$$

The displacement gradient tensor follows from

$$\frac{\partial \boldsymbol{u}}{\partial \boldsymbol{X}} = \nabla \boldsymbol{u} = \begin{bmatrix} \lambda_1 - 1 & 0 & 0 \\ 0 & \lambda_2 - 1 & 0 \\ 0 & 0 & \lambda_3 - 1 \end{bmatrix}$$

The deformation gradient is calculated from $\boldsymbol{F} = \dfrac{\partial \boldsymbol{x}}{\partial \boldsymbol{X}} = \begin{bmatrix} \lambda_1 & 0 & 0 \\ 0 & \lambda_2 & 0 \\ 0 & 0 & \lambda_3 \end{bmatrix}$

Note that these results satisfy the relation $\boldsymbol{F}=\boldsymbol{I}+\nabla\boldsymbol{u}$.

Consider next a few comments on some special deformations. A *homogeneous deformation* is one where the deformation gradient tensor is independent of the coordinates. The previous two examples of simple shear and constant extension were both homogeneous deformations, and many experimental testing procedures attempt to maintain this type of deformation within the specimen under study. Note that if we only have rigid body translation, the motion is $\boldsymbol{x} = \boldsymbol{X} + \boldsymbol{c}$, where $\boldsymbol{c}$ is the constant translation, and the deformation gradient reduces to the unit tensor $\boldsymbol{F} = \boldsymbol{I}$. It should be apparent that the deformation gradient will not be changed if we add a uniform translation to the motion since the gradient of the scalar addition will always vanish. For the special case of rigid body rotation (about the origin) using a constant rotation tensor $\boldsymbol{R}$, the motion becomes $\boldsymbol{x} = \boldsymbol{R}\boldsymbol{X}$, and thus the deformation gradient is given by $\boldsymbol{F} = \boldsymbol{R}$.

EXAMPLE 3.5.3 DISPLACEMENTS AND DEFORMATION GRADIENT TENSOR FOR NONHOMOGENEOUS DEFORMATION

Consider now a nonhomogeneous deformation case with motion

$$x_1 = AX_1X_2$$
$$x_2 = BX_2^2$$
$$x_3 = CX_3$$

where A, B, and C are constants. Determine the displacement field, displacement gradient tensor, and the deformation gradient tensor for this motion.

Solution: From relation (3.5.1), $\boldsymbol{u}(\boldsymbol{X},t) = \boldsymbol{x} - \boldsymbol{X}$, and so for this motion

$$u_1 = x_1 - X_1 = (AX_2 - 1)X_1$$
$$u_2 = x_2 - X_2 = (BX_2 - 1)X_2$$
$$u_3 = x_3 - X_3 = (C - 1)X_3$$

The displacement gradient tensor follows from

$$\frac{\partial \boldsymbol{u}}{\partial \boldsymbol{X}} = \nabla \boldsymbol{u} = \begin{bmatrix} AX_2 - 1 & AX_1 & 0 \\ 0 & 2BX_2 - 1 & 0 \\ 0 & 0 & C-1 \end{bmatrix}$$

The deformation gradient is calculated from $\boldsymbol{F} = \dfrac{\partial \boldsymbol{x}}{\partial \boldsymbol{X}} = \begin{bmatrix} AX_2 & AX_1 & 0 \\ 0 & 2BX_2 & 0 \\ 0 & 0 & C \end{bmatrix}$

Both the displacement gradient and deformation gradient depend on material coordinates and thus the deformation is referred to as nonhomogeneous.

Next let us examine the concept of *sequential deformations* whereby the total deformation is carried out through a sequence of two or more separate steps. Consider the case of a two-step process composed of deformation (1) followed by deformation (2). Using the general result (3.5.6), we can express the first step as $d\boldsymbol{x}^{(1)} = \boldsymbol{F}^{(1)} d\boldsymbol{x}$. Likewise, the second deformation step is then given by $d\boldsymbol{x}^{(2)} = \boldsymbol{F}^{(2)} d\boldsymbol{x}^{(1)} = \boldsymbol{F}^{(2)}\boldsymbol{F}^{(1)} d\boldsymbol{X}$. We can then conclude that for a two-step sequential deformation process of $\boldsymbol{F}^{(1)}$ followed by $\boldsymbol{F}^{(2)}$, the overall deformation gradient is given by the product $\boldsymbol{F}^{(2)}\boldsymbol{F}^{(1)}$. The overall deformation can then be written as

$$d\boldsymbol{x} = \boldsymbol{F}^{(2)}\boldsymbol{F}^{(1)} d\boldsymbol{X} \tag{3.5.10}$$

It is important to note that the product is carried out in reverse order to the deformation steps, and changing this order will result in a totally different overall final deformation. This concept can be shown graphically in Fig. 3.5 which illustrates two 2-step sequential deformations where the order has been switched. It is clearly evident that by changing the sequence order, the final deformations are quite different.

Based on our previous work, we can consider a *spatial deformation gradient tensor* $\boldsymbol{F}^{-1}$:

$$\boldsymbol{F}^{-1} = \frac{\partial \boldsymbol{X}}{\partial \boldsymbol{x}}, \quad F_{ij}^{-1} = \frac{\partial X_i}{\partial x_j} \tag{3.5.11}$$

that maps the spatial line element $d\boldsymbol{x}$ to the material element $d\boldsymbol{X}$ defined by

$$d\boldsymbol{X} = \frac{\partial \boldsymbol{X}}{\partial \boldsymbol{x}} d\boldsymbol{x} = \boldsymbol{F}^{-1} d\boldsymbol{x} \tag{3.5.12}$$

The tensor $\boldsymbol{F}^{-1}$ can rightfully be called *inverse of the deformation gradient* $\boldsymbol{F}$ since

$$F_{ij}F_{jk}^{-1} = \frac{\partial x_i}{\partial X_j}\frac{\partial X_j}{\partial x_k} = \frac{\partial x_i}{\partial x_k} = \delta_{ik} \Rightarrow \boldsymbol{F}\boldsymbol{F}^{-1} = \boldsymbol{I} \tag{3.5.13}$$

It should be noted that the deformation gradient involves both the spatial coordinates $\boldsymbol{x}$ and the material coordinates through the gradient operation $\partial / \partial X$. The same

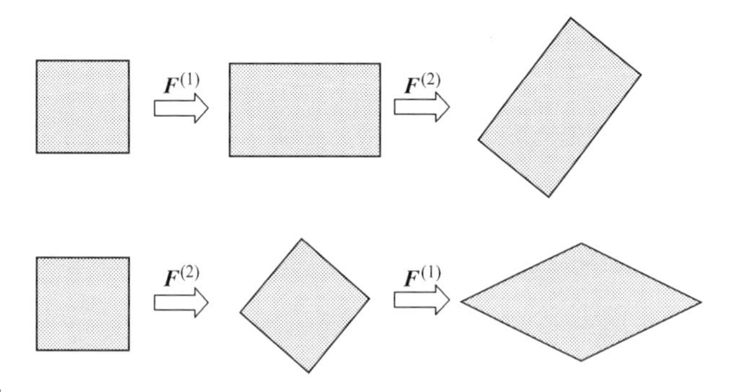

FIGURE 3.5

Two-step sequential deformations with different sequences ($\boldsymbol{F}^{(1)}$—horizontal stretch; $\boldsymbol{F}^{(2)}$ 45°—CCW rotation).

can be said concerning the spatial or inverse deformation gradient F^{-1}. Because of this, both F and F^{-1} are called *two-point tensors* being coupled to both configurations. Sometimes, these tensors are written as

$$\begin{aligned} F &= \frac{\partial x}{\partial X_j} \otimes E_j = \frac{\partial x_i}{\partial X_j} e_i \otimes E_j \\ F^{-1} &= \frac{\partial X}{\partial x_j} \otimes e_j = \frac{\partial X_i}{\partial x_j} E_i \otimes e_j \end{aligned} \tag{3.5.14}$$

where e_i are the spatial basis vectors and E_i are the basis vectors in the material configuration.

We next consider how the deformation gradient tensor transforms under the objectivity test established in Section 2.9. The deformation gradients in the two reference frames shown previously in Fig. 2.4 would simply be

$$F = \frac{\partial x}{\partial X}, \quad F^* = \frac{\partial x^*}{\partial X}$$

Using the chain rule gives

$$F^* = \frac{\partial x^*}{\partial X} = \frac{\partial x^*}{\partial x}\frac{\partial x}{\partial X} = QF \tag{3.5.15}$$

and thus the deformation gradient F satisfies the objectivity test for vectors (2.9.2) but not for second-order tensors as given by (2.9.3). This is expected as F is a function singly of the vector x, and the reference position X is to remain the same within the objectivity study. Following analogous steps for the inverse F^{-1}:

$$F^{*-1} = \frac{\partial X}{\partial x^*} = \frac{\partial X}{\partial x}\frac{\partial x}{\partial x^*} = F^{-1}Q^T \tag{3.5.16}$$

and this leads to a similar conclusion that F^{-1} does not satisfy the objectivity test for second-order tensors. The deformation gradient and its inverse play key roles in describing the *local* deformation in the neighborhood of a continuum particle; however, we will need to construct additional second-order tensors that more properly describe straining deformation and satisfy the objectivity test.

3.6 LAGRANGIAN AND EULERIAN STRAIN TENSORS

From our previous kinematical discussions, the motion relation $x = \chi(X, t)$ can produce movements of continuum particles of unlimited variety and complexity that would include rigid body translation and rotation along with other movements that will cause *relative changes in position* between two or more particles. These relative motions or deformations are what we commonly refer to as *strain*. Rigid body motions will generally be of little interest since they are not related to internal force or stress distributions. Fig. 3.6A illustrates some of these basic concepts graphically

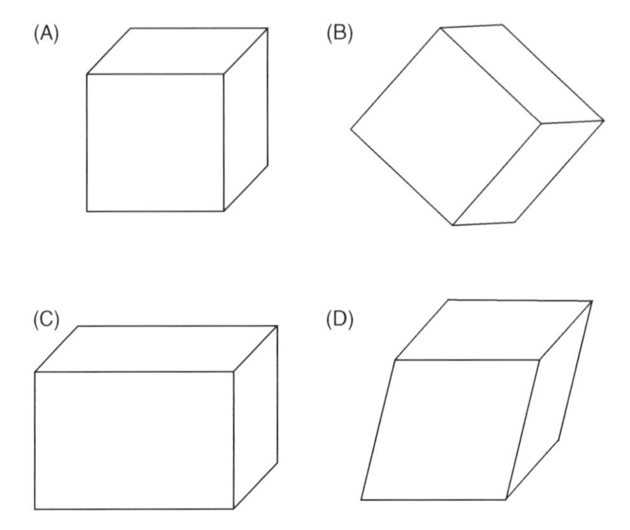

FIGURE 3.6

Examples of motion and deformation: (A) reference element; (B) rigid body motion; (C) extensional deformation; (D) shearing deformation.

by starting with a Cartesian differential element in say the reference configuration. This element can then undergo several different types of motion or deformation. Fig. 3.6B shows the element after rigid body motion including translation and rotation, and no relative distance changes between any particle pairs have occurred. However, in Fig. 3.6C and D extensional and shearing deformations have obviously created relative distance changes among continuum particles, and it is for these cases that we wish to develop quantitative measures.

The major question is then how do we decide on creating such strain measures. This choice is not unique and over the years many schemes have been defined. For modern continuum mechanics theory, we want strain measures that satisfy the principle of objectivity as presented in Section 2.9. Although the deformation gradient plays an important role in the analysis of deformation, it is not suitable by itself to be a measure of strain since it does not satisfy objectivity for second-order tensors. A simple measure is the *distance between two material points* since we have previously shown that this scalar value will be the same for all observers and hence satisfy objectivity.

Thus, we can start in the reference configuration and define the *Lagrangian or Green strain tensor*, by considering the *change in the square of the length of the vector* $d\boldsymbol{X}$ shown in Fig. 3.4.

$$\begin{aligned} |d\boldsymbol{x}|^2 - |d\boldsymbol{X}|^2 &= (\boldsymbol{F}\,d\boldsymbol{X})\cdot(\boldsymbol{F}\,d\boldsymbol{X}) - d\boldsymbol{X}\cdot d\boldsymbol{X} \\ &= F_{ij}\,dX_j F_{ik}\,dX_k - dX_k\,dX_k \\ &= (F_{ij}F_{ik} - \delta_{jk})dX_j\,dX_k \\ &= d\boldsymbol{X}\left(\boldsymbol{F}^{\mathrm{T}}\boldsymbol{F} - \boldsymbol{I}\right)d\boldsymbol{X} \end{aligned} \tag{3.6.1}$$

Defining the Lagrangian stain tensor $\boldsymbol{E}$ by

$$\boldsymbol{E} = \frac{1}{2}\left(\boldsymbol{F}^{\mathrm{T}}\boldsymbol{F} - \boldsymbol{I}\right), \quad E_{ij} = \frac{1}{2}\left(F_{ki}F_{kj} - \delta_{ij}\right) \tag{3.6.2}$$

we then have

$$|d\boldsymbol{x}|^2 - |d\boldsymbol{X}|^2 = 2d\boldsymbol{X}\boldsymbol{E}\,d\boldsymbol{X} = 2E_{ij}\,\mathrm{d}X_i\,\mathrm{d}X_j \tag{3.6.3}$$

Using relation (3.5.8), we can express the Lagrangian stain in terms of the displacement gradients

$$\begin{aligned} \boldsymbol{E} &= \frac{1}{2}[(\nabla\boldsymbol{u}) + (\nabla\boldsymbol{u})^{\mathrm{T}} + (\nabla\boldsymbol{u})^{\mathrm{T}}(\nabla\boldsymbol{u})] \\ E_{ij} &= \frac{1}{2}\left(u_{i,j} + u_{j,i} + u_{k,i}u_{k,j}\right) \end{aligned} \tag{3.6.4}$$

Notice that in all forms the Lagrangian strain is a symmetric second-order tensor, $E_{ij} = E_{ji}$.

We can repeat the past few steps but instead of using the reference configuration we use the current configuration and define the *Eulerian or Almansi strain tensor*, by considering the *change in the square of the length of the vector* $d\boldsymbol{x}$ shown in Fig. 3.4.

$$\begin{aligned} |d\boldsymbol{x}|^2 - |d\boldsymbol{X}|^2 &= d\boldsymbol{x}\cdot d\boldsymbol{x} - \left(\boldsymbol{F}^{-1}\,d\boldsymbol{x}\right)\cdot\left(\boldsymbol{F}^{-1}\,d\boldsymbol{x}\right) \\ &= d\boldsymbol{x}\left(\boldsymbol{I} - \left(\boldsymbol{F}^{-1}\right)^{\mathrm{T}}\boldsymbol{F}^{-1}\right)d\boldsymbol{x} \\ &= (\delta_{jk} - F_{ij}^{-1}F_{ik}^{-1})\,dx_j\,dx_k \end{aligned} \tag{3.6.5}$$

where we have used the spatial deformation gradient tensor defined by (3.5.9). Defining the Eulerian stain tensor $\boldsymbol{e}$ by

$$\boldsymbol{e} = \frac{1}{2}\left(\boldsymbol{I} - \left(\boldsymbol{F}^{-1}\right)^{\mathrm{T}}\boldsymbol{F}^{-1}\right), \quad e_{ij} = \frac{1}{2}(\delta_{jk} - F_{ij}^{-1}F_{ik}^{-1}) \tag{3.6.6}$$

we then have

$$|d\boldsymbol{x}|^2 - |d\boldsymbol{X}|^2 = 2\,d\boldsymbol{x}\,\boldsymbol{e}\,d\boldsymbol{x} = 2e_{ij}\,dx_i\,dx_j \tag{3.6.7}$$

We can define the spatial or Eulerian displacement gradient tensor by

$$\nabla\boldsymbol{u}^* = \frac{\partial\boldsymbol{u}^*}{\partial\boldsymbol{x}} = \frac{\partial u_i^*}{\partial x_j}\boldsymbol{e}_i \otimes \boldsymbol{e}_j = u_{i,j}^*\boldsymbol{e}_i \otimes \boldsymbol{e}_j \tag{3.6.8}$$

Using the chain rule gives $\nabla\boldsymbol{u} = \nabla\boldsymbol{u}^*\boldsymbol{F}$, and thus $\boldsymbol{F}^{-1} = \boldsymbol{I} - \nabla\boldsymbol{u}^*$. These results allow us to express the Eulerian stain in terms of the spatial displacement gradients

$$\begin{aligned} \boldsymbol{e} &= \frac{1}{2}[(\nabla\boldsymbol{u}^*) + (\nabla\boldsymbol{u}^*)^{\mathrm{T}} + (\nabla\boldsymbol{u}^*)^{\mathrm{T}}(\nabla\boldsymbol{u}^*)] \\ e_{ij} &= \frac{1}{2}\left(u_{i,j}^* + u_{j,i}^* + u_{k,i}^*u_{k,j}^*\right) \end{aligned} \tag{3.6.9}$$

and similar to our previous Lagrangian form, the Eulerian strain is a symmetric second-order tensor, $e_{ij} = e_{ji}$.

EXAMPLE 3.6.1 LAGRANGIAN AND EULERIAN STRAIN TENSORS FOR SIMPLE SHEAR AND EXTENSIONAL DEFORMATIONS

For the previous deformations of simple shear and extension specified in Examples 3.1.1 and 3.5.2, determine the Lagrangian and Eulerian strain tensors.

Solution: For the simple shear case

$$\begin{matrix} x_1 = X_1 + \gamma X_2 \\ x_2 = X_2 \\ x_3 = X_3 \end{matrix} \quad \Rightarrow \boldsymbol{F} = \frac{\partial \boldsymbol{x}}{\partial \boldsymbol{X}} = \begin{bmatrix} 1 & \gamma & 0 \\ 0 & 1 & 0 \\ 0 & 0 & 1 \end{bmatrix} \Rightarrow$$

$$\boldsymbol{F}^{\mathrm{T}}\boldsymbol{F} = \begin{bmatrix} 1 & 0 & 0 \\ \gamma & 1 & 0 \\ 0 & 0 & 1 \end{bmatrix} \begin{bmatrix} 1 & \gamma & 0 \\ 0 & 1 & 0 \\ 0 & 0 & 1 \end{bmatrix} = \begin{bmatrix} 1 & \gamma & 0 \\ \gamma & 1+\gamma^2 & 0 \\ 0 & 0 & 1 \end{bmatrix}$$

$$\boldsymbol{E} = \frac{1}{2}\left(\boldsymbol{F}^{\mathrm{T}}\boldsymbol{F} - \boldsymbol{I}\right) = \frac{1}{2}\begin{bmatrix} 0 & \gamma & 0 \\ \gamma & \gamma^2 & 0 \\ 0 & 0 & 0 \end{bmatrix}$$

For the Eulerian forms

$$\begin{matrix} X_1 = x_1 - \gamma x_2 \\ X_2 = x_2 \\ X_3 = x_3 \end{matrix} \quad \Rightarrow \boldsymbol{F}^{-1} = \frac{\partial \boldsymbol{X}}{\partial \boldsymbol{x}} = \begin{bmatrix} 1 & -\gamma & 0 \\ 0 & 1 & 0 \\ 0 & 0 & 1 \end{bmatrix} \Rightarrow$$

$$(\boldsymbol{F}^{-1})^{\mathrm{T}}\boldsymbol{F}^{-1} = \begin{bmatrix} 1 & 0 & 0 \\ -\gamma & 1 & 0 \\ 0 & 0 & 1 \end{bmatrix} \begin{bmatrix} 1 & -\gamma & 0 \\ 0 & 1 & 0 \\ 0 & 0 & 1 \end{bmatrix} = \begin{bmatrix} 1 & -\gamma & 0 \\ -\gamma & 1+\gamma^2 & 0 \\ 0 & 0 & 1 \end{bmatrix}$$

$$\boldsymbol{e} = \frac{1}{2}\left(\boldsymbol{I} - (\boldsymbol{F}^{-1})^{\mathrm{T}}\boldsymbol{F}^{-1}\right) = \frac{1}{2}\begin{bmatrix} 0 & \gamma & 0 \\ \gamma & -\gamma^2 & 0 \\ 0 & 0 & 0 \end{bmatrix}$$

For the extensional motion example

$$\begin{matrix} x_1 = \lambda_1 X_1 \\ x_2 = \lambda_2 X_2 \Rightarrow \boldsymbol{F} = \dfrac{\partial \boldsymbol{x}}{\partial \boldsymbol{X}} = \begin{bmatrix} \lambda_1 & 0 & 0 \\ 0 & \lambda_2 & 0 \\ 0 & 0 & \lambda_3 \end{bmatrix} \Rightarrow \\ X_3 = \lambda_3 X_3 \end{matrix}$$

$$\boldsymbol{F}^{\mathrm{T}}\boldsymbol{F} = \begin{bmatrix} \lambda_1 & 0 & 0 \\ 0 & \lambda_2 & 0 \\ 0 & 0 & \lambda_3 \end{bmatrix} \begin{bmatrix} \lambda_1 & 0 & 0 \\ 0 & \lambda_2 & 0 \\ 0 & 0 & \lambda_3 \end{bmatrix} = \begin{bmatrix} \lambda_1^2 & 0 & 0 \\ 0 & \lambda_2^2 & 0 \\ 0 & 0 & \lambda_3^2 \end{bmatrix}$$

$$E = \frac{1}{2}\left(F^T F - I\right) = \frac{1}{2}\begin{bmatrix} \lambda_1^2 - 1 & 0 & 0 \\ 0 & \lambda_2^2 - 1 & 0 \\ 0 & 0 & \lambda_3^2 - 1 \end{bmatrix}$$

and for the Eulerian forms

$$\begin{aligned} &X_1 = x_1 / \lambda_1 \\ &X_2 = x_2 / \lambda_2 \Rightarrow F^{-1} = \frac{\partial F}{\partial x} = \begin{bmatrix} \lambda_1^{-1} & 0 & 0 \\ 0 & \lambda_2^{-1} & 0 \\ 0 & 0 & \lambda_3^{-1} \end{bmatrix} \Rightarrow \\ &X_3 = x_3 / \lambda_3 \end{aligned}$$

$$(F^{-1})^T F^{-1} = \begin{bmatrix} \lambda_1^{-1} & 0 & 0 \\ 0 & \lambda_2^{-1} & 0 \\ 0 & 0 & \lambda_3^{-1} \end{bmatrix}\begin{bmatrix} \lambda_1^{-1} & 0 & 0 \\ 0 & \lambda_2^{-1} & 0 \\ 0 & 0 & \lambda_3^{-1} \end{bmatrix} = \begin{bmatrix} \lambda_1^{-2} & 0 & 0 \\ 0 & \lambda_2^{-2} & 0 \\ 0 & 0 & \lambda_3^{-2} \end{bmatrix}$$

$$e = \frac{1}{2}\left(I - \left(F^{-1}\right)^T F^{-1}\right) = \frac{1}{2}\begin{bmatrix} 1-\lambda_1^{-2} & 0 & 0 \\ 0 & 1-\lambda_2^{-2} & 0 \\ 0 & 0 & 1-\lambda_3^{-2} \end{bmatrix}$$

Let us now apply the Polar Decomposition Theorem from Section 2.16. Recall that this mathematical result allows for the multiplicative decomposition of any nonsingular second-order tensor. Since the deformation gradient F is nonsingular (see relation (3.1.3), we can use this theorem to write

$$F = RU = VR \tag{3.6.10}$$

where R is a proper orthogonal tensor and U and V are symmetric positive definite tensors, such that

$$\begin{aligned} U^2 &= F^T F \\ V^2 &= FF^T \end{aligned} \tag{3.6.11}$$

The proper orthogonal tensor represents the rigid body rotation, and so the tensors U and V must characterize pure strain and are often referred to as the *right and left stretch tensors*. Note that the rotation tensor can be found from one of the relations $R = FU^{-1} = V^{-1}F$. Since $V = RUR^T$, U and V must have the same principal values $\{\lambda_i\}$ called the *principal stretches*. From another point of view, if we consider the extensional deformation given in Example 3.6.1:

$$U^2 = F^T F = \begin{bmatrix} \lambda_1^2 & 0 & 0 \\ 0 & \lambda_2^2 & 0 \\ 0 & 0 & \lambda_3^2 \end{bmatrix} \Rightarrow U = \begin{bmatrix} \lambda_1 & 0 & 0 \\ 0 & \lambda_2 & 0 \\ 0 & 0 & \lambda_3 \end{bmatrix} = V$$

and thus $\boldsymbol{U}$ and $\boldsymbol{V}$ are directly related to the stretch ratios λ_i from extensional deformation.

Going back to the fundamental definition of the deformation gradient

$$d\boldsymbol{x} = \boldsymbol{F}\,d\boldsymbol{X} = (\boldsymbol{R}\boldsymbol{U})d\boldsymbol{X} = \boldsymbol{R}(\boldsymbol{U}\,d\boldsymbol{X})$$

which can be thought of as a two-step process: pure stretch $\boldsymbol{U}d\boldsymbol{X}$ followed by a rotation $\boldsymbol{R}(\boldsymbol{U}d\boldsymbol{X})$. Likewise, we can use the second form in relation (3.6.10):

$$d\boldsymbol{x} = \boldsymbol{F}\,d\boldsymbol{X} = (\boldsymbol{V}\boldsymbol{R})d\boldsymbol{X} = \boldsymbol{V}(\boldsymbol{R}\,d\boldsymbol{X})$$

which is the two-step process: rotation $\boldsymbol{R}d\boldsymbol{X}$ followed by a stretch $\boldsymbol{V}(\boldsymbol{R}d\boldsymbol{X})$. We thus can see the *sequential or serial mapping* that results from the multiplicative decomposition from the Polar Decomposition Theorem.

Another useful set of strain tensors come from the specification of changes of differential line elements in the reference and spatial configurations. Consider two differential line elements in the reference configuration $d\boldsymbol{X}^{(1)}$ and $d\boldsymbol{X}^{(2)}$ which are mapped into the corresponding elements $d\boldsymbol{x}^{(1)}$ and $d\boldsymbol{x}^{(2)}$ in the current spatial configuration. We can thus write

$$\begin{aligned} d\boldsymbol{x}^{(1)} \cdot d\boldsymbol{x}^{(2)} &= \boldsymbol{F}\,d\boldsymbol{X}^{(1)} \cdot \boldsymbol{F}\,d\boldsymbol{X}^{(2)} \\ &= d\boldsymbol{X}^{(1)}\left(\boldsymbol{F}^{\mathrm{T}}\boldsymbol{F}\right)d\boldsymbol{X}^{(2)} \\ &= d\boldsymbol{X}^{(1)}\boldsymbol{C}\,d\boldsymbol{X}^{(2)} \end{aligned} \tag{3.6.12}$$

where

$$\boldsymbol{C} = \boldsymbol{F}^{\mathrm{T}}\boldsymbol{F}, \quad C_{ij} = F_{ki}F_{kj} \tag{3.6.13}$$

is called the *right Cauchy–Green or Green strain tensor.* It can be shown that this tensor is symmetric and positive definite which implies that it has real positive eigenvalues (see Exercise 2.20).

Next, we can reverse the mapping and work from spatial to reference configurations and write

$$\begin{aligned} d\boldsymbol{X}^{(1)} \cdot d\boldsymbol{X}^{(2)} &= \boldsymbol{F}^{-1}\,d\boldsymbol{x}^{(1)} \cdot \boldsymbol{F}^{-1}\,d\boldsymbol{x}^{(2)} \\ &= d\boldsymbol{x}^{(1)}\left((\boldsymbol{F}^{-1})^{\mathrm{T}}\boldsymbol{F}^{-1}\right)d\boldsymbol{x}^{(2)} \\ &= d\boldsymbol{x}^{(1)}\boldsymbol{B}^{-1}\,d\boldsymbol{x}^{(2)} \end{aligned} \tag{3.6.14}$$

where $\boldsymbol{B}^{-1} = (\boldsymbol{F}^{-1})^{\mathrm{T}}\boldsymbol{F}^{-1}$, and this leads to the definition of $\boldsymbol{B}$ as

$$\boldsymbol{B} = \boldsymbol{F}\boldsymbol{F}^{\mathrm{T}}, \quad B_{ij} = F_{ik}F_{jk} \tag{3.6.15}$$

which is called the *left Cauchy–Green or Finger strain tensor.* Again this tensor is symmetric and positive definite.

Note that $tr\,\boldsymbol{C} = tr\left(\boldsymbol{F}^{\mathrm{T}}\boldsymbol{F}\right) = tr\left(\boldsymbol{F}\boldsymbol{F}^{\mathrm{T}}\right) = tr\,\boldsymbol{B}$ and this result also holds for arbitrary powers of these tensors. Using results (2.11.4) and (2.11.5) implies that the invariants of $\boldsymbol{C}$ and $\boldsymbol{B}$ are the same. Referring back to our definition (3.6.11) of the right and left stretch tensors $\boldsymbol{U}$ and $\boldsymbol{V}$, we can write $\boldsymbol{U}^2 = \boldsymbol{C}$ and $\boldsymbol{V}^2 = \boldsymbol{B}$.

EXAMPLE 3.6.2 RIGHT AND LEFT CAUCHY-GREEN STRAIN TENSORS FOR SIMPLE SHEAR AND EXTENSIONAL DEFORMATIONS

For the previous deformations of simple shear and extension specified in Examples 3.1.1 and 3.5.2, determine the right and left Cauchy–Green strain tensors.

Solution: For the simple shear case

$$\begin{aligned} x_1 &= X_1 + \gamma X_2 \\ x_2 &= X_2 \\ x_3 &= X_3 \end{aligned} \Rightarrow \boldsymbol{F} = \frac{\partial \boldsymbol{x}}{\partial \boldsymbol{X}} = \begin{bmatrix} 1 & \gamma & 0 \\ 0 & 1 & 0 \\ 0 & 0 & 1 \end{bmatrix} \Rightarrow$$

$$\boldsymbol{C} = \boldsymbol{F}^{\mathrm{T}}\boldsymbol{F} = \begin{bmatrix} 1 & 0 & 0 \\ \gamma & 1 & 0 \\ 0 & 0 & 1 \end{bmatrix} \begin{bmatrix} 1 & \gamma & 0 \\ 0 & 1 & 0 \\ 0 & 0 & 1 \end{bmatrix} = \begin{bmatrix} 1 & \gamma & 0 \\ \gamma & 1+\gamma^2 & 0 \\ 0 & 0 & 1 \end{bmatrix}$$

$$\boldsymbol{B} = \boldsymbol{F}\boldsymbol{F}^{\mathrm{T}} = \begin{bmatrix} 1 & \gamma & 0 \\ 0 & 1 & 0 \\ 0 & 0 & 1 \end{bmatrix} \begin{bmatrix} 1 & 0 & 0 \\ \gamma & 1 & 0 \\ 0 & 0 & 1 \end{bmatrix} = \begin{bmatrix} 1+\gamma^2 & \gamma & 0 \\ \gamma & 1 & 0 \\ 0 & 0 & 1 \end{bmatrix}$$

For the extensional motion example

$$\begin{aligned} x_1 &= \lambda_1 X_1 \\ x_2 &= \lambda_2 X_2 \Rightarrow \\ x_3 &= \lambda_3 X_3 \end{aligned} \quad \boldsymbol{F} = \frac{\partial \boldsymbol{x}}{\partial \boldsymbol{X}} = \begin{bmatrix} \lambda_1 & 0 & 0 \\ 0 & \lambda_2 & 0 \\ 0 & 0 & \lambda_3 \end{bmatrix} \Rightarrow$$

$$\boldsymbol{C} = \boldsymbol{F}^{\mathrm{T}}\boldsymbol{F} = \begin{bmatrix} \lambda_1 & 0 & 0 \\ 0 & \lambda_2 & 0 \\ 0 & 0 & \lambda_3 \end{bmatrix} \begin{bmatrix} \lambda_1 & 0 & 0 \\ 0 & \lambda_2 & 0 \\ 0 & 0 & \lambda_3 \end{bmatrix} = \begin{bmatrix} \lambda_1^2 & 0 & 0 \\ 0 & \lambda_2^2 & 0 \\ 0 & 0 & \lambda_3^2 \end{bmatrix} = \boldsymbol{B}$$

EXAMPLE 3.6.3 ROTATION AND RIGHT AND LEFT STRETCH TENSORS

For the following motion, calculate the deformation gradient $\boldsymbol{F}$, the rotation tensor $\boldsymbol{R}$, and the right and left stretch tensors $\boldsymbol{U}$ and $\boldsymbol{V}$:

$$\begin{aligned} x_1 &= X_1 \\ x_2 &= 2X_3 \\ x_3 &= 4X_2 \end{aligned}$$

Solution: Starting with the deformation gradient

$$\boldsymbol{F} = \frac{\partial \boldsymbol{x}}{\partial \boldsymbol{X}} = \begin{bmatrix} 1 & 0 & 0 \\ 0 & 0 & 2 \\ 0 & 4 & 0 \end{bmatrix} \Rightarrow$$

$$U^2 = F^{\mathrm{T}}F = \begin{bmatrix} 1 & 0 & 0 \\ 0 & 0 & 4 \\ 0 & 2 & 0 \end{bmatrix}\begin{bmatrix} 1 & 0 & 0 \\ 0 & 0 & 2 \\ 0 & 4 & 0 \end{bmatrix} = \begin{bmatrix} 1 & 0 & 0 \\ 0 & 16 & 0 \\ 0 & 0 & 4 \end{bmatrix} \Rightarrow U = \begin{bmatrix} 1 & 0 & 0 \\ 0 & 4 & 0 \\ 0 & 0 & 2 \end{bmatrix}$$

$$V^2 = FF^{\mathrm{T}} = \begin{bmatrix} 1 & 0 & 0 \\ 0 & 0 & 2 \\ 0 & 4 & 0 \end{bmatrix}\begin{bmatrix} 1 & 0 & 0 \\ 0 & 0 & 4 \\ 0 & 2 & 0 \end{bmatrix} = \begin{bmatrix} 1 & 0 & 0 \\ 0 & 4 & 0 \\ 0 & 0 & 16 \end{bmatrix} \Rightarrow V = \begin{bmatrix} 1 & 0 & 0 \\ 0 & 2 & 0 \\ 0 & 0 & 4 \end{bmatrix}$$

Note that square roots were an easy computation for this simple case.

$$R = FU^{-1} = \begin{bmatrix} 1 & 0 & 0 \\ 0 & 0 & 2 \\ 0 & 4 & 0 \end{bmatrix}\begin{bmatrix} 1 & 0 & 0 \\ 0 & 1/4 & 0 \\ 0 & 0 & 1/2 \end{bmatrix} = \begin{bmatrix} 1 & 0 & 0 \\ 0 & 0 & 1 \\ 0 & 1 & 0 \end{bmatrix}$$

It is easily checked that the orthogonality condition $RR^{\mathrm{T}} = I$ is satisfied.

We now wish to develop the relationship between the Lagrangian strain tensor E and the so-called *physical strain components*. It should be kept in mind that our development of the various strains tensors E, e, C, and B were all done for arbitrary finite deformations. Small deformation simplifications will be explored in the next section. The Lagrangian strain components E_{ij} can be related to particular physical extensional and shear strain behaviors. In order to do this, consider the orthogonal triad of vectors $\{dX^{(1)}, dX^{(2)}, dX^{(3)}\}$ located at position X in the reference configuration. Under general motion, these vectors are mapped to another triad $\{dx^{(1)}, dx^{(2)}, dx^{(3)}\}$ which are not necessarily orthogonal (see Fig. 3.7).

We define the *extensional strain components* $\gamma^{(i)}$ as the change in length divided by the original length of each fiber $dX^{(i)}$:

$$\gamma^{(i)} = \frac{|dx^{(i)}| - |dX^{(i)}|}{|dX^{(i)}|} \tag{3.6.16}$$

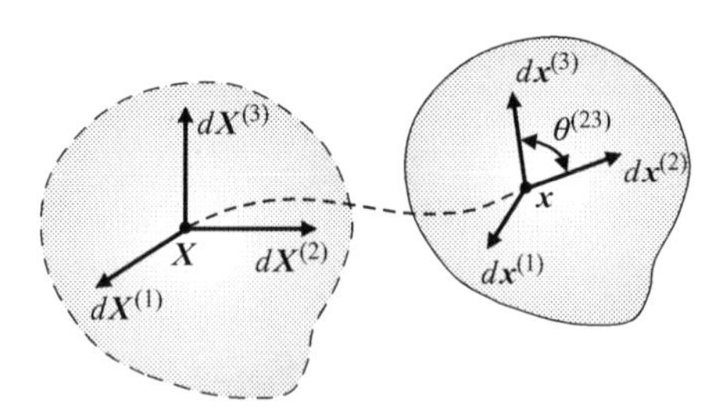

FIGURE 3.7

Deformation of an orthogonal triad.

Likewise, the *shear strain components* $\gamma^{(ij)}$ are the change in angle between fibers $d\boldsymbol{X}^{(i)}$ and $d\boldsymbol{X}^{(j)}$:

$$\gamma^{(ij)} = \frac{\pi}{2} - \theta^{(ij)} \tag{3.6.17}$$

where $\theta^{(ij)}$ is the angle between fibers $d\boldsymbol{x}^{(i)}$ and $d\boldsymbol{x}^{(j)}$ in the current spatial configuration.

Using the basic Lagrangian strain relation (3.6.3), we can write

$$\begin{aligned}\left|d\boldsymbol{x}^{(i)}\right|^2 - \left|d\boldsymbol{X}^{(i)}\right|^2 &= 2\,d\boldsymbol{X}^{(i)}\boldsymbol{E}\,d\boldsymbol{X}^{(i)} \\ &= 2\left|d\boldsymbol{X}^{(i)}\right|\boldsymbol{e}_i\boldsymbol{E}\left|d\boldsymbol{X}^{(i)}\right|\boldsymbol{e}_i \\ &= 2\boldsymbol{e}_i\boldsymbol{E}\boldsymbol{e}_i\left|d\boldsymbol{X}^{(i)}\right|^2 \\ &= 2E_{\underline{ii}}\left|d\boldsymbol{X}^{(i)}\right|^2 \qquad (no\ sum\ on\ i)\end{aligned} \tag{3.6.18}$$

Relation (3.6.16) can be rewritten as $\left|d\boldsymbol{x}^{(i)}\right| = \left(\gamma^{(i)}+1\right)\left|d\boldsymbol{X}^{(i)}\right|$, and combining this with (3.6.18) gives $\left(\gamma^{(i)}+1\right)^2 = 1+2E_{\underline{ii}}$. Solving for the extensional strain then yields the final result

$$\gamma^{(i)} = \sqrt{1+2E_{\underline{ii}}} - 1 \tag{3.6.19}$$

Next, let us consider the angle changes for the shear strain

$$\sin\gamma^{(ij)} = \cos\theta^{(ij)} = \frac{d\boldsymbol{x}^{(i)}\cdot d\boldsymbol{x}^{(j)}}{\left|d\boldsymbol{x}^{(i)}\right|\left|d\boldsymbol{x}^{(j)}\right|}, \quad i \neq j \tag{3.6.20}$$

From (3.5.4) $d\boldsymbol{x}^{(i)} = \boldsymbol{F}\,d\boldsymbol{X}^{(i)}$ and combining with (3.6.16) and (3.6.2), we can express (3.6.20) as

$$\begin{aligned}\sin\gamma^{(ij)} &= \frac{\boldsymbol{F}\,d\boldsymbol{X}^{(i)}\cdot\boldsymbol{F}\,d\boldsymbol{X}^{(j)}}{\left(\gamma^{(i)}+1\right)\left|d\boldsymbol{X}^{(i)}\right|\left(\gamma^{(j)}+1\right)\left|d\boldsymbol{X}^{(j)}\right|} \\ &= \frac{d\boldsymbol{X}^{(i)}\boldsymbol{F}^{\mathrm{T}}\boldsymbol{F}\,d\boldsymbol{X}^{(j)}}{\left(\gamma^{(i)}+1\right)\left(\gamma^{(j)}+1\right)\left|d\boldsymbol{X}^{(i)}\right|\left|d\boldsymbol{X}^{(j)}\right|} \\ &= \frac{2E_{ij}}{\sqrt{1+2E_{\underline{ii}}}\sqrt{1+2E_{\underline{jj}}}}\end{aligned} \tag{3.6.21}$$

Relations (3.6.19) and (3.6.21) then provide the relationships for the physical strain components in terms of the Lagrangian strain tensor components.

We can follow similar steps to relate the physical strains in terms of the Eulerian strains, and the results become

$$\begin{aligned}\gamma^{(i)} &= \frac{1}{\sqrt{1-2e_{\underline{ii}}}} - 1 \\ \sin\gamma^{(ij)} &= \frac{2e_{ij}}{\sqrt{1-2e_{\underline{ii}}}\sqrt{1-2e_{\underline{jj}}}} \quad (i \neq j)\end{aligned} \tag{3.6.22}$$

EXAMPLE 3.6.4 PHYSICAL STRAIN COMPONENTS FOR SIMPLE SHEAR AND EXTENSIONAL DEFORMATIONS

For the previous deformations of simple shear and extension specified in Examples 3.1.1 and 3.5.2, determine the physical strain components $\gamma^{(i)}$ and $\sin\gamma^{(ij)}$.

Solution: For the simple shear case,

$$\begin{array}{l} x_1 = X_1 + \gamma X_2 \\ x_2 = X_2 \\ x_3 = X_3 \end{array} \Rightarrow \boldsymbol{F} = \frac{\partial \boldsymbol{x}}{\partial \boldsymbol{X}} = \begin{bmatrix} 1 & \gamma & 0 \\ 0 & 1 & 0 \\ 0 & 0 & 1 \end{bmatrix} \Rightarrow \boldsymbol{E} = \frac{1}{2}\left(\boldsymbol{F}^{\mathrm{T}}\boldsymbol{F} - \boldsymbol{I}\right) = \frac{1}{2}\begin{bmatrix} 0 & \gamma & 0 \\ \gamma & \gamma^2 & 0 \\ 0 & 0 & 0 \end{bmatrix}$$

$$\gamma^{(i)} = \sqrt{1 + 2E_{\underline{ii}}} - 1 = \begin{bmatrix} 0 \\ \sqrt{1+\gamma^2} - 1 \\ 0 \end{bmatrix}$$

$$\sin\gamma^{(ij)} = \frac{2E_{ij}}{\sqrt{1 + 2E_{\underline{ii}}}\sqrt{1 + 2E_{\underline{jj}}}} = \begin{bmatrix} 0 \\ \dfrac{\gamma}{\sqrt{1+\gamma^2}} \\ 0 \end{bmatrix} \quad (i \neq j)$$

As expected for simple shear there is no extension for reference fibers along the X_1 and X_3 directions, and no angles changes with X_1,X_3 and X_2,X_3 fiber pairs.

For the extensional motion example,

$$\begin{array}{l} x_1 = \lambda_1 X_1 \\ x_2 = \lambda_2 X_2 \\ x_3 = \lambda_3 X_3 \end{array} \Rightarrow \boldsymbol{F} = \frac{\partial \boldsymbol{x}}{\partial \boldsymbol{X}} = \begin{bmatrix} \lambda_1 & 0 & 0 \\ 0 & \lambda_2 & 0 \\ 0 & 0 & \lambda_3 \end{bmatrix} \Rightarrow \boldsymbol{E} = \frac{1}{2}\left(\boldsymbol{F}^{\mathrm{T}}\boldsymbol{F} - \boldsymbol{I}\right) = \frac{1}{2}\begin{bmatrix} \lambda_1^2 - 1 & 0 & 0 \\ 0 & \lambda_2^2 - 1 & 0 \\ 0 & 0 & \lambda_3^2 - 1 \end{bmatrix}$$

$$\gamma^{(i)} = \sqrt{1 + 2E_{\underline{ii}}} - 1 = \begin{bmatrix} \lambda_1 - 1 \\ \lambda_2 - 1 \\ \lambda_3 - 1 \end{bmatrix}$$

$$\sin\gamma^{(ij)} = \frac{2E_{ij}}{\sqrt{1 + 2E_{\underline{ii}}}\sqrt{1 + 2E_{\underline{jj}}}} = \begin{bmatrix} 0 \\ 0 \\ 0 \end{bmatrix} \quad (i \neq j)$$

As expected for extensional motion, there will be no angle changes for fibers along reference coordinate directions.

We have already seen that our kinematic tensor fields may be expressed in terms of reference or spatial coordinates using the basis vectors $\boldsymbol{E}_i$ or $\boldsymbol{e}_i$. As with the case of the deformation gradient, there are also some two-point tensors associated with both reference frames. These transformations between reference and spatial or current representations are commonly called *push-forward* and *pull-back operations* and would employ using the basic motion relations $\boldsymbol{x} = \chi(\boldsymbol{X}, t)$ and $\boldsymbol{X} = \chi^{-1}(\boldsymbol{x}, t)$. The push-forward operation transforms a reference tensor field $(\boldsymbol{X}, t)$ to the current configuration $(\boldsymbol{x}, t)$. An example of this process is the transformation of the reference Lagrangian strain tensor $\boldsymbol{E}$ to the spatial Eulerian–Almansi strain tensor $\boldsymbol{e}$. Starting with the fundamental definition of $\boldsymbol{e}$,

$$\begin{aligned} \boldsymbol{e} &= \frac{1}{2}\left(\boldsymbol{I} - (\boldsymbol{F}^{-1})^{\mathrm{T}} \boldsymbol{F}^{-1}\right) = (\boldsymbol{F}^{-1})^{\mathrm{T}} \frac{1}{2}\left[\boldsymbol{F}^{\mathrm{T}}\left(\boldsymbol{I} - (\boldsymbol{F}^{-1})^{\mathrm{T}} \boldsymbol{F}^{-1}\right)\boldsymbol{F}\right]\boldsymbol{F}^{-1} \\ &= (\boldsymbol{F}^{-1})^{\mathrm{T}} \frac{1}{2}\left(\boldsymbol{F}^{\mathrm{T}}\boldsymbol{F} - \boldsymbol{I}\right)\boldsymbol{F}^{-1} = (\boldsymbol{F}^{-1})^{\mathrm{T}} \boldsymbol{E}\boldsymbol{F}^{-1} \end{aligned} \tag{3.6.23}$$

The pull-back operation is then the inverse of this process, and for the previous example we can easily invert things to get $\boldsymbol{E} = \boldsymbol{F}^{\mathrm{T}}\boldsymbol{e}\boldsymbol{F}$.

Before ending this section, let us explore the objectivity test for the various strain tensors developed. Recall that the fundamental test is based on a change of reference between two frames $\boldsymbol{x}^*$ and $\boldsymbol{x}$, such that $\boldsymbol{x}^* = \boldsymbol{c}(t) + \boldsymbol{Q}(t)\boldsymbol{x}$. Using this, we found earlier that the deformation gradient transformed as $\boldsymbol{F}^* = \boldsymbol{Q}\boldsymbol{F}$, and the inverse followed the relation $\boldsymbol{F}^{*-1} = \boldsymbol{F}^{-1}\boldsymbol{Q}^{\mathrm{T}}$. Starting with the Lagrangian strain tensor $\boldsymbol{E} = \frac{1}{2}\left(\boldsymbol{F}^{\mathrm{T}}\boldsymbol{F} - \boldsymbol{I}\right)$, we have

$$\begin{aligned} \boldsymbol{E}^* &= \frac{1}{2}\left(\boldsymbol{F}^{*\mathrm{T}}\boldsymbol{F}^* - \boldsymbol{I}^*\right) = \frac{1}{2}\left((\boldsymbol{Q}\boldsymbol{F})^{\mathrm{T}}\boldsymbol{Q}\boldsymbol{F} - \boldsymbol{I}\right) = \frac{1}{2}\left(\boldsymbol{F}^{\mathrm{T}}\boldsymbol{Q}^{\mathrm{T}}\boldsymbol{Q}\boldsymbol{F} - \boldsymbol{I}\right) \\ &= \frac{1}{2}\left(\boldsymbol{F}^{\mathrm{T}}\boldsymbol{F} - \boldsymbol{I}\right) = \boldsymbol{E} \end{aligned} \tag{3.6.24}$$

Likewise, the Eulerian strain $\boldsymbol{e} = \frac{1}{2}\left(\boldsymbol{I} - (\boldsymbol{F}^{-1})^{\mathrm{T}}\boldsymbol{F}^{-1}\right)$ follows the transformation:

$$\begin{aligned} \boldsymbol{e}^* &= \frac{1}{2}\left(\boldsymbol{I}^* - (\boldsymbol{F}^{*-1})^{\mathrm{T}}\boldsymbol{F}^{*-1}\right) = \frac{1}{2}\left(\boldsymbol{I} - (\boldsymbol{F}^{-1}\boldsymbol{Q}^{\mathrm{T}})^{\mathrm{T}}\boldsymbol{F}^{-1}\boldsymbol{Q}^{\mathrm{T}}\right) \\ &= \frac{1}{2}\left(\boldsymbol{I} - \boldsymbol{Q}(\boldsymbol{F}^{-1})^{\mathrm{T}}\boldsymbol{F}^{-1}\boldsymbol{Q}^{\mathrm{T}}\right) = \boldsymbol{Q}\frac{1}{2}\left(\boldsymbol{I} - (\boldsymbol{F}^{-1})^{\mathrm{T}}\boldsymbol{F}^{-1}\right)\boldsymbol{Q}^{\mathrm{T}} = \boldsymbol{Q}\boldsymbol{e}\boldsymbol{Q}^{\mathrm{T}} \end{aligned} \tag{3.6.25}$$

For the right Cauchy–Green strain tensor $\boldsymbol{C} = \boldsymbol{F}^{\mathrm{T}}\boldsymbol{F}$:

$$\boldsymbol{C}^* = \boldsymbol{F}^{*\mathrm{T}}\boldsymbol{F}^* = (\mathbf{QF})^{\mathrm{T}}\boldsymbol{Q}\boldsymbol{F} = \boldsymbol{F}^{\mathrm{T}}\boldsymbol{Q}^{\mathrm{T}}\boldsymbol{Q}\boldsymbol{F} = \boldsymbol{F}^{\mathrm{T}}\boldsymbol{F} = \boldsymbol{C} \tag{3.6.26}$$

and finally for the left Cauchy–Green strain $\boldsymbol{B} = \boldsymbol{F}\boldsymbol{F}^{\mathrm{T}}$:

$$\boldsymbol{B}^* = \boldsymbol{F}^*\boldsymbol{F}^{*\mathrm{T}} = \boldsymbol{Q}\boldsymbol{F}(\mathbf{QF})^{\mathrm{T}} = \boldsymbol{Q}\boldsymbol{F}\boldsymbol{F}^{\mathrm{T}}\boldsymbol{Q}^{\mathrm{T}} = \boldsymbol{Q}\boldsymbol{B}\boldsymbol{Q}^{\mathrm{T}} \tag{3.6.27}$$

Thus, we conclude that the Eulerian and left Cauchy–Green tensors are spatially based and satisfy the standard objective relation. However, the Lagrangian and right Cauchy–Green tensors do not satisfy the standard relation, but because they are based in reference coordinates they will transform as a scalar field.

We have now developed several second-order tensors $\boldsymbol{F}, \boldsymbol{E}, \boldsymbol{e}, \boldsymbol{C}$, and $\boldsymbol{B}$ that can be used to characterize finite strain deformations. As previously mentioned, there is no single unique strain tensor, and over the years a sizeable number of strain measures have been constructed. We could continue our study by reviewing some of these additional schemes; however, the current strain tensor set is sufficient for most continuum mechanics theories. Since we wish to complete the basic continuum formulation as soon as possible in order to move into the vast array of constitutive models, we will not further explore other strain measures.

3.7 CHANGES IN LINE, AREA, AND VOLUME ELEMENTS

It will be useful to develop relations for the changes in line, area, and volume elements under the motion from reference to spatial configurations. We simplify things somewhat by choosing a cubical reference element with sides aligned with the reference coordinate frame as shown in Fig. 3.8. Under general motion $\boldsymbol{x} = \chi(\boldsymbol{X},t)$, this element is mapped to a new element with totally different dimensions and shape in the spatial configuration. This scheme will allow us to determine element changes in the line edge, surface area, and volume. The deformation gradient and other strain tensor components can be used to find these types of changes. We basically seek a relational form that will give the spatial or current value in terms of a simple multiplier times the corresponding reference value.

Starting with line elements, we can use many of the relations from our previous discussion on physical extensional strain components. Starting with (3.6.16) and using (3.6.19) and (3.6.2) gives

$$\left|d\boldsymbol{x}^{(i)}\right| = \left(\gamma^{(i)}+1\right)\left|d\boldsymbol{X}^{(i)}\right| = \sqrt{1+2E_{\underline{i}\underline{i}}}\left|d\boldsymbol{X}^{(i)}\right| = \sqrt{F_{ki}F_{k\underline{i}}}\left|d\boldsymbol{X}^{(i)}\right| = \sqrt{C_{\underline{i}\underline{i}}}\left|d\boldsymbol{X}^{(i)}\right| \tag{3.7.1}$$

Next consider the area behavior of one face of the differential cubical element in Fig. 3.8. In the reference configuration, we choose the differential area $d\boldsymbol{A}$ in the

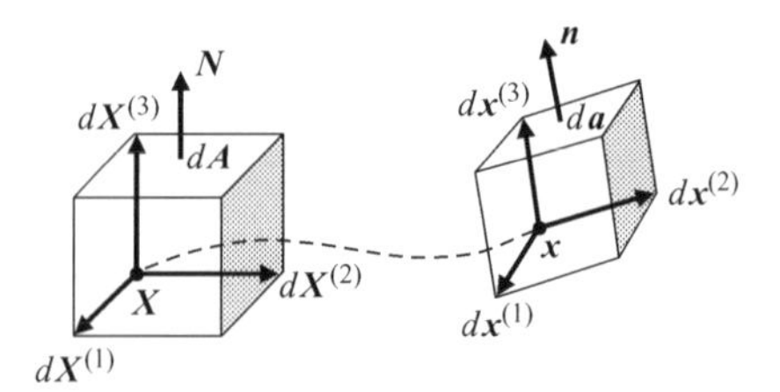

FIGURE 3.8

Line, area, and volume deformation.

$(X^{(1)}, X^{(2)})$-plane. This area is spanned by the vectors $d\boldsymbol{X}^{(1)}$ and $d\boldsymbol{X}^{(2)}$ and thus the area and unit normal vector are given by

$$dA = \left|d\boldsymbol{X}^{(1)} \times d\boldsymbol{X}^{(2)}\right|, \quad \boldsymbol{N} = \frac{d\boldsymbol{X}^{(1)} \times d\boldsymbol{X}^{(2)}}{\left|d\boldsymbol{X}^{(1)} \times d\boldsymbol{X}^{(2)}\right|} \tag{3.7.2}$$

These variables then define what is normally called an *oriented element area in the reference configuration* $d\boldsymbol{A} = \boldsymbol{N}\,dA = d\boldsymbol{X}^{(1)} \times d\boldsymbol{X}^{(2)}$. Under the motion, the reference vectors $d\boldsymbol{X}^{(1)}$ and $d\boldsymbol{X}^{(2)}$ are mapped to the spatial vectors $d\boldsymbol{x}^{(1)}$ and $d\boldsymbol{x}^{(2)}$ in the deformed configuration. The corresponding oriented element area then becomes $d\boldsymbol{a} = \boldsymbol{n}\,da = d\boldsymbol{x}^{(1)} \times d\boldsymbol{x}^{(2)}$. Shifting to index notation

$$\begin{aligned} da_i &= n_i\,da = \varepsilon_{ijk}\,dx_j^{(1)}\,dx_k^{(2)} \\ &= \varepsilon_{ijk} F_{jl}\,dX_l^{(1)} F_{km}\,dX_m^{(2)} \\ &= \varepsilon_{ijk} F_{jl} F_{km}\,dX_l^{(1)}\,dX_m^{(2)} \end{aligned}$$

Multiplying both sides by F_{in} gives

$$da_i F_{in} = \varepsilon_{ijk} F_{in} F_{jl} F_{km}\,dX_l^{(1)}\,dX_m^{(2)}$$

and then using property (2.5.10)

$$da_i F_{in} = (\det \boldsymbol{F})\varepsilon_{nlm}\,dX_l^{(1)}\,dX_m^{(2)} = J(d\boldsymbol{X}^{(1)} \times d\boldsymbol{X}^{(2)})_n = JN_n\,dA$$

where the Jacobian $J = \det \boldsymbol{F}$. Finally, multiplying both sides by F_{nm}^{-1} gives the final result

$$\begin{aligned} da_i F_{in} F_{nm}^{-1} &= JF_{nm}^{-1} N_n\,dA \Rightarrow da_m = JF_{nm}^{-1} N_n\,dA \\ \boldsymbol{n}\,da &= J(\boldsymbol{F}^{-1})^{\mathrm{T}} \boldsymbol{N}\,dA \end{aligned} \tag{3.7.3}$$

This relation is often called *Nanson's formula*.

Our last step in this section is to explore the volume changes that occur during general motion. We again start with the element shown in the reference configuration in Fig. 3.8. The volume of this element is spanned by the vectors $(d\boldsymbol{X}^{(1)}, d\boldsymbol{X}^{(2)}, d\boldsymbol{X}^{(3)})$. It is well known that the volume of such a shape is given by the scalar triple product (2.10.4):

$$dV = d\boldsymbol{X}^{(1)} \cdot d\boldsymbol{X}^{(2)} \times d\boldsymbol{X}^{(2)} = dX^{(1)}\,dX^{(2)}\,dX^{(2)}$$

Likewise, the volume of the element in the deformed configuration is given by $dv = d\boldsymbol{x}^{(1)} \cdot d\boldsymbol{x}^{(2)} \times d\boldsymbol{x}^{(2)}$, and again shifting to index notation

$$\begin{aligned} dv &= \varepsilon_{imp} F_{ij}\,dX_j^{(1)} F_{mn}\,dX_n^{(2)} F_{pq}\,dX_q^{(3)} \\ &= \varepsilon_{imp} F_{ij} F_{mn} F_{pq}\,dX_j^{(1)}\,dX_n^{(2)}\,dX_q^{(3)} \\ &= (\det F)\varepsilon_{jnq}\,dX_j^{(1)}\,dX_n^{(2)}\,dX_q^{(3)} \\ &= J\,\mathrm{d}V \end{aligned} \tag{3.7.4}$$

where we have used similar reduction steps as done with the area analysis.

The *volume dilatation* ϑ is defined by

$$\vartheta = \frac{dv - dv}{dv} = \det \boldsymbol{F} - 1 = J - 1 \tag{3.7.5}$$

Hence, we can say that a motion is *volume preserving* or *isochoric* if det $\boldsymbol{F}$ = 1. It turns out that incompressible materials can only undergo isochoric motions.

EXAMPLE 3.7.1 LINE, AREA, AND VOLUME CHANGES FOR SIMPLE SHEAR AND EXTENSIONAL DEFORMATIONS

Using again our previous deformations of simple shear and extension specified in Examples 3.1.1 and 3.5.2, determine the line, area, and volume changes.

Solution: For the simple shear case,

$$\begin{aligned} x_1 &= X_1 + \gamma X_2 \\ x_2 &= X_2 \\ x_3 &= X_3 \end{aligned} \Rightarrow \boldsymbol{F} = \frac{\partial \boldsymbol{x}}{\partial \boldsymbol{X}} = \begin{bmatrix} 1 & \gamma & 0 \\ 0 & 1 & 0 \\ 0 & 0 & 1 \end{bmatrix} \Rightarrow \boldsymbol{F}^{-1} = \frac{\partial \boldsymbol{X}}{\partial \boldsymbol{x}} = \begin{bmatrix} 1 & -\gamma & 0 \\ 0 & 1 & 0 \\ 0 & 0 & 1 \end{bmatrix}, \quad J = 1$$

$$\boldsymbol{C} = \boldsymbol{F}^{\mathrm{T}} \boldsymbol{F} = \begin{bmatrix} 1 & \gamma & 0 \\ \gamma & 1+\gamma^2 & 0 \\ 0 & 0 & 1 \end{bmatrix}$$

$$\left| d\boldsymbol{x}^{(i)} \right| = \sqrt{C_{\underline{ii}}} \left| d\boldsymbol{X}^{(i)} \right| = \begin{bmatrix} 1 \\ \sqrt{1+\gamma^2} \\ 1 \end{bmatrix} \left| d\boldsymbol{X}^{(i)} \right|$$

$$\boldsymbol{n}\, da = J(\boldsymbol{F}^{-1})^{\mathrm{T}} \boldsymbol{N}\, dA = \begin{bmatrix} 1 & 0 & 0 \\ -\gamma & 1 & 0 \\ 0 & 0 & 1 \end{bmatrix} \begin{bmatrix} N_1 \\ N_2 \\ N_3 \end{bmatrix} dA = \begin{bmatrix} N_1 \\ -\gamma N_1 + N_2 \\ N_3 \end{bmatrix} dA$$

$dv = J\, dV = dV$ ... no volumetric change

For the extensional motion example,

$$\begin{aligned} x_1 &= \lambda_1 X_1 \\ x_2 &= \lambda_2 X_2 \Rightarrow \\ x_3 &= \lambda_3 X_3 \end{aligned} \boldsymbol{F} = \frac{\partial \boldsymbol{x}}{\partial \boldsymbol{X}} = \begin{bmatrix} \lambda_1 & 0 & 0 \\ 0 & \lambda_2 & 0 \\ 0 & 0 & \lambda_3 \end{bmatrix} \Rightarrow \boldsymbol{F}^{-1} = \frac{\partial \boldsymbol{X}}{\partial \boldsymbol{x}} = \begin{bmatrix} \lambda_1^{-1} & 0 & 0 \\ 0 & \lambda_2^{-1} & 0 \\ 0 & 0 & \lambda_3^{-1} \end{bmatrix}, \quad J = \lambda_1 \lambda_2 \lambda_3$$

$$\boldsymbol{C} = \boldsymbol{F}^{\mathrm{T}} \boldsymbol{F} = \begin{bmatrix} \lambda_1^2 & 0 & 0 \\ 0 & \lambda_2^2 & 0 \\ 0 & 0 & \lambda_3^2 \end{bmatrix}$$

$$\left| d\boldsymbol{x}^{(i)} \right| = \sqrt{C_{\underline{ii}}} \left| d\boldsymbol{X}^{(i)} \right| = \begin{bmatrix} \lambda_1 \\ \lambda_2 \\ \lambda_3 \end{bmatrix} \left| d\boldsymbol{X}^{(i)} \right| \text{ ... uniform extensional line changes}$$

$$\boldsymbol{n}da = J(\boldsymbol{F}^{-1})^{\mathrm{T}}\boldsymbol{N}dA = \begin{bmatrix} \lambda_1^{-1} & 0 & 0 \\ 0 & \lambda_2^{-1} & 0 \\ 0 & 0 & \lambda_3^{-1} \end{bmatrix}\begin{bmatrix} N_1 \\ N_2 \\ N_3 \end{bmatrix} dA = \begin{bmatrix} \lambda_1^{-1}N_1 \\ \lambda_2^{-1}N_2 \\ \lambda_3^{-1}N_3 \end{bmatrix} dA$$

$$dv = J\,dV = \lambda_1\lambda_2\lambda_3\,dV$$

3.8 SMALL DEFORMATION KINEMATICS AND STRAIN TENSORS

In many classical continuum mechanics applications, the deformations under study are small and the need to distinguish between reference and spatial configurations is unnecessary. This situation occurs in classical linear elasticity, plasticity, viscoelasticity, and other such theories. Many of these models will be discussed in detail in Chapter 6. This situation greatly simplifies much of our previous kinematical relations, and we now wish to look at the details of these simplifications. In general, our finite strain relations were nonlinear in the displacement gradients, and we expect that for small deformations this nonlinearity will disappear.

With the removal of the difference between reference $\boldsymbol{X}$ coordinates and spatial or current $\boldsymbol{x}$ coordinates, one might think that the deformation gradient tensor becomes the unit tensor, and hence is essentially eliminated from the kinematical study. However, this would eliminate all of the strain tensors previous developed. Thus, we start with relation (3.5.8) $\boldsymbol{F} = \boldsymbol{I} + \nabla\boldsymbol{u}$ under the condition that the displacement gradient tensor is small $\nabla\boldsymbol{u} \ll \boldsymbol{I}$. Likewise, the inverse deformation gradient was given by $\frac{\partial u_i}{\partial X_j} = \frac{\partial u_i}{\partial x_k}\frac{\partial x_k}{\partial X_j} = \frac{\partial u_i}{\partial x_k}\left(\frac{\partial u_k}{\partial X_j} + \delta_{kj}\right) \approx \frac{\partial u_i}{\partial x_k}\delta_{kj} = \frac{\partial u_i}{\partial x_j}$, and so for small spatial displacement gradients, $\nabla\boldsymbol{u}^* \ll \boldsymbol{I}$. We now use these criteria to simplify the basic kinematical strain tensors.

Starting with the Lagrangian strain tensor given by (3.6.4),

$$E_{ij} = \frac{1}{2}\left(u_{i,j} + u_{j,i} + u_{k,i}u_{k,j}\right)$$

We make the argument that for small deformations, the displacement gradient product $u_{k,i}u_{k,j}$ is small in comparison to the displacement gradient itself, and so it can be dropped from the relation giving

$$E_{ij} = \frac{1}{2}\left(u_{i,j} + u_{j,i}\right)$$

In similar fashion, the Eulerian strain given by (3.6.9) will reduce to

$$e_{ij} = \frac{1}{2}\left(u^*_{i,j} + u^*_{j,i}\right)$$

Furthermore, differences between spatial and reference derivatives disappear following the simple analysis:

$$\frac{\partial u_i}{\partial X_j} = \frac{\partial u_i}{\partial x_k}\frac{\partial x_k}{\partial X_j} = \frac{\partial u_i}{\partial x_k}\left(\frac{\partial u_k}{\partial X_j} + \delta_{kj}\right) \approx \frac{\partial u_i}{\partial x_k}\delta_{kj} = \frac{\partial u_i}{\partial x_j}$$

Thus, the Lagrangian and Eulerian strain tensors both reduce to the infinitesimal strain

$$\varepsilon_{ij} = \frac{1}{2}\left(u_{i,j} + u_{j,i}\right) \tag{3.8.1}$$

and the derivatives in (3.8.1) can be taken with respect to either reference or spatial as they are indistinguishable. This result reduces the right and left Cauchy–Green tensor as

$$\begin{aligned} C_{ij} &= F_{ki}F_{kj} = 2\varepsilon_{ij} + \delta_{ij} \\ B_{ij} &= F_{ik}F_{jk} = 2\varepsilon_{ij} + \delta_{ij} \end{aligned} \tag{3.8.2}$$

For small deformations, since the distinction between reference and spatial position is dropped, the displacement vector is then used to represent position. The velocity and acceleration relations then reduce to

$$\begin{aligned} \boldsymbol{v}(\boldsymbol{x},t) &= \frac{D\boldsymbol{x}}{Dt} \approx \frac{\partial \boldsymbol{u}}{\partial t} \\ \boldsymbol{a}(\boldsymbol{x},t) &= \frac{\partial \boldsymbol{v}}{\partial t} + \boldsymbol{v}\cdot\nabla\boldsymbol{v} \approx \frac{\partial \boldsymbol{v}}{\partial t} = \frac{\partial^2 \boldsymbol{u}}{\partial t^2} \end{aligned} \tag{3.8.3}$$

EXAMPLE 3.8.1 PHYSICAL STRAIN COMPONENTS FOR SMALL STRAIN APPROXIMATION

Determine the general physical strain components given by (3.6.19) and (3.6.21) for the case of small strain approximation.

Solution: From (3.6.19) and (3.8.1) $\Rightarrow \gamma^{(i)} = \sqrt{1+2E_{\underline{ii}}} - 1 = \sqrt{1+2\varepsilon_{\underline{ii}}} - 1 \approx \varepsilon_{\underline{ii}}$.
Using (3.6.21) and (3.8.1) $\Rightarrow$

$$\sin\gamma^{(ij)} = \frac{2E_{ij}}{\sqrt{1+2E_{\underline{ii}}}\sqrt{1+2E_{\underline{jj}}}} = \frac{2\varepsilon_{ij}}{\sqrt{1+2\varepsilon_{\underline{ii}}}\sqrt{1+2\varepsilon_{\underline{jj}}}} \approx \frac{2\varepsilon_{ij}}{\left(1+\varepsilon_{\underline{ii}}\right)\left(1+\varepsilon_{\underline{jj}}\right)} \approx \frac{2\varepsilon_{ij}}{1+\varepsilon_{\underline{ii}}+\varepsilon_{\underline{jj}}} \approx 2\varepsilon_{ij} \quad (i \neq j)$$

and for small deformations $\sin\gamma^{(ij)} \approx \gamma^{(ij)} \Rightarrow \gamma^{(ij)} = \frac{\pi}{2} - \theta^{(ij)} \approx 2\varepsilon_{ij} \quad (i \neq j)$

Thus, for small deformations, the physical extensional strain is $\varepsilon_{\underline{ii}}$ and the change in the right angle between two orthogonal fibers is $2\varepsilon_{ij}\ (i \neq j)$.

3.9 PRINCIPAL AXES FOR STRAIN TENSORS

From our previous developments, the Lagrangian strain $\boldsymbol{E}$, the Eulerian strain $\boldsymbol{e}$, the right and left Cauchy–Green strains $\boldsymbol{C}$ and $\boldsymbol{B}$, and the infinitesimal strain $\boldsymbol{\varepsilon}$ were all symmetric second-order tensors. For each of these, we can therefore apply our previous principal value theory from Section 2.11.

Focusing our attention to the Lagrangian strain tensor defined by relation (3.6.2) $\boldsymbol{E}=\frac{1}{2}\left(\boldsymbol{F}^{\mathrm{T}}\boldsymbol{F}-\boldsymbol{I}\right)$, we can then conclude that there will exist at least three principal directions and at most three principal values for the tensor $\boldsymbol{E}$. The principal values (eigenvalues) E_1, E_2, and E_3 are referred to as *principal strains* and come from the roots of the characteristic equation

$$E^3-I_E E^2+II_E E-III_E=0 \tag{3.9.1}$$

where the fundamental invariants of $\boldsymbol{E}$ are given by

$$\begin{aligned}
I_E &= E_{ii}=E_{11}+E_{22}+E_{33}\\
II_E &= \frac{1}{2}\left(E_{ii}E_{jj}-E_{ij}E_{ij}\right)=\begin{vmatrix} E_{11} & E_{12}\\ E_{21} & E_{22}\end{vmatrix}+\begin{vmatrix} E_{22} & E_{23}\\ E_{32} & E_{33}\end{vmatrix}+\begin{vmatrix} E_{11} & E_{13}\\ E_{31} & E_{33}\end{vmatrix}\\
III_E &= \det \boldsymbol{E}
\end{aligned} \tag{3.9.2}$$

Using the special principal coordinate system for strain, we can conclude that tensor $\boldsymbol{E}$ written as a matrix will take the diagonal form

$$E_{ij}=\begin{bmatrix} E_1 & 0 & 0\\ 0 & E_2 & 0\\ 0 & 0 & E_3\end{bmatrix} \tag{3.9.3}$$

and using this representation, the invariants can be expressed in simpler form using the principal values:

$$\begin{aligned}
I_E &= E_1+E_2+E_3\\
II_E &= E_1E_2+E_2E_3+E_3E_1\\
III_E &= E_1E_2E_3
\end{aligned} \tag{3.9.4}$$

Obviously, all of this theoretical structure would hold for any of the other symmetric strain tensors. Note that for the infinitesimal strain case, the absence of the off-diagonal terms in the principal coordinate system implies zero angle change deformation (see Example 3.8.1).

3.10 SPHERICAL AND DEVIATORIC STRAIN TENSORS

Again from our previous developments in Section 2.12, any of our strain tensors $\boldsymbol{E}$, $\boldsymbol{e}$, $\boldsymbol{C}$, $\boldsymbol{B}$, and $\boldsymbol{\varepsilon}$ can be decomposed into spherical and deviatoric parts. For the infinitesimal strain $\boldsymbol{\varepsilon}$, the spherical part is given by

$$\tilde{\varepsilon}_{ij}=\frac{1}{3}\varepsilon_{kk}\delta_{ij} \tag{3.10.1}$$

while the deviatoric part is specified as

$$\hat{\varepsilon}_{ij} = \varepsilon_{ij} - \frac{1}{3}\varepsilon_{kk}\delta_{ij} \tag{3.10.2}$$

Note again that the sum of these two parts then gives the original tensor ε_{ij}.

For the infinitesimal strain, the spherical part is associated with *volumetric deformation* while the deviatoric part measures *shape or shear changes*. It can be shown that the principal directions of the strain deviator are the same as those of the strain tensor, and the deviator principal values are $\{\varepsilon_1 + p, \varepsilon_2 + p, \varepsilon_3 + p\}$, where $p = \varepsilon_{kk}/3$. Also note that $I_{\hat{\varepsilon}} = 0$, and spherical strain tensor is isotropic.

3.11 STRAIN COMPATIBILITY

Strain compatibility is a very interesting topic as it generates additional continuum field equations that come from unexpected places. While most of our field equations originate from Euclidean geometry, force analysis, balance or conservation principles, and thermodynamic theories, strain compatibility arises from a more abstract mathematical concept. The idea starts with a careful review of the various strain tensor forms that have been previously developed. If the motions or displacements are known, then the strain tensors, $\boldsymbol{E}$, $\boldsymbol{e}$, $\boldsymbol{C}$, $\boldsymbol{B}$, and $\boldsymbol{\varepsilon}$ can be determined by differentiation (assuming the motions/displacements are differentiable functions). However, the question also comes up of determining the displacements when the strains are given. Because our previous strain definitions (3.6.4), (3.6.9), (3.6.13), (3.6.15), and (3.8.1) all yield symmetric second-order tensors, there are *six* independent partial differential equations for *three* unknown displacement components. Such a system of equations is *over-determined* and will not have a *single-valued* solution for the displacements. One would expect, however, that a solution would exist if the strains satisfy certain additional conditions, and these are called *integrability* or *compatibility conditions*.

We will now only pursue in detail the infinitesimal strain $\boldsymbol{\varepsilon}$ case which has the linear relation form with the displacement gradients $\varepsilon_{ij} = \frac{1}{2}(u_{i,j} + u_{j,i})$. Writing these equations out in a scalar format yields the six *strain–displacement* relations

$$\varepsilon_{ij} = \frac{1}{2}(u_{i,j} + u_{j,i}) \Rightarrow \begin{matrix} \varepsilon_{11} = \dfrac{\partial u_1}{\partial x_1}, & \varepsilon_{12} = \dfrac{1}{2}\left(\dfrac{\partial u_1}{\partial x_2} + \dfrac{\partial u_2}{\partial x_1}\right) \\ \varepsilon_{22} = \dfrac{\partial u_2}{\partial x_2}, & \varepsilon_{23} = \dfrac{1}{2}\left(\dfrac{\partial u_2}{\partial x_3} + \dfrac{\partial u_3}{\partial x_2}\right) \\ \varepsilon_{33} = \dfrac{\partial u_3}{\partial x_3}, & \varepsilon_{31} = \dfrac{1}{2}\left(\dfrac{\partial u_3}{\partial x_1} + \dfrac{\partial u_1}{\partial x_3}\right) \end{matrix} \tag{3.11.1}$$

Thus, if we specify continuous, single-valued displacements u_1, u_2, and u_3, then through differentiation the resulting strain field can be determined and will be equally well-behaved. However, the converse is not necessarily true, that is, given the six strain components, integration of the strain–displacement relations (3.11.1) will not necessarily produce continuous, single-valued displacements. This should not be totally surprising since we are trying to solve six equations for only three unknown displacement components. In order to ensure continuous, single-valued displacements, the strains must satisfy additional relations.

Before proceeding with the mathematics to develop these equations, it is instructive to consider a geometric interpretation of this concept (Sadd, 2014). A two-dimensional example is shown in Fig. 3.9A whereby a continuum is first divided up into a series of elements. For simple visualization, consider only four such elements. In the undeformed configuration shown in Fig. 3.9B, these elements of course fit together perfectly. Next let us arbitrarily specify the strain of each of the four elements and attempt to reconstruct the continuum back together. For Fig. 3.9C, the elements have been carefully strained taking into consideration neighboring elements so that the system fits together, thus yielding continuous, single-valued displacements. However, for Fig. 3.9D, the

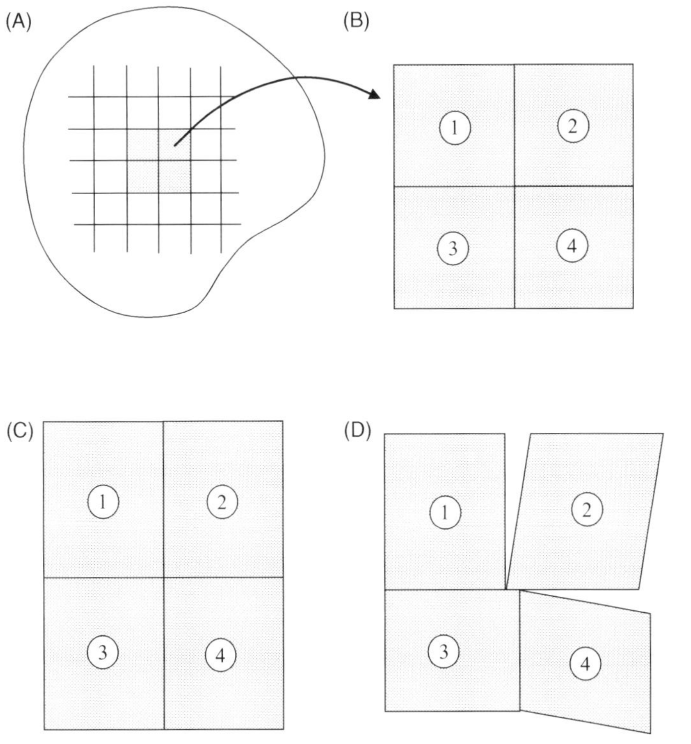

FIGURE 3.9

Physical interpretation of strain compatibility: (A) discretized continuum; (B) undeformed configuration; (C) deformed configuration continuous displacements; (D) deformed configuration discontinuous displacements.

elements have been individually deformed regardless of neighboring deformations. It is observed for this case that the system will not fit together without voids and gaps, and this situation produces a discontinuous displacement field. So we again conclude that the strain components must be somehow related to yield continuous, single-valued displacements. We now pursue these particular relations.

The development process is based on eliminating the displacements from the strain–displacement relations. Working in index notation, we start by differentiating (3.11.1) twice with respect to x_k and x_l:

$$\varepsilon_{ij,kl} = \frac{1}{2}(u_{i,jkl} + u_{j,ikl})$$

Through simple interchange of subscripts, we can generate the following additional relations:

$$\begin{aligned} \varepsilon_{kl,ij} &= \frac{1}{2}(u_{k,lij} + u_{l,kij}) \\ \varepsilon_{jl,ik} &= \frac{1}{2}(u_{j,lik} + u_{l,jik}) \\ \varepsilon_{ik,jl} &= \frac{1}{2}(u_{i,kjl} + u_{k,ijl}) \end{aligned}$$

Under the assumption of continuous displacements, we can interchange the order of differentiation on u_i and the displacements can be eliminated from the preceding set to get

$$\varepsilon_{ij,kl} + \varepsilon_{kl,ij} - \varepsilon_{ik,jl} - \varepsilon_{jl,ik} = 0 \tag{3.11.2}$$

These are called the well-known *Saint Venant compatibility equations*. Although the system would generally have 81 individual equations, most are either simple identities or repetitions and only six are meaningful. It turns out that one can set $k = l$ without loss in generality, and this leads to determine the six nonzero relations given by

$$\begin{aligned} &\frac{\partial^2 \varepsilon_{11}}{\partial x_2^2} + \frac{\partial^2 \varepsilon_{22}}{\partial x_1^2} = 2\frac{\partial^2 \varepsilon_{12}}{\partial x_1 \partial x_2} \\ &\frac{\partial^2 \varepsilon_{22}}{\partial x_3^2} + \frac{\partial^2 \varepsilon_{33}}{\partial x_2^2} = 2\frac{\partial^2 \varepsilon_{23}}{\partial x_2 \partial x_3} \\ &\frac{\partial^2 \varepsilon_{33}}{\partial x_1^2} + \frac{\partial^2 \varepsilon_{11}}{\partial x_3^2} = 2\frac{\partial^2 \varepsilon_{31}}{\partial x_3 \partial x_1} \\ &\frac{\partial^2 \varepsilon_{11}}{\partial x_2 \partial x_3} = \frac{\partial}{\partial x_1}\left(-\frac{\partial \varepsilon_{23}}{\partial x_1} + \frac{\partial \varepsilon_{31}}{\partial x_2} + \frac{\partial \varepsilon_{12}}{\partial x_3}\right) \\ &\frac{\partial^2 \varepsilon_{22}}{\partial x_3 \partial x_1} = \frac{\partial}{\partial x_2}\left(-\frac{\partial \varepsilon_{31}}{\partial x_2} + \frac{\partial \varepsilon_{12}}{\partial x_3} + \frac{\partial \varepsilon_{23}}{\partial x_1}\right) \\ &\frac{\partial^2 \varepsilon_{33}}{\partial x_3 \partial x_1} = \frac{\partial}{\partial x_3}\left(-\frac{\partial \varepsilon_{12}}{\partial x_3} + \frac{\partial \varepsilon_{23}}{\partial x_1} + \frac{\partial \varepsilon_{31}}{\partial x_2}\right) \end{aligned} \tag{3.11.3}$$

It should be noted that not all six of these equations are independent, and it can be shown that they are equivalent to three independent fourth-order relations (see Exercise 3.22). However, in applications, it is usually more convenient to use the second-order equations (3.11.3). Most applications of these equations are found in linear elasticity for the case where the displacements are not included in the basic formulation. We will discuss this situation later in Chapter 6. Note that all linear strain fields will satisfy these equations and thus are automatically compatible.

Using relation (2.5.5), we can show that the compatibility relations (3.11.2) with $l = k$ can be expressed by

$$\eta_{ij} = \varepsilon_{ikl}\varepsilon_{jmp}\varepsilon_{lp,km} = 0 \tag{3.11.4}$$

where η_{ij} is sometimes referred to as the *incompatibility tensor* (Asaro and Lubarda, 2006). Note the notation is a little less than ideal where $\varepsilon_{...}$ is the alternating symbol while $\varepsilon_{..}$ is the small strain tensor. This form can also be written in vector notation as

$$\nabla \times \boldsymbol{\varepsilon} \times \nabla = 0 \tag{3.11.5}$$

In the previous development of the compatibility relations, we assumed that the displacements were continuous and thus the resulting equations (3.11.3) are actually only a necessary condition. It is not very difficult to prove that they are also sufficient (Sadd, 2014), but with the caveat that the region must be simply connected. Thus, the compatibility equations (3.11.3) are necessary and sufficient conditions for single-valued continuous displacements in simply connected domains. Such domains refer to regions of space for which all simple closed curves drawn in the region can be continuously shrunk to a point without going outside the region. Domains not having this property are called *multiply connected*. For two-dimensional regions, the presence of one or more holes will make the region multiply connected.

Note that these compatibility conditions do not ensure uniqueness of the displacements; and in actuality the displacements are not unique since we can always superpose a rigid body motion which changes the displacements of points but has no effect on the strains. If the displacements are explicitly included in the problem formulation, then necessarily the solution will produce single-valued displacements and there is no need to use the compatibility equations. For multiply-connected regions, additional conditions must be satisfied by the strains for single-valued displacements (see Fung, 1965).

Compatibility for the nonlinear finite strain case is much more difficult to handle since the relations are nonlinear partial differential equations. For this case, it is necessary to use general tensor theory, and the interested reader is referred to Truesdell and Toupin (1960) or Malvern (1969).

EXAMPLE 3.11.1 COMPATIBLE AND INCOMPATIBLE INFINITESIMAL STRAIN TENSORS

Show that the strain field (a) is compatible, whereas strain (b) is not.

$$\text{a. } \varepsilon_{ij} = \begin{bmatrix} Ax_2^2 & 0 & 0 \\ 0 & -Ax_1^2 & 0 \\ 0 & 0 & Bx_3 \end{bmatrix}, \quad \text{b. } \varepsilon_{ij} = \begin{bmatrix} 0 & Ax_1x_2 & 0 \\ Ax_1x_2 & 0 & 0 \\ 0 & 0 & Bx_3 \end{bmatrix}, \quad (A \text{ and } B \text{ small constants})$$

Solution:

(a) From the first compatibility relation $\frac{\partial^2 \varepsilon_{11}}{\partial x_2^2} + \frac{\partial^2 \varepsilon_{21}}{\partial x_1^2} = 2\frac{\partial^2 \varepsilon_{12}}{\partial x_1 \partial x_2} \Rightarrow$ $2A - 2A = 0$. All other equations give $0 = 0$, and so this strain is compatible.

(b) From the first compatibility relation $\frac{\partial^2 \varepsilon_{11}}{\partial x_2^2} + \frac{\partial^2 \varepsilon_{22}}{\partial x_1^2} = 2\frac{\partial^2 \varepsilon_{12}}{\partial x_1 \partial x_2} \Rightarrow$ $0 + 0 = 2A$, which is not satisfied, and so the strain is not compatible.

EXAMPLE 3.11.2 DETERMINATION OF STRAINS FROM DISPLACEMENTS, AND DISPLACEMENTS FROM STRAINS FOR SMALL DEFORMATION THEORY

(a) For small deformation theory, find the strains from the given displacement field $u_1 = Ax_1^2 + Bx_1x_2^2$, $u_2 = Cx_1x_2^2$, $u_3 = 0$, with A, B, and C constants.

(b) Given the following strain field, $\varepsilon_{ij} = \begin{bmatrix} Ax_1^2 & 0 & 0 \\ 0 & Bx_2^2 & 0 \\ 0 & 0 & 0 \end{bmatrix}$, with A and B constants. Determine the corresponding displacements assuming they only depend on x_1 and x_2, and that $u_3 = 0$ (two-dimensional problem).

Solution: (a) To get strains, simply use (3.11.1) and differentiate the displacements

$$\varepsilon_{11} = \frac{\partial u_1}{\partial x_1} = 2Ax_1 + Bx_2^2, \quad \varepsilon_{12} = \frac{1}{2}\left(\frac{\partial u_1}{\partial x_2} + \frac{\partial u_2}{\partial x_1}\right) = \frac{1}{2}\left(2Bx_1x_2 + Cx_2^2\right)$$

$$\varepsilon_{22} = \frac{\partial u_2}{\partial x_2} = 2Cx_1x_2, \quad \varepsilon_{23} = \frac{1}{2}\left(\frac{\partial u_2}{\partial x_3} + \frac{\partial u_3}{\partial x_2}\right) = 0$$

$$\varepsilon_{33} = \frac{\partial u_3}{\partial x_3} = 0, \quad \varepsilon_{31} = \frac{1}{2}\left(\frac{\partial u_3}{\partial x_1} + \frac{\partial u_1}{\partial x_3}\right) = 0$$

Solution: (b) We have to integrate the strain–displacement relations to determine displacements

$\varepsilon_{11} = \frac{\partial u_1}{\partial x_1} = Ax_1^2 \Rightarrow u_1 = \frac{A}{3}x_1^3 + f(x_2)$, integration creates arbitrary function $f(x_2)$

$\varepsilon_{22} = \frac{\partial u_2}{\partial x_2} = Bx_2^2 \Rightarrow u_2 = \frac{B}{3}x_2^3 + g(x_1)$, integration creates arbitrary function $g(x_1)$

$$\varepsilon_{12} = \frac{1}{2}\left(\frac{\partial u_1}{\partial x_2} + \frac{\partial u_2}{\partial x_1}\right) = 0 \Rightarrow \frac{\mathrm{d}f(x_2)}{\mathrm{d}x_2} + \frac{\mathrm{d}g(x_1)}{\mathrm{d}x_1} = 0 \Rightarrow \frac{\mathrm{d}f(x_2)}{\mathrm{d}x_2} = -\frac{\mathrm{d}g(x_1)}{\mathrm{d}x_1} = \text{constant}$$

The last relation must equal a constant since x_1 and x_2 are independent variables.

$\therefore f(x_2) = ax_2 + b, g(x_1) = -ax_1 + c$, with a, b, and c all arbitrary constants.

Thus, the displacements are given by $u_1 = \frac{A}{3}x_1^3 + ax_2 + b,\ u_2 = \frac{B}{2}x_2^2 - ax_1 + c$,

Note that portions from $f(x_2)$ and $g(x_1)$ are actually rigid body motion terms.

3.12 ROTATION TENSOR

Let us now explore the kinematical concept of continuum rotation. We will look at this special type of motion from both small and large deformation theory. Starting first with small deformations, relation (3.6.3) may be written as

$$|d\mathbf{x}|^2 - |d\mathbf{X}|^2 = 2\varepsilon_{ij}\, dX_i\, dX_j \tag{3.12.1}$$

and so a body moves rigidly (preserving relative distances) if and only if $\varepsilon_{ij} = 0$. This fact implies that

$$u_{i,j} = -u_{j,i} \tag{3.12.2}$$

Next combining previous relations (3.5.5) and (3.5.8) yields

$$dx_i = F_{ij}\, dX_j = (\delta_{ij} + u_{i,j})\, dX_j \tag{3.12.3}$$

Expanding the displacement gradient into symmetric and antisymmetric parts as per relation (2.3.4)

$$u_{i,j} = \frac{1}{2}(u_{i,j} + u_{j,i}) + \frac{1}{2}(u_{i,j} - u_{j,i}) \tag{3.12.4}$$

and defining

$$\varepsilon_{ij} = \frac{1}{2}\left(u_{i,j} + u_{j,i}\right)$$
$$\omega_{ij} = \frac{1}{2}\left(u_{i,j} - u_{j,i}\right) \tag{3.12.5}$$

we can write

$$u_{i,j} = \varepsilon_{ij} + \omega_{ij} \tag{3.12.6}$$

The symmetric part of $u_{i,j}$ is the infinitesimal strain tensor ε_{ij} previously defined by (3.8.1), whereas the antisymmetric part ω_{ij} is called the *infinitesimal rotation tensor*. In light of relation (3.12.2), the skew-symmetric tensor ω_{ij} represents kinematical motion without strain and is thus associated with rigid body rotational motion. Combining (3.12.3) and (3.12.6) allows us to write

$$d\boldsymbol{x} = (\boldsymbol{I} + \omega + \varepsilon)d\boldsymbol{X} \tag{3.12.7}$$

We can then conclude that every infinitesimal deformation consists locally of the *sum of a translation, rotation*, and *pure strain*. The order of these individual motions is arbitrary but the decomposition is unique.

From the discussion in Section 2.3 and Exercise 2.13, the rotation tensor ω_{ij} can be associated with an axial or dual vector ω_i defined by

$$\omega_i = -\frac{1}{2}\varepsilon_{ijk}\omega_{jk} = \frac{1}{2}\varepsilon_{ijk}u_{k,j} \Rightarrow \boldsymbol{\omega} = \frac{1}{2}\nabla \times \boldsymbol{u} \tag{3.12.8}$$

where $\boldsymbol{\omega}$ is called the *rotation vector* with components

$$\omega_1 = \omega_{32} = \frac{1}{2}\left(\frac{\partial u_3}{\partial x_2} - \frac{\partial u_2}{\partial x_3}\right)$$
$$\omega_2 = \omega_{13} = \frac{1}{2}\left(\frac{\partial u_1}{\partial x_3} - \frac{\partial u_3}{\partial x_1}\right) \tag{3.12.9}$$
$$\omega_3 = \omega_{21} = \frac{1}{2}\left(\frac{\partial u_2}{\partial x_1} - \frac{\partial u_1}{\partial x_2}\right)$$

For the large or finite strain case, we have already (Section 3.6) presented most of the relations associated with the rotation tensor. For example, starting with the basic relation $d\boldsymbol{x} = \boldsymbol{F}\,d\boldsymbol{X}$, we used the Polar Decomposition Theorem to write

$$d\boldsymbol{x} = \boldsymbol{RU}\,d\boldsymbol{X} = \boldsymbol{VR}\,d\boldsymbol{X} \tag{3.12.10}$$

The orthogonal tensor $\boldsymbol{R}$ represented the rotation tensor, whereas $\boldsymbol{U}$ and $\boldsymbol{V}$ were the right and left stretch tensors that encompass pure strain. Relation (3.12.10) can be thought of as a two-step process of pure stretch $\boldsymbol{U}\,\mathrm{d}\boldsymbol{X}$ followed by a rotation $\boldsymbol{R}(\boldsymbol{U}\,\mathrm{d}\boldsymbol{X})$; or rotation $\boldsymbol{R}\,\mathrm{d}\boldsymbol{X}$ followed by a stretch $\boldsymbol{V}(\boldsymbol{R}\,\mathrm{d}\boldsymbol{X})$. This format then implies a *sequential or serial mapping* that results from the *multiplicative decomposition* of the deformation into pure strain and rigid body motion. This of course is totally different than the additive decomposition for the small deformation case (3.12.7).

EXAMPLE 3.12.1 SEPARATION OF STRAIN AND ROTATION TENSORS FOR SMALL DEFORMATION

Determine the strain and rotation tensors for the following small deformation case with a displacement field given by

$$u_1 = Ax_1^2x_2 - \omega_0 x_2 + u_0, \quad u_2 = Ax_1x_2^2 + \omega_0 x_1 + v_0, \quad u_3 = Bx_2^2x_3, \quad A, B, \omega_0, u_0, v_0 \text{ all constants}$$

Solution:

$$\varepsilon_{ij} = \frac{1}{2}\left(u_{i,j} + u_{j,i}\right) = \begin{bmatrix} 2Ax_1x_2 & A(x_1^2 + x_2^2)/2 & 0 \\ A(x_1^2 + x_2^2)/2 & 2Ax_1x_2 & 2Bx_2x_3 \\ 0 & 2Bx_2x_3 & Bx_2^2 \end{bmatrix}$$

$$\omega_{ij} = \frac{1}{2}\left(u_{i,j} - u_{j,i}\right) = \begin{bmatrix} 0 & A(x_1^2 - x_2^2)/2 - \omega_0 & 0 \\ -A(x_1^2 - x_2^2)/2 + \omega_0 & 0 & -2Bx_2x_3 \\ 0 & 2Bx_2x_3 & 0 \end{bmatrix}$$

$$= \begin{bmatrix} 0 & A(x_1^2 - x_2^2)/2 & 0 \\ -A(x_1^2 - x_2^2)/2 & 0 & -2Bx_2x_3 \\ 0 & 2Bx_2x_3 & 0 \end{bmatrix} + \begin{bmatrix} 0 & -\omega_0 & 0 \\ \omega_0 & 0 & 0 \\ 0 & 0 & 0 \end{bmatrix}$$

Note that the second term in the rotation tensor is actually a rigid body rotation about the x_3-axis. An example of calculating the rotation and strain tensors for a finite deformation was previously given in Example 3.6.3.

3.13 RATE OF STRAIN TENSORS

Since many materials have rate-dependent behaviors, we will be interested in extending our analysis into time rates of change of various kinematical tensors. This will bring in added complications, as we have already seen in Section 3.3. There we presented both a material time derivative holding particles constant, and a spatial time derivate holding spatial coordinates constant. Thus, we will have to keep track of the proper kinematical representation that we are interested in calculating the time rate of change.

We start with the material time derivative of the differential element $d\boldsymbol{x}$:

$$\begin{aligned} \frac{D}{Dt}(d\boldsymbol{x}) &= \frac{D}{Dt}(\boldsymbol{x}(\boldsymbol{X} + \mathrm{d}\boldsymbol{X}, t) - \boldsymbol{x}(\boldsymbol{X}, t)) \\ &= \frac{D}{Dt}\boldsymbol{x}(\boldsymbol{X} + d\boldsymbol{X}, t) - \frac{D}{Dt}\boldsymbol{x}(\boldsymbol{X}, t) \\ &= \boldsymbol{v}(\boldsymbol{X} + d\boldsymbol{X}, t) - \boldsymbol{v}(\boldsymbol{X}, t) = \nabla \boldsymbol{v}(\boldsymbol{X}, t)\, d\boldsymbol{X} \end{aligned} \tag{3.13.1}$$

where $\boldsymbol{v}(\boldsymbol{X}, t)$ is the velocity vector in material coordinates, and $\nabla \boldsymbol{v}(\boldsymbol{X}, t) = \dfrac{\partial \boldsymbol{v}}{\partial \boldsymbol{X}}$ is the *velocity gradient tensor with respect to the Lagrangian description*. Likewise, we could use the Eulerian or spatial description and rewrite (3.13.1) as

$$\frac{D}{Dt}(d\boldsymbol{x}) = \nabla \boldsymbol{v}(\boldsymbol{x}, t)\, d\boldsymbol{x} \tag{3.13.2}$$

where $\nabla \boldsymbol{v}(\boldsymbol{x}, t) = \dfrac{\partial \boldsymbol{v}}{\partial \boldsymbol{x}}$ is the *spatial velocity gradient tensor*.

Using the spatial description, it is common to define $\boldsymbol{L}(\boldsymbol{x}, t) = \nabla \boldsymbol{v}(\boldsymbol{x}, t) = \dfrac{\partial \boldsymbol{v}}{\partial \boldsymbol{x}}$ and decompose the velocity gradient into symmetric and antisymmetric parts as

$$\boldsymbol{L}(\boldsymbol{x}, t) = \nabla \boldsymbol{v}(\boldsymbol{x}, t) = \boldsymbol{D} + \boldsymbol{W} \tag{3.13.3}$$

where

$$\begin{aligned} \boldsymbol{D} &= \frac{1}{2}\left(\nabla \boldsymbol{v} + (\nabla \boldsymbol{v})^{\mathrm{T}}\right) \\ \boldsymbol{W} &= \frac{1}{2}\left(\nabla \boldsymbol{v} - (\nabla \boldsymbol{v})^{\mathrm{T}}\right) \end{aligned} \tag{3.13.4}$$

$\boldsymbol{D}$ is called the *rate of deformation tensor* and $\boldsymbol{W}$ is known as the *spin tensor*. In hydrodynamics, the quantity $2\boldsymbol{W}$ is commonly referred to as the *vorticity tensor*. Recall our previous discussions on the dual vector associated with antisymmetric tensors. Using this theory, we can define the dual vector $\boldsymbol{w}$ to the spin tensor $\boldsymbol{W}$ by the standard relations

$$w_i = -\frac{1}{2}\varepsilon_{ijk} W_{jk} = \frac{1}{2}\varepsilon_{ijk} v_{k,j} = \frac{1}{2}\varepsilon_{ijk} L_{kj} \tag{3.13.5}$$

where $\boldsymbol{w}$ is called the *angular velocity vector*. Again in classical fluid dynamics, $2\boldsymbol{w}$ is often referred to as the *vorticity vector*. Note that the discussion here makes no assumption on the magnitudes of the velocity gradient components, and so all results have no smallness restrictions.

Next let us consider the material time derivative of the squared length $(ds)^2$:

$$\begin{aligned} \frac{D}{Dt}[(ds)^2] &= \frac{D}{Dt}[d\boldsymbol{x} \cdot d\boldsymbol{x}] = 2\, d\boldsymbol{x} \cdot \frac{D}{Dt}(d\boldsymbol{x}) \\ &= 2\, d\boldsymbol{x} \cdot \nabla \boldsymbol{v}(\boldsymbol{x}, t)\, d\boldsymbol{x} = 2\, d\boldsymbol{x}(\boldsymbol{D} + \boldsymbol{W})\, d\boldsymbol{x} \\ &= 2\, d\boldsymbol{x}\, \boldsymbol{D}\, d\boldsymbol{x} = 2 D_{ij}\, dx_i\, dx_j \end{aligned} \tag{3.13.6}$$

where we have used (3.13.2) and (3.13.3), and the spin tensor $\boldsymbol{W}$ drops out since it is antisymmetric. Note that relation (3.13.6) indicates that only the rate of deformation tensor governs the material time derivative of $(ds)^2$.

Let us move on to time rates of several of our strain tensors. Consider first the material time derivatives of the deformation gradient and its inverse:

$$\dot{F}_{ij} = \frac{D}{Dt}\frac{\partial x_i}{\partial X_j} = \frac{\partial}{\partial X_j}\frac{Dx_i}{Dt} = \frac{\partial v_i}{\partial X_j} = \frac{\partial v_i}{\partial x_k}\frac{\partial x_k}{\partial X_j} = L_{ik} F_{kj}$$
$$\text{or } \dot{\boldsymbol{F}} = \boldsymbol{L}\boldsymbol{F} \tag{3.13.7}$$

$$\frac{D}{Dt}\left(F_{ik}F_{kj}^{-1}=\delta_{ij}\right)\Rightarrow F_{ik}\dot{F}_{kj}^{-1}+\dot{F}_{ik}F_{kj}^{-1}=0\Rightarrow F_{ik}\dot{F}_{kj}^{-1}=-\dot{F}_{ik}F_{kj}^{-1}\Rightarrow$$
$$F_{li}^{-1}\left(F_{ik}\dot{F}_{kj}^{-1}=-\dot{F}_{ik}F_{kj}^{-1}\right)\Rightarrow \dot{F}_{lj}^{-1}=-F_{li}^{-1}L_{im}F_{mk}F_{kj}^{-1}=-F_{li}^{-1}L_{ij} \tag{3.13.8}$$
$$\text{or } \dot{\boldsymbol{F}}^{-1}=-\boldsymbol{F}^{-1}\boldsymbol{L}$$

Using property (2.5.11), we can determine the time rate of the Jacobian J:

$$\frac{D}{Dt}J=\frac{D}{Dt}\left(\det F_{ij}\right)=\det(\boldsymbol{F})F_{ij}^{-1}\dot{F}_{ji}=JF_{ij}^{-1}L_{jk}F_{ki}=JL_{kk}$$
$$\text{or } \dot{J}=Jv_{k,k}=J\,div\,\boldsymbol{v} \tag{3.13.9}$$

Since the Jacobian provides a local measure of the volume change, (3.13.9) would give the time rate of volume change.

Next, consider the material time derivative of the Lagrangian strain tensor $\boldsymbol{E}$:

$$\dot{\boldsymbol{E}}=\frac{D}{Dt}\left[\frac{1}{2}\left(\boldsymbol{F}^{\mathrm{T}}\boldsymbol{F}-\boldsymbol{I}\right)\right]=\frac{1}{2}\left(\dot{\mathbf{F}}^{\mathrm{T}}\boldsymbol{F}+\boldsymbol{F}^{\mathrm{T}}\dot{\mathbf{F}}\right)$$
$$=\frac{1}{2}\left((\boldsymbol{LF})^{\mathrm{T}}\boldsymbol{F}+\boldsymbol{F}^{\mathrm{T}}\boldsymbol{LF}\right)=\frac{1}{2}\boldsymbol{F}^{\mathrm{T}}(\boldsymbol{L}^{\mathrm{T}}+\boldsymbol{L})\boldsymbol{F}=\boldsymbol{F}^{\mathrm{T}}\boldsymbol{DF} \tag{3.13.10}$$

where we have used (3.13.7). This result clearly indicates that the rate of deformation tensor $\boldsymbol{D}$ is not the same as time rate of change of $\boldsymbol{E}$. However, for the infinitesimal strain case with $\boldsymbol{F}=\boldsymbol{I}+\nabla\boldsymbol{u}\approx\boldsymbol{I}$:

$$\dot{\varepsilon}=\boldsymbol{D} \tag{3.13.11}$$

Continuing on with time rates, consider next the right Cauchy–Green strain $\boldsymbol{C}$:

$$\boldsymbol{C}=\boldsymbol{F}^{\mathrm{T}}\boldsymbol{F}=2\boldsymbol{E}+\boldsymbol{I}\Rightarrow\dot{\mathbf{C}}=2\dot{\mathbf{E}}=2\boldsymbol{F}^{\mathrm{T}}\boldsymbol{DF} \tag{3.13.12}$$

and finally the left Cauchy–Green stain $\boldsymbol{B}$

$$\dot{\mathbf{B}}=\frac{D}{Dt}\left(\boldsymbol{FF}^{\mathrm{T}}\right)=\dot{\mathbf{F}}\boldsymbol{F}^{\mathrm{T}}+\boldsymbol{F}\dot{\mathbf{F}}^{\mathrm{T}}=\boldsymbol{LFF}^{\mathrm{T}}+\boldsymbol{F}(\boldsymbol{LF})^{\mathrm{T}}$$
$$=\boldsymbol{LFF}^{\mathrm{T}}+\boldsymbol{FF}^{\mathrm{T}}\boldsymbol{L}^{\mathrm{T}}=\boldsymbol{LB}+\boldsymbol{BL}^{\mathrm{T}} \tag{3.13.13}$$

EXAMPLE 3.13.1 RATE EXAMPLES

A continuum material undergoes the following time-dependent motion:

$$x_1=2tX_1+4t^2X_2$$
$$x_2=4tX_2$$
$$x_3=X_3$$

Determine the following rate variables: $\boldsymbol{v}(\boldsymbol{X},t)$, $\boldsymbol{v}(\boldsymbol{x},t)$, $\boldsymbol{L}$, $\boldsymbol{D}$, $\boldsymbol{W}$, $\boldsymbol{w}$, $\dot{\boldsymbol{E}}$, $\dot{J}$.

Solution: The inverse motion is easily found $X_1=(x_1-tx_2)/2t, X_2=x_2/4t$, $X_3=x_3$ and the deformation gradient becomes $\boldsymbol{F}=\begin{bmatrix}2t & 4t^2 & 0\\ 0 & 4t & 0\\ 0 & 0 & 1\end{bmatrix}\Rightarrow$

$J=\det\boldsymbol{F}=8t^2$.

The velocity in material coordinates is $\boldsymbol{v}(\boldsymbol{X},t)=\dfrac{D}{Dt}\boldsymbol{x}(\boldsymbol{X},t)=\begin{bmatrix}2X_1+8tX_2\\4X_2\\0\end{bmatrix}$

and using the inverse motion results, $\boldsymbol{v}(\boldsymbol{x},t)=\begin{bmatrix}(x_1-tx_2)/t+2x_2\\x_2/t\\0\end{bmatrix}$

$=\begin{bmatrix}x_1/t+x_2\\x_2/t\\0\end{bmatrix}$

The velocity gradient tensor follows from the basic definition $\boldsymbol{L}=\dfrac{\partial\boldsymbol{v}}{\partial\boldsymbol{x}}$

$=\begin{bmatrix}1/t & 1 & 0\\0 & 1/t & 0\\0 & 0 & 0\end{bmatrix}.$

The rate of deformation and spin tensors are thus $\boldsymbol{D}=\dfrac{1}{2}(\boldsymbol{L}+\boldsymbol{L}^{\mathrm{T}})$

$=\begin{bmatrix}1/t & 1/2 & 0\\1/2 & 1/t & 0\\0 & 0 & 0\end{bmatrix}$, $\boldsymbol{W}=\dfrac{1}{2}(\boldsymbol{L}-\boldsymbol{L}^{\mathrm{T}})=\begin{bmatrix}0 & 1/2 & 0\\-1/2 & 0 & 0\\0 & 0 & 0\end{bmatrix}$

For the angular velocity vector,

$$w_i=-\frac{1}{2}\varepsilon_{ijk}W_{jk}=-\frac{1}{2}(\varepsilon_{i12}W_{12}+\varepsilon_{i21}W_{21})=-\frac{1}{4}(\varepsilon_{i12}-\varepsilon_{i21})=-\frac{1}{2}\begin{bmatrix}0\\0\\1\end{bmatrix}=-\frac{1}{2}\boldsymbol{e}_3$$

The rate of strain tensor is given by

$$\dot{\boldsymbol{E}}=\boldsymbol{F}^{\mathrm{T}}\boldsymbol{D}\boldsymbol{F}=\begin{bmatrix}2t & 0 & 0\\4t^2 & 4t & 0\\0 & 0 & 1\end{bmatrix}\begin{bmatrix}1/t & 1/2 & 0\\1/2 & 1/t & 0\\0 & 0 & 0\end{bmatrix}\begin{bmatrix}2t & 4t^2 & 0\\0 & 4t & 0\\0 & 0 & 1\end{bmatrix}$$

$$=\begin{bmatrix}2 & t & 0\\6t & 2t^2+4 & 0\\0 & 0 & 0\end{bmatrix}\begin{bmatrix}2t & 4t^2 & 0\\0 & 4t & 0\\0 & 0 & 1\end{bmatrix}=\begin{bmatrix}4t & 12t^2 & 0\\12t^2 & 32t^3+16t & 0\\0 & 0 & 0\end{bmatrix}$$

and finally the Jacobian rate $\dot{J}=Jv_{k,k}=JL_{kk}=8t^2(2/t)=16t$.

Before completing this section, let us check on the objectivity properties of some of the deformation rate tensors we have developed. Recall for a second-order tensor $\boldsymbol{A}$, the objectivity test expects that under change in reference frame $\boldsymbol{x}^*=\boldsymbol{c}(t)+\boldsymbol{Q}(t)\boldsymbol{x}$, the tensor will obey the transformation relation $\boldsymbol{A}^*=\boldsymbol{Q}\boldsymbol{A}\boldsymbol{Q}^{\mathrm{T}}$. First, consider the spatial velocity gradient $\boldsymbol{L}(\boldsymbol{x},t)=\dfrac{\partial \boldsymbol{v}}{\partial \boldsymbol{x}}$. From (3.13.7) and using the transformation properties of $\boldsymbol{F}$,

$$\begin{aligned} \boldsymbol{L}^* &= \dot{\boldsymbol{F}}^* \boldsymbol{F}^{*-1} = (\dot{\boldsymbol{Q}}\boldsymbol{F} + \boldsymbol{Q}\dot{\boldsymbol{F}})(\boldsymbol{Q}\boldsymbol{F})^{-1} \\ &= (\dot{\boldsymbol{Q}}\boldsymbol{F} + \boldsymbol{Q}\dot{\boldsymbol{F}})\boldsymbol{F}^{-1}\boldsymbol{Q}^{\mathrm{T}} \\ &= \dot{\boldsymbol{Q}}\boldsymbol{F}\boldsymbol{F}^{-1}\boldsymbol{Q}^{\mathrm{T}} + \boldsymbol{Q}\dot{\boldsymbol{F}}\boldsymbol{F}^{-1}\boldsymbol{Q}^{\mathrm{T}} \\ &= \dot{\boldsymbol{Q}}\boldsymbol{Q}^{\mathrm{T}} + \boldsymbol{Q}\boldsymbol{L}\boldsymbol{Q}^{\mathrm{T}} \end{aligned} \tag{3.13.14}$$

Next, consider the rate of deformation tensor $\boldsymbol{D}$:

$$\begin{aligned} \boldsymbol{D}^* &= \frac{1}{2}\left(\boldsymbol{L}^* + \boldsymbol{L}^{*\mathrm{T}}\right) \\ &= \frac{1}{2}\left(\dot{\boldsymbol{Q}}\boldsymbol{Q}^{\mathrm{T}} + \boldsymbol{Q}\boldsymbol{L}\boldsymbol{Q}^{\mathrm{T}} + (\dot{\boldsymbol{Q}}\boldsymbol{Q}^{\mathrm{T}} + \boldsymbol{Q}\boldsymbol{L}\boldsymbol{Q}^{\mathrm{T}})^{\mathrm{T}}\right) \\ &= \frac{1}{2}\left((\dot{\boldsymbol{Q}}\boldsymbol{Q}^{\mathrm{T}} + \boldsymbol{Q}\dot{\boldsymbol{Q}}^{\mathrm{T}}) + \boldsymbol{Q}\boldsymbol{L}\boldsymbol{Q}^{\mathrm{T}} + \boldsymbol{Q}\boldsymbol{L}^{\mathrm{T}}\boldsymbol{Q}^{\mathrm{T}}\right) \\ &= \frac{1}{2}\left(\overline{(\boldsymbol{Q}\boldsymbol{Q}^{\mathrm{T}})} + \boldsymbol{Q}(\boldsymbol{L} + \boldsymbol{L}^{\mathrm{T}})\boldsymbol{Q}^{\mathrm{T}}\right) = \boldsymbol{Q}\boldsymbol{D}\boldsymbol{Q}^{\mathrm{T}} \end{aligned} \tag{3.13.15}$$

and then the spin tensor $\boldsymbol{W}$:

$$\begin{aligned} \boldsymbol{W}^* &= \frac{1}{2}\left(\boldsymbol{L}^* - \boldsymbol{L}^{*\mathrm{T}}\right) \\ &= \frac{1}{2}\left(\dot{\boldsymbol{Q}}\boldsymbol{Q}^{\mathrm{T}} + \boldsymbol{Q}\boldsymbol{L}\boldsymbol{Q}^{\mathrm{T}} - (\dot{\boldsymbol{Q}}\boldsymbol{Q}^{\mathrm{T}} + \boldsymbol{Q}\boldsymbol{L}\boldsymbol{Q}^{\mathrm{T}})^{\mathrm{T}}\right) \\ &= \frac{1}{2}\left((\dot{\boldsymbol{Q}}\boldsymbol{Q}^{\mathrm{T}} - \boldsymbol{Q}\dot{\boldsymbol{Q}}^{\mathrm{T}}) + \boldsymbol{Q}\boldsymbol{L}\boldsymbol{Q}^{\mathrm{T}} - \boldsymbol{Q}\boldsymbol{L}^{\mathrm{T}}\boldsymbol{Q}^{\mathrm{T}}\right) \\ &= \frac{1}{2}\left(2\dot{\boldsymbol{Q}}\boldsymbol{Q}^{\mathrm{T}} + \boldsymbol{Q}(\boldsymbol{L} - \boldsymbol{L}^{\mathrm{T}})\boldsymbol{Q}^{\mathrm{T}}\right) = \dot{\boldsymbol{Q}}\boldsymbol{Q}^{\mathrm{T}} + \boldsymbol{Q}\boldsymbol{W}\boldsymbol{Q}^{\mathrm{T}} \end{aligned} \tag{3.13.16}$$

Finally, let us check on the strain rate tensor $\dot{\boldsymbol{E}}$:

$$\begin{aligned} \dot{\boldsymbol{E}}^* &= \boldsymbol{F}^{*\mathrm{T}}\boldsymbol{D}^*\boldsymbol{F}^* \\ &= (\boldsymbol{Q}\boldsymbol{F})^{\mathrm{T}}\boldsymbol{Q}\boldsymbol{D}\boldsymbol{Q}^{\mathrm{T}}(\boldsymbol{Q}\boldsymbol{F}) \\ &= \boldsymbol{F}^{\mathrm{T}}\boldsymbol{Q}^{\mathrm{T}}\boldsymbol{Q}\boldsymbol{D}\boldsymbol{Q}^{\mathrm{T}}\boldsymbol{Q}\boldsymbol{F} = \boldsymbol{F}^{\mathrm{T}}\boldsymbol{D}\boldsymbol{F} = \dot{\boldsymbol{E}} \end{aligned} \tag{3.13.17}$$

We thus conclude that only the rate of deformation tensor $\boldsymbol{D}$ satisfies the standard objective relation. However, as we pointed out in Section 3.6, the Lagrangian strain is based on reference coordinates and thus will transform as a scalar field.

3.14 OBJECTIVE TIME DERIVATIVES

Recall that in Section 2.9, we explored the principle of objectivity or frame indifference and indicated that we wish to have our general tensor variables and relations satisfy this principle. This can be again summarized that under the change in reference frame $x^* = c(t) + Q(t)x$, objective first- and second -order tensors a and A must satisfy the relations

$$\begin{aligned} a^* &= Qa \\ A^* &= QAQ^{\mathrm{T}} \end{aligned} \tag{3.14.1}$$

In that section, we found that the velocity vector is in general not objective, since it transforms as

$$v^* = \dot{c} + \dot{Q}x + Qv \tag{3.14.2}$$

Since the velocity is part of the material time derivative as per relation (3.3.3), this derivative is not objective or frame indifferent. Because several rate-dependent constitutive relations use such time derivatives, this will produce an unpleasant situation. In order to correct this problem, various *objective time derivatives* have been developed.

Let us first consider the material time derivative of an objective vector a:

$$\dot{a}^* = Q\dot{a} + \dot{Q}a \tag{3.14.3}$$

Clearly $\dot{a}$ is no longer objective because of the extra term $\dot{Q}a$ in relation (3.14.3). We now wish to modify the ordinary material time derivative to another form which will satisfy the objectivity test. Recall from the previous relation (3.13.16) that the spin tensor satisfied the relation $W^* = QWQ^{\mathrm{T}} + \dot{Q}Q^{\mathrm{T}}$ and $W = -W^{\mathrm{T}}$. Now from (3.14.1) and (3.14.3),

$$\begin{gathered} \dot{a}^* = Q\dot{a} + \dot{Q}Q^{\mathrm{T}}a^* = \\ Q\dot{a} + (W^* - QWQ^{\mathrm{T}})a^* \Rightarrow \dot{a}^* - W^*a^* = Q(\dot{a} - Wa) \end{gathered} \tag{3.14.4}$$

and thus a modified time derivative form for vectors which satisfies objectivity can be written as

$$\overset{\circ}{a} = \dot{a} - Wa \tag{3.14.5}$$

Going through a similar process for second-order tensors A gives the objective time derivative

$$\overset{\circ}{A} = \dot{A} - WA + AW \tag{3.14.6}$$

Relations (3.14.5) and (3.14.6) are known as *Zarembra–Jaumann objective time rates*. These may be thought of as the rate of change as seen by an observer who is rotating with the media. Several other such rates have been defined in the literature most of which are special forms of the so-called *Lie derivative*. We will not pursue any further relations of this type here, and the interested reader can find more information in Haupt (2002) and Holzapfel (2006).

3.15 CURRENT CONFIGURATION AS REFERENCE CONFIGURATION

Recall back in Section 3.1, we indicated that the choice of the reference configuration was completely arbitrary. We now wish to redevelop some of the basic kinematics for the case where *the current configuration is used as the reference configuration*. This concept provides a convenient method to determine the *past history of the deformation*, and this is an important ingredient in some constitutive equations to be developed later. Let x_i and ξ_i be the positions at times t and τ of particle X which was at position X_α in the reference configuration.

$$\boldsymbol{x} = \chi(\boldsymbol{X}, t), \quad \xi = \chi(\boldsymbol{X}, \tau) \tag{3.15.1}$$

Using the standard definition, the deformation gradients at times t and τ are given by

$$F_{i\alpha}(t) = \frac{\partial x_i}{\partial X_\alpha}, \quad F_{i\alpha}(\tau) = \frac{\partial \xi_i}{\partial X_\alpha} \tag{3.15.2}$$

Now, if instead, we take the current configuration at time t to be the reference configuration, then we can write the *relative motion* as

$$\xi = \chi(\chi^{-1}(\boldsymbol{x}, t), \tau) = \xi(\boldsymbol{x}, t, \tau) \tag{3.15.3}$$

and the *relative deformation gradient tensor* becomes

$$\begin{aligned} F_t(\tau)_{ij} &= \frac{\partial \xi_i}{\partial x_j} = F_{i\alpha}(\tau) F^{-1}_{\alpha j}(t) \\ \boldsymbol{F}_t(\tau) &= \boldsymbol{F}(\tau)\boldsymbol{F}^{-1}(t) \end{aligned} \tag{3.15.4}$$

For this case, the right and left Cauchy–Green strain tensors are then written as

$$\begin{aligned} C_t(\tau)_{ij} &= F_t(\tau)_{ki} F_t(\tau)_{kj}, \quad \boldsymbol{C}_t(\tau) = \boldsymbol{F}_t(\tau)^{\mathrm{T}} \boldsymbol{F}_t(\tau) \\ B_t(\tau)_{ij} &= F_t(\tau)_{ik} F_t(\tau)_{jk}, \quad \boldsymbol{B}_t(\tau) = \boldsymbol{F}_t(\tau) \boldsymbol{F}_t(\tau)^{\mathrm{T}} \end{aligned} \tag{3.15.5}$$

Note that

$$\begin{aligned} \boldsymbol{C}_t(\tau) &= \boldsymbol{F}_t(\tau)^{\mathrm{T}} \boldsymbol{F}_t(\tau) \\ &= \left(\boldsymbol{F}(\tau)\boldsymbol{F}^{-1}(t)\right)^{\mathrm{T}} \boldsymbol{F}(\tau)\boldsymbol{F}^{-1}(t) \\ &= \boldsymbol{F}^{-1}(t)^{\mathrm{T}} \boldsymbol{F}(\tau)^{\mathrm{T}} \boldsymbol{F}(\tau)\boldsymbol{F}^{-1}(t) \\ &= \boldsymbol{F}^{-1}(t)^{\mathrm{T}} \boldsymbol{C}(\tau)\boldsymbol{F}^{-1}(t) \end{aligned} \tag{3.15.6}$$

Applying our objectivity test to relation (3.15.6) reveals that

$$\boldsymbol{C}_t^*(\tau) = \boldsymbol{Q}\boldsymbol{C}_t(\tau)\boldsymbol{Q}^{\mathrm{T}} \tag{3.15.7}$$

and thus the relative right Cauchy–Green strain tensor $\boldsymbol{C}_t(\tau)$ is objective, but recall that $\boldsymbol{C}(t)$ was not. It can be shown that for the left Cauchy–Green strain tensor both $\boldsymbol{B}(t)$ and $\boldsymbol{B}_t(\tau)$ are objective.

Now if time τ is in the range $-\infty < \tau < t$, then $\boldsymbol{C}_t(\tau)$ represents the entire past history of the deformation. This concept is also commonly written in an alternative form by letting $\tau = t - s$, and thus $\boldsymbol{C}_t(t-s)$ is the past strain history for s ranging $0 < s < \infty$. Other various strain and rate of strain tensors follow similar definitions with respect to this referencing scheme. We will put some of these tensors to use later in our studies on constitutive equation theories in Chapter 8.

EXAMPLE 3.15.1 RELATIVE DEFORMATION IN SIMPLE SHEAR AND EXTENSION

For the following time-dependent motions of simple shear and extension, determine the relative deformation gradient and relative right and left Cauchy–Green strain tensors:

Simple shear	Extension motion
$x_1 = X_1 + \gamma(t)X_2$	$x_1 = \lambda_1(t)X_1$
$x_2 = X_2$	$x_2 = \lambda_2(t)X_2$
$x_3 = X_3$	$x_3 = \lambda_3(t)X_3$

Solution: For the simple shear case,

$$\begin{aligned} x_1 &= X_1 + \gamma(t)X_2 \\ x_2 &= X_2 \\ x_3 &= X_3 \end{aligned} \quad \Rightarrow \boldsymbol{F}(t) = \frac{\partial \boldsymbol{x}}{\partial \boldsymbol{X}} = \begin{bmatrix} 1 & \gamma(t) & 0 \\ 0 & 1 & 0 \\ 0 & 0 & 1 \end{bmatrix} \Rightarrow \boldsymbol{F}^{-1}(t) = \begin{bmatrix} 1 & -\gamma(t) & 0 \\ 0 & 1 & 0 \\ 0 & 0 & 1 \end{bmatrix}$$

$$\boldsymbol{F}_t(\tau) = \boldsymbol{F}(\tau)\boldsymbol{F}^{-1}(t) = \begin{bmatrix} 1 & \gamma(\tau) & 0 \\ 0 & 1 & 0 \\ 0 & 0 & 1 \end{bmatrix} \begin{bmatrix} 1 & -\gamma(t) & 0 \\ 0 & 1 & 0 \\ 0 & 0 & 1 \end{bmatrix} = \begin{bmatrix} 1 & \gamma(\tau)-\gamma(t) & 0 \\ 0 & 1 & 0 \\ 0 & 0 & 1 \end{bmatrix}$$

$$\begin{aligned} \boldsymbol{C}_t(\tau) = \boldsymbol{F}_t(\tau)^{\mathrm{T}}\boldsymbol{F}_t(\tau) &= \begin{bmatrix} 1 & 0 & 0 \\ \gamma(\tau)-\gamma(t) & 1 & 0 \\ 0 & 0 & 1 \end{bmatrix} \begin{bmatrix} 1 & \gamma(\tau)-\gamma(t) & 0 \\ 0 & 1 & 0 \\ 0 & 0 & 1 \end{bmatrix} \\ &= \begin{bmatrix} 1 & \gamma(\tau)-\gamma(t) & 0 \\ \gamma(\tau)-\gamma(t) & 1+[\gamma(\tau)-\gamma(t)]^2 & 0 \\ 0 & 0 & 1 \end{bmatrix} \end{aligned}$$

$$\begin{aligned} \boldsymbol{B}_t(\tau) = \boldsymbol{F}_t(\tau)\boldsymbol{F}_t(\tau)^{\mathrm{T}} &= \begin{bmatrix} 1 & \gamma(\tau)-\gamma(t) & 0 \\ 0 & 1 & 0 \\ 0 & 0 & 1 \end{bmatrix} \begin{bmatrix} 1 & 0 & 0 \\ \gamma(\tau)-\gamma(t) & 1 & 0 \\ 0 & 0 & 1 \end{bmatrix} \\ &= \begin{bmatrix} 1+[\gamma(\tau)-\gamma(t)]^2 & \gamma(\tau)-\gamma(t) & 0 \\ \gamma(\tau)-\gamma(t) & 1 & 0 \\ 0 & 0 & 1 \end{bmatrix} \end{aligned}$$

For the extensional deformation,

$$\begin{aligned} x_1 &= \lambda_1(t)X_1 \\ x_2 &= \lambda_2(t)X_2 \\ x_3 &= \lambda_3(t)X_3 \end{aligned} \quad \Rightarrow \boldsymbol{F}(t) = \begin{bmatrix} \lambda_1(t) & 0 & 0 \\ 0 & \lambda_2(t) & 0 \\ 0 & 0 & \lambda_3(t) \end{bmatrix}$$

$$\Rightarrow \boldsymbol{F}^{-1}(t) = \begin{bmatrix} \lambda_1^{-1}(t) & 0 & 0 \\ 0 & \lambda_2^{-1}(t) & 0 \\ 0 & 0 & \lambda_3^{-1}(t) \end{bmatrix}$$

$$\boldsymbol{F}_t(\tau) = \boldsymbol{F}(\tau)\boldsymbol{F}^{-1}(t) = \begin{bmatrix} \lambda_1(\tau) & 0 & 0 \\ 0 & \lambda_2(\tau) & 0 \\ 0 & 0 & \lambda_3(\tau) \end{bmatrix} \begin{bmatrix} \lambda_1^{-1}(t) & 0 & 0 \\ 0 & \lambda_2^{-1}(t) & 0 \\ 0 & 0 & \lambda_3^{-1}(t) \end{bmatrix}$$

$$= \begin{bmatrix} \dfrac{\lambda_1(\tau)}{\lambda_1(t)} & 0 & 0 \\ 0 & \dfrac{\lambda_2(\tau)}{\lambda_2(t)} & 0 \\ 0 & 0 & \dfrac{\lambda_3(\tau)}{\lambda_3(t)} \end{bmatrix}$$

$$\boldsymbol{C}_t(\tau) = \boldsymbol{F}_t(\tau)^{\mathrm{T}}\boldsymbol{F}_t(\tau) = \begin{bmatrix} \left(\dfrac{\lambda_1(\tau)}{\lambda_1(t)}\right)^2 & 0 & 0 \\ 0 & \left(\dfrac{\lambda_2(\tau)}{\lambda_2(t)}\right)^2 & 0 \\ 0 & 0 & \left(\dfrac{\lambda_3(\tau)}{\lambda_3(t)}\right)^2 \end{bmatrix} = \boldsymbol{B}_t(\tau)$$

3.16 RIVLIN–ERICKSEN TENSORS

Nonlinear rate-dependent constitutive equations often use functional forms that incorporate higher-order time derivatives of strain tensors. These forms are constructed to be objective strain rates and are called *Rivlin–Ericksen tensors* $\boldsymbol{A}^{(n)}$. Truesdell and Noll (1965) defined these tensors by the relation

$$\boldsymbol{A}^{(n)} = \frac{D^n}{D\tau^n}[\boldsymbol{C}_t(\tau)]_{\tau=t} \tag{3.16.1}$$

and another equivalent form often given (Malvern, 1969; Eringen, 1967) is related to the expression

$$\frac{D^n}{Dt^n}\left[(ds)^2\right] = d\boldsymbol{x}\,\boldsymbol{A}^{(n)}\,d\boldsymbol{x} \tag{3.16.2}$$

which can be compared with (3.13.6) for the case with $n = 1$. These forms produce the set of tensors

$$\begin{gathered} \boldsymbol{A}^{(1)} = \boldsymbol{L} + \boldsymbol{L}^{\mathrm{T}} = 2\boldsymbol{D} \\ \boldsymbol{A}^{(2)} = \frac{D}{Dt}\boldsymbol{A}^{(1)} + \boldsymbol{A}^{(1)}\boldsymbol{L} + \boldsymbol{L}^{\mathrm{T}}\boldsymbol{A}^{(1)} \\ \vdots \\ \boldsymbol{A}^{(n)} = \frac{D}{Dt}\boldsymbol{A}^{(n-1)} + \boldsymbol{A}^{(n-1)}\boldsymbol{L} + \boldsymbol{L}^{\mathrm{T}}\boldsymbol{A}^{(n-1)} \end{gathered} \tag{3.16.3}$$

Note from (3.6.12), $(ds)^2 = d\boldsymbol{X}\,\boldsymbol{C}\,d\boldsymbol{X}$, and so

$$\frac{D^n}{Dt^n}\left[(ds)^2\right] = d\boldsymbol{X}\frac{D^n\boldsymbol{C}}{Dt^n}d\boldsymbol{X} \tag{3.16.4}$$

Now from definition (3.16.2),

$$\begin{aligned} \frac{D^n}{Dt^n}\left[(ds)^2\right] &= d\boldsymbol{x}\,\boldsymbol{A}^{(n)}\,d\boldsymbol{x} = (d\boldsymbol{X}\,\boldsymbol{F}^T)\boldsymbol{A}^{(n)}(\boldsymbol{F}\,\mathrm{d}\boldsymbol{X}) \\ &= d\boldsymbol{X}(\boldsymbol{F}^T\boldsymbol{A}^{(n)}\boldsymbol{F})d\boldsymbol{X} \end{aligned} \tag{3.16.5}$$

Comparing results (3.16.4) and (3.16.5) implies

$$\frac{D^n\boldsymbol{C}}{Dt^n} = \boldsymbol{F}^{\mathrm{T}}\boldsymbol{A}^{(n)}\boldsymbol{F} \quad \text{or} \quad \boldsymbol{A}^{(n)} = (\boldsymbol{F}^{\mathrm{T}})^{-1}\frac{D^n\boldsymbol{C}}{Dt^n}\boldsymbol{F}^{-1} \tag{3.16.6}$$

which matches with form (3.16.1).

3.17 CURVILINEAR CYLINDRICAL AND SPHERICAL COORDINATE RELATIONS

In order to formulate and solve many continuum mechanics problems, it is necessary to use curvilinear coordinates typically cylindrical or spherical systems. We now wish to develop a few of the kinematical expressions for these coordinate systems. Starting first with small deformation theory, the strain–displacement relation was given by equation (3.8.1):

$$\boldsymbol{\varepsilon} = \frac{1}{2}\left[\nabla\boldsymbol{u} + (\nabla\boldsymbol{u})^{\mathrm{T}}\right] \tag{3.17.1}$$

The desired curvilinear relations can be determined using the appropriate forms for the displacement gradient term $\nabla\boldsymbol{u}$.

The cylindrical coordinate system previously defined in Fig. 2.6 establishes new components for the displacement vector and strain tensor

$$\begin{gathered} \boldsymbol{u} = u_r\boldsymbol{e}_r + u_\theta\boldsymbol{e}_\theta + u_z\boldsymbol{e}_z \\ \boldsymbol{\varepsilon} = \begin{bmatrix} \varepsilon_{rr} & \varepsilon_{r\theta} & \varepsilon_{rz} \\ \varepsilon_{r\theta} & \varepsilon_{\theta\theta} & \varepsilon_{\theta z} \\ \varepsilon_{rz} & \varepsilon_{\theta z} & \varepsilon_{zz} \end{bmatrix} \end{gathered} \tag{3.17.2}$$

Notice that the symmetry of the strain tensor is preserved in this orthogonal curvilinear system. Using results (2.18.18) and (2.18.10), the derivative operation $\nabla \boldsymbol{u}$ in cylindrical coordinates can be expressed by

$$
\begin{aligned}
\nabla \boldsymbol{u} = \frac{\partial u_r}{\partial r}\boldsymbol{e}_r\boldsymbol{e}_r + \frac{\partial u_\theta}{\partial r}\boldsymbol{e}_r\boldsymbol{e}_\theta + \frac{\partial u_z}{\partial r}\boldsymbol{e}_r\boldsymbol{e}_z \\
+ \frac{1}{r}\left(\frac{\partial u_r}{\partial \theta} - u_\theta\right)\boldsymbol{e}_\theta\boldsymbol{e}_r + \frac{1}{r}\left(u_r + \frac{\partial u_\theta}{\partial \theta}\right)\boldsymbol{e}_\theta\boldsymbol{e}_\theta + \frac{1}{r}\frac{\partial u_z}{\partial \theta}\boldsymbol{e}_\theta\boldsymbol{e}_z \\
+ \frac{\partial u_r}{\partial z}\boldsymbol{e}_z\boldsymbol{e}_r + \frac{\partial u_\theta}{\partial z}\boldsymbol{e}_z\boldsymbol{e}_\theta + \frac{\partial u_z}{\partial z}\boldsymbol{e}_z\boldsymbol{e}_z
\end{aligned}
\tag{3.17.3}
$$

Placing this result into the strain–displacement form (3.17.1) gives the desired relations in cylindrical coordinates. The individual scalar equations are given by

$$
\begin{aligned}
\varepsilon_{rr} &= \frac{\partial u_r}{\partial r}, \quad \varepsilon_{\theta\theta} = \frac{1}{r}\left(u_r + \frac{\partial u_\theta}{\partial \theta}\right), \quad \varepsilon_{zz} = \frac{\partial u_z}{\partial z} \\
\varepsilon_{r\theta} &= \frac{1}{2}\left(\frac{1}{r}\frac{\partial u_r}{\partial \theta} + \frac{\partial u_\theta}{\partial r} - \frac{u_\theta}{r}\right) \\
\varepsilon_{\theta z} &= \frac{1}{2}\left(\frac{\partial u_\theta}{\partial z} + \frac{1}{r}\frac{\partial u_z}{\partial \theta}\right) \\
\varepsilon_{zr} &= \frac{1}{2}\left(\frac{\partial u_r}{\partial z} + \frac{\partial u_z}{\partial r}\right)
\end{aligned}
\tag{3.17.4}
$$

For spherical coordinates defined by Fig. 2.7, the displacement vector and strain tensor can be written as

$$
\begin{aligned}
\boldsymbol{u} &= u_R\boldsymbol{e}_R + u_\phi\boldsymbol{e}_\phi + u_\theta\boldsymbol{e}_\theta \\
\boldsymbol{\varepsilon} &= \begin{bmatrix} \varepsilon_{RR} & \varepsilon_{R\phi} & \varepsilon_{R\theta} \\ \varepsilon_{R\phi} & \varepsilon_{\phi\phi} & \varepsilon_{\phi\theta} \\ \varepsilon_{R\theta} & \varepsilon_{\phi\theta} & \varepsilon_{\theta\theta} \end{bmatrix}
\end{aligned}
\tag{3.17.5}
$$

Following identical procedures as used for the cylindrical equation development, the strain–displacement relations for spherical coordinates become

$$
\begin{aligned}
\varepsilon_{RR} &= \frac{\partial u_R}{\partial R}, \quad \varepsilon_{\phi\phi} = \frac{1}{R}\left(u_R + \frac{\partial u_\phi}{\partial \phi}\right) \\
\varepsilon_{\theta\theta} &= \frac{1}{R\sin\phi}\left(\frac{\partial u_\theta}{\partial \theta} + \sin\phi\, u_R + \cos\phi\, u_\phi\right) \\
\varepsilon_{R\phi} &= \frac{1}{2}\left(\frac{1}{R}\frac{\partial u_R}{\partial \phi} + \frac{\partial u_\phi}{\partial R} - \frac{u_\phi}{R}\right) \\
\varepsilon_{\phi\theta} &= \frac{1}{2R}\left(\frac{1}{\sin\phi}\frac{\partial u_\phi}{\partial \theta} + \frac{\partial u_\theta}{\partial \phi} - \cot\phi\, u_\theta\right) \\
\varepsilon_{\theta R} &= \frac{1}{2}\left(\frac{1}{R\sin\phi}\frac{\partial u_R}{\partial \theta} + \frac{\partial u_\theta}{\partial R} - \frac{u_\theta}{R}\right)
\end{aligned}
\tag{3.17.6}
$$

It is observed that these relations in curvilinear systems contain additional terms that do not include derivatives of individual displacement components. For example, in spherical coordinates, a simple uniform radial displacement u_R will give rise to transverse extensional strains $\varepsilon_{\phi\phi} = \varepsilon_{\theta\theta} = \frac{u_R}{R}$. This deformation can be simulated by blowing up a spherical balloon and observing the separation of points on the balloon's surface. Such terms were not found in the Cartesian forms, and their appearance is related to the curvature of the spatial coordinate system. Clearly, the curvilinear forms (3.17.4) and (3.17.6) are more complicated than the corresponding Cartesian relations (3.11.1). However, for particular problems, the curvilinear relations when combined with other field equations will allow analytical solutions to be developed that could not be found using a Cartesian formulation.

In regard to finite deformation theory, curvilinear forms are more challenging to generate, and we will only give a few of the relations for cylindrical coordinates. If we choose cylindrical coordinates for both the reference (R, Θ, Z) and current configurations (r, θ, z), the motion can be expressed by

$$\begin{aligned} r &= r(R,\Theta,Z,t) \\ \theta &= \theta(R,\Theta,Z,t) \\ z &= z(R,\Theta,Z,t) \end{aligned} \tag{3.17.7}$$

Using the results in either Sections 2.18 or 2.19, we can write

$$\begin{aligned} d\boldsymbol{x} &= dr\boldsymbol{e}_r + r\,d\theta\boldsymbol{e}_\theta + dz\boldsymbol{e}_z \\ d\boldsymbol{X} &= dR\boldsymbol{E}_R + R\,d\Theta\boldsymbol{E}_\Theta + dZ\boldsymbol{E}_Z \end{aligned} \tag{3.17.8}$$

where we are using $(\boldsymbol{e}_r, \boldsymbol{e}_\theta, \boldsymbol{e}_z)$ as the basis in the current configuration and $(\boldsymbol{E}_R, \boldsymbol{E}_\Theta, \boldsymbol{E}_Z)$ in the reference configuration. Employing the usual relation $d\boldsymbol{x} = \boldsymbol{F}\,d\boldsymbol{X}$, we find

$$\begin{aligned} dr\boldsymbol{e}_r + r\,d\theta\boldsymbol{e}_\theta + dz\boldsymbol{e}_z &= \boldsymbol{F}(dR\boldsymbol{E}_R + R\,d\Theta\boldsymbol{E}_\Theta + dZ\boldsymbol{E}_Z) \Rightarrow \\ dr &= dR(\boldsymbol{e}_r\boldsymbol{F}\boldsymbol{E}_R) + R\,d\Theta(\boldsymbol{e}_r\boldsymbol{F}\boldsymbol{E}_\Theta) + dZ(\boldsymbol{e}_r\boldsymbol{F}\boldsymbol{E}_Z) \\ r\,d\theta &= dR(\boldsymbol{e}_\theta\boldsymbol{F}\boldsymbol{E}_R) + R\,d\Theta(\boldsymbol{e}_\theta\boldsymbol{F}\boldsymbol{E}_\Theta) + dZ(\boldsymbol{e}_\theta\boldsymbol{F}\boldsymbol{E}_Z) \\ dz &= dR(\boldsymbol{e}_z\boldsymbol{F}\boldsymbol{E}_R) + R\,d\Theta(\boldsymbol{e}_z\boldsymbol{F}\boldsymbol{E}_\Theta) + dZ(\boldsymbol{e}_z\boldsymbol{F}\boldsymbol{E}_Z) \end{aligned} \tag{3.17.9}$$

Thus, $\boldsymbol{e}_r\boldsymbol{F}\boldsymbol{E}_R = F_{11} = \frac{\partial r}{\partial R}, \boldsymbol{e}_r\boldsymbol{F}\boldsymbol{E}_\Theta = F_{12} = \frac{1}{R}\frac{\partial r}{\partial \Theta}, \boldsymbol{e}_r\boldsymbol{F}\boldsymbol{E}_Z = F_{13} = \frac{\partial r}{\partial Z}, \ldots$, and so the final form for the deformation gradient tensor is

$$\boldsymbol{F} = \begin{bmatrix} \frac{\partial r}{\partial R} & \frac{1}{R}\frac{\partial r}{\partial \Theta} & \frac{\partial r}{\partial Z} \\ r\frac{\partial \theta}{\partial R} & \frac{r}{R}\frac{\partial \theta}{\partial \Theta} & r\frac{\partial \theta}{\partial Z} \\ \frac{\partial z}{\partial R} & \frac{1}{R}\frac{\partial z}{\partial \Theta} & \frac{\partial z}{\partial Z} \end{bmatrix} \tag{3.17.10}$$

If we were to use Cartesian coordinates for reference (X, Y, Z) and cylindrical coordinates for the current configuration (r, θ, z), the deformation gradient would become

$$\boldsymbol{F} = \begin{bmatrix} \dfrac{\partial r}{\partial X} & \dfrac{\partial r}{\partial Y} & \dfrac{\partial r}{\partial Z} \\ r\dfrac{\partial \theta}{\partial X} & r\dfrac{\partial \theta}{\partial Y} & r\dfrac{\partial \theta}{\partial Z} \\ \dfrac{\partial z}{\partial X} & \dfrac{\partial z}{\partial Y} & \dfrac{\partial z}{\partial Z} \end{bmatrix} \quad (3.17.11)$$

Knowing these forms for the deformation gradient allows for the calculation of the right and left Cauchy–Green strain tensors. Lai et al. (2010) provided further details on these calculations. Appendices A and B also give additional summary details on curvilinear forms for a variety of continuum mechanics relations and variables.

REFERENCES

Asaro, R.J., Lubarda, V.A., 2006. Mechanics of Solids and Materials. Cambridge University Press, New York.
Eringen, A.C., 1967. Mechanics of Continua. John Wiley, New York.
Fung, Y.C., 1965. Foundations of Solid Mechanics. Prentice-Hall, Englewood Cliffs, NJ.
Haupt, P., 2002. Continuum Mechanics and Theory of Materials, Second ed. Springer, Berlin.
Holzapfel, G.A., 2006. Nonlinear Solid Mechanics: A Continuum Approach for Engineering. John Wiley, West Sussex.
Lai, W.M., Rubin, D., Krempl, E., 2010. Introduction to Continuum Mechanics, fourth ed. Butterworth Heinemann/Elsevier, Amsterdam.
Malvern, L.E., 1969. Introduction to the Mechanics of a Continuous Medium. Prentice-Hall, Englewood Cliffs, NJ.
Sadd, M.H., 2014. Elasticity: Theory Applications and Numerics, Third ed. Elsevier, Waltham, MA.
Truesdell, C.A., Noll, W., 1965. Nonlinear field theories. Flugge, S. (Ed.), Encyclopedia of Physics, III, Third ed. Springer, Berlin.
Truesdell, C.A., Toupin, R.A., 1960. Classical Field Theories. Flugge, S. (Ed.), Encyclopedia of Physics, III, 1 ed. Berlin, Springer.

EXERCISES

3.1 Use MATLAB Code C-3 to make two-dimensional deformation plots of reference unit squares and circles for the following material motions:

(a) $x_1 = (\sqrt{3}/2)X_1 + 0.5X_2 + 1.5$, $x_2 = -0.5X_1 + (\sqrt{3}/2)X_2 + 1.5$
(b) $x_1 = 2X_1 - 0.5X_2 + 1.5$, $x_2 = 0.5X_1 + 1.2X_2 + 1.5$
(c) $x_1 = 2X_1 + 1.5X_2 + 2$, $x_2 = 0.5X_1 + 2$

3.2 For the following temperature and motion fields, determine the material time derivative of the temperature

$$\text{(a)}\ \begin{aligned} \theta &= x_1 + 2x_2 + 4tx_3 \\ x_1 &= 2tX_1 + 3tX_2 \\ x_2 &= X_1 + 2tX_2 \\ x_3 &= X_3 \end{aligned} \qquad \text{(b)}\ \begin{aligned} \theta &= 2tx_1 + 4x_2 + x_3 \\ x_1 &= X_1 + 3tX_2 \\ x_2 &= 2tX_1 + 3t^2X_2 \\ x_3 &= X_3 \end{aligned} \qquad \text{(c)}\ \begin{aligned} \theta &= x_1 + 4t^2x_2 + x_3 \\ x_1 &= X_1 + 2X_2 \\ x_2 &= 2X_1 + 3tX_2 \\ x_3 &= 2X_3 \end{aligned}$$

3.3 For the following motions, determine the velocity and acceleration fields in material and spatial coordinates:

$$\text{(a)}\ \begin{aligned} x_1 &= X_1 + 3t^2X_2 \\ x_2 &= X_1 + 2tX_2 \\ x_3 &= X_3 \end{aligned} \qquad \text{(b)}\ \begin{aligned} x_1 &= 2tX_1 + 4X_2 \\ x_2 &= X_1 + 2t^3X_2 \\ x_3 &= X_3 \end{aligned} \qquad \text{(c)}\ \begin{aligned} x_1 &= tX_1 + 3X_2 \\ x_2 &= 2t^2X_1 + X_2 \\ x_3 &= X_3 \end{aligned}$$

3.4 Calculate the acceleration from the following spatial velocity fields:

$$\text{(a)}\quad \boldsymbol{v} = 2x_1x_2\boldsymbol{e}_1 + 3x_2\boldsymbol{e}_2 + \boldsymbol{e}_3$$

$$\text{(b)}\quad \boldsymbol{v} = 4x_1t\boldsymbol{e}_1 + 2x_2\boldsymbol{e}_2 + 2t^2\boldsymbol{e}_3$$

$$\text{(c)}\quad \boldsymbol{v} = 2x_1t\boldsymbol{e}_1 + 4x_2t^2\boldsymbol{e}_2 + 3\boldsymbol{e}_3$$

3.5 Determine the displacement field, displacement gradient, and deformation gradient tensors for the following material motions:

$$\text{(a)}\ \begin{aligned} x_1 &= X_1 + 3X_2 \\ x_2 &= 2X_1 + 4X_2 \\ x_3 &= 4X_1X_3 \end{aligned} \qquad \text{(b)}\ \begin{aligned} x_1 &= X_1 + 3X_2X_3 \\ x_2 &= 2X_1 + 4X_2 + X_3 \\ x_3 &= 6X_1X_2 \end{aligned} \qquad \text{(c)}\ \begin{aligned} x_1 &= 3X_1X_3 + 2X_2 \\ x_2 &= 2X_1 + 4X_2X_3 \\ x_3 &= 4X_1 + 3X_3 \end{aligned}$$

3.6 Consider the deformations of simple shear and two-dimensional extension given by the two deformation gradients

$$\boldsymbol{F}^{(1)} = \begin{bmatrix} 1 & \gamma & 0 \\ 0 & 1 & 0 \\ 0 & 0 & 1 \end{bmatrix}, \quad \boldsymbol{F}^{(2)} = \begin{bmatrix} \lambda_1 & 0 & 0 \\ 0 & \lambda_2 & 0 \\ 0 & 0 & 1 \end{bmatrix}.$$ Using the sequential deformation concept, determine the overall relation between spatial and reference differential elements for the cases $d\boldsymbol{x} = \boldsymbol{F}^{(1)}\boldsymbol{F}^{(2)}\,d\boldsymbol{X}$ and $d\boldsymbol{x} = \boldsymbol{F}^{(2)}\boldsymbol{F}^{(1)}\,d\boldsymbol{X}$.

3.7 For the following motions, determine the deformation gradient $\boldsymbol{F}$. Next, calculate the inverse motion $\boldsymbol{X} = \chi^{-1}(\boldsymbol{x},t)$, and then using $\boldsymbol{F}^{-1} = \dfrac{\partial \boldsymbol{X}}{\partial \boldsymbol{x}}$ find the inverse deformation gradient tensor. Finally, verify that $\boldsymbol{F}\boldsymbol{F}^{-1} = \boldsymbol{I}$.

$$\text{(a)}\ \begin{aligned} x_1 &= X_1 + 2X_2 \\ x_2 &= 2X_1 + X_2 \\ x_3 &= X_3 \end{aligned} \qquad \text{(b)}\ \begin{aligned} x_1 &= 4X_1 + X_2 \\ x_2 &= 4X_2 + X_3 \\ x_3 &= X_3 \end{aligned} \qquad \text{(c)}\ \begin{aligned} x_1 &= X_1 + X_2 \\ x_2 &= 2X_1 + 4X_2 \\ x_3 &= 4X_1 + 3X_3 \end{aligned}$$

3.8 *Rigid body motion* preserves the relative distances between particles in the body. Show that a body moves rigidly if $\boldsymbol{F}^{\mathrm{T}}\boldsymbol{F} = \boldsymbol{I}$, and hence $\det \boldsymbol{F} = 1$. Also justify that such a motion must be of the form $\boldsymbol{x} = \boldsymbol{Q}(t)\boldsymbol{X} + \boldsymbol{c}(t)$ where $\boldsymbol{Q}^{\mathrm{T}}\boldsymbol{Q} = \boldsymbol{I}$.

3.9 Using the sequential deformation concept, consider the case where the deformation gradient can be decomposed into volumetric and distortional parts such that $\boldsymbol{F} = \boldsymbol{F}_v\boldsymbol{F}_d$. Since there is no volume change for the distortional deformation, $\det \boldsymbol{F}_d = 1$. Show that this condition is satisfied by the form $\boldsymbol{F}_d = J^{-1/3}\boldsymbol{F}$, and that $J = \det \boldsymbol{F}_v$.

3.10 Determine the Lagrangian, Eulerian, and right and left Cauchy–Green strain tensors for the following motions: (a) $x_1 = X_1 + 3X_2$, $x_2 = 2X_1 + 4X_2$, $x_3 = X_3$ (b) $x_1 = X_1 + 3X_2$, $x_2 = 2X_1 + 4X_3$, $x_3 = X_3$

3.11 Explicitly verify that the invariants of the right and left Cauchy–Green strain tensors $\boldsymbol{C}$ and $\boldsymbol{B}$ are the same.

3.12 Using the Polar Decomposition Theorem, explicitly justify relations (3.6.11) and $\boldsymbol{R} = \boldsymbol{F}\boldsymbol{U}^{-1} = \boldsymbol{V}^{-1}\boldsymbol{F}$, $\boldsymbol{V} = \boldsymbol{R}\boldsymbol{U}\boldsymbol{R}^{\mathrm{T}}$. Also justify that $\boldsymbol{U}$ and $\boldsymbol{V}$ must have the same principal values.

3.13 For the following motions, calculate the deformation gradient $\boldsymbol{F}$, the rotation tensor $\boldsymbol{R}$, and the right and left stretch tensors $\boldsymbol{U}$ and $\boldsymbol{V}$. For each case, also verify that $\boldsymbol{R}\boldsymbol{R}^{\mathrm{T}} = \boldsymbol{I}$:

$$\text{(a)}\ \begin{aligned} x_1 &= 2X_1 \\ x_2 &= 4X_3 \\ x_3 &= 4X_2 \end{aligned} \quad \text{(b)}\ \begin{aligned} x_1 &= 2X_3 \\ x_2 &= 4X_1 \\ x_3 &= 4X_2 \end{aligned} \quad \text{(c)}\ \begin{aligned} x_1 &= 4X_1 \\ x_2 &= 4X_2 \\ x_3 &= 2X_3 \end{aligned}$$

3.14 For the motions previously given in Exercise 3.10, determine the physical strain components $\gamma^{(i)}$ and $\sin\gamma^{(ij)}$.

3.15 For a unit length reference fiber $d\boldsymbol{X} = \frac{1}{\sqrt{2}}(\boldsymbol{E}_1 + \boldsymbol{E}_2)$, determine the fiber's new length and orientation after simple shearing and extensional deformation given in Examples 3.5.1 and 3.5.2.

3.16 Develop relations (3.6.22) for the physical strains in terms of the Eulerian strains.

3.17 As per the relations developed in Section 3.7, determine the line, area, and volume changes for the motions given in Exercise 3.10.

3.18 Assuming small deformations for the following set of displacements, calculate the strain tensor $\boldsymbol{\varepsilon}$ and the rotation tensor $\boldsymbol{\omega}$: (a) $\boldsymbol{u} = x_1x_2\boldsymbol{e}_1 + 3x_2\boldsymbol{e}_2 + x_3^2\boldsymbol{e}_3$

(b) $\boldsymbol{u} = 2x_1x_3^2\boldsymbol{e}_1 + 2x_1x_2\boldsymbol{e}_2 + 2x_3^3\boldsymbol{e}_3$ (c) $\boldsymbol{u} = 2\frac{x_1}{x_2}\boldsymbol{e}_1 + 4x_2x_3\boldsymbol{e}_2 + 3x_1\boldsymbol{e}_3$

3.19 Show that for small strain, the volume dilatation defined by relation (3.7.5) may be written as $\vartheta = \dfrac{dv - dV}{dV} = \varepsilon_{kk} = I_{\varepsilon}$.

3.20 Justify that the first invariant of the deviatoric small strain tensor is zero. In light of the results from Exercise 3.19, what does the vanishing of the dilatation imply?

3.21 For small strain, show that the principal directions of the strain deviator are the same as those of the strain tensor, and the deviator principal values are $\{\varepsilon_1 - \varepsilon_{kk}/3,\ \varepsilon_2 - \varepsilon_{kk}/3,\ \varepsilon_3 - \varepsilon_{kk}/3\}$.

3.22 Using relation (2.5.5), show that the compatibility relations (3.11.2) with $l = k$ can be expressed by $\eta_{ij} = \varepsilon_{ikl}\varepsilon_{jmp}\varepsilon_{lp,km} = 0$, which can also be written in vector notation as $\boldsymbol{\nabla} \times \boldsymbol{e} \times \boldsymbol{\nabla} = \mathbf{0}$.

3.23 Show that the six compatibility equations (3.11.3) may also be represented by the three independent fourth-order equations

$$\frac{\partial^4 \varepsilon_{11}}{\partial x_2^2 \partial x_3^2} = \frac{\partial^3}{\partial x_1 \partial x_2 \partial x_3}\left(-\frac{\partial \varepsilon_{23}}{\partial x_1} + \frac{\partial \varepsilon_{31}}{\partial x_2} + \frac{\partial \varepsilon_{12}}{\partial x_3}\right)$$

$$\frac{\partial^4 \varepsilon_{22}}{\partial x_3^2 \partial x_1^2} = \frac{\partial^3}{\partial x_1 \partial x_2 \partial x_3}\left(-\frac{\partial \varepsilon_{31}}{\partial x_2} + \frac{\partial \varepsilon_{12}}{\partial x_3} + \frac{\partial \varepsilon_{23}}{\partial x_1}\right)$$

$$\frac{\partial^4 \varepsilon_{33}}{\partial x_1^2 \partial x_2^2} = \frac{\partial^3}{\partial x_1 \partial x_2 \partial x_3}\left(-\frac{\partial \varepsilon_{12}}{\partial x_3} + \frac{\partial \varepsilon_{23}}{\partial x_1} + \frac{\partial \varepsilon_{31}}{\partial x_2}\right)$$

3.24 Determine if the following strain fields are compatible (A and B constants):

(a) $\varepsilon_{11} = Ax_2x_3,\quad \varepsilon_{22} = -Ax_1x_3,\quad \varepsilon_{33} = Bx_3,\quad \varepsilon_{12} = \varepsilon_{13} = \varepsilon_{23} = 0$

(b) $\varepsilon_{11} = Ax_2^3,\quad \varepsilon_{22} = Ax_1^3,\quad \varepsilon_{12} = \dfrac{3}{2}Ax_1x_2(x_1 + x_2),\quad \varepsilon_{33} = \varepsilon_{13} = \varepsilon_{23} = 0$

(c) $\varepsilon_{12} = A\dfrac{x_1}{x_1^2 + x_2^2},\quad \varepsilon_{13} = -A\dfrac{x_2}{x_1^2 + x_2^2},\quad \varepsilon_{11} = \varepsilon_{22} = \varepsilon_{33} = \varepsilon_{23} = 0$

3.25 Consider a two-dimensional small deformation case with strains given by $\varepsilon_{11} = Ax_1, \varepsilon_{22} = Ax_2, \varepsilon_{12} = Bx_1$. Following the integration scheme demonstrated in Example 3.11.2, determine the in-plane displacements u_1 and u_2.

3.26 Show for the two-dimensional small deformation case, the displacements $u_1 = -\omega_o x_2 + u_o, u_2 = \omega_o x_1 + v_o$ (ω_o, u_o, v_o all constants) represent rigid body motion and will yield zero strains.

3.27 Determine the following rate variables: $\boldsymbol{v}(\boldsymbol{X},t)$, $\boldsymbol{v}(\boldsymbol{x},t)$, $\boldsymbol{L}$, $\boldsymbol{D}$, $\boldsymbol{W}$, $\boldsymbol{w}$, $\dot{\boldsymbol{E}}$, $\dot{J}$ for the following material motions:

(a) $x_1 = 2t^2X_1 + 4tX_2,\quad x_2 = 4X_2,\quad x_3 = 2tX_3$

(b) $x_1 = 2tX_1 + 4t^2X_2,\quad x_2 = 4X_2 - 2tX_1,\quad x_3 = X_3$

(c) $x_1 = 2tX_1 + 4X_2,\quad x_2 = 4t^2X_2 - 2tX_1,\quad x_3 = 6t^2X_3$

3.28 Justify that the objective time derivative for a second-order tensor is given by (3.14.6)

$$\overset{\circ}{\boldsymbol{A}} = \dot{\boldsymbol{A}} - \boldsymbol{W}\boldsymbol{A} + \boldsymbol{A}\boldsymbol{W}$$

3.29 Apply the objectivity test to the relative right Cauchy–Green strain tensor to show that

$$\boldsymbol{C}_t^*(\tau) = \boldsymbol{Q}\boldsymbol{C}_t(\tau)\boldsymbol{Q}^{\mathrm{T}}$$

Next explore the corresponding result for the case of the left Cauchy–Green strain $\boldsymbol{B}_t(\tau)$.

3.30 For the following time-dependent motions, determine the relative deformation gradient and the relative right and left Cauchy–Green strain tensors

$$\text{(a)}\ \begin{array}{l} x_1 = 2X_1 + 4tX_2 \\ x_2 = 4X_2 \\ x_3 = 2tX_3 \end{array} \qquad \text{(b)}\ \begin{array}{l} x_1 = 2tX_1 + X_2 \\ x_2 = 4X_2 - 2tX_1 \\ x_3 = X_3 \end{array} \qquad \text{(c)}\ \begin{array}{l} x_1 = 2tX_1 + 4X_2 \\ x_2 = -4X_2 + 2tX_1 \\ x_3 = 6t^2X_3 \end{array}$$

3.31 Using Cartesian coordinates for reference and cylindrical coordinates for the current configuration, consider the finite deformation defined by

$$r^2 = 2AX, \quad \theta = BY, \quad z = \frac{Z}{AB}$$

where A and B are arbitrary constants. First determine the deformation geometry of where a reference rectangular domain: $X_1 \le X \le X_2, Y_1 \le Y \le Y_2$ gets mapped into the current configuration. Next, using the appropriate form from Section 3.17, determine the deformation gradient for this case and show that the motion is isochoric.

CHAPTER

4 Force and Stress

Chapter 3 investigated the kinematics of continuum deformation regardless of the force or stress distribution that produced the motion or deformation. In this chapter, we now wish to examine how these forces and stresses can be quantitatively described. Following the classical continuum mechanics model, we assume a continuously distributed internal force system composed of body and surface forces. Each of these will be associated with a continuous density function that will represent the force per unit volume or per unit surface area. For surface forces, this will lead to the definition and use of the *stress or traction vector* and *stress tensor*. Each of these provides a quantitative method to describe both boundary and internal force distributions within a continuum. Since stress is related to force per unit area, we will have to keep track of whether we wish to use reference or current areas for large deformation problems. Stress is very important in continuum mechanics applications, because many materials exhibit some type of failure condition based on this variable. It should be noted that the developments in this chapter will not require a material constitutive assumption and thus they will apply to a broad class of material behavior.

4.1 BODY AND SURFACE FORCES

When a continuum material is subjected to applied external loadings, internal forces are induced inside the body. Following our continuum philosophy, we assume that these internal forces are distributed continuously within the material. In order to study such forces, it is convenient to categorize them into two major groups, commonly referred to as *body forces* and *surface forces*.

Body forces are proportional to the body's mass and are reacted with an agent *outside* the body. Examples of these include gravitational-weight forces, magnetic forces, and inertial forces. Fig. 4.1A shows an example body force of an object's self-weight producing a particular deformation. Using continuum mechanics principles, a *body force density* (force per unit mass) $\boldsymbol{b}(\boldsymbol{x}, t)$ can be defined such that the total resultant body force $\boldsymbol{F}_R$ of an entire object can be written as a volume integral over the body B:

$$\boldsymbol{F}_R = \iiint_B \rho \boldsymbol{b}(\boldsymbol{x}, t)\, dV \tag{4.1.1}$$

where ρ is the mass density of the material. Later in the text, we will often use the *body force per unit volume* defined by $\boldsymbol{F} = \rho \boldsymbol{b}$.

Continuum Mechanics Modeling of Material Behavior. http://dx.doi.org/10.1016/B978-0-12-811474-2.00004-6

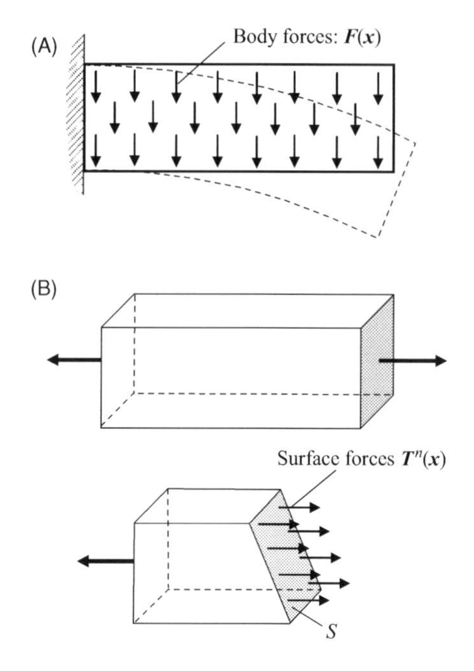

FIGURE 4.1

Examples of body and surfaces forces: (A) cantilever beam under self-weight loading; (B) sectioned axially loaded structure.

Surface forces always act on a surface and result from physical contact with another body. Fig. 4.1B illustrates surface forces existing in a structural solid section that has been created by sectioning the body into two pieces. For this particular case, the surface S is a virtual one in the sense that it was artificially created to investigate the nature of the internal forces at this location in the body. Again, the resultant surface force $\boldsymbol{F}_S$ over an entire surface S can be expressed as the integral of a *surface force density function* $\boldsymbol{t}(\boldsymbol{x}, t)$:

$$\boldsymbol{F}_S = \iint_S \boldsymbol{t}(\boldsymbol{x},t)\,dS \tag{4.1.2}$$

The surface force density is normally referred to as the *traction or stress vector* and will be discussed in more detail in the next section. We assume *a priori* that the body force density $\boldsymbol{b}$ and the traction $\boldsymbol{t}$ are objective vectors. These density variables can be functions of both position and time, but in the future we will generally drop writing the time dependence.

In the development of classical continuum mechanics, distributions of body or surface couples are normally not included. Theories which include such force system distributions have been constructed in an effort to extend the classical theory for applications in micromechanical modeling. Such approaches include *micropolar* or *couple-stress theory* (see Eringen, 1968) and will be further explored in Chapter 9.

4.2 CAUCHY STRESS PRINCIPLE: STRESS VECTOR

In order to quantify the nature of the internal distribution of forces within a continuum, consider a general body subject to arbitrary (concentrated and distributed) external loadings as shown in Fig. 4.2. For now, we will assume that the body is in its current deformed configuration. To investigate the internal forces, a section is made through the body as shown. On this section, consider a small area Δa with a unit normal vector $\boldsymbol{n}$. In general, the resultant of the surface forces acting on Δa would be a force $\Delta \boldsymbol{F}$ and a moment $\Delta \boldsymbol{M}$. We now wish to determine the pointwise value of the force per unit area and thus the *stress* or *traction vector* is defined by

$$\boldsymbol{t}(\boldsymbol{x}, t, \boldsymbol{n}) = \lim_{\Delta a \to 0} \frac{\Delta \boldsymbol{F}}{\Delta a} \tag{4.2.1}$$

For the resultant moment case, we assume that in the limit

$$\lim_{\Delta a \to 0} \frac{\Delta \boldsymbol{M}}{\Delta a} = 0 \tag{4.2.2}$$

which is consistent with our earlier discussion that no resultant surface couple will be included in classical continuum mechanics.

Notice that the traction vector depends on both the spatial location and the unit normal vector to the surface under study. Thus, even though we may be investigating the same point, the traction vector will still vary as a function of the orientation of the surface normal. This concept is often called the *Cauchy Stress Principle* which can be stated as *the stress or traction vector at any given place and time has a same value*

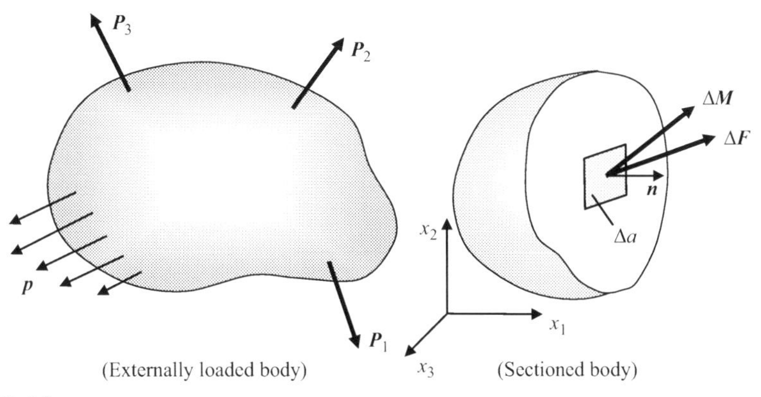

FIGURE 4.2

Stress vector concept.

on all parts of material having a common tangent plane and lying on the same side of it. Note also the expected action–reaction principle (Newton's third law) which is often expressed as

$$t(\boldsymbol{x},\boldsymbol{n}) = -\boldsymbol{t}(\boldsymbol{x},-\boldsymbol{n}) \tag{4.2.3}$$

4.3 CAUCHY STRESS TENSOR

Consider now the special case where the small area Δa (shown in Fig. 4.2) coincides with each of the three coordinate planes with unit normal vectors pointing along the positive coordinate axes. This concept is shown in Fig. 4.3 where the three coordinate surfaces partition off a cube of material. For this case, the traction vector on each face can be written as

$$\begin{aligned}
\boldsymbol{t}(\boldsymbol{x},\boldsymbol{n}=\boldsymbol{e}_1) &= T_{11}\boldsymbol{e}_1 + T_{12}\boldsymbol{e}_2 + T_{13}\boldsymbol{e}_3 \\
\boldsymbol{t}(\boldsymbol{x},\boldsymbol{n}=\boldsymbol{e}_2) &= T_{21}\boldsymbol{e}_1 + T_{22}\boldsymbol{e}_2 + T_{23}\boldsymbol{e}_3 \\
\boldsymbol{t}(\boldsymbol{x},\boldsymbol{n}=\boldsymbol{e}_3) &= T_{31}\boldsymbol{e}_1 + T_{32}\boldsymbol{e}_2 + T_{33}\boldsymbol{e}_3
\end{aligned} \tag{4.3.1}$$

where $\boldsymbol{e}_1$, $\boldsymbol{e}_2$, and $\boldsymbol{e}_3$ are the coordinate unit vectors, and the nine quantities T_{11}, T_{22}, T_{33}, T_{12}, T_{21}, T_{23}, T_{32}, T_{31}, and T_{13} are the components of the traction vector on each of the three coordinate planes as illustrated. These nine components are called the *Cauchy stress components*, with T_{11}, T_{22}, and T_{33} being referred to as *normal stresses* and T_{12}, T_{21}, T_{23}, T_{32}, T_{31}, and T_{13} called the *shearing stresses*. The components of stress T_{ij} are commonly written in a matrix format as

$$T_{ij} = \begin{bmatrix} T_{11} & T_{12} & T_{13} \\ T_{21} & T_{22} & T_{23} \\ T_{31} & T_{32} & T_{33} \end{bmatrix} \tag{4.3.2}$$

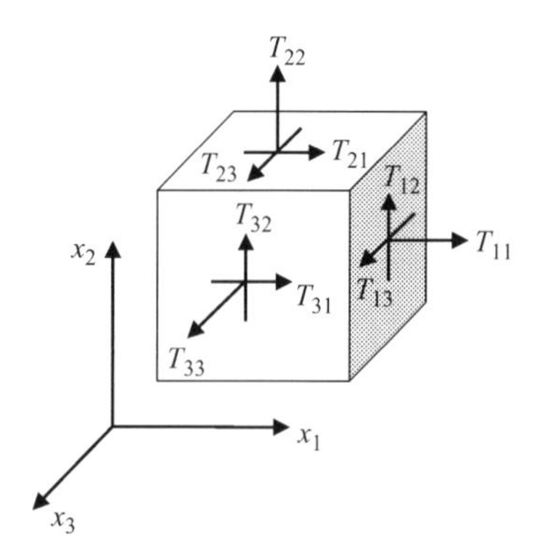

FIGURE 4.3

Traction components on coordinate surfaces—Cauchy stresses.

We will shortly show that this stress is a second-order tensor obeying the appropriate transformation law $(2.8.1)_3$. Since it represents the force per unit area in the deformed or current configuration, T_{ij} is sometimes referred to as the *true stress*. It should be mentioned that in Chapter 5, we will formally prove that the Cauchy stress is a symmetric tensor.

The positive directions of each stress component are illustrated in Fig. 4.3. Regardless of the coordinate system, positive normal stress always acts in tension out of the face, and only one subscript would be necessary for definition since it always acts normal to the surface. The shear stress, however, requires two subscripts, the first representing the plane of action and the second designating the direction of the stress. Similar to shear strain, the sign of the shear stress depends on coordinate system orientation. For example, on a plane with a normal in the positive x_1-direction, positive T_{12} acts in the positive x_2-direction. Similar definitions would follow for the other shear stress components. Proper formulation of continuum mechanics problems will require knowledge of these basic definitions, directions, and sign conventions for particular stress components. Notice that relations (4.3.1) imply that on coordinate boundary surfaces, the components of the traction vector are directly related to particular stress components. This fact will simplify certain boundary conditions and thus lead to a more direct problem solution.

Consider next the tetrahedron element bounded by an oblique plane of arbitrary orientation and three coordinate planes as shown in Fig. 4.4. Appropriate traction loadings exist on each of these surfaces. The unit normal to the oblique surface is given by

$$\boldsymbol{n} = n_1\boldsymbol{e}_1 + n_2\boldsymbol{e}_2 + n_3\boldsymbol{e}_3 \tag{4.3.3}$$

where n_1, n_2, and n_3 are the direction cosines of the unit vector $\boldsymbol{n}$ relative to the given coordinate system. The area of the oblique surface will be denoted by da and thus the areas of each of the coordinate surfaces are given by n_1da, n_2da, and n_3da. We

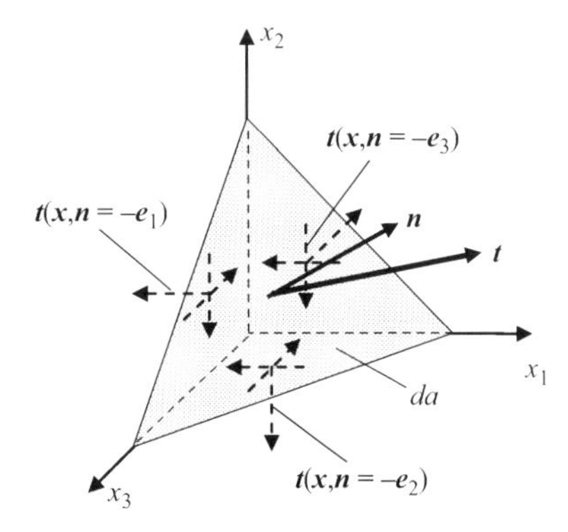

FIGURE 4.4

Tractions on tetrahedron element.

now consider the dynamic equilibrium of the element interior to the oblique and coordinate planes and take the limit as the volume shrinks to zero. Invoking Newton's law, the force balance between the various surface tractions must equal the element's mass times acceleration:

$$\boldsymbol{t}\,da - n_1\,da\,\boldsymbol{t}(\boldsymbol{n}=\boldsymbol{e}_1) - n_2\,da\,\boldsymbol{t}(\boldsymbol{n}=\boldsymbol{e}_2) - n_3\,da\,\boldsymbol{t}(n=\boldsymbol{e}_3) = m\boldsymbol{a} \tag{4.3.4}$$

Note that we have used the fact that $\boldsymbol{t}(\boldsymbol{n}=-\boldsymbol{e}_i)=-\boldsymbol{t}(\boldsymbol{n}=\boldsymbol{e}_i)$. Since the elements mass is proportional to $(da)^{3/2}$, in the limit as element volume goes to zero ($da \to 0$), the inertia term on the right-hand side of relation (4.3.4) will vanish. Thus, using relations (4.3.1), we can solve for the traction vector in terms of the stress components

$$\begin{aligned}\boldsymbol{t} &= (T_{11}n_1 + T_{21}n_2 + T_{31}n_3)\boldsymbol{e}_1\\ &+ (T_{12}n_1 + T_{22}n_2 + T_{32}n_3)\boldsymbol{e}_2\\ &+ (T_{13}n_1 + T_{23}n_2 + T_{33}n_3)\boldsymbol{e}_3\end{aligned} \tag{4.3.5}$$

or in index and direct notation

$$t_i = T_{ji}n_j, \quad \boldsymbol{t} = \boldsymbol{T}^T\boldsymbol{n} \tag{4.3.6}$$

Relations (4.3.5) and (4.3.6) thus indicate that the stress or traction vector is linearly related to the unit normal vector through a linear transformation using the transpose of the stress tensor. This result is often called the *Cauchy stress formula*. This result provides a simple and direct method to calculate the forces on oblique planes and surfaces and will prove to be very useful to specify general boundary conditions during the formulation and solution of solid mechanics problems.

Next, let us formally show that the Cauchy stress is a second-order tensor. We start with the assumption that the traction and normal vectors are first-order tensors and thus $t'_i = Q_{ij}t_j$ and $n'_i = Q_{ij}n_j$. Starting from (4.3.6) in the primed frame

$$\begin{aligned}&t'_i = T'_{ji}n'_j \Rightarrow Q_{ij}t_j = T'_{ji}Q_{jk}n_k \Rightarrow Q_{ij}T_{kj}n_k = T'_{ji}Q_{jk}n_k \Rightarrow\\ &Q_{im}(Q_{ij}T_{kj} - T'_{ji}Q_{jk})n_k = 0 \Rightarrow T_{km} = Q_{jk}Q_{im}T'_{ji}\end{aligned}$$

The final expression is just the standard tensor transformation law for second-order tensors.

The distinction between the traction vector and stress tensor should be carefully understood. Although each quantity has the same units of force per unit area, they are fundamentally different since the traction is a vector, whereas the stress is a second-order tensor. Components of traction can be defined on any surface, but particular stress components only exist on coordinate surfaces as shown in Fig. 4.3 for the Cartesian case. Eq. (4.3.6) establishes the relation between the two variables, thereby indicating that each traction component can be expressed as a linear combination of particular stress components. Further discussion on this topic will be given later in Chapter 6 when boundary condition development is presented.

EXAMPLE 4.3.1 FOR THE GIVEN STATE OF STRESS $T_{ij}=\begin{bmatrix}1 & 2 & 2\\ 2 & 3 & 4\\ 2 & 4 & 6\end{bmatrix}$, DETERMINE THE TRACTION VECTOR ON A PLANES WITH UNIT NORMAL VECTORS $n^{(1)}=\frac{1}{\sqrt{3}}(e_1+e_2+e_3)$ AND $n^{(2)}=\frac{1}{\sqrt{2}}(e_1+e_2)$.

Solution: Using (4.3.6)

$$t_i^{(1)}=T_{ji}n_j^{(1)}=\frac{1}{\sqrt{3}}\begin{bmatrix}1 & 2 & 2\\ 2 & 3 & 4\\ 2 & 4 & 6\end{bmatrix}\begin{bmatrix}1\\ 1\\ 1\end{bmatrix}=\frac{1}{\sqrt{3}}\begin{bmatrix}5\\ 9\\ 12\end{bmatrix},$$

$$t_i^{(2)}=T_{ji}n_j^{(2)}=\frac{1}{\sqrt{2}}\begin{bmatrix}1 & 2 & 2\\ 2 & 3 & 4\\ 2 & 4 & 6\end{bmatrix}\begin{bmatrix}1\\ 1\\ 0\end{bmatrix}=\frac{1}{\sqrt{2}}\begin{bmatrix}3\\ 5\\ 6\end{bmatrix}$$

The Cauchy stress components follow the standard transformation rules for second-order tensors established in Section 2.8. Applying transformation relation $(2.8.1)_3$ for the stress gives

$$T'_{ij}=Q_{ip}Q_{jq}T_{pq} \tag{4.3.7}$$

where the rotation matrix $Q_{ij}=\cos(x'_i,x_j)$. Therefore, given the stress in one coordinate system, we can determine the new components in any other rotated system. This will be a very handy tool for many problems in continuum mechanics. For the general three-dimensional case, the rotation matrix may be chosen in the form

$$Q_{ij}=\begin{bmatrix}l_1 & m_1 & n_1\\ l_2 & m_2 & n_2\\ l_3 & m_3 & n_3\end{bmatrix} \tag{4.3.8}$$

Using this notational scheme, the specific transformation relations for the stress then become

$$\begin{aligned}
T'_{11}&=T_{11}l_1^2+T_{22}m_1^2+T_{33}n_1^2+2(T_{12}l_1m_1+T_{23}m_1n_1+T_{31}n_1l_1)\\
T'_{22}&=T_{11}l_2^2+T_{22}m_2^2+T_{33}n_2^2+2(T_{12}l_2m_2+T_{23}m_2n_2+T_{31}n_2l_2)\\
T'_{33}&=T_{11}l_3^2+T_{22}m_3^2+T_{33}n_3^2+2(T_{12}l_3m_3+T_{23}m_3n_3+T_{31}n_3l_3)\\
T'_{12}&=T_{11}l_1l_2+T_{22}m_1m_2+T_{33}n_1n_2+T_{12}(l_1m_2+m_1l_2)+T_{23}(m_1n_2+n_1m_2)+T_{31}(n_1l_2+l_1n_2)\\
T'_{23}&=T_{11}l_2l_3+T_{22}m_2m_3+T_{33}n_2n_3+T_{12}(l_2m_3+m_2l_3)+T_{23}(m_2n_3+n_2m_3)+T_{31}(n_2l_3+l_2n_3)\\
T'_{31}&=T_{11}l_3l_1+T_{22}m_3m_1+T_{33}n_3n_1+T_{12}(l_3m_1+m_3l_1)+T_{23}(m_3n_1+n_3m_1)+T_{31}(n_3l_1+l_3n_1)
\end{aligned} \tag{4.3.9}$$

For the two-dimensional case, the transformation matrix reduces to

$$Q_{ij}=\begin{bmatrix}\cos\theta & \sin\theta\\ -\sin\theta & \cos\theta\end{bmatrix} \tag{4.3.10}$$

with θ being the counterclockwise angle defined in Fig. 2.9. Under this transformation, the in-plane stress components transform according to

$$\begin{aligned} T'_{11} &= T_{11}\cos^2\theta + T_{22}\sin^2\theta + 2T_{12}\sin\theta\cos\theta \\ T'_{22} &= T_{11}\sin^2\theta + T_{22}\cos^2\theta - 2T_{12}\sin\theta\cos\theta \\ T'_{12} &= -T_{11}\sin\theta\cos\theta + T_{22}\sin\theta\cos\theta + T_{12}(\cos^2\theta - \sin^2\theta) \end{aligned} \tag{4.3.11}$$

which is commonly rewritten in terms of the double angle

$$\begin{aligned} T'_{11} &= \frac{T_{11}+T_{22}}{2} + \frac{T_{11}-T_{22}}{2}\cos 2\theta + T_{12}\sin 2\theta \\ T'_{22} &= \frac{T_{11}+T_{22}}{2} - \frac{T_{11}-T_{22}}{2}\cos 2\theta - T_{12}\sin 2\theta \\ T'_{12} &= \frac{T_{22}-T_{11}}{2}\sin 2\theta + T_{12}\cos 2\theta \end{aligned} \tag{4.3.12}$$

Both two- and three-dimensional stress transformation equations can be easily incorporated within MATLAB to provide numerical solution to problems of interest.

4.4 PRINCIPAL STRESSES AND AXES FOR CAUCHY STRESS TENSOR

We can again use the previous developments from Section 2.11 to discuss the issues of principal stresses and directions. As mentioned, it will be shown later that the Cauchy stress is a symmetric tensor. Using this fact, appropriate theory has been developed to determine the principal axes and values for the stress. For any given Cauchy stress tensor, we can establish the principal value problem and solve the characteristic equation to explicitly determine the principal values and directions. The general characteristic equation for this case becomes

$$\det[T_{ij} - T\delta_{ij}] = -T^3 + I_T T^2 - II_T T + III_T = 0 \tag{4.4.1}$$

where T are the *principal stresses* and the fundamental invariants are given by

$$\begin{aligned} I_T &= T_{ii} = T_{11} + T_{22} + T_{33} \\ II_T &= \frac{1}{2}(T_{ii}T_{jj} - T_{ij}T_{ij}) = \begin{vmatrix} T_{11} & T_{12} \\ T_{21} & T_{22} \end{vmatrix} + \begin{vmatrix} T_{22} & T_{23} \\ T_{32} & T_{33} \end{vmatrix} + \begin{vmatrix} T_{11} & T_{13} \\ T_{31} & T_{33} \end{vmatrix} \\ III_T &= \det \boldsymbol{T} \end{aligned} \tag{4.4.2}$$

In the principal coordinate system, the stress matrix will take the special diagonal form

$$T_{ij} = \begin{bmatrix} T_1 & 0 & 0 \\ 0 & T_2 & 0 \\ 0 & 0 & T_3 \end{bmatrix} \tag{4.4.3}$$

The fundamental invariants can be expressed in terms of the three principal stressess T_1, T_2, and T_3 as

$$\begin{aligned} I_T &= T_1 + T_2 + T_3 \\ II_T &= T_1T_2 + T_2T_3 + T_3T_1 \\ III_T &= T_1T_2T_3 \end{aligned} \tag{4.4.4}$$

A comparison of the general and principal stress states are shown in Fig. 4.5. Notice that for the principal coordinate system, all shearing stresses will vanish and thus the state will only have normal stresses.

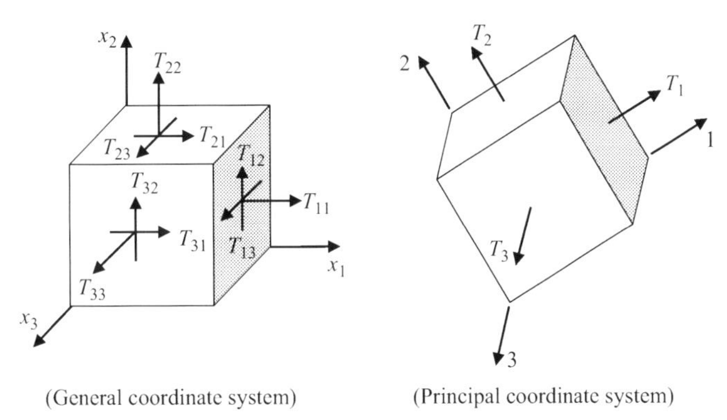

FIGURE 4.5

Comparison of general and principal stress states.

EXAMPLE 4.4.1 CAUCHY STRESS PRINCIPAL VALUE PROBLEM

Determine the principal stresses and directions for the given state of stress,

$$T_{ij} = \begin{bmatrix} 3 & 1 & 1 \\ 1 & 0 & 2 \\ 1 & 2 & 0 \end{bmatrix}.$$

Solution: The principal stress problem is started by calculating the three invariants, giving the result $I_T = 3$, $II_T = -6$, $III_T = -8$. This yields the following characteristic equation:

$$-T^3 + 3T^2 + 6T - 8 = 0$$

The roots of this equation are found to be $T = 4, 1, -2$. Back-substituting the first root into the fundamental system (2.11.1) gives

$$\begin{aligned} -n_1^{(1)} + n_2^{(1)} + n_3^{(1)} &= 0 \\ n_1^{(1)} - 4n_2^{(1)} + 2n_3^{(1)} &= 0 \\ n_1^{(1)} + 2n_2^{(1)} - 4n_3^{(1)} &= 0 \end{aligned}$$

Solving this system, the normalized principal direction is found to be $\boldsymbol{n}^{(1)} = (2, 1, 1)/\sqrt{6}$. In a similar fashion, the other two principal directions are $\boldsymbol{n}^{(2)} = (-1, 1, 1)/\sqrt{3}$, $\boldsymbol{n}^{(3)} = (0, -1, 1)/\sqrt{2}$.

We now wish to go back to investigate another issue related to stress and traction transformation that makes use of principal stresses. Consider the general traction vector $\boldsymbol{t}$ acting on an arbitrary surface as shown in Fig. 4.6. The issue of interest is to determine the traction vector's normal and shear components N and S. The normal component is simply the traction's projection in the direction of the unit normal vector $\boldsymbol{n}$, whereas the shear component is found by Pythagorean theorem:

$$N = \boldsymbol{t} \cdot \boldsymbol{n}$$
$$S = \left(|\boldsymbol{t}|^2 - N^2\right)^{1/2} \tag{4.4.5}$$

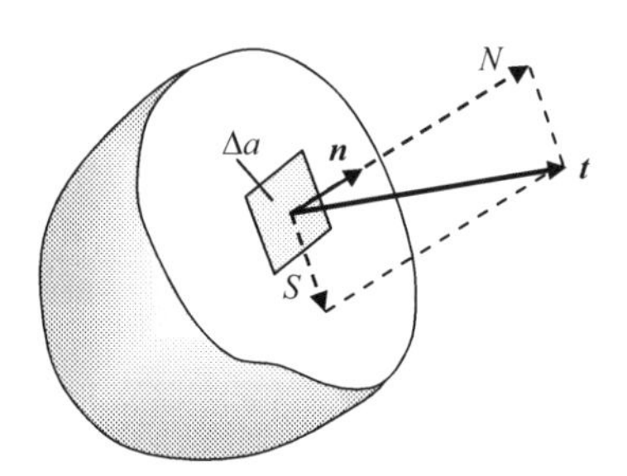

FIGURE 4.6

Traction vector decomposition.

Using the relationship for the traction vector (4.3.6) into $(4.4.5)_1$ gives

$$\begin{aligned} N &= \boldsymbol{t} \cdot \boldsymbol{n} = t_i n_i = T_{ji} n_j n_i \\ &= T_1 n_1^2 + T_2 n_2^2 + T_3 n_3^2 \end{aligned} \tag{4.4.6}$$

where in order to simplify the expressions, we have used the principal axes for the stress tensor. In a similar manner,

$$\begin{aligned} |\boldsymbol{t}|^2 &= \boldsymbol{t} \cdot \boldsymbol{t} = t_i t_i = T_{ji} n_j T_{ki} n_k \\ &= T_1^2 n_1^2 + T_2^2 n_2^2 + T_3^2 n_3^2 \end{aligned} \tag{4.4.7}$$

Using these results in relations (4.4.5) yields

$$\begin{aligned} N &= T_1 n_1^2 + T_2 n_2^2 + T_3 n_3^2 \\ S^2 + N^2 &= T_1^2 n_1^2 + T_2^2 n_2^2 + T_3^2 n_3^2 \end{aligned} \tag{4.4.8}$$

In addition, we also add the condition that the vector $\boldsymbol{n}$ has unit magnitude

$$1 = n_1^2 + n_2^2 + n_3^2 \tag{4.4.9}$$

Relations (4.4.8) and (4.4.9) can be viewed as three linear algebraic equations for the unknowns n_1^2, n_2^2, n_3^2. Solving this system gives the following result:

$$\begin{aligned}
n_1^2 &= \frac{S^2 + (N - T_2)(N - T_3)}{(T_1 - T_2)(T_1 - T_3)} \\
n_2^2 &= \frac{S^2 + (N - T_3)(N - T_1)}{(T_2 - T_3)(T_2 - T_1)} \\
n_3^2 &= \frac{S^2 + (N - T_1)(N - T_2)}{(T_3 - T_1)(T_3 - T_2)}
\end{aligned} \tag{4.4.10}$$

Without loss in generality, we can rank the principal stresses as $T_1 > T_2 > T_3$. Noting that the expressions given by (4.4.10) must be greater than or equal to zero, we can conclude the following:

$$\begin{aligned}
S^2 + (N - T_2)(N - T_3) &\geq 0 \\
S^2 + (N - T_3)(N - T_1) &\leq 0 \\
S^2 + (N - T_1)(N - T_2) &\geq 0
\end{aligned} \tag{4.4.11}$$

For the equality case, Eqs. (4.4.11) represent three circles in an S–N coordinate system, and Fig. 4.7 illustrates the location of each circle. These results were originally generated by Otto Mohr over a century ago, and the circles are commonly called *Mohr's circles of stress*. The three inequalities given in (4.4.11) imply that all admissible values of N and S lie in the shaded regions bounded by the three circles. Note that for the ranked principal stresses, the largest shear component is easily determined as $S_{max} = ½|T_1 - T_3|$. Although these circles can be effectively used for two-dimensional stress transformation, the general tensorial-based equations (4.3.7) or (4.3.9) are normally used for general transformation computations.

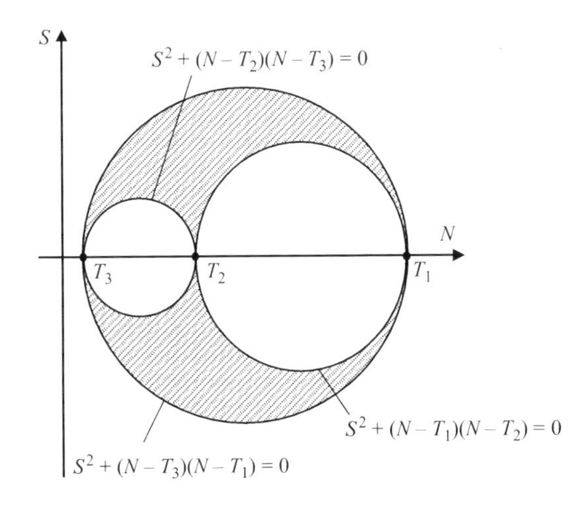

FIGURE 4.7

Mohr's circles of stress.

4.5 SPHERICAL, DEVIATORIC, OCTAHEDRAL, AND VON MISES STRESS

Similar to our earlier discussion on strain, it is often convenient to decompose the stress into two parts called the *spherical* and *deviatoric stress tensors*. Analogous to relations given in Chapter 3, the spherical Cauchy stress is defined by

$$\tilde{T}_{ij} = \frac{1}{3} T_{kk} \delta_{ij} \tag{4.5.1}$$

whereas the deviatoric stress becomes

$$\hat{T}_{ij} = T_{ij} - \frac{1}{3} T_{kk} \delta_{ij} \tag{4.5.2}$$

Note that the total stress is then simply the sum

$$T_{ij} = \tilde{T}_{ij} + \hat{T}_{ij} \tag{4.5.3}$$

The spherical stress is an isotropic tensor, being the same in all coordinate systems (as per discussion in Section 2.15). Using the fact that the characteristic equation for the deviatoric stress reduces to

$$\hat{T}^3 + II_{\hat{T}} \hat{T} - III_{\hat{T}} = 0 \tag{4.5.4}$$

it can be shown that the principal directions of the deviatoric stress are the same as those of the stress tensor itself (see Exercise 4.13).

We next briefly explore a couple of particular stress components or combinations that have been defined in the literature and are commonly used in formulating failure theories related to inelastic deformation. It has been found that ductile materials normally exhibit inelastic yielding failures that can be characterized by these particular stresses. Consider first the normal and shear stresses (tractions) that act on a special plane whose normal makes equal angles with the three principal axes. This plane is commonly referred to as the *octahedral plane*. Determination of these stresses is straight forward if we use the principal axes of stress. Since the unit normal vector to the octahedral plane makes equal angles with the principal axes, its components are given by $n_i = \pm(1,1,1)/\sqrt{3}$. Referring to Fig. 4.6 and using the results of the previous section, relations (4.4.8) give the desired normal and shear stresses as

$$\begin{aligned} N = \sigma_{oct} &= \frac{1}{3}(T_1 + T_2 + T_3) = \frac{1}{3} T_{kk} = \frac{1}{3} I_T \\ S = \tau_{oct} &= \frac{1}{3}\left[(T_1 - T_2)^2 + (T_2 - T_3)^2 + (T_3 - T_1)^2\right]^{1/2} \\ &= \frac{1}{3}\left(2I_T^2 - 6II_T\right)^{1/2} \end{aligned} \tag{4.5.5}$$

It can be shown that the octahedral shear stress τ_{oct} is directly related to the *distortional strain energy* (defined in Chapter 6) which is often used in failure theories for ductile materials.

Another specially defined stress also related to the distortional strain energy failure criteria is known as the *effective* or *von Mises stress* and is given by the expression

$$\begin{aligned}\sigma_e = \sigma_{von\ Mises} &= \sqrt{\frac{3}{2}\hat{T}_{ij}\hat{T}_{ij}} \\ &= \frac{1}{\sqrt{2}}\left[(T_{11}-T_{22})^2+(T_{22}-T_{33})^2+(T_{33}-T_{11})^2+6\left(T_{12}^2+T_{23}^2+T_{31}^2\right)\right]^{1/2} \\ &= \frac{1}{\sqrt{2}}\left[(T_1-T_2)^2+(T_2-T_3)^2+(T_3-T_1)^2\right]^{1/2}\end{aligned} \quad (4.5.6)$$

Note that although the von Mises stress is not really a particular stress or traction component in the usual sense, it is directly related to the octahedral shear stress by the relation $\sigma_e = \left(3/\sqrt{2}\right)\tau_{oct}$. If at some point in the structure, the von Mises stress equals the tensile yield stress, then the material is considered to be at the failure condition. Because of this fact, many finite-element computer codes commonly plot von Mises stress distributions based on the numerically generated stress field. It should be noted that the von Mises and octahedral shear stresses involve only the *differences* in the principal stresses and not the individual values. Thus, increasing each principal stress by the same amount will not change the value of σ_e or τ_{oct}. This result also implies that these values are independent of the hydrostatic stress. The interested reader on failure theories is referred to Ugural and Fenster (2003) for further details on this topic. It should be pointed out that the spherical, octahedral, and von Mises stresses are all expressible in terms of the stress invariants and thus are independent of the coordinate system used to calculate them.

4.6 STRESS DISTRIBUTIONS AND CONTOUR LINES

Over the years, the stress analysis community has developed a large variety of schemes to help visualize and understand the nature of the stress distribution in solids. While stress has been the common pursuit, similar graphical analysis has also been made of strain and displacement distribution. Much of this effort is aimed at determining the magnitude and location of maximum stresses within the structure. Simple schemes involve just plotting the distribution of particular stress components along chosen directions within the body under study. Other methods focus on constructing contour plots of principal stress, maximum shear stress, von Mises stress, and other stress variables or combinations. Some techniques have been constructed to compare with optical experimental methods that provide photographic data of particular stress variables (Shukla and Dally, 2010).

We now will briefly explore some of these schemes as they relate to two-dimensional *plane stress distributions* defined in the x_1–x_2 plane by the field:

$$T_{11} = T_{11}(x_1, x_2),\quad T_{22} = T_{22}(x_1, x_2),\quad T_{12} = T_{12}(x_1, x_2),\quad T_{13} = T_{23} = T_{33} = 0 \quad (4.6.1)$$

Note for this case, the principal stresses and maximum shear stress are given in Exercise 4.6. By passing polarized light through transparent model samples under load, the experimental method of *photoelasticity* can provide full field photographic stress data of particular stress combinations. The method can generate *isochromatic* fringe patterns that represent lines of constant difference in the principal stresses, that is, $T_1 - T_2 =$ constant, which would also be lines of maximum shearing stress. An example of an isochromatic fringe pattern is shown in Fig. 4.8 which illustrates a disk under opposite diametrical compressive loadings. Photoelasticity can also generate another series of fringe lines called *isoclinics* along which the principal stresses have a constant orientation. Still another set of contour lines often used in optical experimental stress analysis are *isopachic* contours, which are lines of $T_{11} + T_{22} = T_1 + T_2 =$ constant. These contours are related to the out-of-plane strain and displacement.

An additional, useful set of lines are *isostatics*; they are also referred to as *stress trajectories*. Such lines are oriented along the direction of a particular principal stress. For the two-dimensional plane stress case, the principal stresses T_1 and T_2 give rise to two families of stress trajectories that form an orthogonal network composed of lines free of shear stress. These trajectories have proved to be useful aids for understanding load paths, that is, how external loadings move through a solid continuum structure to the reaction points (Kelly and Tosh, 2000). Stress trajectories are also related to structural optimization, and Michell structures composed of frameworks of continuous members in tension and compression (Dewhurst, 2001). Additional two-dimensional contour plots of stress function derivatives have also been proposed in the literature (Rathkjen, 1997).

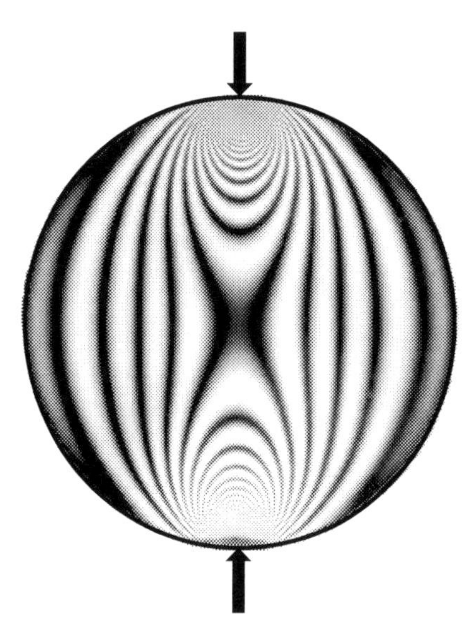

(Courtesy of Dynamic Photomechanics Laboratory, University of Rhode Island)

FIGURE 4.8

Photoelastic isochromatic contours for a disk under diametrical compression.

Considering a particular stress trajectory, the orientation angle θ_p with respect to the x_1-axis can be found using the relation

$$\tan 2\theta_p = \frac{2T_{12}}{T_{11} - T_{22}} \tag{4.6.2}$$

Now replacing x_1, x_2 with x, y, for a given trajectory specified by $y(x)$, $\tan\theta_p = \frac{dy}{dx}$ and combining these results with a standard trigonometric identity gives

$$\tan 2\theta_p = \frac{2\tan\theta_p}{1-\tan^2\theta_p} = \frac{2\frac{dy}{dx}}{1-\left(\frac{dy}{dx}\right)^2} = \frac{2T_{12}}{T_{11} - T_{22}}$$

This relation is easily solved for the trajectory slope

$$\frac{dy}{dx} = -\frac{T_{11} - T_{22}}{2T_{12}} \pm \sqrt{1 + \left(\frac{T_{11} - T_{22}}{2T_{12}}\right)^2} \tag{4.6.3}$$

So given in-plane stress components, the differential equation (4.6.3) can be integrated to generate the stress trajectories, $y(x)$. Although some special cases can be done analytically (Molleda et al., 2005), most stress distributions will generate complicated forms that require numerical integration (Breault, 2012). A particular example will now be explored, and several of the previously discussed stress contours and lines are generated and plotted.

EXAMPLE 4.6.1 ELASTIC STRESS DISTRIBUTIONS IN A DISK UNDER DIAMETRICAL COMPRESSION

Consider a specific two-dimensional problem of a circular disk loaded by equal but opposite concentrated forces along a given diameter as shown in Fig. 4.9A. Sadd (2014) provides the linear elastic solution to this problem, and with respect to the given axes the in-plane stresses are found to be

$$\begin{aligned}
T_{11} &= -\frac{2P}{\pi}\left[\frac{(R-y)x^2}{r_1^4} + \frac{(R+y)x^2}{r_2^4} - \frac{1}{2R}\right] \\
T_{22} &= -\frac{2P}{\pi}\left[\frac{(R-y)^3}{r_1^4} + \frac{(R+y)^3}{r_2^4} - \frac{1}{2R}\right] \\
T_{12} &= \frac{2P}{\pi}\left[\frac{(R-y)^2 x}{r_1^4} - \frac{(R+y)^2 x}{r_2^4}\right]
\end{aligned} \tag{4.6.4}$$

where $r_{1,2} = \sqrt{x^2 + (R \mp y)^2}$. Numerical results are presented for the case with unit radius and unit loading ($R = 1$, $P = 1$). For this case, Fig. 4.9 illustrates

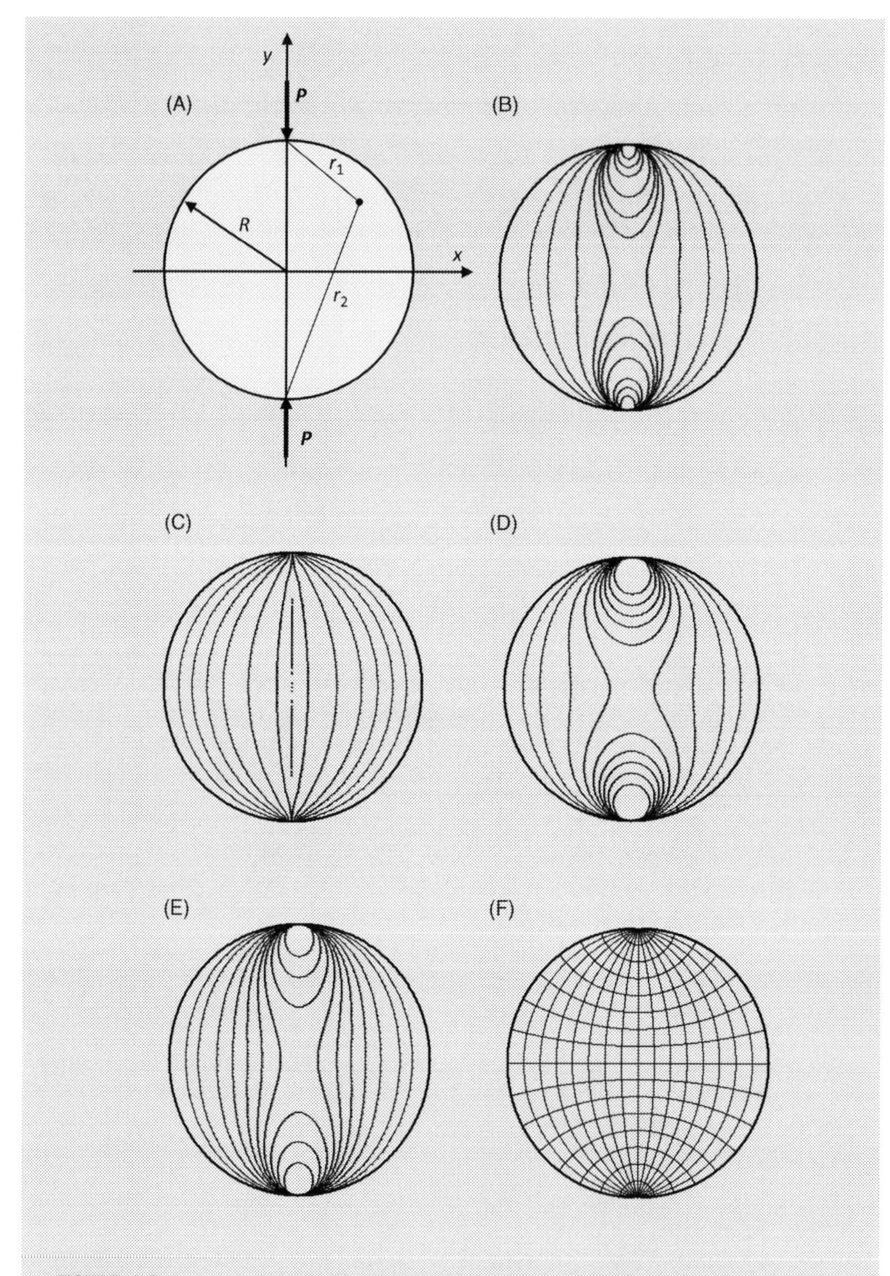

FIGURE 4.9

Elastic stress contours for diametrically loaded disk problem: (A) disk problem; (B) maximum shear stress contours (isochromatic lines); (C) maximum principal stress contours; (D) sum of principal stress contours (isopachic lines); (E) von Mises stress contours; (F) stress trajectories (isostatic lines).

several contour distributions and the stress trajectories that have been previously discussed. It should be apparent that each of these contour distributions is, in general, different from one another and each will convey particular information about the nature of the stress field under study.

4.7 REFERENCE CONFIGURATION PIOLA–KIRCHHOFF STRESS TENSORS

As discussed in Chapter 3, under finite deformations the reference area and current deformed area will not be the same (see Section 3.7). This difference must then be reconciled in the definitions of the stress tensor. Recall that the Cauchy stress previously discussed was an Eulerian variable defined as the force per unit area in the current deformed configuration. This stress tensor is widely used in many fields of continuum mechanics. However, there are particular cases where it will be advantageous to formulate the problem in the reference configuration in which certain variables will be known. For example, choosing the initial configuration as reference, we might know some features about the problem geometry when things begin. We now wish to explore the stress in the Lagrangian context, that is, the force per unit area in the reference configuration, and this will lead to two new and different stress tensors.

We start by reviewing the situation shown in Fig. 4.10 which illustrates the traction and normal vectors on an arbitrary differential area in both the reference and current configurations. As previously described in Section 4.2, the traction vector in the current configuration is denoted by $\boldsymbol{t}$ and the area on which it acts is da with unit normal vector $\boldsymbol{n}$. In the reference configuration, this differential area is dA with unit normal $\boldsymbol{N}$, and the traction vector in this configuration is designated as $\boldsymbol{T}^R$. Note that traction $\boldsymbol{T}^R$ is often referred to as a *pseudo*-vector since it represents the force in the current configuration but is evaluated per unit area in the undeformed reference configuration. It is important to realize what is different and what is the same between

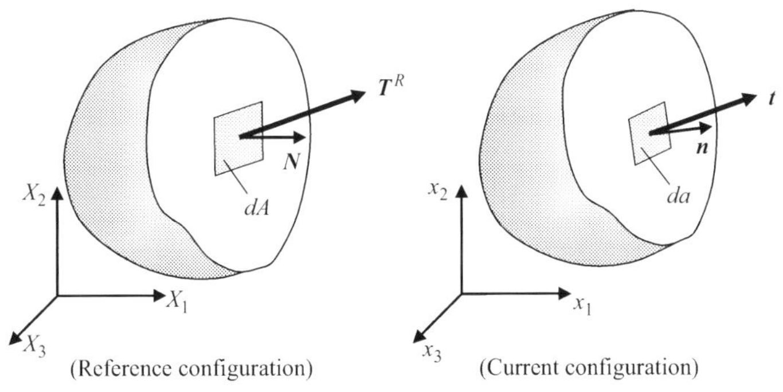

FIGURE 4.10

Reference and current differential areas and tractions.

the two configurations. Clearly, the areas and the unit normals are different since there is kinematic deformation between the two configurations. However, the total force acting on each area is to be the same and thus we can write

$$\boldsymbol{T}^R\, dA = \boldsymbol{t}\, da \tag{4.7.1}$$

Reviewing this relation, we note that the two traction vectors $\boldsymbol{t}$ and $\boldsymbol{T}^R$ act in the same direction with different magnitudes. Following a similar (but not identical) scheme used for the Cauchy stress tensor (4.3.6), we define a *Lagrangian stress tensor* T_{ij}^o as

$$T_i^R = T_{ij}^o N_j \tag{4.7.2}$$

T_{ij}^o is commonly called the *first Piola–Kirchhoff (PK1) stress tensor*. Note that relation (4.7.2) differs from (4.3.6) by using the form T_{ij}^o and not the transpose T_{ji}^o.

Combining (4.3.6) and (4.7.2) in (4.7.1) then expresses the common force relation as

$$T_{ij}^o N_j\, dA = T_{ji} n_j\, da \tag{4.7.3}$$

Going back to Section 3.7, the changes in the areas were given by Nanson's formula (3.7.3):

$$\boldsymbol{n}\, da = J(\boldsymbol{F}^{-1})^T\, \boldsymbol{N}\, dA \tag{4.7.4}$$

where $J = \det F$ is the Jacobian of the deformation. Using this result in (4.7.3) yields

$$T_{ij}^o N_j\, dA = T_{ji} J F_{kj}^{-1} N_k\, dA \Rightarrow (T_{ik}^o - T_{ji} J F_{kj}^{-1}) N_k = 0$$

thus resulting in the equation relating the PK1 stress to the Cauchy stress:

$$\begin{aligned} T_{ij}^o &= J F_{jk}^{-1} T_{ki} = J F_{jk}^{-1} T_{ik} \\ \boldsymbol{T}^o &= J \boldsymbol{T} \left(\boldsymbol{F}^{-1} \right)^T \end{aligned} \tag{4.7.5}$$

This relation can be easily inverted to express the Cauchy stress in terms of the PK1 stress:

$$T_{ij} = J^{-1} F_{jk} T_{ik}^o, \quad \boldsymbol{T} = J^{-1} \boldsymbol{T}^o \boldsymbol{F}^T \tag{4.7.6}$$

We have used the fact that the Cauchy stress will be shown to be symmetric. Exploring the symmetry of the PK1 stress tensor, we take the transpose of (4.7.5) to get

$$T_{ji}^o = J F_{ik}^{-1} T_{kj} = J F_{ik}^{-1} T_{jk} \neq T_{ij}^o$$

and thus the first Piola–Kirchhoff stress tensor T_{ij}^o is not, in general, symmetric even with a symmetric Cauchy stress. Note that some authors define a *nominal stress* $\boldsymbol{N}$ as the transpose of the PK1 stress, that is, $\boldsymbol{N} = \boldsymbol{T}^{o^T}$.

A second reference stress tensor may be developed by evaluating the force in the reference configuration using a pull-back operation as discussed in Section 3.6. Similar to the relation $d\boldsymbol{X} = \boldsymbol{F}^{-1}\,d\boldsymbol{x}$, we can express the force acting on dA as

$$\begin{aligned} \boldsymbol{SN}\,dA = \boldsymbol{F}^{-1}\boldsymbol{t}\,da = \boldsymbol{F}^{-1}(\boldsymbol{T}^{o}\boldsymbol{N}\,dA) \Rightarrow \\ \boldsymbol{S} = \boldsymbol{F}^{-1}\boldsymbol{T}^{o} \end{aligned} \tag{4.7.7}$$

where we have used relation (4.7.3), and $\boldsymbol{S}$ is the new stress called the *second Piola–Kirchhoff stress tensor* (PK2). Using (4.7.5), we can express PK2 in terms of the Cauchy stress as

$$S_{ij} = JF_{il}^{-1}F_{jk}^{-1}T_{lk}, \quad \boldsymbol{S} = J\boldsymbol{F}^{-1}\boldsymbol{T}(\boldsymbol{F}^{-1})^{T} \tag{4.7.8}$$

Checking the symmetry property of S_{ij} gives

$$S_{ji} = JF_{jl}^{-1}F_{ik}^{-1}T_{lk} = JF_{ik}^{-1}F_{jl}^{-1}T_{kl} = S_{ij}$$

and thus the second Piola–Kirchhoff stress is symmetric if the Cauchy stress is symmetric. It is a simple exercise to determine the Cauchy stress in terms of the PK2 stress:

$$T_{ij} = J^{-1}F_{ik}F_{jl}S_{kl}, \quad \boldsymbol{T} = J^{-1}\boldsymbol{FSF}^{T} \tag{4.7.9}$$

For small deformations $\nabla\boldsymbol{u} \approx \mathrm{O}(\varepsilon)$, where $\varepsilon \ll 1$. This situation implies the following:

$$\begin{aligned} &\boldsymbol{F} = \boldsymbol{I} + \nabla\boldsymbol{u} \approx \boldsymbol{I} + \mathrm{O}(\varepsilon) \\ &\boldsymbol{F}^{-1} = \boldsymbol{I} - \nabla\boldsymbol{u}^{*} \approx \boldsymbol{I} - \mathrm{O}(\varepsilon) \\ &J = \det\boldsymbol{F} \approx 1 + \mathrm{O}(\varepsilon) \\ &\boldsymbol{T}^{o} = J\boldsymbol{T}\left(\boldsymbol{F}^{-1}\right)^{T} \approx \boldsymbol{T} \\ &\boldsymbol{S} = J\boldsymbol{F}^{-1}\boldsymbol{T}\left(\boldsymbol{F}^{-1}\right)^{T} \approx \boldsymbol{T} \end{aligned}$$

Thus, for small deformations, both Piola–Kirchhoff stress tensors reduce to the Cauchy stress.

EXAMPLE 4.7.1 DETERMINATION OF PIOLA–KIRCHHOFF STRESS TENSORS

For a Cauchy stress state $T_{ij} = \begin{bmatrix} 3 & 1 & 0 \\ 1 & 2 & 0 \\ 0 & 0 & 0 \end{bmatrix}$, under simple shearing deformation, determine the PK1 and PK2 stress tensors. Next, calculate the pseudo-stress vector $\boldsymbol{T}^{R}$ acting on a plane with unit normal $\boldsymbol{e}_1$ associated with PK1 and PK2.

Solution: For simple shearing deformation,

$$\boldsymbol{F} = \frac{\partial\boldsymbol{x}}{\partial\boldsymbol{X}} = \begin{bmatrix} 1 & \gamma & 0 \\ 0 & 1 & 0 \\ 0 & 0 & 1 \end{bmatrix}, \quad \boldsymbol{F}^{-1} = \frac{\partial\boldsymbol{X}}{\partial\boldsymbol{x}} = \begin{bmatrix} 1 & -\gamma & 0 \\ 0 & 1 & 0 \\ 0 & 0 & 1 \end{bmatrix}, \quad J = 1$$

For PK1, use (4.7.5) ⇒

$$\boldsymbol{T}^{o} = J\boldsymbol{T}(\boldsymbol{F}^{-1})^{T} = \begin{bmatrix} 3 & 1 & 0 \\ 1 & 2 & 0 \\ 0 & 0 & 0 \end{bmatrix} \begin{bmatrix} 1 & 0 & 0 \\ -\gamma & 1 & 0 \\ 0 & 0 & 1 \end{bmatrix} = \begin{bmatrix} 3-\gamma & 1 & 0 \\ 1-2\gamma & 2 & 0 \\ 0 & 0 & 0 \end{bmatrix}$$

For PK2, use (4.7.7) ⇒

$$\boldsymbol{S} = \boldsymbol{F}^{-1}\boldsymbol{T}^{o} = \begin{bmatrix} 1 & -\gamma & 0 \\ 0 & 1 & 0 \\ 0 & 0 & 1 \end{bmatrix} \begin{bmatrix} 3-\gamma & 1 & 0 \\ 1-2\gamma & 2 & 0 \\ 0 & 0 & 0 \end{bmatrix} = \begin{bmatrix} 3-2\gamma+2\gamma^{2} & 1-2\gamma+ & 0 \\ 1-2\gamma & 2 & 0 \\ 0 & 0 & 0 \end{bmatrix}$$

Note that PK2 is symmetric, whereas PK1 is not.
Now using (4.7.2) with $\boldsymbol{N} = \boldsymbol{e}_1$, traction associated with PK1 is thus

$$\boldsymbol{T}^{R} = \boldsymbol{T}^{o}\boldsymbol{N} = \begin{bmatrix} 3-\gamma & 1 & 0 \\ 1-2\gamma & 2 & 0 \\ 0 & 0 & 0 \end{bmatrix} \begin{bmatrix} 1 \\ 0 \\ 0 \end{bmatrix} = \begin{bmatrix} 3-\gamma \\ 1-2\gamma \\ 0 \end{bmatrix}$$

and similarly from (4.7.7), the traction related to PK2 is

$$\boldsymbol{SN} = \begin{bmatrix} 3-2\gamma+2\gamma^{2} & 1-2\gamma+ & 0 \\ 1-2\gamma & 2 & 0 \\ 0 & 0 & 0 \end{bmatrix} \begin{bmatrix} 1 \\ 0 \\ 0 \end{bmatrix} = \begin{bmatrix} 3-2\gamma+2\gamma^{2} \\ 1-2\gamma \\ 0 \end{bmatrix}$$

4.8 OTHER STRESS TENSORS

Over the years, several other stress tensors have been defined by the continuum mechanics community. Normally these definitions have particular advantages in special applications. In contrast to the Cauchy and Piola–Kirchhoff stresses, most of these special stress components do not have a direct physical interpretation.

KIRCHHOFF STRESS

The *Kirchhoff stress tensor* is defined in terms of the Cauchy stress by

$$\tau = J\boldsymbol{T} \tag{4.8.1}$$

This stress is sometimes used in plasticity theories and it will occasionally simplify particular equation forms.

BIOT STRESS

The Biot or Jaumann stress tensor is defined as

$$\boldsymbol{T}^{B} = \boldsymbol{R}^{T}\boldsymbol{T}^{o} \tag{4.8.2}$$

where $\boldsymbol{R}$ is the rotation tensor from the polar decomposition of the deformation gradient, $\boldsymbol{F} = \boldsymbol{RU}$, and $\boldsymbol{T}^{o}$ is the first Piola–Kirchhoff stress. This tensor form is useful because it is the energy conjugate (see Section 5.6) to the right stretch tensor $\boldsymbol{U}$. From (4.7.7), $\boldsymbol{T}^o = \boldsymbol{FS}$ and thus (4.8.2) can be written as

$$\boldsymbol{T}^B = \boldsymbol{R}^T \boldsymbol{FS} = \boldsymbol{US} \tag{4.8.3}$$

COROTATIONAL CAUCHY STRESS

For some applications, we may be interested in the stress at an intermediate rotated configuration. The *corotational Cauchy stress* $\boldsymbol{T}^C$ is related to this concept and is given by

$$\boldsymbol{T}^C = \boldsymbol{R}^T \boldsymbol{TR} \tag{4.8.4}$$

where $\boldsymbol{R}$ is an orthogonal rotation tensor. Using (4.7.9), relation (4.8.4) can be expressed in terms of the PK2 stress and the right stretch tensors

$$\boldsymbol{T}^C = J^{-1}\boldsymbol{R}^T \boldsymbol{FSF}^T \boldsymbol{R} = J^{-1}\boldsymbol{USU} \tag{4.8.5}$$

Although a few more specialized stress examples exist, they are not widely used and we will end our discussion on this topic.

4.9 OBJECTIVITY OF STRESS TENSORS

Recalling our objectivity discussions from Sections 1.2, 2.9, 3.5, and 3.6, we now wish to do an objectivity check on the various second-order stress tensors previously presented. For second-order tensors, the $\boldsymbol{Q}$-objectivity test was specified by relation (2.9.3), and thus a general tensor $\boldsymbol{A}$ would have to satisfy the relation $\boldsymbol{A}^* = \boldsymbol{QAQ}^T$. Also collecting some previous results from Section 3.5, we found that the deformation gradient and its inverse satisfied relations $\boldsymbol{F}^* = \boldsymbol{QF}$ and $\boldsymbol{F}^{*-1} = \boldsymbol{F}^{-1}\boldsymbol{Q}^T$.

Starting with the Cauchy stress, we begin with relation (4.3.6), $\boldsymbol{t} = \boldsymbol{T}^T\boldsymbol{n}$. We assume that from their fundamental definitions, vectors $\boldsymbol{n}$ and $\boldsymbol{t}$ are objective and transform as first-order tensors, and thus

$$\begin{aligned} &\boldsymbol{t} = \boldsymbol{T}^T\boldsymbol{n} \Rightarrow \boldsymbol{Q}^T\boldsymbol{t}^* = \boldsymbol{T}^T\boldsymbol{Q}^T\boldsymbol{n}^* \Rightarrow \boldsymbol{t}^* = \boldsymbol{QT}^T\boldsymbol{Q}^T\boldsymbol{n}^* \\ &\text{but } \ \boldsymbol{t}^* = \boldsymbol{T}^{*T}\boldsymbol{n}^* \Rightarrow \boldsymbol{T}^{*T}\boldsymbol{n}^* = \boldsymbol{QT}^T\boldsymbol{Q}^T\boldsymbol{n}^* \Rightarrow \\ &\boldsymbol{T}^{*T} = \boldsymbol{QT}^T\boldsymbol{Q}^T \Rightarrow \boldsymbol{T}^* = \boldsymbol{QTQ}^T \end{aligned} \tag{4.9.1}$$

and thus the Cauchy stress tensor is objective.

Next, let us explore the first Piola–Kirchhoff stress $\boldsymbol{T}^o$. From (4.7.5),

$$\begin{aligned} &\boldsymbol{T}^{o^*} = J^*\boldsymbol{T}^*\left(\boldsymbol{F}^{*^{-1}}\right)^T \Rightarrow \boldsymbol{T}^{o^*} = J\boldsymbol{QTQ}^T\left(\boldsymbol{F}^{-1}\boldsymbol{Q}^T\right)^T \Rightarrow \\ &\boldsymbol{T}^{o^*} = J\boldsymbol{QTQ}^T\boldsymbol{Q}\left(\boldsymbol{F}^{-1}\right)^T = J\boldsymbol{QT}\left(\boldsymbol{F}^{-1}\right)^T = \boldsymbol{QT}^o \end{aligned} \tag{4.9.2}$$

We conclude that the first Piola–Kirchhoff stress tensor satisfies the objectivity test for first-order tensors but not for second-order tensors.

Moving on to the PK2 stress $\boldsymbol{S}$, we start with relation (4.7.7):

$$\boldsymbol{S}^* = \boldsymbol{F}^{*-1}\boldsymbol{T}^{*o} = \boldsymbol{F}^{-1}\boldsymbol{Q}^T\boldsymbol{Q}\boldsymbol{T}^o = \boldsymbol{F}^{-1}\boldsymbol{T}^o = \boldsymbol{S} \tag{4.9.3}$$

and thus the second Piola–Kirchhoff stress tensor is the same in both reference frames, and so it is not an objective second-order tensor.

In regard to the Kirchhoff stress $\boldsymbol{\tau} = J\boldsymbol{T}$, since the Cauchy stress is objective and J is a scalar, the Kirchhoff stress is objective. For the Biot stress, using (4.8.3) $\boldsymbol{T}^B = \boldsymbol{U}\boldsymbol{S}$, we can argue that since $\boldsymbol{S}$ and $\boldsymbol{U}^2$ remain the same under the frame change, then $\boldsymbol{T}^B$ will also remain the same and thus will not be an objective second-order tensor. Finally, in light of (4.8.5), the corotational Cauchy stress also remains the same under the $\boldsymbol{Q}$-transformation and thus does not satisfy objectivity for a second-order tensor.

4.10 CYLINDRICAL AND SPHERICAL COORDINATE CAUCHY STRESS FORMS

As discussed in Chapter 3, in order to solve many continuum mechanics problems, formulation must be done in curvilinear coordinates typically using cylindrical or spherical systems. Thus, following similar methods as used with the strain tensor relations in Chapter 3, we now wish to develop expressions for a few of the basic stress and traction concepts for cylindrical and spherical coordinates.

Cylindrical coordinates were originally presented in Fig. 2.6, and for such a system, the Cauchy stress components are defined on the differential element shown in Fig. 4.11. For this coordinate system, the stress matrix is given by

$$\boldsymbol{T} = \begin{bmatrix} T_{rr} & T_{r\theta} & T_{rz} \\ T_{\theta r} & T_{\theta\theta} & T_{\theta z} \\ T_{zr} & T_{z\theta} & T_{zz} \end{bmatrix} \tag{4.10.1}$$

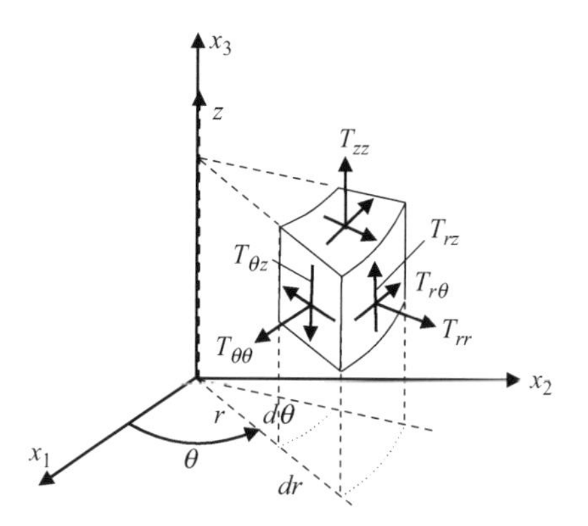

FIGURE 4.11

Stress components in cylindrical coordinates.

Now the stress can be expressed in terms of the traction components as

$$\boldsymbol{T} = \boldsymbol{e}_r \boldsymbol{t}_r + \boldsymbol{e}_\theta \boldsymbol{t}_\theta + \boldsymbol{e}_z \boldsymbol{t}_z \tag{4.10.2}$$

where

$$\begin{aligned} \boldsymbol{t}_r &= T_{rr}\boldsymbol{e}_r + T_{r\theta}\boldsymbol{e}_\theta + T_{rz}\boldsymbol{e}_z \\ \boldsymbol{t}_\theta &= T_{\theta r}\boldsymbol{e}_r + T_{\theta\theta}\boldsymbol{e}_\theta + T_{\theta z}\boldsymbol{e}_z \\ \boldsymbol{t}_z &= T_{zr}\boldsymbol{e}_r + T_{z\theta}\boldsymbol{e}_\theta + T_{zz}\boldsymbol{e}_z \end{aligned} \tag{4.10.3}$$

Next consider these developments for the spherical coordinate system, previously shown in Fig. 2.7. The Cauchy stress components in spherical coordinates are defined on the differential element illustrated in Fig. 4.12, and the stress matrix for this case is

$$\boldsymbol{T} = \begin{bmatrix} T_{RR} & T_{R\phi} & T_{R\theta} \\ T_{\phi R} & T_{\phi\phi} & T_{\phi\theta} \\ T_{\theta R} & T_{\theta\phi} & T_{\theta\theta} \end{bmatrix} \tag{4.10.4}$$

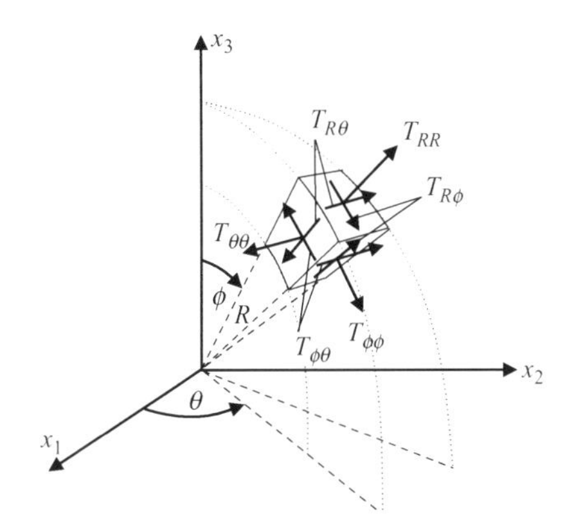

FIGURE 4.12

Stress components in spherical coordinates.

As before, the stress can be expressed in terms of the traction components as

$$\boldsymbol{T} = \boldsymbol{e}_R \boldsymbol{t}_R + \boldsymbol{e}_\phi \boldsymbol{t}_\phi + \boldsymbol{e}_\theta \boldsymbol{t}_\theta \tag{4.10.5}$$

where

$$\begin{aligned} \boldsymbol{t}_R &= T_{RR}\boldsymbol{e}_R + T_{R\phi}\boldsymbol{e}_\phi + T_{R\theta}\boldsymbol{e}_\theta \\ \boldsymbol{t}_\phi &= T_{\phi R}\boldsymbol{e}_R + T_{\phi\phi}\boldsymbol{e}_\phi + T_{\phi\theta}\boldsymbol{e}_\theta \\ \boldsymbol{t}_\theta &= T_{\theta R}\boldsymbol{e}_R + T_{\theta\phi}\boldsymbol{e}_\phi + T_{\theta\theta}\boldsymbol{e}_\theta \end{aligned} \tag{4.10.6}$$

REFERENCES

Breault, B., 2012. Improving load distributions in cellular materials using stress trajectory topology. University of Rhode Island, MS Thesis.

Dewhurst, P., 2001. Analytical solutions and numerical procedures for minimum-weight Michell structures. J. Mech. Phys. Solids 49, 445–467.

Eringen, A.C., 1968. Theory of micropolar elasticity. Liebowitz, H. (Ed.), Fracture, 2, Academic Press, New York, pp. 662–729.

Kelly, D., Tosh, M., 2000. Interpreting load paths and stress trajectories in elasticity. Eng. Comput. 17, 117–135.

Molleda, F., Mora, J., Molleda, F.J., Carrillo, E., Mellor, B.G., 2005. Stress trajectories for mode I fracture. Mater. Characterization 54, 9–12.

Rathkjen, A., 1997. Force lines in plane stress. J. Elasticity 47, 167–178.

Sadd, M.H., 2014. Elasticity theory, applications and numerics, third ed. Elsevier, Waltham, MA.

Shukla, A., Dally, J.W., 2010. Experimental Solid Mechanics. College House Enterprises, Knoxville, TN.

Ugural, A.C., Fenster, S.K., 2003. Advanced strength of materials and applied elasticity. Prentice Hall, Englewood liffs, NJ.

EXERCISES

4.1 For the following Cauchy stress tensors, find the traction vector acting on a plane with the given normal vector $\boldsymbol{n}$. Note that you must first make the normal vector have unit length.

(*a*) $\boldsymbol{T} = \begin{bmatrix} 2 & 1 & 1 \\ 1 & 0 & 2 \\ 1 & 2 & 0 \end{bmatrix}$, $\boldsymbol{n} = \boldsymbol{e}_1 + 2\boldsymbol{e}_2 + \boldsymbol{e}_3$

(*b*) $\boldsymbol{T} = \begin{bmatrix} 1 & 0 & 1 \\ 0 & 2 & 2 \\ 1 & 2 & 0 \end{bmatrix}$, $\boldsymbol{n} = 2\boldsymbol{e}_1 + \boldsymbol{e}_2 + \boldsymbol{e}_3$

(*c*) $\boldsymbol{T} = \begin{bmatrix} 2 & 1 & 0 \\ 1 & 4 & 2 \\ 0 & 2 & 0 \end{bmatrix}$, $\boldsymbol{n} = \boldsymbol{e}_1 + 2\boldsymbol{e}_2$

4.2 For each of the stress tensors in Exercise 4.1, determine the new components in a coordinate frame rotated +90° around the x_1-axis (direction *via* the right-hand rule).

4.3 Explicitly verify the two-dimensional transformation equations (4.3.11) and (4.3.12).

4.4 Show that the general two-dimensional stress transformation relations (4.3.11) can be used to generate relations for the normal and shear stresses in a polar coordinate system in terms of Cartesian components:

$$
\begin{aligned}
T_{rr} &= T_{11}\cos^2\theta + T_{22}\sin^2\theta + 2T_{12}\sin\theta\cos\theta \\
T_{\theta\theta} &= T_{11}\sin^2\theta + T_{22}\cos^2\theta - 2T_{12}\sin\theta\cos\theta \\
T_{r\theta} &= -T_{11}\sin\theta\cos\theta + T_{22}\sin\theta\cos\theta + T_{12}(\cos^2\theta - \sin^2\theta)
\end{aligned}
$$

4.5 Verify that the two-dimensional transformation relations giving Cartesian stresses in terms of polar components are given by

$$
\begin{aligned}
T_{11} &= T_{rr}\cos^2\theta + T_{\theta\theta}\sin^2\theta - 2T_{r\theta}\sin\theta\cos\theta \\
T_{22} &= T_{rr}\sin^2\theta + T_{\theta\theta}\cos^2\theta - 2T_{r\theta}\sin\theta\cos\theta \\
T_{12} &= T_{rr}\sin\theta\cos\theta - T_{\theta\theta}\sin\theta\cos\theta + T_{r\theta}(\cos^2\theta - \sin^2\theta)
\end{aligned}
$$

4.6 A two-dimensional state of *plane Cauchy stress* in the x_1,x_2-plane is defined by

$$T_{ij} = \begin{bmatrix} T_{11} & T_{12} & 0 \\ T_{21} & T_{22} & 0 \\ 0 & 0 & 0 \end{bmatrix}$$

Using general principal value theory, show that for this case the in-plane principal stresses and maximum shear stress are given by

$$
\begin{aligned}
T_{1,2} &= \frac{T_{11}+T_{22}}{2} \pm \sqrt{\left(\frac{T_{11}-T_{22}}{2}\right)^2 + T_{12}^2} \\
\tau_{max} &= \sqrt{\left(\frac{T_{11}-T_{22}}{2}\right)^2 + T_{12}^2}
\end{aligned}
$$

4.7 For the plane stress case in Exercise 4.6, demonstrate the invariant nature of the principal stresses and maximum shear stresses by showing that

$$T_{1,2} = \frac{1}{2}I_T \pm \sqrt{\frac{1}{4}I_T^2 - II_T} \text{ and } \tau_{max} = \sqrt{\frac{1}{4}I_T^2 - II_T}$$

4.8 It was discussed in Section 4.4 that for the case of ranked principal stresses ($T_1 > T_2 > T_3$), the maximum shear stress was given by $S_{max} = (T_1 - T_3)/2$ which was the radius of the largest Mohr's circle shown in Fig. 4.7. For this case, show that the normal stress acting on the plane of maximum shear is given by $N = (T_1 + T_3)/2$. Finally, using relations (4.4.10), show that the components of the unit normal vector to this plane are $n_i = \pm(1,0,1)/\sqrt{2}$. This result implies that the maximum shear stress acts on a plane that bisects the angle between the directions of the largest and smallest principal stress.

4.9 Explicitly show that the stress state given in Example 4.4.1 will reduce to the proper diagonal form under transformation to principal axes. Follow the transformation scheme shown in Example 2.11.1.

4.10 For the case of *pure shear*, where the stress is given by $\boldsymbol{T} = \begin{bmatrix} 0 & \tau & 0 \\ \tau & 0 & 0 \\ 0 & 0 & 0 \end{bmatrix}$, determine the principal stresses and directions, and compute the normal and shear stress on the octahedral plane.

4.11 For linear elastic materials, stress field solutions can often be written in the general form $T_{ij} = Pf_{ij}(\boldsymbol{x})$, where P is a loading parameter and the tensor function f_{ij} specifies only the field distribution. Show, in general, that for this case, the principal stresses will be a linear form in P, that is, $T_{1,2,3} = Pg_{1,2,3}(\boldsymbol{x})$. Next demonstrate that the principal directions will not depend on P.

4.12 Determine the spherical and deviatoric stress tensors for the Cauchy stress states given in Exercise 4.1.

4.13 Show that the principal directions of the deviatoric Cauchy stress tensor $\hat{\boldsymbol{T}}$ coincide with the principal directions of the stress tensor $\boldsymbol{T}$. Also show that the principal values of the deviatoric stress $\hat{T}_i$ can be expressed in terms of the principal values T_i by the relation $\hat{T}_i = T_i - \frac{1}{3}T_{kk}$. This problem is analogous to Exercise 3.21.

4.14 Show that the second invariant of the Cauchy stress deviator tensor may be written in the following forms:

$$
\begin{aligned}
II_{\hat{T}} &= -\frac{1}{2}\left(\hat{T}_{11}^2 + \hat{T}_{22}^2 + \hat{T}_{33}^2\right) - T_{12}^2 - T_{23}^2 - T_{31}^2 \\
&= -\frac{1}{6}\left[\left(\hat{T}_{11} - \hat{T}_{22}\right)^2 + \left(\hat{T}_{22} - \hat{T}_{33}\right)^2 + \left(\hat{T}_{33} - \hat{T}_{11}\right)^2\right] - T_{12}^2 - T_{23}^2 - T_{31}^2 \\
&= -\frac{1}{6}\left[(T_{11} - T_{22})^2 + (T_{22} - T_{33})^2 + (T_{33} - T_{11})^2\right] T_{12}^2 - T_{23}^2 - T_{31}^2
\end{aligned}
$$

4.15 Show that the second and third invariants of the Cauchy deviatoric stress tensor can be written as

$$
\begin{aligned}
II_{\hat{T}} &= -\frac{1}{2}\left(T_{ij}T_{ij} - \frac{2}{3}I_T^2 + \frac{1}{3}I_T^2\right) = II_T - \frac{1}{3}I_T^2 \\
III_{\hat{T}} &= \frac{1}{3}\operatorname{tr}\hat{\boldsymbol{T}}^3 = \frac{2}{27}I_T^3 - \frac{1}{3}I_T II_T + III_T
\end{aligned}
$$

For the third invariant, make use of the Cayley–Hamilton Theorem.

4.16 Explicitly verify relations (4.5.5) for the octahedral Cauchy stress components. Also assuming that $\boldsymbol{T}$ is symmetric, show that they can be expressed in terms of the general stress components by

$$
\sigma_{oct} = \frac{1}{3}(T_{11} + T_{22} + T_{33})
$$

$$
\tau_{oct} = \frac{1}{3}\left[(T_{11} - T_{22})^2 + (T_{22} - T_{33})^2 + (T_{33} - T_{11})^2 + 6T_{12}^2 + 6T_{23}^2 + 6T_{31}^2\right]^{1/2}
$$

4.17 Determine the von Mises and octahedral stresses for the Cauchy stress states given in Exercise 4.1. Use results in Exercise 4.16 for the octahedral stresses.

4.18 The plane stress solution for a semi-infinite linear elastic solid under a concentrated point loading P (Flamant problem) is given by

$$T_{11} = -\frac{2Px_1^2x_2}{\pi\left(x_1^2+x_2^2\right)^2}$$

$$T_{22} = -\frac{2Px_2^3}{\pi\left(x_1^2+x_2^2\right)^2}$$

$$T_{12} = -\frac{2Px_1x_2^2}{\pi\left(x_1^2+x_2^2\right)^2}$$

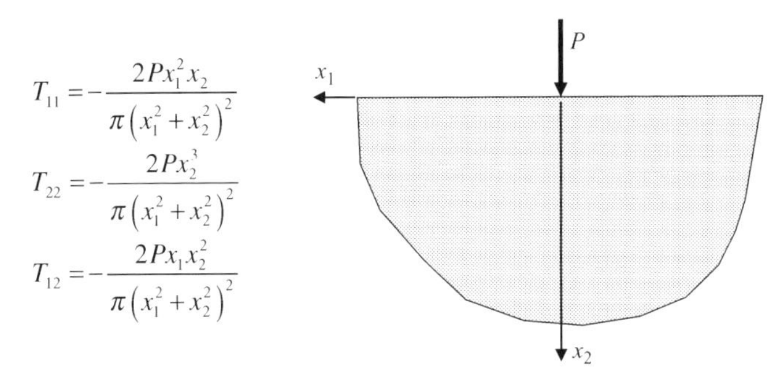

Calculate the maximum shear stress at any point in the body (see Exercise 4.6) and then use MATLAB or equivalent (see example code C-5) to plot contours of $\tau_{\max}$.

4.19 A linear elastic plane stress solution $T_{11} = -\frac{3Px_1x_2}{2c^3} + \frac{N}{2c}, T_{22} = 0, T_{12} = -\frac{3P}{4c}\left(1-\frac{x_2^2}{c^2}\right)$, where P and N are constants, has been determined for the illustrated rectangular domain. Determine the boundary stresses and their resultants (including both forces and moments) that act on each of the four boundaries. These boundary conditions should indicate that this is the solution to a particular beam problem.

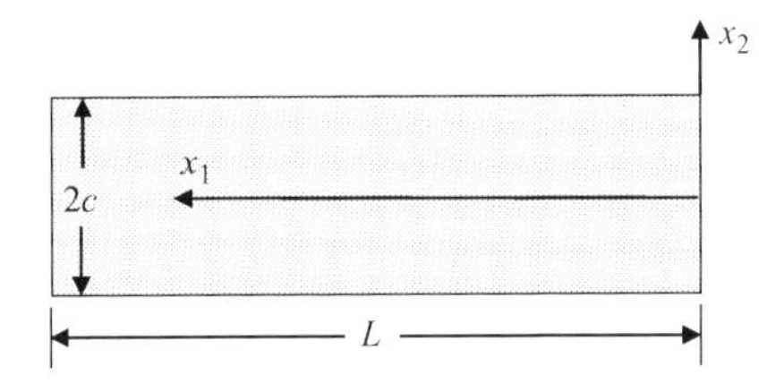

4.20 For each of the stress tensors in Exercise 4.1 under extensional deformation

$$x_1 = \lambda_1 X_1, \quad x_2 = \lambda_2 X_2, \quad x_3 = \lambda_3 X_3$$

determine the first and second Piola–Kirchhoff stress tensors. Check that PK2 is symmetric.

4.21 Consider the isochoric extensional deformation $x_1 = 2X_1, x_2 = X_2/\sqrt{2}, x_3 = X_3/\sqrt{2}$ of a long bar whose axis lies along the x_1-direction and has unit square cross-section in the reference configuration. First, determine the change in cross-sectional area from the reference to the current configuration. Next, if the Cauchy stress is to be taken as uniaxial,

$\boldsymbol{T} = T_{11}\boldsymbol{e}_1\boldsymbol{e}_1$, calculate the first and second Piola–Kirchhoff stress tensors. Compare the 11-components of each of the three stress tensors.

4.22 Consider uniaxial extensional deformation of a material only along the x_1-direction. Transverse deformation is constrained and so $x_1 = \lambda X_1, x_2 = X_2, x_3 = X_3$. If the Cauchy stress is to be taken as uniaxial, $\boldsymbol{T} = T_{11}\boldsymbol{e}_1\boldsymbol{e}_1$, calculate the first and second Piola–Kirchhoff stress tensors.

CHAPTER

General Conservation or Balance Laws

5

We now wish to explore a number of fundamental physical laws that apply to continuum material behavior. These laws have been developed from many years of past research and are appropriate for deformations, loadings, and rate effects found in typical engineering applications where relativistic and nuclear behaviors can be neglected. These relations commonly represent some type of *conservation principle* and are also often referred to as *balance laws*. They are applicable to all continuum materials regardless of whether they are solids, fluids, elastic, plastic, etc. Our presentation will include *conservation of mass, linear and angular momentum*, as well as *energy* (first law of thermodynamics). In addition, we will also present the *Clausius–Duhem inequality* which is a form of the *second law of thermodynamics*. The balance laws will be initially formulated in *integral form* and then later reduced to *differential field equations* that apply to all continuum points within a body under study. While these relations will include many previously defined tensor fields, they will also introduce a few new variables in our study.

5.1 GENERAL CONSERVATION PRINCIPLES AND THE REYNOLDS TRANSPORT THEOREM

Fundamentally balance principles start as axioms involving integral relations over material body configurations. Consider first the time rate of change of certain integrals. Clearly for a *fixed region of space* R, and with $\boldsymbol{G}(\boldsymbol{x}, t)$ being an arbitrary tensor field

$$\frac{\partial}{\partial t}\int_R \boldsymbol{G}\,dv = \int_R \frac{\partial}{\partial t}\boldsymbol{G}\,dv \tag{5.1.1}$$

The time derivative passes through the volume integral sign since the limits of integration are time independent.

However, referring to our previous discussion in Section 3.1, we next reconsider this time rate of change over a *fixed group of continuum particles* that occupy the region of space R_m at some particular point in time. For this case, not only does the integrand change with time, but so does the spatial volume over which the integral is taken. We thus wish to define a material time derivative of a volume integral in such a way that it measures the rate of change of the total amount of some quantity carried by the given mass system in R_m. Hence, we want to consider

$$\frac{D}{Dt}\int_{R_m} \boldsymbol{G}\,dV$$

Continuum Mechanics Modeling of Material Behavior. http://dx.doi.org/10.1016/B978-0-12-811474-2.00005-8

Note that $R = R_m$ at the instant of time under consideration, and the differential volume elements dv and dV are related through the Jacobian determinant by relation (3.7.4), $dv = J\,dV$. Some authors refer to R as the *control volume* and R_m as the *material volume*.

Thus, we can write

$$\begin{aligned}\frac{D}{Dt}\int_{R_m}\boldsymbol{G}\,dv &= \int_R \frac{D}{Dt}\boldsymbol{G}\,dv + \int_R \boldsymbol{G}\frac{D}{Dt}dv = \int_R \dot{\boldsymbol{G}}\,dv + \int_{R_m}\boldsymbol{G}\dot{J}\,dV \\ &= \int_R \dot{\boldsymbol{G}}\,dv + \int_{R_m}\boldsymbol{G}v_{k,k}J\,dV = \int_R \dot{\boldsymbol{G}}\,dv + \int_R \boldsymbol{G}v_{k,k}\,dv \\ &= \int_R \frac{\partial}{\partial t}\boldsymbol{G}\,dv + \int_R (\boldsymbol{G}_{,k}v_k + \boldsymbol{G}v_{k,k})\,dv = \int_R \frac{\partial}{\partial t}\boldsymbol{G}\,dv + \int_R (\boldsymbol{G}v_k)_{,k}\,dv\end{aligned} \tag{5.1.2}$$

where we have used (3.13.9) and (3.3.3), with v_k being the velocity field. Next, using the Divergence Theorem on the second integral produces

$$\frac{D}{Dt}\int_R \boldsymbol{G}\,dV = \int_R \frac{\partial}{\partial t}\boldsymbol{G}\,dv + \int_{\partial R}\boldsymbol{G}v_k n_k\,ds \tag{5.1.3}$$

where n_k is the unit outward normal vector to the surface ∂R which encloses the region R, as shown in Fig. 5.1. Result (5.1.3) is often called the *Reynolds Transport Theorem*, and we see from this result that the material time rate of change is given by two terms: one related to the simple time rate of change within the region R, whereas another term related to a particular amount entering (flux) through the boundary ∂R.

Most of our physical laws will be first stated in a general global equation of balance or conservation in the form

$$\frac{D}{Dt}\int_R \rho\Psi\,dV = -\int_{\partial R}\boldsymbol{I}\cdot\boldsymbol{n}\,ds + \int_R \rho\boldsymbol{S}\,dv \tag{5.1.4}$$

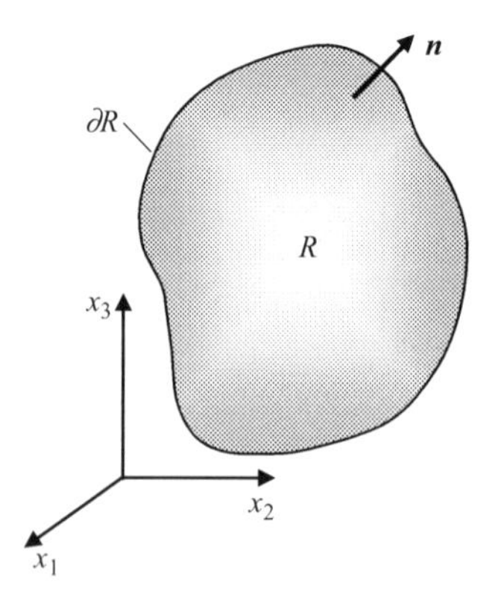

FIGURE 5.1

Typical control volume.

where $\boldsymbol{\Psi}$ is some tensor field, $\boldsymbol{I}$ is an influx term due to transfer through the boundary ∂R, $\boldsymbol{S}$ is an internal source term, and the minus sign is needed since $\boldsymbol{n}$ is the outward normal. We now proceed to several specific conservation or balance relations.

5.2 CONSERVATION OF MASS

Considering an arbitrary portion of a material R_m, the global principle of conservation of mass can be simply stated that total mass in R_m should remain constant at all times. Thus, we can write

$$\frac{D}{Dt}\int_{R_m} \rho\, dV = 0 \tag{5.2.1}$$

Incorporating the Reynolds Transport Theorem (5.1.2) or (5.1.3) gives the results

$$\frac{D}{Dt}\int_{R_m} \rho\, dV = \int_R \frac{\partial}{\partial t}\rho\, dv + \int_{\partial R} \rho v_k n_k\, ds = \int_R \left(\frac{\partial \rho}{\partial t} + (\rho v_k)_{,k}\right) dv = 0 \tag{5.2.2}$$

Note that the first form of (5.2.2) indicates that the rate of change of mass in R is equal to the density change plus the total rate of mass entering the region through the boundary ∂R. Employing the Localization Theorem (2.17.16) implies that the common integrand in (5.2.2) must be zero:

$$\begin{gathered}\frac{\partial \rho}{\partial t} + (\rho v_k)_{,k} = 0 \\ \frac{\partial \rho}{\partial t} + \nabla \cdot (\rho \boldsymbol{v}) = 0\end{gathered} \tag{5.2.3}$$

Relation (5.2.3) is known as the *differential statement of the conservation of mass* and is a point-wise relation that applies at all continuum points within R. It is also often referred to as the *continuity equation* in fluid mechanics. This relation can be written in the alternative form as

$$\begin{gathered}\frac{D\rho}{Dt} + \rho v_{k,k} = 0 \\ \frac{D\rho}{Dt} + \rho \nabla \cdot \boldsymbol{v} = 0\end{gathered} \tag{5.2.4}$$

Note that if the material is *incompressible*, then $D\rho / Dt = 0$, and thus

$$v_{k,k} = 0, \quad \nabla \cdot \boldsymbol{v} = 0 \tag{5.2.5}$$

For a fixed group of continuum particles, this conservation principle requires the mass to be the same in all configurations. Thus, we may use the reference and current configurations and write

$$\int_{R_o} \rho_o(\boldsymbol{X}, t)\, dV = \int_R \rho(\boldsymbol{x}, t)\, dv \tag{5.2.6}$$

where R_o and ρ_o are the reference volume and density, respectively, and R and ρ are the respective current values. Using (3.7.4) in (5.2.6) gives

$$\int_{R_o} (\rho_o - \rho J)\, dV = 0 \tag{5.2.7}$$

which is again true for all regions R_o, and so

$$\rho_o - \rho J = 0 \Rightarrow \frac{D}{Dt}(\rho J) = 0 \tag{5.2.8}$$

where we have used the fact that in the reference configuration $J_o = 1$. Relations (5.2.7) and (5.2.8) are sometimes referred to as the *Lagrangian or reference form of the conservation of mass*. Notice that expanding (5.2.8) gives the usual form (5.2.4).

$$\frac{D}{Dt}(\rho J) = \dot{\rho} J + \rho \dot{J} = J(\dot{\rho} + \rho v_{k,k}) = 0 \Rightarrow \dot{\rho} + \rho v_{k,k} = 0$$

Note that with the conservation of mass, the Reynolds Transport Theorem then gives the general result that for any tensor field $\boldsymbol{G}$:

$$\frac{D}{Dt}\int_R \rho \boldsymbol{G}\, dV = \int_R \rho \frac{D}{Dt}\boldsymbol{G}\, dv \tag{5.2.9}$$

EXAMPLE 5.2.1 CONSERVATION OF MASS CHECK

A continuum has a mass density given by $\rho = \dfrac{\rho_o}{a + bt}$, and a motion specified by $x_1 = (a + bt)X_1, x_2 = X_2, x_3 = X_3$, where ρ_o, a, and b are constants.

(a) Verify that the differential form of the conservation of mass is satisfied.

(b) Verify the global mass balance on a particular Cartesian unit cubical element defined by R: $\{1 \le x_1 \le 2; 0 \le x_2 \le 1; 0 \le x_3 \le 1\}$.

Solution: First compute the velocity field: $v_1 = \dfrac{Dx_1}{Dt} = bX_1 = \dfrac{bx_1}{a + bt}, v_2 = v_3 = 0$

(a) Differential mass balance check:

$$\frac{\partial \rho}{\partial t} + (\rho v_k),_k = 0 \Rightarrow \frac{\partial \rho}{\partial t} + \rho v_1,_1 = 0 \Rightarrow -\frac{\rho_o b}{(a+bt)^2} + \frac{\rho_o}{a+bt}\frac{b}{a+bt} = 0$$

(b) For global mass balance check on particular element, use (5.2.1) and (5.2.2). With $v_2 = v_3 = 0$, only need to check x_1-faces for surface integrals $\Rightarrow$

$$\frac{D}{Dt}\int_{R_m} \rho dV = \int_R \frac{\partial}{\partial t}\rho\, dv + \int_{\partial R} \rho v_1 n_1\, ds$$

$$= \int_0^1\int_0^1\int_1^2 \frac{\partial}{\partial t}\frac{\rho_o}{a+bt} dx_1\, dx_2\, dx_3 + \frac{\rho_o}{a+bt}\frac{b(1)}{a+bt}(-1)\int_0^1\int_0^1 dx_2\, dx_3$$

$$+ \frac{\rho_o}{a+bt}\frac{b(2)}{a+bt}(1)\int_0^1\int_0^1 dx_2\, dx_3$$

$$= -\frac{\rho_o b}{(a+bt)^2} + \frac{\rho_o b}{(a+bt)^2} = 0$$

5.3 CONSERVATION OF LINEAR MOMENTUM

The principle of conservation of linear momentum is basically a statement of Newton's second law for a collection of particles. It can be stated as *the time rate of change of the total linear momentum of a given group of continuum particles equals the sum of all the external forces acting on the group*. This concept is valid provided that Newton's third law of action–reaction governs the internal forces between the particles and thus all internal forces will cancel each other when summed over the entire system. Thus, consider a fixed group of continuum particles instantaneously occupying a region of space R and acted upon by external surfaces forces $\boldsymbol{t}$ and body forces $\boldsymbol{b}$ as shown in Fig. 5.2. Using previous force relations (4.1.1) and (4.1.2), we may express this concept as

$$\frac{D}{Dt}\int_R \rho \boldsymbol{v}\, dv = \int_{\partial R} \boldsymbol{t}\, ds + \int_R \rho \boldsymbol{b}\, dv \tag{5.3.1}$$

Shifting to index notation and introducing the Cauchy stress tensor

$$\begin{aligned}\frac{D}{Dt}\int_R \rho v_i\, dv &= \int_{\partial R} t_i\, ds + \int_R \rho b_i\, dv \\ &= \int_{\partial R} T_{ji} n_j\, ds + \int_R \rho b_i\, dv \\ &= \int_R T_{ji,j}\, dv + \int_R \rho b_i\, dv\end{aligned} \tag{5.3.2}$$

where we have used the Divergence Theorem. Employing the Reynolds Transport Theorem (5.1.2) and the conservation of mass (5.2.4) allows the evaluation

$$\begin{aligned}\frac{D}{Dt}\int_R \rho v_i\, dv &= \int_R \frac{D}{Dt}(\rho v_i) dv + \int_R \rho v_i v_{k,k}\, dv \\ &= \int_R \left(\frac{D\rho}{Dt} v_i + \rho \frac{Dv_i}{Dt}\right) dv + \int_R \rho v_i v_{k,k}\, dv \\ &= \int_R \rho \frac{Dv_i}{Dt} dv + \int_R v_i \left(\frac{D\rho}{Dt} + \rho v_{k,k}\right) dv \\ &= \int_R \rho \frac{Dv_i}{Dt} dv = \int_R \rho a_i\, dv\end{aligned} \tag{5.3.3}$$

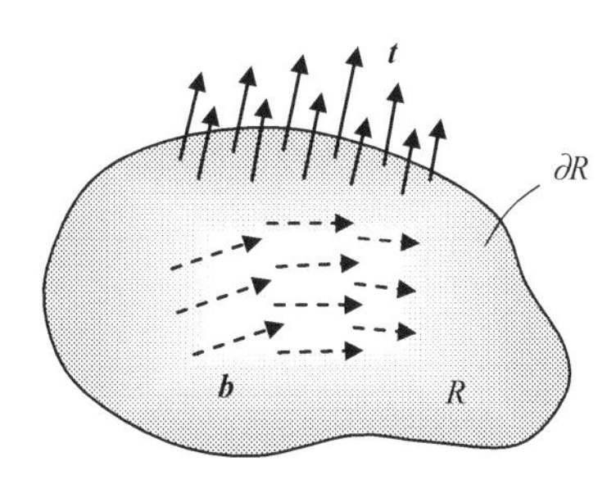

FIGURE 5.2

Body and surface forces acting on arbitrary portion of a continuum.

where a_i is the acceleration field. This result would also come more directly from (5.2.9).

Combining (5.3.2) and (5.3.3) yields

$$\int_R (T_{ji,j} + \rho b_i - \rho a_i)\, dv = 0 \tag{5.3.4}$$

Again using the Localization Theorem implies that the integrand must vanish

$$T_{ji,j} + \rho b_i = \rho a_i \tag{5.3.5}$$

or in a vector format

$$\nabla \boldsymbol{T}^T + \rho \boldsymbol{b} = \rho \boldsymbol{a} = \rho\left(\frac{\partial \boldsymbol{v}}{\partial t} + \boldsymbol{v} \cdot \nabla \boldsymbol{v}\right) \tag{5.3.6}$$

or in a scalar format

$$\begin{aligned}
\frac{\partial T_{11}}{\partial x_1} + \frac{\partial T_{21}}{\partial x_2} + \frac{\partial T_{31}}{\partial x_3} + \rho b_1 &= \rho a_1 \\
\frac{\partial T_{12}}{\partial x_1} + \frac{\partial T_{22}}{\partial x_2} + \frac{\partial T_{32}}{\partial x_3} + \rho b_2 &= \rho a_2 \\
\frac{\partial T_{13}}{\partial x_1} + \frac{\partial T_{23}}{\partial x_2} + \frac{\partial T_{33}}{\partial x_3} + \rho b_3 &= \rho a_3
\end{aligned} \tag{5.3.7}$$

Relations (5.3.5)–(5.3.7) are known as *Cauchy's equations of motion*—a differential form of the conservation of linear momentum. For the case of small deformations, the acceleration expression (3.8.3) $\boldsymbol{a}(\boldsymbol{x},t) = \dfrac{\partial^2 \boldsymbol{u}}{\partial t^2}$ is then used in the equations of motion. When the problem is in static equilibrium where there is no or negligible acceleration, Eqs. (5.3.5) reduce to the *equations of equilibrium*:

$$T_{ji,j} + \rho b_i = 0 \tag{5.3.8}$$

The previous equations of motion or equilibrium were expressed in terms of the Cauchy stress in the current configuration. We now wish to develop the conservation of linear momentum in the reference configuration by employing the first Piola–Kirchhoff (PK1) stress tensor $\boldsymbol{T}^o$ previously defined in relation (4.7.5), $\boldsymbol{T}^o = J\boldsymbol{T}(\boldsymbol{F}^{-1})^T$ or by the inverted relation (4.7.6) $\boldsymbol{T} = J^{-1}\boldsymbol{T}^o\boldsymbol{F}^T$. From our previous conservation of mass work

$$\rho_o\, dV = \rho\, dv \tag{5.3.9}$$

and from (4.7.1) and (4.7.2), we can write

$$t_i\, ds = T_i^R\, dS = T_{ij}^o N_j\, dS \tag{5.3.10}$$

Using these results, relations (5.3.2) and (5.3.3) can be transformed to the reference configuration as

$$\int_{R_o} \rho_o \frac{D}{Dt} v_i \, dV = \int_{\partial R_o} T_i^R \, dS + \int_{R_o} \rho_o B_i \, dV$$
$$= \int_{\partial R_o} T_{ij}^o N_j \, dS + \int_{R_o} \rho_o B_i \, dV \tag{5.3.11}$$

where B_i is the body force density in the reference configuration and all variables are functions of the material coordinates $\boldsymbol{X}$. Using the Divergence Theorem again on the surface integral term gives

$$\int_{R_o} \left(T_{ij,j}^o + \rho_o B_i - \rho_o \frac{Dv_i}{Dt} \right) dV = 0 \tag{5.3.12}$$

Finally invoking the Localization Theorem gives the differential form

$$\frac{\partial T_{ij}^o}{\partial X_j} + \rho_o B_i = \rho_o a_i \tag{5.3.13}$$

Result (5.3.13) then represents the desired form, the *conservation of linear momentum in the reference configuration*. This result can also be derived from the current configuration form (5.3.5) by eliminating the Cauchy stress using (4.7.6) and employing some detailed tensor analysis steps (see Exercise 5.11).

EXAMPLE 5.3.1 BODY FORCE: PRISMATIC BAR UNDER SELF-WEIGHT

Consider a simple static example of a uniform prismatic bar being loaded under its own self-weight as shown in Fig. 5.3. The bar is fixed at the top end, and assume a one-dimensional stress field with T_{11} being the only nonzero stress. Develop the equation of equilibrium for this case and integrate the single equation to determine the nonzero stress component.

Solution: The body force density (per unit mass) for this problem with the given coordinate system would be given by $b_1 = -g, b_2 = b_3 = 0$, where g is the local acceleration of gravity. For this case, the only nonzero equilibrium equation becomes

$$\frac{\partial T_{11}}{\partial x_1} + \rho b_1 = 0 \Rightarrow \frac{\partial T_{11}}{\partial x_1} - \rho g = 0$$

Note that since the bar is uniform, the mass density ρ is constant. This simple differential equation is easily integrated to give the stress result $T_{11} = \rho g x_1$, where the constant of integration has been set to zero to satisfy the boundary condition $T_{11}(0) = 0$. This is a rare case where we can solve for the stress without incorporating a material constitutive equation.

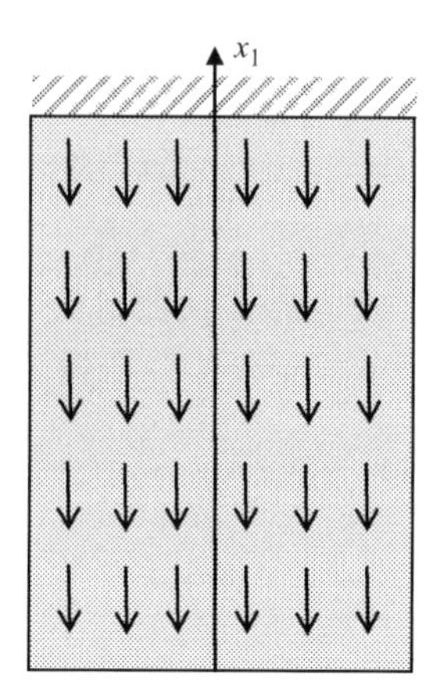

FIGURE 5.3

Bar under self-weight.

5.4 CONSERVATION OF MOMENT OF MOMENTUM

The angular or moment of momentum principle is another form of Newton's second law. For our application, it states that *the time rate of change of the total angular momentum of a group of particles must equal the sum of all external moments acting on the system with respect to some arbitrary point in space*. This concept is again valid provided that Newton's third law of action–reaction governs the internal forces between the particles and thus the moments of all internal forces will cancel each other.

So we again consider a fixed group of continuum particles instantaneously occupying a region of space R and acted upon by external surfaces forces $\boldsymbol{t}$ and body forces $\boldsymbol{b}$ as was shown in Fig. 5.2. Neglecting any distributed body of surface couples, we then can write

$$\frac{D}{Dt}\int_R \boldsymbol{r}\times\rho\boldsymbol{v}\,dv = \int_{\partial R}\boldsymbol{r}\times\boldsymbol{t}\,ds + \int_R \boldsymbol{r}\times\rho\boldsymbol{b}\,dv \tag{5.4.1}$$

where $\boldsymbol{r}$ is the position vector with respect to some arbitrary point. Working in index notation

$$\begin{aligned}
&\frac{D}{Dt}\int_R \varepsilon_{ijk}x_j\rho v_k\,dv = \int_{\partial R}\varepsilon_{ijk}x_jt_k\,ds + \int_R \varepsilon_{ijk}x_j\rho b_k\,dv\\
&\int_R \varepsilon_{ijk}\frac{D}{Dt}(x_jv_k)\rho\,dv = \int_R \varepsilon_{ijk}(x_jT_{lk})_{,l}\,dv + \int_R \varepsilon_{ijk}x_j\rho b_k\,dv\\
&\int_R \varepsilon_{ijk}[v_jv_k\rho + x_ja_k\rho - T_{jk} - x_jT_{lk,l} + x_j\rho b_k]dv = 0\\
&\int_R \varepsilon_{ijk}(v_jv_k\rho - T_{jk} - [T_{lk,l} + \rho b_k - \rho a_k]x_j)\,dv = 0
\end{aligned} \tag{5.4.2}$$

Now since $\varepsilon_{ijk}v_jv_k = 0$ (symmetric times antisymmetric) and $T_{lk,l} + \rho b_k - \rho a_k = 0$ from the equations of motion (5.3.5), we get

$$\int_R \varepsilon_{ijk}T_{jk}\,dv = 0 \rightarrow \varepsilon_{ijk}T_{jk} = 0 \tag{5.4.3}$$

Since ε_{ijk} is antisymmetric, then T_{jk} must be symmetric, and this proves the *symmetry of the Cauchy stress tensor.* Thus,

$$\begin{aligned} &T_{ij} = T_{ji}, \boldsymbol{T} = \boldsymbol{T}^{\mathrm{T}} \Rightarrow \\ &T_{12} = T_{21} \\ &T_{23} = T_{32} \\ &T_{31} = T_{13} \end{aligned} \tag{5.4.4}$$

Relation (5.4.4), sometimes called *Cauchy's Second Law of Motion*, is the result from the balance of the moment of momentum. Recall that this result hinges on the assumption of no distributed body or surface couples. We will later explore some specialized continuum mechanics theories for *polar materials* where we wish to include such internal loadings, and for such cases the Cauchy stress will no longer be symmetric.

5.5 CONSERVATION OF LINEAR MOMENTUM EQUATIONS IN CYLINDRICAL AND SPHERICAL COORDINATES

Similar to Sections 3.17 and 4.10, we now wish to list the Cauchy equations of motion (5.3.7) in curvilinear cylindrical and spherical coordinate systems. In direct notation, these equations read

$$\nabla \boldsymbol{T} + \rho \boldsymbol{b} = \rho \boldsymbol{a} \tag{5.5.1}$$

where we have used the symmetry of the stress tensor. The various curvilinear components of the traction vector and stress tensor have been previously defined in Section 4.10. The spatial acceleration term follows from relation (3.4.3):

$$\boldsymbol{a}(\boldsymbol{x},t) = \frac{\partial \boldsymbol{v}}{\partial t} + \boldsymbol{v} \cdot \nabla \boldsymbol{v} \tag{5.5.2}$$

Section 2.18 outlined the basic procedures to determine the vector differential operator and various gradient operations necessary to construct the specific equations. Here, we only provide a listing of the final results. Appendix A includes these results along with other field equation forms.

Equations of motion in cylindrical coordinates:

$$\begin{aligned} &\frac{\partial T_{rr}}{\partial r} + \frac{1}{r}\frac{\partial T_{r\theta}}{\partial \theta} + \frac{\partial T_{rz}}{\partial z} + \frac{1}{r}(T_{rr} - T_{\theta\theta}) + \rho b_r \\ &= \rho\left(\frac{\partial v_r}{\partial t} + v_r \frac{\partial v_r}{\partial r} + \frac{v_\theta}{r}\frac{\partial v_r}{\partial \theta} + v_z \frac{\partial v_r}{\partial z} - \frac{v_\theta^2}{r}\right) \\ &\frac{\partial T_{r\theta}}{\partial r} + \frac{1}{r}\frac{\partial T_{\theta\theta}}{\partial \theta} + \frac{\partial T_{\theta z}}{\partial z} + \frac{2}{r}T_{r\theta} + \rho b_\theta \\ &= \rho\left(\frac{\partial v_\theta}{\partial t} + v_r \frac{\partial v_\theta}{\partial r} + \frac{v_\theta}{r}\frac{\partial v_\theta}{\partial \theta} + v_z \frac{\partial v_\theta}{\partial z} + \frac{v_r v_\theta}{r}\right) \\ &\frac{\partial T_{rz}}{\partial r} + \frac{1}{r}\frac{\partial T_{\theta z}}{\partial \theta} + \frac{\partial T_{zz}}{\partial z} + \frac{1}{r}T_{rz} + \rho b_z \\ &= \rho\left(\frac{\partial v_z}{\partial t} + v_r \frac{\partial v_z}{\partial r} + \frac{v_\theta}{r}\frac{\partial v_z}{\partial \theta} + v_z \frac{\partial v_z}{\partial z}\right) \end{aligned} \tag{5.5.3}$$

Equations of motion in spherical coordinates:

$$
\begin{aligned}
&\frac{\partial T_{RR}}{\partial R}+\frac{1}{R}\frac{\partial T_{R\phi}}{\partial \phi}+\frac{1}{R\sin\phi}\frac{\partial T_{R\theta}}{\partial \theta}+\frac{1}{R}(2T_{RR}-T_{\phi\phi}-T_{\theta\theta}+T_{R\phi}\cot\phi)+\rho b_R \\
&=\rho\left(\frac{\partial v_R}{\partial t}+v_R\frac{\partial v_R}{\partial R}+\frac{v_\phi}{R}\frac{\partial v_R}{\partial \phi}+\frac{v_\theta}{R\sin\phi}\frac{\partial v_R}{\partial \theta}-\frac{v_\phi^2+v_\theta^2}{R}\right) \\
&\frac{\partial T_{R\phi}}{\partial R}+\frac{1}{R}\frac{\partial T_{\phi\phi}}{\partial \phi}+\frac{1}{R\sin\phi}\frac{\partial T_{\phi\theta}}{\partial \theta}+\frac{1}{R}[(T_{\phi\phi}-T_{\theta\theta})\cot\phi+3T_{R\phi}]+\rho b_\phi \\
&=\rho\left(\frac{\partial v_\phi}{\partial t}+v_R\frac{\partial v_\phi}{\partial R}+\frac{v_\phi}{R}\frac{\partial v_\phi}{\partial \phi}+\frac{v_\theta}{R\sin\phi}\frac{\partial v_\phi}{\partial \theta}+\frac{v_R v_\phi}{R}-\frac{v_\theta^2\cot\phi}{R}\right) \\
&\frac{\partial T_{R\theta}}{\partial R}+\frac{1}{R}\frac{\partial T_{\phi\theta}}{\partial \phi}+\frac{1}{R\sin\phi}\frac{\partial T_{\theta\theta}}{\partial \theta}+\frac{1}{R}(2T_{\phi\theta}\cot\phi+3T_{R\theta})+\rho b_\theta \\
&=\rho\left(\frac{\partial v_\theta}{\partial t}+v_R\frac{\partial v_\theta}{\partial R}+\frac{v_\phi}{R}\frac{\partial v_\theta}{\partial \phi}+\frac{v_\theta}{R\sin\phi}\frac{\partial v_\theta}{\partial \theta}+\frac{v_\theta v_R}{R}+\frac{v_\phi v_\theta\cot\phi}{R}\right)
\end{aligned}
\tag{5.5.4}
$$

It is again noticeable that these cylindrical and spherical forms are considerably more complicated in form than the corresponding Cartesian relations. However, for many problems, these curvilinear equations will allow solutions to be generated that would not be possible using a Cartesian formulation.

5.6 CONSERVATION OF ENERGY

Our final conservation principle involves the energy balance, and this has the potential to bring into play many possible energy forms including mechanical, thermal, chemical, electrical, magnetic, and others. We will generally limit our study here to problems involving only mechanical and thermal phenomena and thus will develop various *thermomechanical material models*. However, continuum mechanics theories can easily be constructed in a similar manner to handle much broader physical behaviors.

The principle of conservation of energy may be stated as *the time rate of change of the kinetic and internal energy of a given group of continuum particles is equal to the sum of the rate of change of work done by the external forces, and the energy entering the system through the boundary*. The *internal energy* represents all energy associated with the microscopic modes of continuum motion which are not accounted for in the bulk mechanical energy terms. For most applications, the energy entering through the boundary will be in the form of *thermal energy*. Note that we can also include internal distributed sources as energy entering the system. This energy principle is usually referred to as the *First Law of Thermodynamics*. A more complete description of thermodynamics can be found in Tadmor et al. (2012).

Using our control volume shown in Fig. 5.1 to hold our fixed group of particles, the *kinetic energy* of the group is given by

$$
K=\int_R \frac{1}{2}v^2\rho\, dv \tag{5.6.1}
$$

where $v^2 = \boldsymbol{v} \cdot \boldsymbol{v}$ is the square of the velocity field within R. The *internal energy* may be written as

$$E = \int_R \varepsilon \rho \, dv \tag{5.6.2}$$

where $\varepsilon = \varepsilon(\boldsymbol{x}, t)$ is the *internal energy density per unit mass* and is a thermodynamic state variable. The rate of working of the external surface and body forces, often referred to as the *external mechanical power*, can be expressed by

$$P_{ext} = \int_{\partial R} \boldsymbol{t} \cdot \boldsymbol{v} \, ds + \int_R \boldsymbol{b} \cdot \boldsymbol{v} \rho \, dv \tag{5.6.3}$$

Relation (5.6.3) can be rewritten in several alternative forms. Working on the surface integral term

$$\begin{aligned} \int_{\partial R} \boldsymbol{t} \cdot \boldsymbol{v} \, ds &= \int_{\partial R} t_i v_i \, ds = \int_{\partial R} T_{ij} n_j v_i \, ds \\ &= \int_R (T_{ij} v_i)_{,j} \, dv = \int_R (T_{ij,j} v_i + T_{ij} v_{i,j}) \, dv \end{aligned} \tag{5.6.4}$$

where we have used the Divergence Theorem to change the surface integral to volume integral. Next we note that using (3.13.3), $T_{ij} v_{i,j} = T_{ij}(D_{ij} + W_{ij}) = T_{ij} D_{ij}$, since W_{ij} is antisymmetric. Thus, the external mechanical power relation can be written as

$$\begin{aligned} P_{ext} &= \int_R [(T_{ij,j} + \rho b_i) v_i + T_{ij} D_{ij}] \, dv \\ &= \int_R [\rho a_i v_i + T_{ij} D_{ij}] \, dv \end{aligned} \tag{5.6.5}$$

The term $\int_R T_{ij} D_{ij} \, dv$ is commonly referred to as the *stress power*. The two variables T_{ij} and D_{ij} are often called *energetic conjugates*, since the integral of the double dot product gives the system energy due to the stress and strain fields. Introducing the first and second Piola–Kirchhoff stress tensors T_{ij}^o and S_{ij}, the stress power can be expressed in the following equivalent relations in the reference configuration:

$$\int_R T_{ij} D_{ij} \, dv = \int_{R_o} T_{ij}^o \dot{F}_{ij} \, dV = \int_{R_o} S_{ij} \dot{E}_{ij} \, dV \tag{5.6.6}$$

where $\dot{F}_{ij}$ is the material time rate of the deformation gradient tensor and $\dot{E}_{ij}$ is the material time rate of the Lagrangian strain (3.6.2). Notice that new conjugate stress-deformation rate pairs are developed in these alternative forms.

For our thermomechanical study, the energy entering system R both through the boundary ∂R and from internal sources may be written as

$$Q = -\int_{\partial R} \boldsymbol{q} \cdot \boldsymbol{n} \, ds + \int_R h \rho \, dv \tag{5.6.7}$$

with $\boldsymbol{q}$ being the *rate of heat flux per unit area* and h is the *specific energy source (supply) per unit mass*. A common example of a source term would be from a radioactive material generating heat in a distributed fashion.

The general energy balance statement then reads

$$\dot{K} + \dot{E} = P_{ext} + Q \tag{5.6.8}$$

and substituting in the previous specific results for the various energy pieces gives

$$\frac{D}{Dt}\int_R \frac{1}{2}v^2 \rho\, dv + \frac{D}{Dt}\int_R \varepsilon\rho\, dv = \int_R [\rho a_i v_i + T_{ij}D_{ij}]dv - \int_{\partial R} q_i n_i\, ds + \int_R h\rho\, dv \tag{5.6.9}$$

Using the Reynolds Transport and Divergence Theorems, this result can be expressed as

$$\int_R \rho \frac{D}{Dt}\left(\frac{1}{2}v^2 + \varepsilon\right)dv = \int_R [\rho a_i v_i + T_{ij}D_{ij}]dv - \int_R q_{i,i}\, dv + \int_R h\rho\, dv \tag{5.6.10}$$

Now the term $\rho a_i v_i = \rho \frac{Dv_i}{Dt} v_i = \frac{1}{2}\rho \frac{Dv^2}{Dt}$, and thus (5.6.8) reduces to

$$\int_R (\rho\dot{\varepsilon} - T_{ij}D_{ij} + q_{i,i} - \rho h)\, dv = 0 \tag{5.6.11}$$

Again employing the Localization Theorem (2.17.16), the integrand itself must vanish

$$\begin{aligned} &\rho\dot{\varepsilon} - T_{ij}D_{ij} + q_{i,i} - \rho h = 0 \\ &\rho\dot{\varepsilon} - tr(\boldsymbol{TD}) + \nabla \cdot \boldsymbol{q} - \rho h = 0 \end{aligned} \tag{5.6.12}$$

which is the *differential form of the energy balance equation* for thermomechanical continuum mechanics. Clearly, differential relation (5.6.12) is written in terms of local spatial variables, but could be restated in the reference configuration using (5.6.6) for the stress power and simply converting the other scalar variables to reference values. Since relation (5.6.12) involves additional unknowns of internal energy ε, and the heat flux vector $\boldsymbol{q}$ (usually the source term h is given), the energy equation requires additional relations among ε, $\boldsymbol{q}$, and the temperature θ. These addition relations are normally involved with *thermodynamical constitutive equations* sometimes called *equations of state*. These will be discussed later in Chapter 7.

EXAMPLE 5.6.1 STRESS POWER CALCULATION FOR SIMPLE SHEAR

Determine the stress power term $T_{ij}D_{ij}$, if the continuum undergoes simple shearing motion and has the particular Cauchy plane stress form

$$\begin{aligned} x_1 &= X_1 + \gamma(t)X_2 \\ x_2 &= X_2 \\ x_3 &= X_3 \end{aligned} \qquad T_{ij} = \begin{bmatrix} T_{11} & T_{12} & 0 \\ T_{12} & T_{22} & 0 \\ 0 & 0 & 0 \end{bmatrix}$$

Solution: For simple shearing deformation, the velocity field becomes $v_1 = \dot{\gamma}(t)x_2, v_2 = v_3 = 0$, and thus the velocity gradient and rate of deformation tensors are

$$\boldsymbol{L} = \frac{\partial \boldsymbol{v}}{\partial \boldsymbol{x}} = \begin{bmatrix} 0 & \dot{\gamma}(t) & 0 \\ 0 & 0 & 0 \\ 0 & 0 & 0 \end{bmatrix} \Rightarrow \boldsymbol{D} = \frac{1}{2}\left(\boldsymbol{L} + \boldsymbol{L}^T\right) = \frac{1}{2}\begin{bmatrix} 0 & \dot{\gamma}(t) & 0 \\ \dot{\gamma}(t) & 0 & 0 \\ 0 & 0 & 0 \end{bmatrix}$$

The stress power term is then

$$T_{ij}D_{ij} = tr(\boldsymbol{TD}) = \frac{1}{2}tr\left\{\begin{bmatrix} T_{11} & T_{12} & 0 \\ T_{12} & T_{22} & 0 \\ 0 & 0 & 0 \end{bmatrix}\begin{bmatrix} 0 & \dot{\gamma}(t) & 0 \\ \dot{\gamma}(t) & 0 & 0 \\ 0 & 0 & 0 \end{bmatrix}\right\}$$

$$= \frac{1}{2}tr\begin{bmatrix} T_{12}\dot{\gamma}(t) & T_{11}\dot{\gamma}(t) & 0 \\ T_{22}\dot{\gamma}(t) & T_{12}\dot{\gamma}(t) & 0 \\ 0 & 0 & 0 \end{bmatrix} = T_{12}\dot{\gamma}(t)$$

5.7 SECOND LAW OF THERMODYNAMICS—ENTROPY INEQUALITY

The previous section developed the basic energy balance or conservation of energy principle commonly known as the first law of thermodynamics. The differential form (5.6.12) can be viewed as a measure of the interconvertibility of heat and work while maintaining a proper energy balance. However, the expression provides no restrictions on the direction of any such interconvertibility processes, and this is an important issue in thermomechanical behavior of materials. When considering thermal effects with dissipation phenomena, the direction of energy transfer must satisfy certain criteria, and this introduces *irreversible processes*. For example, a process in which friction changes mechanical energy into heat energy cannot be reversed. Another common observable restriction is that heat only flows from warmer regions to cooler regions and not the other way. Collectively, such restrictions are connected to the *second law of thermodynamics*. Of course various restrictions relate to the second law in different ways, and we wish to establish a mathematical relationship applicable to the thermomechanical behavior of continuum materials. One particular mathematical statement associated with the second law is the *Clausius–Duhem entropy inequality*, and this relation has broad applications for many materials we wish to study. As we shall see in later chapters, this inequality will place restrictions on the material response functions. More detailed background on this topic can be found in Haupt (2002), Holzapfel (2006), Asaro and Lubarda (2006), and Tadmor et al. (2012).

Basic to our development in this section is the definition of *entropy*. This rather abstract variable can be interpreted as a measure of the microscopic randomness or disorder of the continuum system. In classical thermodynamics, it is commonly defined as a state function related to heat transfer. For a reversible process, the *entropy per unit mass*, $s(\boldsymbol{x}, t)$ is commonly defined by the relation

$$ds = \left(\frac{\delta q}{\theta}\right)_{rev} \tag{5.7.1}$$

where θ is the absolute temperature (Kelvin-scale, always positive) and δq is the heat input per unit mass over a *reversible process*. Since relation (5.7.1) is an exact differential, we may write it between two states 1 and 2, or for an entire cycle as

$$\Delta s = s_2 - s_1 = \int_1^2 \left(\frac{\delta q}{\theta}\right)_{rev} \quad \text{or} \quad \oint ds = \oint \left(\frac{\delta q}{\theta}\right)_{rev} = 0 \tag{5.7.2}$$

However, for irreversible processes (the real world), observations indicate that

$$\oint \left(\frac{\delta q}{\theta}\right)_{irrev} < 0 \tag{5.7.3}$$

Since we interpret $\delta q / \theta$ as the entropy input from the heat input δq, we conclude that over an irreversible cycle, the *net entropy input is negative.* However, as entropy is assumed to be a state variable, it must return to its initial value at the end of any cycle. Because of this, the negative entropy input shown in (5.7.3) implies that entropy has been created inside the system. In other words, *dissipative irreversible processes produce a positive internal entropy production.* Therefore, for an irreversible change of state $1 \rightarrow 2$, the entropy increase will be greater than the entropy input by heat transfer

$$\Delta s > \int_1^2 \left(\frac{\delta q}{\theta}\right)_{irrev} \tag{5.7.4}$$

The previous relations form the fundamental basis for the second law.

However, for use in continuum mechanics, the second law is normally rephrased in a different form. First, consider a fixed group (closed system) of continuum particles R_m that occupying spatial region R:

$$\textit{Entropy input rate} = \int_R \frac{\rho h}{\theta} dv - \int_{\partial R} \frac{\boldsymbol{q} \cdot \boldsymbol{n}}{\theta} ds \tag{5.7.5}$$

with again $\boldsymbol{q}$ being the *rate of heat flux per unit area* and h the *specific energy source per unit mass*. Note that the ds term in the surface integral is the differential surface area and not the differential entropy. Now according to relation (5.7.4), the rate of entropy increase in R must be greater than or equal to (for the reversible case) the entropy input rate, and thus

$$\frac{D}{Dt}\int_R s\rho\, dv \geq \int_R \frac{\rho h}{\theta} dv - \int_{\partial R} \frac{\boldsymbol{q} \cdot \boldsymbol{n}}{\theta} ds \tag{5.7.6}$$

Using our usual procedures on the integral formulations, we apply relation (5.2.9) and the Divergence Theorem to get

$$\int_R \left(\dot{s}\rho - \frac{\rho h}{\theta} + \left(\frac{q_i}{\theta}\right)_{,i} \right) dv \geq 0 \tag{5.7.7}$$

and this implies

$$\dot{s} \geq \frac{h}{\theta} - \frac{1}{\rho}\left(\frac{q_i}{\theta}\right)_{,i}$$
$$\dot{s} \geq \frac{h}{\theta} - \frac{1}{\rho}\nabla \cdot \left(\frac{\boldsymbol{q}}{\theta}\right) \tag{5.7.8}$$

Relations (5.7.7) and (5.7.8) are known as the *integral and differential forms of the Clausius–Duhem inequality*. They represent forms of the second law of thermodynamics for continuum mechanics applications.

It is easily shown that (5.7.8) can also be expressed as

$$\dot{s} \geq \frac{h}{\theta} - \frac{1}{\rho\theta}(\nabla \cdot \boldsymbol{q}) + \frac{1}{\rho\theta^2}(\boldsymbol{q} \cdot \nabla\theta) \tag{5.7.9}$$

By using the energy equation (5.6.12), the entropy inequality can be expressed as

$$\rho(\theta\dot{s} - \dot{\varepsilon}) + T_{ij}D_{ij} - \frac{1}{\theta}(\boldsymbol{q} \cdot \nabla\theta) \geq 0 \tag{5.7.10}$$

which is sometimes referred to as the *reduced Clausius–Duhem or Dissipation inequality*. In some applications, the *free energy* $\Psi = \varepsilon - s\theta$, is introduced, and relation (5.7.10) would become

$$-\rho(\dot{\Psi} + s\dot{\theta}) + T_{ij}D_{ij} - \frac{1}{\theta}(\boldsymbol{q} \cdot \nabla\theta) \geq 0 \tag{5.7.11}$$

It should also be pointed out that the observable and accepted concept that heat only flows from regions of higher temperature to lower temperature implies that

$$\boldsymbol{q} \cdot \nabla\theta \leq 0 \tag{5.7.12}$$

with equality only if $\nabla\theta = 0$. This relation is sometimes called the *classical heat conduction inequality*. Using this relation with (5.7.9), we can argue the stronger statement

$$\dot{s} \geq \frac{h}{\theta} - \frac{1}{\rho\theta}(\nabla \cdot \boldsymbol{q}) \tag{5.7.13}$$

which is known as the *Clausius–Planck inequality*. Another form of this relation can be found using (5.7.10):

$$\rho(\theta\dot{s} - \dot{\varepsilon}) + T_{ij}D_{ij} \geq 0 \tag{5.7.14}$$

5.8 SUMMARY OF CONSERVATION LAWS, GENERAL PRINCIPLES, AND UNKNOWNS

Since so many laws and principles have been developed in this chapter, we now provide a listing of these basic relations along with the associated unknowns appearing in the equations. Table 5.1 illustrates these relations and summarizes the

Table 5.1 Summary of general equations and unknowns

Equation	No. of Equations	Unknowns (No Repeats)	No. of Unknowns
Conservation of mass: $\frac{D\rho}{Dt} + \rho\nabla\cdot v = 0$	1	Density, velocity	4
Conservation of linear momentum: $\nabla T^T + \rho b = \rho a = \frac{\partial v}{\partial t} + v\cdot\nabla v$	3	Cauchy stress	6
Conservation of angular momentum: $T = T^T$	—	—	—
Conservation of energy: $\rho\dot{\varepsilon} - \mathrm{tr}(TD) + \nabla\cdot q - \rho h = 0$	1	Internal energy, heat flux vector	4
Clausius–Duhem inequality: $\dot{s} \geq \frac{h}{\theta} - \frac{1}{\rho}\nabla\cdot\left(\frac{q}{\theta}\right)$	—	Entropy, temperature	2
Strain–displacement: $E = \frac{1}{2}[(\nabla u) + (\nabla u)^T + (\nabla u)^T(\nabla u)]$	6	Strain or strain rate	6
Strain Rate–velocity: $D = \frac{1}{2}(\nabla v + (\nabla v)^T)$			
Total	11		22

number of equations and unknowns for each general principle. We have also included a fundamental kinematic relation from Chapter 3 that relates either the strain and displacements or strain rates and velocities. The conservation of angular momentum simply gives symmetry of the Cauchy stress tensor (for nonpolar materials). This result is automatically incorporated into all other relations, and so this system of equations (5.4.4) is generally not included in the table total. Likewise, although the Clausius–Duhem inequality will place restrictions on material behavior, it is often not considered a governing equation and is used only on an irregular basis. Under these conditions, the thermomechanical system coming from conservation principles and kinematics totals to 11 governing equations with 22 unknowns including mass density, velocity (or equivalently motion or displacement), stress, internal energy, heat flux, and temperature. Entropy would be added to the unknown listing if the second law is included. For the reduced problem of a nonthermal mechanical system, the energy equation is dropped and thus the number of equations reduces to 10, while the unknowns reduce to 16 including mass density, velocity (or equivalently motion or displacement), and stress. Note that the body force and heat source variables are assumed to be given *a priori*.

For both mechanical and thermomechanical problems, our number counting finds insufficient numbers of equations to solve for all of the unknown variables. For the mechanical model, we need six more equations, while the thermomechanical case needs 10 additional relations. These additional equations are needed to properly close the mathematical system. Up until this point, our field equations have been applicable to all continuum materials irrespective of their material properties. We have not brought into consideration any specific material behavior relations among either mechanical or thermodynamic variables. From common experience, we know that different materials will behave differently even under identical loading conditions. Therefore, the missing equations are to be found from *material response relations or constitutive relations*. These will typically involve stress–strain and/or stress–strain rate mechanical relations, and thermodynamic equations involving heat flux–temperature and internal energy specification. The next few chapters will develop these relations in detail, and all of this will eventually be included into overall system models in order to formulate complete continuum theories for a large variety of material types.

REFERENCES

Asaro, R.J., Lubarda, V.A., 2006. Mechanics of Solids and Materials. Cambridge University Press, New York.

Haupt, P., 2002. Continuum Mechanics and Theory of Materials, Second ed Springer, Berlin.

Holzapfel, G.A., 2006. Nonlinear Solid Mechanics: A Continuum Approach for Engineering. John Wiley & Sons, West Sussex.

Tadmor, E.B., Miller, R.E., Elliott, R.S., 2012. Continuum Mechanics and Thermodynamics. Cambridge University Press, Cambridge.

EXERCISES

5.1 Explicitly justify relation (5.2.9).

5.2 Using the results in Section 2.18, develop the conservation of mass equation for cylindrical coordinates

$$\frac{\partial \rho}{\partial t}+\frac{1}{r}\left[\frac{\partial(r\rho v_r)}{\partial r}+\frac{\partial(\rho v_\theta)}{\partial \theta}+r\frac{\partial(\rho v_z)}{\partial z}\right]=0$$

5.3 Show that if the spatial velocity field is expressible by $\boldsymbol{v}=-\nabla\Phi$, then $\nabla\times\boldsymbol{v}=0$ and if the density is a constant (incompressible case), then conservation of mass implies that $\nabla^2\Phi=0$.

5.4 For the incompressible case, check if the given velocity fields satisfy conservation of mass

(a) $v_1=a(t)x_1, v_2=b(t)x_2, v_3=c(t)x_3$, with $a(t)+b(t)+c(t)=0$

(b) $v_1=kx_1x_2, v_2=-\frac{k}{2}x_2^2, v_3=0 \quad k=constant$

(c) $v_1=2k(x_1^2-x_2^2), v_2=-2kx_1x_2, v_3=kx_1x_2, \quad k=constant$

5.5 For two-dimensional flow of an incompressible fluid, the velocity field can be represented by a scalar *Lagrange stream function* $\psi(x_1, x_2)$ as $v_1 = \frac{\partial \psi}{\partial x_2}, v_2 = -\frac{\partial \psi}{\partial x_1}$. Show that this representation satisfies conservation of mass for any ψ with appropriate continuous derivatives.

5.6 In an *ideal nonviscous fluid*, there can be no shear stress, and so the Cauchy stress tensor must be hydrostatic (spherical), i.e. $T_{ij} = -p\delta_{ij}$. For this case, show that the equations of motion reduce to *Euler's equations of motion*

$$-\frac{1}{\rho}\nabla p + \boldsymbol{b} = \frac{\partial \boldsymbol{v}}{\partial t} + \boldsymbol{v} \cdot \nabla \boldsymbol{v}$$

5.7 Show that the following stress fields in the absence of body forces satisfy the equations of equilibrium (a, b, α all constants):

$$(a) \quad \boldsymbol{T} = \begin{bmatrix} \frac{a}{2}x_1^2 & -ax_1x_2 & 0 \\ -ax_1x_2 & \frac{a}{2}x_2^2 & 0 \\ 0 & 0 & 0 \end{bmatrix};$$

$$(b) \quad \boldsymbol{T} = \begin{bmatrix} 0 & -\alpha x_3 & \alpha x_2 \\ -\alpha x_3 & 0 & 0 \\ \alpha x_2 & 0 & 0 \end{bmatrix};$$

$$(c) \quad \boldsymbol{T} = \begin{bmatrix} ax_2 & bx_3 & 0 \\ bx_3 & ax_1 & 0 \\ 0 & 0 & 0 \end{bmatrix}$$

5.8 For the plane stress case defined by relation (4.6.1), assume that the nonzero stresses can be defined in terms of the *Airy stress function*, $\phi(x_1, x_2)$:

$$T_{11} = \frac{\partial^2 \phi}{\partial x_2^2}, T_{22} = \frac{\partial^2 \phi}{\partial x_1^2}, T_{12} = -\frac{\partial^2 \phi}{\partial x_1 \partial x_2}$$

Show that this representation identically satisfies the equilibrium equations (with no body forces) and hence is a *self-equilibrated form*.

5.9 Express the cylindrical equations of motion (5.5.3) for the simplified case of *axisymmetry* where all stresses and velocities are independent of the angle θ, and $v_\theta = 0$.

5.10 As a continuation of Exercise 5.9, express the cylindrical equations of motion (5.5.3) for the simplified case of *axisymmetry* where all stresses and velocities are independent of coordinates θ and z, with $v_\theta = v_z = 0$. Note for this case all partial derivatives become *d/dr*. Can any of the resulting ordinary differential equations be integrated using standard methods?

5.11 Using (4.7.6), eliminate the Cauchy stress from the equations of motion (5.3.5), and develop (5.3.13), the conservation of linear momentum equations in the reference configuration.

5.12 Recall the energetically conjugate form in the stress power relation $\int_R T_{ij}D_{ij}\,dv$.

Develop the corresponding forms using each of the two Piola–Kirchhoff stress tensors as given in relation (5.6.6).

5.13 Calculate the stress power term $T_{ij}D_{ij}$, if the continuum undergoes the following motion and has the particular Cauchy stress form:

$$(a)\quad \begin{aligned} x_1 &= 2tX_1 + 4t^2X_2 \\ x_2 &= 4tX_2 \\ x_3 &= X_3 \end{aligned} \qquad T_{ij} = \begin{bmatrix} T_{11} & T_{12} & 0 \\ T_{12} & T_{22} & 0 \\ 0 & 0 & 0 \end{bmatrix} \qquad (b)\quad \begin{aligned} x_1 &= 2tX_1 \\ x_2 &= tX_1 + 2tX_2 \\ x_3 &= X_3 \end{aligned} \qquad T_{ij} = \begin{bmatrix} T_{11} & T_{12} & 0 \\ T_{12} & T_{22} & 0 \\ 0 & 0 & 0 \end{bmatrix}$$

5.14 Consider the case of an ideal nonviscous fluid (see Exercise 5.6) with no internal heat sources and assume that the transfer of heat is governed by *Fourier's law* $\boldsymbol{q} = -k\nabla\theta$, where k is a constant known as the material *thermal conductivity*. Show that the energy equation reduces to

$$\rho\dot{\varepsilon} = k\nabla^2\theta - p(\nabla\cdot\boldsymbol{v})$$

5.15 Starting with the Clausius–Duhem inequality (5.7.8), develop the equivalent forms (5.7.9) and (5.7.10).

5.16 For a reversible process, the Clausius–Planck inequality (5.7.14) now becomes an equality, $\rho(\theta\dot{s} - \dot{\varepsilon}) + T_{ij}D_{ij} = 0$. Using the energy equation for this case with zero source term, show that

$$\rho\dot{s} = -\frac{q_{i,i}}{\theta}$$

5.17 For an isothermal process (constant temperature distribution), show that the Clausius–Duhem inequality (5.7.11) reduces to $\dot{\Psi} \le \frac{1}{\rho}T_{ij}D_{ij}$.

CHAPTER

6 Constitutive relations and formulation of classical linear theories of solids and fluids

Chapters 6–9 will explore numerous types of constitutive equations for a broad class of materials commonly found in engineering and scientific applications. Constitutive relations characterize a continuum material's macroscopic response to applied mechanical, thermal, or other types of loadings. Such relations are often based on the material's internal constitution and commonly result in idealized material models such as *elastic solids* or *viscous fluids*. These models do not come directly from general principles, but rather are normally developed from observed behaviors found in collected experimental data. As shown in Fig. 1.5, we also wish to combine various constitutive laws with our previous kinematical, stress, and general principle relations in order to create a closed system of field equations that contains sufficient numbers of relations to solve for all model unknowns (see discussion in Section 5.8 and Table 5.1).

For each theoretical model, the complete formulation will require various manipulations of the governing equations and the creation of proper boundary and/or initial conditions appropriate to the model. Following these steps, we will then explore some particular basic solutions to problems of interest. We will generally limit the number of solutions presented for each model, in order to spend more time on expanding the number of material theories developed. This will provide a much broader background on how continuum mechanics is applied to a very large group of materials with quite different behaviors. Many of the classical continuum theories (elasticity, plasticity, fluid mechanics, rheology, etc.) have separate courses and textbooks entirely devoted to each of them. We will thus try to avoid duplication of some of this material and keep our presentations brief.

Because of our interest in the wide application of continuum mechanics, we will divide this study into four separate chapters. This chapter will focus on the classical linear theories of solids and fluids, neglecting thermal effects. This material will provide the foundation for further, more advanced study. Subsequent chapters will then explore more complex problems dealing with multiple constitutive fields, thermomechanical effects, nonlinear theories, materials with microstructure, and damage mechanics.

Continuum Mechanics Modeling of Material Behavior. http://dx.doi.org/10.1016/B978-0-12-811474-2.00006-X

6.1 INTRODUCTION TO CONSTITUTIVE EQUATIONS

As previously mentioned, constitutive equations describe the macroscopic behavior resulting from the material's *internal constitution*. Such internal composition is primarily related to the material makeup at several orders of length scale below that being used in the continuum study, thus commonly focusing on *microstructure or nanostructure*. Because of the wide variety of materials we wish to study, internal microstructure can vary significantly. For example, such microstructure in solids normally includes crystalline, polycrystalline, and amorphous (molecular) types as illustrated in Fig. 6.1. Many solids also have a composite microstructure with fiber or particulate reinforcement embedded in a somewhat uniform matrix material as shown in Fig. 6.2. Additionally, some solids have porous and/or cellular microstructure as illustrated in Fig. 6.3. Fluids can also have complex internal structure such as: liquid mixtures, flows of polymerics, slurries, biological fluids, etc. The classical linear theories to be discussed in this chapter commonly neglect much (but not all) of these microstructures. More sophisticated continuum theories that incorporate some of these details will be discussed in Chapter 9.

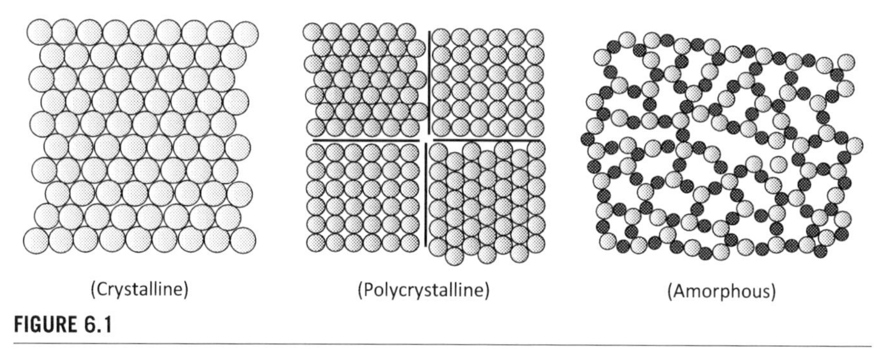

FIGURE 6.1

Typical microstructures in solids.

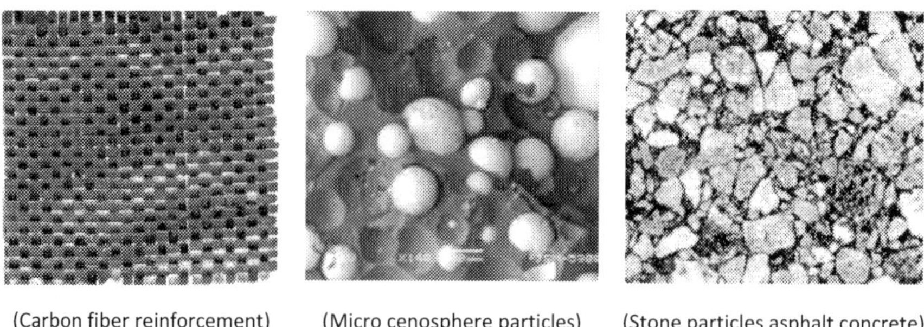

FIGURE 6.2

Typical microstructures in composite materials.

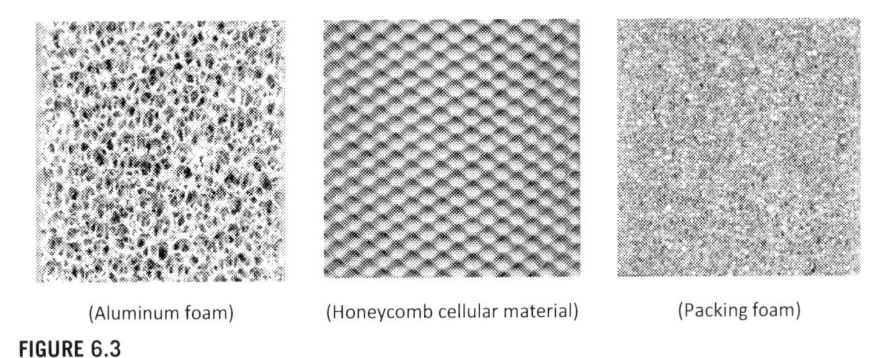

FIGURE 6.3

Microstructures in porous and cellular materials.

For solids, internal material microstructure is also related to two important constitutive issues commonly referred to as *homogeneity* and *isotropy*. If the material is *homogeneous*, the constitutive behavior will not depend on *spatial position* within the body. For many materials, this is a reasonable assumption and will simplify the resulting constitutive equations and subsequent overall formulation. However, there are some important applications (geomechanics and functionally graded materials) where nonhomogeneous behavior must be included in the model. The second issue, the concept of *isotropy*, has to do with *directional dependence* on the constitutive properties. A material with no directional dependence is referred to as *isotropic*, whereas *anisotropic* materials have some form of constitutive dependency on direction. Again assuming an isotropic model will suffice for many applications and will reduce the complexity of the resulting model equations. However, for many materials with significant internal directional microstructure like fiber composites and wood, an anisotropic constitutive model is necessary. Note that fluid media will normally have sufficient internal mixing behavior to eliminate inhomogeneity and anisotropy.

In addition to microstructure, many materials also exhibit complicated rate- and temperature-dependent behavior. With this vast amount of variation, it is generally difficult to establish one constitutive law that can accurately model a real material over a wide range of behavior that includes mechanical, thermal, rate-dependent and microstructural effects. Thus, we commonly formulate specific equations describing the *macroscopic behavior of ideal material types* (e.g. elastic solids, viscoelastic materials, plastic materials, viscous fluids, etc.), each of which will yield a mathematical formulation designed to *approximate real material behavior* over a *restricted range* of deformations, loading rates, and temperatures. In this chapter, we will consider only linear mechanical theories and not explore thermal effects. Subsequent chapters will cover many other continuum material theories that include much broader modeling.

Our study of solid behaviors in this chapter will comprise both elastic and inelastic responses, whereas our fluid modeling will encompass inviscid and Newtonian viscous flows. More precise definitions of these terms will be provided in later sections. Since it is well known that some materials exhibit both solid- and fluid-like behaviors together, we also wish to develop linear viscoelastic continuum theories for such materials. For the nonthermal case, we will be developing six constitutive equations characterizing the particular material's response to applied mechanical loadings. Such material response relations provide the link between applied forces and the material deformation and typically establish a functional relationship between stress and strain or strain rate. For each material model, we will develop the constitutive law, present the general model formulations of the governing field equations, and then present analytical solutions to a few basic problems of interest. We begin each case, with particular experimental evidence to help motivate the constitutive formulation and understand the nature of the modeling. Much more sophisticated and general procedures for constitutive equation development will be given in later chapters.

6.2 LINEAR ELASTIC SOLIDS

Linear elasticity is perhaps one of the most studied fields in continuum mechanics. For over a century, numerous books have been written on the subject and a very large body of research papers has been published on formulation and solution strategies in this field. We will now explore the basics of this formulation and present a few common solutions to boundary value problems within the static theory. Our emphasis will be on the application of continuum mechanics principles to the elasticity formulation. More complete coverage of this topic can be found in Sadd (2014), Barber (2010), Timoshenko and Goodier (1970), and Kachanov et al. (2003) provides an extensive compendium of elasticity solutions.

6.2.1 CONSTITUTIVE LAW

For many centuries, experimental testing has been employed to characterize the behavior of real materials. One such technique to determine the mechanical response is the *simple tension test* in which a specially prepared cylindrical or flat stock sample is axially loaded in a testing machine. This creates a simple and uniform one-dimensional deformation field. Strain is determined by the change in length between prescribed reference marks on the sample and is usually measured by a clip gage. Load data collected from a load cell is divided by the cross-sectional area in the test section to calculate the stress. Axial stress and strain data are recorded and plotted using standard experimental techniques. Typical qualitative data for three types of common structural metals (mild steel, aluminum, cast iron) are shown in Fig. 6.4. It is observed that each material exhibits an initial linear stress–strain response for small deformation. This is followed by a change to nonlinear behavior that can lead to large deformation, finally ending with sample failure.

For each material, the initial linear response ends at a point normally referred to as the *proportional limit*. Another observation in this initial region is that if the loading is removed, the sample will return to its original shape, and the strain will

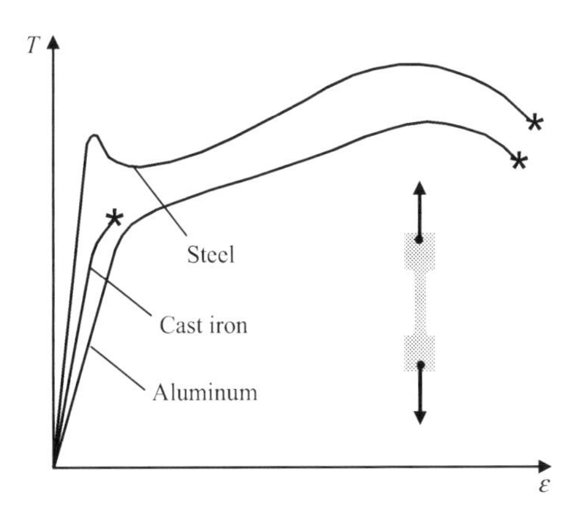

FIGURE 6.4

Typical uniaxial stress–strain curves for three structural metals.

disappear. This characteristic is the primary descriptor of elastic behavior. However, at some further point on the stress–strain curve, unloading will not bring the sample back to zero strain, and some permanent inelastic or plastic deformation will result. The point at which this nonelastic behavior begins is called the *elastic limit*. Although some materials exhibit somewhat different elastic and proportional limits, many times these values are taken to be approximately the same. Another demarcation on the stress–strain curve is referred to as the *yield point*, defined by the location where large inelastic deformation begins. Such inelastic behavior will be discussed in Section 6.6.

Since mild steel and aluminum are ductile materials, their stress–strain response indicates extensive plastic deformation, and during this period the sample dimensions will be changing. In particular, the sample's cross-sectional area will undergo significant reduction and the stress calculation using division by the original area will now be in error. This accounts for the reduction in the stress at large strain. If we were to calculate the load divided by the true area, the *true stress* would continue to increase until failure. On the other hand, cast iron is a brittle material, and thus its stress–strain response does not show large plastic deformation. For this material, very little nonelastic or nonlinear behavior is observed.

We may thus summarize that for many solids under small deformations, the relation between the applied load and deformation is linear; upon removal of the loads, the deformations disappear; and the rate of loading did not affect these observations. These characteristics lead to the formulation of the ideal material model referred to as *linear infinitesimal elasticity theory*. For the one-dimensional axial loading case, we can then write the constitutive response between the stress T and strain ε by the simple relation $T = E\varepsilon$, Twhere E is the slope of the uniaxial stress–strain curve. We will now use this simple concept to develop the general three-dimensional forms of the linear elastic constitutive model.

First, we can generalize things a little and say that for elastic materials, the stress is only a function of the current strain and thus the general constitutive equation would read

$$\begin{aligned} \boldsymbol{T} &= \boldsymbol{f}(\boldsymbol{E}) \\ T_{ij} &= f_{ij}(E_{kl}) \end{aligned} \tag{6.2.1}$$

where $\boldsymbol{T}$ is the Cauchy stress defined in Section 4.3, $\boldsymbol{E}$ is Lagrangian strain defined in Section 3.6, and $\boldsymbol{f}$ is a general tensor-valued material response function. We choose the condition $\boldsymbol{f}(\boldsymbol{0}) = \boldsymbol{0}$, so that when the strain vanishes the stress also becomes zero. Now we wish to limit this model to small deformations and would then drop the distinction between Lagrangian and Euler descriptions and only make use of the small strain tensor ε_{ij} defined by Eq. (3.8.1). Furthermore, the general response function in (6.2.1) must reduce to a general linear form, and thus the response relation is given by

$$T_{ij} = C_{ijkl}\varepsilon_{kl} \tag{6.2.2}$$

where C_{ijkl} is a *fourth-order elasticity tensor* whose components include all the material parameters necessary to characterize the material. This relation is called the *generalized Hooke's law.*

EXAMPLE 6.2.1 PROVE THAT C_{ijkl} IS A FOURTH-ORDER TENSOR

Given that the Cauchy stress T_{ij} and the small strain tensor ε_{ij} are the components of a second-order tensor, explicitly show that the elasticity tensor is a fourth-order tensor.

Solution: From relation (6.2.2) $\Rightarrow T'_{ij} = C'_{ijkl}\varepsilon'_{kl}$.

Since the stress and strain are second-order tensors, $T'_{ij} = Q_{im}Q_{jn}T_{mn}$ and $\varepsilon'_{kl} = Q_{kr}Q_{ls}\varepsilon_{rs}$.

Thus, we get $Q_{im}Q_{jn}T_{mn} = C'_{ijkl}Q_{kr}Q_{ls}\varepsilon_{rs} \Rightarrow Q_{im}Q_{jn}C_{mnrs}\varepsilon_{rs} = C'_{ijkl}Q_{kr}Q_{ls}\varepsilon_{rs}$ $\Rightarrow Q_{im}Q_{jn}C_{mnrs} = C'_{ijkl}Q_{kr}Q_{ls}$. Multiply both sides by $Q_{pr}Q_{qs} \Rightarrow$ $Q_{im}Q_{jn}C_{mnrs}Q_{pr}Q_{qs} = C'_{ijkl}Q_{kr}Q_{ls}Q_{pr}Q_{qs} = C'_{ijkl}\delta_{kp}\delta_{lq} = C'_{ijpq}$ $C'_{ijpq} = Q_{im}Q_{jn}Q_{pr}Q_{qs}C_{mnrs}$ and thus the elasticity tensor $\boldsymbol{C}$ is a fourth-order tensor.

Based on the symmetry of the stress and strain tensors, the elasticity tensor must have the properties (see Exercise 6.1)

$$\begin{aligned} C_{ijkl} &= C_{jikl} \\ C_{ijkl} &= C_{ijlk} \end{aligned} \tag{6.2.3}$$

In general, the fourth-order tensor C_{ijkl} has 81 components. However, relations (6.2.3) reduce the number of independent components to 36, and this provides an alternative way to express Hooke's law in the contracted *Voigt matrix notational form*

$$\begin{bmatrix} T_{11} \\ T_{22} \\ T_{33} \\ T_{23} \\ T_{31} \\ T_{12} \end{bmatrix} = \begin{bmatrix} C_{11} & C_{12} & \cdot & \cdot & \cdot & C_{16} \\ C_{21} & \cdot & \cdot & \cdot & \cdot & \cdot \\ \cdot & \cdot & \cdot & \cdot & \cdot & \cdot \\ \cdot & \cdot & \cdot & \cdot & \cdot & \cdot \\ \cdot & \cdot & \cdot & \cdot & \cdot & \cdot \\ C_{61} & \cdot & \cdot & \cdot & \cdot & C_{66} \end{bmatrix} \begin{bmatrix} \varepsilon_{11} \\ \varepsilon_{22} \\ \varepsilon_{33} \\ 2\varepsilon_{23} \\ 2\varepsilon_{31} \\ 2\varepsilon_{12} \end{bmatrix} \tag{6.2.4}$$

The components of C_{ijkl} or equivalently C_{ij} are called *elastic moduli* and have the same units as stress (force/area).

Next, let us go back to the differential form of energy equation (5.6.12). Neglecting all thermal terms and considering only small deformations, this equation reduces to

$$\rho\dot{\varepsilon} = T_{ij}\dot{\varepsilon}_{ij} \tag{6.2.5}$$

where ε is the internal energy per unit mass. Using our conservation of mass relation (5.2.8) for small deformations (with $J = 1$) gives $\rho = \rho_o$, and thus the mass density remains a constant. For our purely mechanical theory, we define a strain energy density function (per unit volume) as $U = \rho\varepsilon$ and assume that it depends only on the strain, $U = U(\varepsilon_{ij})$. Thus, $\rho\dot{\varepsilon} = \dot{U} = \dfrac{\partial U}{\partial \varepsilon_{ij}}\dot{\varepsilon}_{ij}$, and combining this result with relation (6.2.5) gives

$$T_{ij} = \frac{\partial U}{\partial \varepsilon_{ij}} \tag{6.2.6}$$

Relation (6.2.6) can be thought of as a constitutive form for the stress written in terms of a strain energy function, and this form is commonly referred to as a *hyperelastic* or *Green elastic* material.

Differentiating form (6.2.6), we can write

$$\frac{\partial T_{ij}}{\partial \varepsilon_{kl}} = \frac{\partial^2 U}{\partial \varepsilon_{kl}\,\partial \varepsilon_{ij}} \quad \text{and} \quad \frac{\partial T_{kl}}{\partial \varepsilon_{ij}} = \frac{\partial^2 U}{\partial \varepsilon_{ij}\,\partial \varepsilon_{kl}} \Rightarrow \frac{\partial T_{ij}}{\partial \varepsilon_{kl}} = \frac{\partial T_{kl}}{\partial \varepsilon_{ij}} \tag{6.2.7}$$

Combining Eqs. (6.2.7) with (6.2.2) produces another symmetry relation for the elasticity tensor

$$C_{ijkl} = C_{klij} \quad \text{or equivalently} \quad C_{ij} = C_{ji} \tag{6.2.8}$$

and this result implies that for a general linear elastic solid there are only 21 independent elastic moduli. On occasion we may wish to invert (6.2.2) and write strain in terms of stress

$$\varepsilon_{ij} = S_{ijkl}T_{kl} \tag{6.2.9}$$

where S_{ijkl} is the *elastic compliance tensor* which has identical symmetry properties as those of the elasticity tensor C_{ijkl}.

To go further with the constitutive equation development, we need to decide on the isotropy and homogeneity of the material to be investigated. These issues will reduce the 21 independent elastic moduli to more manageable numbers. For our limited study, we will only consider homogeneous materials and thus all elastic moduli C_{ijkl} will be constants. Further details on nonhomogeneous linear elasticity can be found in Sadd (2014). In regard to the directional dependence, it has been commonly observed that many materials exhibit particular *microstructural symmetries*. Examples of this include the hexagonal crystalline solid shown in Fig. 6.1 and the carbon fiber composite illustrated in Fig. 6.2. Many other crystalline materials and fiber composites exist with various symmetries producing identical constitutive response for *particular* directions within the solid. Some symmetries even follow a curvilinear reference system such as that found in natural wood. These symmetries will generally lead to a reduction in the complexity of the stress–strain constitutive relations, and examples of this will now be shown.

Orientations for which a material has the same stress–strain response can be determined by coordinate transformation (rotation) theory previously developed in Sections 2.7 and 2.8. Such particular transformations are sometimes called the *material symmetry group*. Further details on this topic have been presented by Zheng and Spencer (1993) and Cowin and Mehrabadi (1995). We will have more to say on this topic when we explore more general issues related to nonlinear material behavior in Chapter 8. In order to determine various material symmetries, we review the results from Example 6.2.1 in which we established the fourth-order transformation law for the elasticity tensor

$$C'_{ijkl} = Q_{im}Q_{jn}Q_{kp}Q_{lq}C_{mnpq} \tag{6.2.10}$$

If under a specific transformation $\boldsymbol{Q}$, the material response is to be the same, relation (6.2.10) reduces to

$$C_{ijkl} = Q_{im}Q_{jn}Q_{kp}Q_{lq}C_{mnpq} \tag{6.2.11}$$

This material symmetry relation will provide a system of equations that will allow reduction in the number of independent elastic moduli. We now consider some specific cases of practical interest.

Plane of Symmetry (Monoclinic Material)

We first consider the case of a material with a *plane of symmetry*. Such a medium is commonly referred to as a *monoclinic material*. We consider the case of symmetry with respect to the x_1,x_2-plane as shown in Fig. 6.5.

For this particular symmetry, the required transformation is simply a mirror reflection about the x_1,x_2-plane and is given by

$$Q_{ij} = \begin{bmatrix} 1 & 0 & 0 \\ 0 & 1 & 0 \\ 0 & 0 & -1 \end{bmatrix} \tag{6.2.12}$$

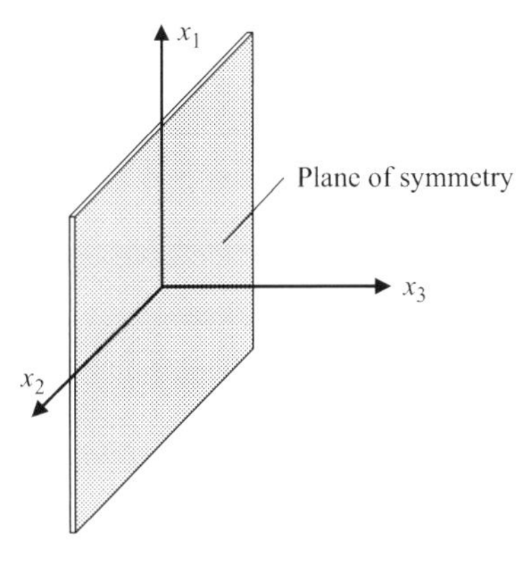

FIGURE 6.5

Plane of symmetry for a monoclinic material.

Note that this transformation is not a simple rotation that preserves the right-handedness of the coordinate system, that is, it is not a proper orthogonal transformation. Nevertheless, it can be used for our symmetry investigations. Using this specific transformation in relation (6.2.11) gives $C_{ijkl} = -C_{ijkl}$ if the index 3 appears an *odd* number of times and thus these particular moduli would have to vanish. In terms of the contracted notation, this gives

$$C_{i4} = C_{i5} = C_{46} = C_{56} = 0, \quad (i = 1, 2, 3) \tag{6.2.13}$$

Thus, the elasticity matrix takes the form

$$C_{ij} = \begin{bmatrix} C_{11} & C_{12} & C_{13} & 0 & 0 & C_{16} \\ \cdot & C_{22} & C_{23} & 0 & 0 & C_{26} \\ \cdot & \cdot & C_{33} & 0 & 0 & C_{36} \\ \cdot & \cdot & \cdot & C_{44} & C_{45} & 0 \\ \cdot & \cdot & \cdot & \cdot & C_{55} & 0 \\ \cdot & \cdot & \cdot & \cdot & \cdot & C_{66} \end{bmatrix} \tag{6.2.14}$$

It is, therefore, observed that *13 independent elastic moduli* are needed to characterize monoclinic materials.

Three Perpendicular Planes of Symmetry (Orthotropic Material)

A material with three mutually perpendicular planes of symmetry is called *orthotropic*. Common examples of such materials include wood and fiber-reinforced composites. In order to investigate the material symmetries for this case, it is convenient to let the symmetry planes correspond to coordinate planes as shown in Fig. 6.6.

The symmetry relations can be determined by using 180° rotations about each of the coordinate axes. Another convenient scheme is to start with the reduced form from the previous monoclinic case and re-apply the same transformation with respect

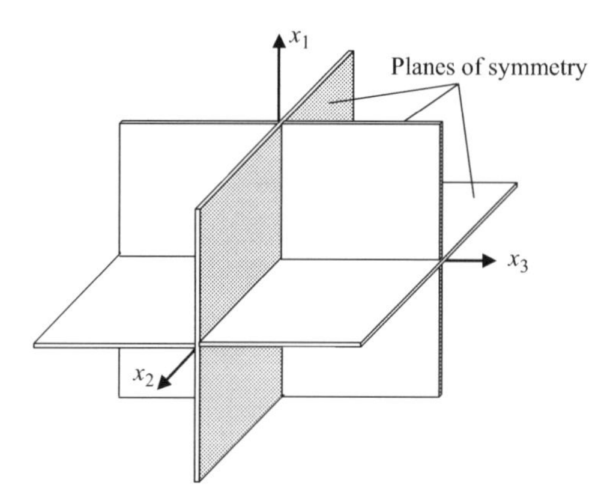

FIGURE 6.6

Three planes of symmetry for an orthotropic material.

to say the x_2,x_3-plane. This results in the additional elastic moduli being reduced to zero

$$C_{16} = C_{26} = C_{36} = C_{45} = 0 \tag{6.2.15}$$

Thus, the elasticity matrix for the orthotropic case reduces to having only *nine independent* elastic moduli given by

$$C_{ij} = \begin{bmatrix} C_{11} & C_{12} & C_{13} & 0 & 0 & 0 \\ \cdot & C_{22} & C_{23} & 0 & 0 & 0 \\ \cdot & \cdot & C_{33} & 0 & 0 & 0 \\ \cdot & \cdot & \cdot & C_{44} & 0 & 0 \\ \cdot & \cdot & \cdot & \cdot & C_{55} & 0 \\ \cdot & \cdot & \cdot & \cdot & \cdot & C_{66} \end{bmatrix} \tag{6.2.16}$$

It should be noted that only two transformations were needed to develop the final reduced constitutive form (6.2.16). The material also must satisfy a third required transformation that the properties would be the same under a reflection of the x_1,x_3-plane; however, this transformation is already identically satisfied. Thus, for some materials, the reduced constitutive form may be developed by only using a portion of the total material symmetries (see Ting, 1996 for more on this topic). On another issue for orthotropic materials, vanishing shear strains imply vanishing shear stresses, and thus the principal axes of stress coincide with the principal axes of strain. This result is of course not true for general anisotropic materials; see for example, the monoclinic constitutive form (6.2.14).

Complete Symmetry (Isotropic Material)

For the case of complete symmetry where all directions have the same constitutive response, the material is referred to as *isotropic*. For this case, the fourth-order elasticity tensor must reduce to an isotropic tensor and this form has been previously given in Table 2.15.1 as

$$C_{ijkl} = \alpha\delta_{ij}\delta_{kl} + \beta\delta_{ik}\delta_{jl} + \gamma\delta_{il}\delta_{jk}, \quad \text{for any constants} \quad \alpha, \beta, \gamma \tag{6.2.17}$$

Invoking the symmetry relations (6.2.3), allows Eq. (6.2.17) to be written as

$$C_{ijkl} = \lambda\delta_{ij}\delta_{kl} + \mu(\delta_{ik}\delta_{jl} + \delta_{il}\delta_{jk}) \tag{6.2.18}$$

where we have redefined some of the general constants associated with (6.2.17). In contracted matrix form, this result would be expressed as

$$C_{ij} = \begin{bmatrix} \lambda+2\mu & \lambda & \lambda & 0 & 0 & 0 \\ \cdot & \lambda+2\mu & \lambda & 0 & 0 & 0 \\ \cdot & \cdot & \lambda+2\mu & 0 & 0 & 0 \\ \cdot & \cdot & \cdot & \mu & 0 & 0 \\ \cdot & \cdot & \cdot & \cdot & \mu & 0 \\ \cdot & \cdot & \cdot & \cdot & \cdot & \mu \end{bmatrix} \tag{6.2.19}$$

Thus, only *two* independent elastic constants exist for isotropic materials. For each of the presented cases, a similar compliance elasticity matrix could be developed.

Focusing now on the isotropic case, using (6.2.18) in Hooke's law (6.2.2) yields

$$T_{ij} = \lambda\varepsilon_{kk}\delta_{ij} + 2\mu\varepsilon_{ij} \tag{6.2.20}$$

where the elastic constant λ is called *Lamé's constant* and μ is referred to as the *shear modulus* or *modulus of rigidity*. Some studies use the notation G for the shear modulus. Eq. (6.2.20) can be written out as six individual scalar equations as

$$\begin{aligned} T_{11} &= \lambda(\varepsilon_{11} + \varepsilon_{22} + \varepsilon_{33}) + 2\mu\varepsilon_{11} \\ T_{22} &= \lambda(\varepsilon_{11} + \varepsilon_{22} + \varepsilon_{33}) + 2\mu\varepsilon_{22} \\ T_{33} &= \lambda(\varepsilon_{11} + \varepsilon_{22} + \varepsilon_{33}) + 2\mu\varepsilon_{33} \\ T_{12} &= 2\mu\varepsilon_{12} \\ T_{23} &= 2\mu\varepsilon_{23} \\ T_{31} &= 2\mu\varepsilon_{31} \end{aligned} \tag{6.2.21}$$

Stress–strain relations (6.2.20) may be inverted to express the strain in terms of the stress. This can easily be done in index notation by first setting the two free indices the same (contraction process) to get

$$T_{kk} = (3\lambda + 2\mu)\varepsilon_{kk} \tag{6.2.22}$$

This relation can be solved for ε_{kk} and substituted back into (6.2.20) to get

$$\varepsilon_{ij} = \frac{1}{2\mu}\left(T_{ij} - \frac{\lambda}{3\lambda + 2\mu}T_{kk}\delta_{ij}\right)$$

which is more commonly written as

$$\varepsilon_{ij} = \frac{1+\nu}{E}T_{ij} - \frac{\nu}{E}T_{kk}\delta_{ij} \tag{6.2.23}$$

where $E = \dfrac{\mu(3\lambda + 2\mu)}{\lambda + \mu}$ and is called *the modulus of elasticity* or *Young's modulus*, and $\nu = \dfrac{\lambda}{2(\lambda + \mu)}$ is referred to as *Poisson's ratio*. It is easy to show that the principal axes of stress and stain are the same for isotropic elastic materials. For cylindrical and spherical coordinates (shown in Figs. 2.6 and 2.7), the basic forms of Hooke's law will remain the same with simple interchange of (x_1,x_2,x_3) to (r,θ,z) or to (R,ϕ,θ).

The previously defined isotropic elastic moduli have simple physical meaning. These can be determined through investigation of particular states of stress commonly used in laboratory materials testing as shown in Fig. 6.7.

Simple Tension

The simple tension test as discussed previously includes a sample subjected to uniaxial tension σ with an approximate state of stress shown in the figure. The strain field is determined from relations (6.2.23). These results give the following information: $E = \sigma / \varepsilon_x$ and is simply the slope of the stress–strain curve, whereas $\nu = -\varepsilon_y / \varepsilon_x = -\varepsilon_z / \varepsilon_x$ is minus the ratio of the transverse strain to the axial strain. Standard measurement systems can easily collect axial stress and transverse and axial strain data, and thus through this one type of test both elastic constants can be determined for materials of interest.

Pure Shear

This test involves a thin-walled circular tube subjected to torsional loading. The approximate state of stress on the surface of the cylindrical sample is shown in the figure. Again using Hooke's law, the corresponding strain field can easily be determined. These results indicate that the shear modulus is given by $\mu = \tau / 2\varepsilon_{xy}$, and this moduli is therefore related to the slope of the shear stress–shear strain curve.

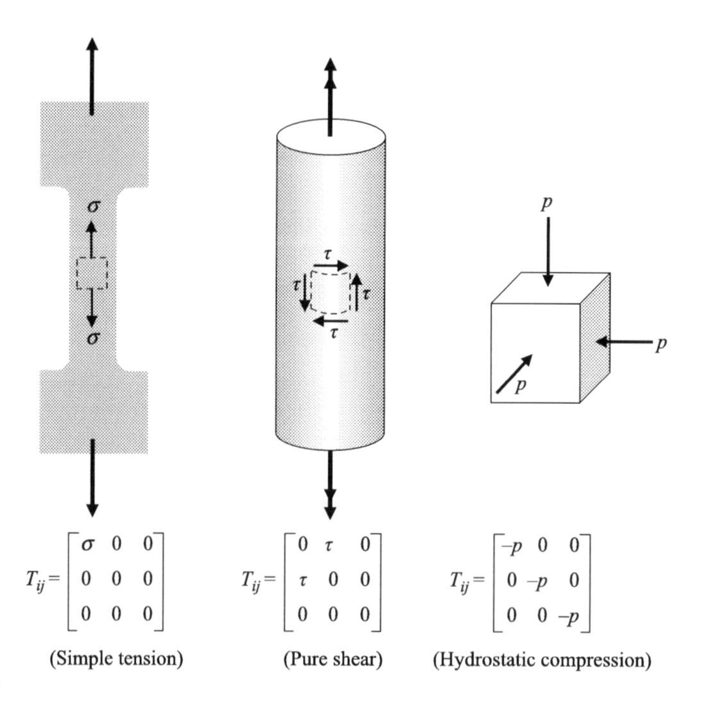

FIGURE 6.7

Special characterization states of stress.

Hydrostatic Compression (or Tension)

The final example is associated with the uniform compression (or tension) loading of a cubical specimen. This type of test could be realizable if the sample was placed in a high-pressure compression chamber. The approximate state of stress is shown in the figure, and the corresponding strains follow from Hooke's law. The dilatation which represents the change in material volume (see Exercise 3.18) is thus given by $\vartheta = \varepsilon_{kk} = -\dfrac{3(1-2\nu)}{E}p$, which can be written as

$$p = -k\vartheta \tag{6.2.24}$$

where $k = \dfrac{E}{3(1-2\nu)}$ is called the *bulk modulus of elasticity*. This additional elastic constant represents the ratio of pressure to the dilatation, which could be referred to as the volumetric stiffness of the material. It can be argued that the elastic constants E, k, and μ must be positive and this will imply that $-1 < \nu < \dfrac{1}{2}$ (see Exercise 6.3). Notice that as Poisson's ratio approaches 0.5, the bulk modulus will become unbounded and the material will not undergo any volumetric deformation and hence will be incompressible.

Our discussion of elastic moduli for isotropic materials has led to the definition of five constants λ, μ, E, ν, and k. However, it should be kept in mind that only two of these are needed to characterize the material. While we have developed a few relationships between various moduli, many other such relations can also be found. In fact, it can be shown that all five elastic constants are interrelated, and if any two are given, the remaining three can be determined using simple formulae (see Sadd, 2014 for details).

EXAMPLE 6.2.2 HYDROSTATIC COMPRESSION OF AN ELASTIC MONOCLINIC AND ISOTROPIC CUBE

In order to demonstrate the difference in linear elastic behavior between isotropic and anisotropic materials, consider a simple example of a cube of both monoclinic and isotropic material under hydrostatic compression p. For this case, the state of stress is given by $T_{ij} = -p\delta_{ij}$, and we wish to determine the corresponding deformations in each type of material.

Solution: for the monoclinic case, hooke's law in compliance form would read

$$\begin{bmatrix} \varepsilon_{11} \\ \varepsilon_{22} \\ \varepsilon_{33} \\ 2\varepsilon_{23} \\ 2\varepsilon_{31} \\ 2\varepsilon_{12} \end{bmatrix} = \begin{bmatrix} S_{11} & S_{12} & S_{13} & 0 & 0 & S_{16} \\ \cdot & S_{22} & S_{23} & 0 & 0 & S_{26} \\ \cdot & \cdot & S_{33} & 0 & 0 & S_{36} \\ \cdot & \cdot & \cdot & S_{44} & S_{45} & 0 \\ \cdot & \cdot & \cdot & \cdot & S_{55} & 0 \\ \cdot & \cdot & \cdot & \cdot & \cdot & S_{66} \end{bmatrix} \begin{bmatrix} -p \\ -p \\ -p \\ 0 \\ 0 \\ 0 \end{bmatrix} \tag{6.2.25}$$

Expanding this matrix relation gives the following deformation field components:

$$\begin{aligned}
\varepsilon_{11} &= -(S_{11}+S_{12}+S_{13})p \\
\varepsilon_{22} &= -(S_{12}+S_{22}+S_{23})p \\
\varepsilon_{33} &= -(S_{13}+S_{23}+S_{33})p \\
\varepsilon_{12} &= -\frac{1}{2}(S_{16}+S_{26}+S_{36})p \\
\varepsilon_{23} &= \varepsilon_{31} = 0
\end{aligned} \tag{6.2.26}$$

The corresponding strains for the isotropic case follow from (6.2.23) And are given by

$$\varepsilon_{11} = \varepsilon_{22} = \varepsilon_{33} = -\left(\frac{1-2\nu}{E}\right)p, \quad \varepsilon_{12} = \varepsilon_{23} = \varepsilon_{31} = 0 \tag{6.2.27}$$

Thus, the response of the monoclinic material is considerably different from isotropic behavior and yields a nonzero shear strain even under uniform hydrostatic stress. Additional examples using simple shear and/or bending deformations can also be used to demonstrate the complexity of anisotropic stress–strain behavior (see Sendeckyj, 1975). It should be apparent that laboratory testing methods attempting to characterize anisotropic materials would have to be more complicated than those used for isotropic solids.

6.2.2 GENERAL FORMULATION

We now wish to establish the general formulation of the linear elasticity model. This process combines the constitutive law with the other governing equations and looks for solution strategies to solve these equations. We will limit our study to only static problems and will thus drop the acceleration term in the equations of motion. Since we are essentially setting up a mathematical boundary value problem, we must also establish appropriate boundary conditions suitable for the theory.

We first start listing all of the governing equations that we have previously developed:

$$\text{Strain-Displacement Relations} \quad \varepsilon_{ij} = \frac{1}{2}(u_{i,j}+u_{j,i}) \tag{6.2.28}$$

$$\text{Compatibility Relations} \quad \varepsilon_{ij,kl}+\varepsilon_{kl,ij}-\varepsilon_{ik,jl}-\varepsilon_{jl,ik}=0 \tag{6.2.29}$$

$$\text{Equilibrium Equations} \quad T_{ij,j}+F_i=0,(F_i=\rho b_i) \tag{6.2.30}$$

$$\text{Constitutive Law} \qquad \begin{aligned} T_{ij} &= (\lambda+\mu)\varepsilon_{kk}\delta_{ij} + 2\mu\varepsilon_{ij} \\ \varepsilon_{ij} &= \frac{1+\nu}{E}T_{ij} - \frac{\nu}{E}T_{kk}\delta_{ij} \end{aligned} \tag{6.2.31}$$

As mentioned in Section 3.11, the compatibility relations ensure that the displacements are continuous and single-valued and are necessary only when the strains are arbitrarily specified. If, however, the displacements are included in the problem formulation, the solution will normally generate single-valued displacements, and strain compatibility will automatically be satisfied. Thus, in discussing the general system of equations of elasticity, the compatibility relations (6.2.29) are normally set aside, to be used only with the stress formulation that we will discuss shortly. Therefore, the general system of elasticity field equations will refer to the 15 relations (6.2.28), (6.2.30), and (6.2.31). This system involves 15 unknowns including three displacements u_i, six strains ε_{ij}, and six stresses T_{ij}. The equation system also includes two elastic material constants (for isotropic materials) and the body force density and these are to be given *a priori* with the problem formulation. It is re-assuring that the number of equations matches the number of unknowns to be determined. However, this general system of equations is of such complexity that solutions via analytical methods are extremely difficult and further simplification is required to solve problems of interest. Before proceeding with the development of such simplifications, it will be useful to first discuss typical boundary conditions connected with elasticity problems and this will lead us to the classification of the fundamental problems.

The common types of boundary conditions for linear elasticity normally include specification of how the body is being *supported* or *loaded*. This concept is mathematically formulated by specifying either the *displacements* or *tractions* at boundary points. Fig. 6.8 illustrates this general idea for three typical cases including tractions, displacements, and a *mixed case* where tractions are specified on boundary S_t and displacements are given on the remaining portion S_u such that the total boundary is given by $S = S_t + S_u$.

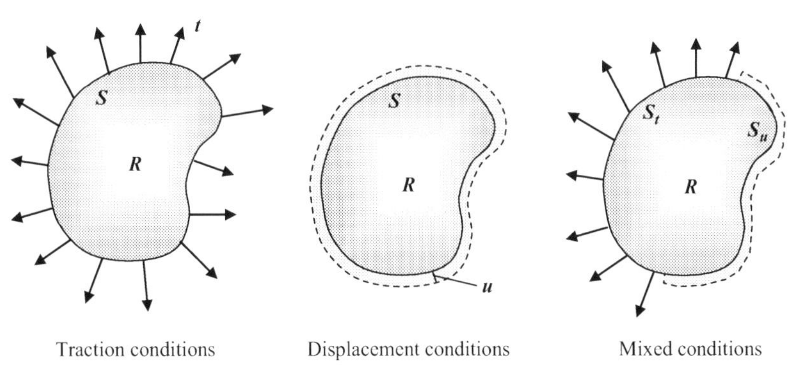

FIGURE 6.8

Typical elasticity boundary conditions.

Traction boundary conditions simplify for the case where the boundary coincides with a coordinate surface. Fig. 6.9 illustrates particular two-dimensional cases where the boundaries coincide with Cartesian or polar coordinate surfaces. Using results from Section 4.3, the traction specification can be reduced to a Cauchy stress specification. For the Cartesian example where x_2 = constant, the normal traction becomes simply the stress component T_{22}, whereas the tangential traction would reduce to T_{12}. For this case, T_{11} exists only *inside* the region, and thus this component of stress cannot be specified on the boundary surface x_2 = constant. A similar situation exists on the vertical boundary x_1 = constant, where the normal traction is now T_{11}, the tangential traction is T_{12}, and the stress component T_{22} exists inside the domain. Similar arguments can be made for polar coordinate boundary surfaces as shown. Drawing the appropriate element along the boundary as illustrated allows for a clear visualization of the particular stress components that act *on* the surface in question. Such a sketch also allows for the determination of the positive directions of these boundary stresses, and this is useful to properly match with boundary loadings that might be prescribed.

We now formulate and classify the *three fundamental boundary-value problems in the theory of linear elasticity* that are related to solving the general system of field equations.

Problem 1 (Traction Problem). Determine the distribution of displacements, strains, and stresses in the interior of an elastic body in equilibrium when body forces are given and the distribution of the tractions are prescribed over the surface of the body:

$$t_i^{(n)}(x_i^{(s)}) = f_i(x_i^{(s)}) \tag{6.2.32}$$

where $x_i^{(s)}$ denotes boundary points and $f_i(x_i^{(s)})$ are the prescribed traction values.

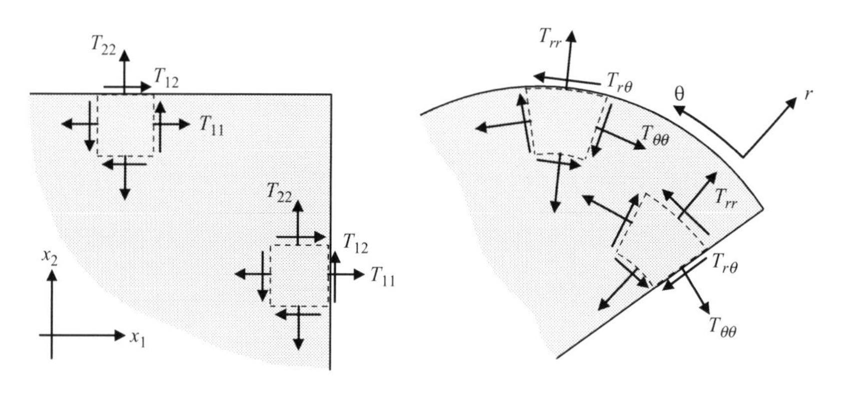

FIGURE 6.9

Boundary stress components on coordinate surfaces.

Problem 2 (Displacement Problem). Determine the distribution of displacements, strains, and stresses in the interior of an elastic body in equilibrium when body forces are given and the distribution of the displacements are prescribed over the surface of the body:

$$u_i(x_i^{(s)}) = g_i(x_i^{(s)}) \tag{6.2.33}$$

where $x_i^{(s)}$ denotes boundary points and $g_i(x_i^{(s)})$ are the prescribed displacement values.

Problem 3 (Mixed Problem). Determine the distribution of displacements, strains, and stresses in the interior of an elastic body in equilibrium when body forces are given and the distribution of the tractions are prescribed as per (6.2.32) over the surface S_t and the distribution of the displacements are prescribed as per (6.2.33) over the surface S_u of the body (see Fig. 6.8).

As mentioned previously, the solution to any of these types of problems is formidable, and further reduction and simplification of the general field equation system is normally required to develop analytical solutions. Based on the description of Problem 1 with only traction boundary conditions, it would appear to be desirable to express the fundamental system solely in terms of stress, thereby reducing the number of unknowns in the system. Likewise, for Problem 2, a displacement-only formulation would appear to simplify the problem. We will now pursue these two specialized formulations and explicitly determine these reduced field equation systems.

6.2.2.1 Stress formulation

For the first fundamental problem in elasticity, the boundary conditions are to be given only in terms of the tractions or stress components. In order to develop solution methods for this case, it will be very helpful to reformulate the general system (6.2.28)–(6.2.31) by eliminating the displacements and strains, thereby casting a new system solely in terms of the stresses. By eliminating the displacements, we must now include the compatibility equations in the fundamental system of field equations. Starting with Hooke's law $(6.2.31)_2$ and eliminate the strains in the compatibility relations (6.2.29) to get

$$\begin{aligned} &T_{ij,kk} + T_{kk,ij} - T_{ik,jk} - T_{jk,ik} = \\ &\frac{\nu}{1+\nu}(T_{mm,kk}\delta_{ij} + T_{mm,ij}\delta_{kk} - T_{mm,jk}\delta_{ik} - T_{mm,ik}\delta_{jk}) \end{aligned} \tag{6.2.34}$$

where we have used the arguments of Section 3.11, that the six independent compatibility relations are found by setting $k = l$ in (6.2.29). Although Eqs. (6.2.34) represent the compatibility in terms of stress, a more useful result is found by incorporating the equilibrium equations into the system. Skipping the details (see Sadd, 2014), the final relation becomes

$$T_{ij,kk} + \frac{1}{1+\nu}T_{kk,ij} = -\frac{\nu}{1-\nu}\delta_{ij}F_{k,k} - F_{i,j} - F_{j,i} \tag{6.2.35}$$

This result is the compatibility relations in terms of the stress and is commonly called the *Beltrami–Michell compatibility equations*. Recall that the six developed relations (6.2.35) actually represent three independent results as per our discussion in Section 3.11. Thus, combining these results with the three equilibrium equations (6.2.30) provides the necessary six relations to solve for the six unknown stress components for the general three-dimensional case. This system constitutes the stress formulation for linear elasticity theory and would be appropriate for use with traction boundary condition problems. Once the stresses have been determined, the strains may be found from Hooke's law $(6.2.31)_2$, and the displacements can be then computed through integration of (6.2.28).

The system of equations for this stress formulation is still rather complex, and analytical solutions are commonly determined by making use of *stress functions*. This concept establishes a representation for the stresses that will automatically satisfy the equilibrium equations. For the two-dimensional case, this concept represents the in-plane stresses in terms of a single function. The representation satisfies equilibrium, and the remaining compatibility equations yield a single partial differential equation (biharmonic equation) in terms of the stress function. Having reduced the system to a single equation then allows many analytical methods to be employed to find solutions of interest. We will explore such solutions in the Problems Solution section.

6.2.2.2 Displacement formulation

We now wish to develop the reduced set of field equations solely in terms of the displacements. This system will be referred to as the *displacement formulation* and would be most useful when combined with displacement-only boundary conditions found in the Problem 2 statement. For this case, we wish to eliminate the strains and stresses from the fundamental system (6.2.28)–(6.2.31). This is easily accomplished by using the strain displacement relations in Hooke's law to give

$$T_{ij} = (\lambda + \mu) u_{k,k} \delta_{ij} + \mu (u_{i,j} + u_{j,i}) \tag{6.2.36}$$

Using these relations in the equilibrium equations gives the result

$$\mu u_{i,kk} + (\lambda + \mu) u_{k,ki} + F_i = 0 \tag{6.2.37}$$

which are the equilibrium equations in terms of the displacements and are referred to as *Navier's or Lamé's equations*. This system can be expressed in vector form as

$$\mu \nabla^2 u + (\lambda + \mu) \nabla (\nabla \cdot u) + F = 0 \tag{6.2.38}$$

Navier's equations are the desired formulation for the displacement problem, and the system represents three equations for the three unknown displacement components. Although this formulation represents a considerable reduction in the number of equations, each relation is difficult to solve since all three displacement components appear in each equation. Similar to the stress formulation, additional mathematical techniques have been developed to further simplify these equations for problem solution. Common methods normally employ the use of *displacement potential functions*. These schemes generally simplify the problem by yielding uncoupled governing equations, and this then allows several analytical methods to be employed to solve problems of interest.

Before moving to the problem solutions, consider a few more details on the strain energy function U that was previously introduced. Boundary tractions and body forces will do work on an elastic solid and this work will be stored inside the material in the form of strain energy. For the elastic case, removal of these loadings will result in the complete recovery of the stored energy. It can be easily shown (Sadd, 2014) that the strain energy function (strain energy per unit volume) can be expressed by

$$U = \frac{1}{2} T_{ij}\varepsilon_{ij} \tag{6.2.39}$$

For the isotropic case, using Hooke's law, the stresses can be eliminated from relation (6.2.39) and the strain energy can be expressed solely in terms of strain

$$\begin{aligned} U(\varepsilon) &= \frac{1}{2}\lambda\varepsilon_{jj}\varepsilon_{kk} + \mu\varepsilon_{ij}\varepsilon_{ij} \\ &= \frac{1}{2}\lambda(\varepsilon_{11}+\varepsilon_{22}+\varepsilon_{33})^2 + \mu(\varepsilon_{11}^2+\varepsilon_{22}^2+\varepsilon_{33}^2+2\varepsilon_{12}^2+2\varepsilon_{23}^2+2\varepsilon_{31}^2) \end{aligned} \tag{6.2.40}$$

Likewise, the strains can be eliminated and the strain energy can be written in terms of stress

$$\begin{aligned} U(\boldsymbol{T}) &= \frac{1+\nu}{2E}T_{ij}T_{ij} - \frac{\nu}{2E}T_{ii}T_{jj} \\ &= \frac{1+\nu}{2E}(T_{11}^2+T_{22}^2+T_{33}^2+2T_{12}^2+2T_{23}^2+2T_{31}^2) - \frac{\nu}{2E}(T_{11}+T_{22}+T_{33})^2 \end{aligned} \tag{6.2.41}$$

These forms for the strain energy will be useful in other parts of the text.

6.2.3 PROBLEM SOLUTIONS

In this section, we shall explore the solution to a few selected two-dimensional problems in isotropic linear elasticity. This will provide some additional insights and details as to how the material model solutions are found and how the stress, strain, and displacement distributions are determined. As is normally the case, analytical solutions are more likely to be found for one- and two-dimensional problems of limited geometric complexity.

For linear elasticity, there exist several two-dimensional formulations including *plane strain, plane stress*, and *antiplane strain*. For now, we will limit our study to only plane stress and focus on problems in the x_1,x_2-plane. Since we will be dealing primarily with a scalar analysis, variables (x_1,x_2,x_3) will be replaced by the usual (x,y,z) notation. Plane stress (in x,y-plane) is a two-dimensional approximation to a three-dimensional world, such that the state of stress is of the form

$$T_{xx} = T_{xx}(x,y), \quad T_{yy} = T_{yy}(x,y), \quad T_{xy} = T_{xy}(x,y), \quad T_{zz} = T_{xz} = T_{yz} = 0 \tag{6.2.42}$$

Thus, only in-plane stresses exist and they are functions only of the in-plane coordinates. This particular assumption is most applicable to bodies that are thin in the out-of-plane direction and are loaded only with in-plane forces. Examples of this type of geometry are plate-like structures with stress free surfaces on $z = \pm h$ as shown in Fig. 6.10.

Neglecting body forces, under plane stress conditions, the equilibrium equations reduce to

$$\begin{aligned} \frac{\partial T_{xx}}{\partial x} + \frac{\partial T_{xy}}{\partial y} = 0 \\ \frac{\partial T_{xy}}{\partial x} + \frac{\partial T_{yy}}{\partial y} = 0 \end{aligned} \tag{6.2.43}$$

The equilibrium equations in terms of the in-plane displacements u, v become

$$\begin{aligned} \mu \nabla^2 u + \frac{E}{2(1-\nu)} \frac{\partial}{\partial x}\left(\frac{\partial u}{\partial x} + \frac{\partial v}{\partial y}\right) = 0 \\ \mu \nabla^2 v + \frac{E}{2(1-\nu)} \frac{\partial}{\partial y}\left(\frac{\partial u}{\partial x} + \frac{\partial v}{\partial y}\right) = 0 \end{aligned} \tag{6.2.44}$$

The compatibility equations reduce to the single relation

$$\frac{\partial^2 \varepsilon_{xx}}{\partial y^2} + \frac{\partial^2 \varepsilon_{yy}}{\partial x^2} = 2\frac{\partial^2 \varepsilon_{xy}}{\partial x \partial y} \tag{6.2.45}$$

and expressing this relation in terms of stress gives the corresponding Beltrami–Michell equation

$$\nabla^2 (T_{xx} + T_{yy}) = 0 \tag{6.2.46}$$

The plane stress problem is then formulated in a two-dimensional region R in the x,y-plane. The displacement formulation is specified by the two governing

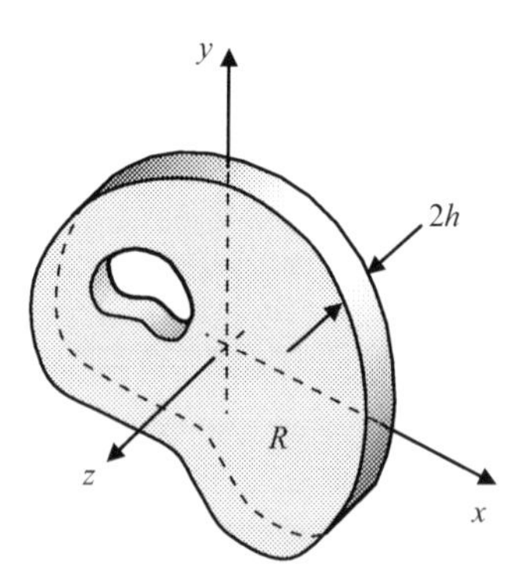

FIGURE 6.10

Thin elastic plate representing plane stress conditions.

Navier relations (6.2.44) with displacement boundary conditions on u and v. The stress formulation includes the three governing equations (6.2.43) and (6.2.46) with appropriate stress or traction boundary conditions. The solution to the plane stress problem then involves determination of the in-plane displacements, strains, and stresses $\{u, v, \varepsilon_{xx}, \varepsilon_{yy}, \varepsilon_{xy}, T_{xx}, T_{yy}, T_{xy}\}$ in R. The out-of-plane strain ε_z can be determined from the in-plane strains via an additional Hooke's law relation.

Even with this two-dimensional reduction in the field equations, the resulting system is still challenging, and additional mathematical help is needed to find solutions. The significant breakthrough for the stress formulation is found by introducing the *Airy stress function* $\phi(x,y)$, such that the stresses can be expressed by the form

$$T_{xx} = \frac{\partial^2 \phi}{\partial y^2}, \quad T_{yy} = \frac{\partial^2 \phi}{\partial x^2}, \quad T_{xy} = -\frac{\partial^2 \phi}{\partial x \partial y} \tag{6.2.47}$$

Using this form, it is observed that the equilibrium equations (6.2.43) will be identically satisfied, and the compatibility relation (6.2.46) becomes

$$\frac{\partial^4 \phi}{\partial x^4} + 2\frac{\partial^4 \phi}{\partial x^2 \partial y^2} + \frac{\partial^4 \phi}{\partial y^4} = \nabla^4 \phi = 0 \tag{6.2.48}$$

The form $\nabla^4 = \nabla^2\nabla^2$ is called the *biharmonic operator*, and relation (6.2.48) is known as the *biharmonic equation*. Thus, the plane stress problem of linear elasticity has been reduced to a single equation in terms of the Airy stress function ϕ. This reduction to a single partial differential equation now offers many solutions to be found to a large variety of problems types.

One common solution method useful for problems in Cartesian geometries is a polynomial/power series representation of the Airy stress function of the form

$$\phi(x, y) = \sum_{m=0}^{\infty}\sum_{n=0}^{\infty} A_{mn} x^m y^n \tag{6.2.49}$$

where A_{mn} are constant coefficients to be determined from problem boundary conditions. It is noted that the three lowest order terms with $m + n \leq 1$ do not contribute to the stresses and are thus are dropped. Terms with $m + n \leq 3$ will automatically satisfy the biharmonic equation for any choice of constants A_{mn}; however, for higher-order terms, the constants A_{mn} will have to be related in order to have the polynomial satisfy the governing equation. It is observed that the second-order terms will produce a constant stress field, third-order terms will give a linear distribution of stress, and so on for higher-order polynomials. Since this method will produce only polynomial stress distributions, the scheme will not satisfy general boundary conditions.

Another more general method that is useful in polar coordinate problems comes from a general solution to the biharmonic equation originally credited to Michell (1899). A final form (commonly called the Michell solution) can be written as

$$
\begin{aligned}
\phi = {} & a_0 + a_1 \log r + a_2 r^2 + a_3 r^2 \log r \\
& + (a_4 + a_5 \log r + a_6 r^2 + a_7 r^2 \log r)\theta \\
& + \left(a_{11} r + a_{12} r \log r + \frac{a_{13}}{r} + a_{14} r^3 + a_{15} r\theta + a_{16} r\theta \log r\right)\cos\theta \\
& + \left(b_{11} r + b_{12} r \log r + \frac{b_{13}}{r} + b_{14} r^3 + b_{15} r\theta + b_{16} r\theta \log r\right)\sin\theta \\
& + \sum_{n=2}^{\infty} (a_{n1} r^n + a_{n2} r^{2+n} + a_{n3} r^{-n} + a_{n4} r^{2-n})\cos n\theta \\
& + \sum_{n=2}^{\infty} (b_{n1} r^n + b_{n2} r^{2+n} + b_{n3} r^{-n} + b_{n4} r^{2-n})\sin n\theta
\end{aligned}
\tag{6.2.50}
$$

where a_n, a_{nm}, and b_{nm} are constants to be determined. Note that this general solution is restricted to the periodic case, which has the most practical applications since it allows the Fourier method to be applied to handle general boundary conditions.

We will now explore a few specific example solutions using these listed methods, and these will focus on a simple Cartesian geometry, and two polar coordinate problems involving stress concentration and stress singularity.

EXAMPLE 6.2.3 SOLUTION TO A RECTANGULAR PLATE UNDER UNIFORM BIAXIAL LOADING

Consider the rectangular plate under uniform biaxial loading N_x and N_y as shown in Fig. 6.11. Determine the plane stress solution for the stresses, strains, and displacements.

Solution: For such Cartesian geometries, we use the polynomial/power series representation (6.2.49). Since a uniform distribution of stress in the plate is expected, based on relations (6.2.47) we are motivated to try a stress function of the general form $\phi(x, y) = Ax^2 + By^2$, where A and B are constants to be determined. Using (6.2.47), the stresses take the general form

$$T_{xx} = \frac{\partial^2 \phi}{\partial y^2} = 2B, \quad T_{yy} = \frac{\partial^2 \phi}{\partial x^2} = 2A, \quad T_{xy} = -\frac{\partial^2 \phi}{\partial x \partial y} = 0 \tag{6.2.51}$$

The boundary conditions on this problem are all traction type and because the boundaries are coordinate surfaces, they will reduce to a stress specification as per our previous discussion (see Fig. 6.9). Thus, the boundary conditions are

$$
\begin{aligned}
& T_{xx}(\pm a, y) = N_x, \quad T_{yy}(x, \pm b) = N_y \\
& T_{xy}(\pm a, y) = T_{xy}(x, \pm b) = 0
\end{aligned}
\tag{6.2.52}
$$

Applying these conditions to the general stress field (6.2.51) determines the constants $A = N_y/2$ and $B = N_x/2$, and thus the stress field is now determined

$$T_{xx} = N_x, \quad T_{yy} = N_y, \quad T_{xy} = 0 \tag{6.2.53}$$

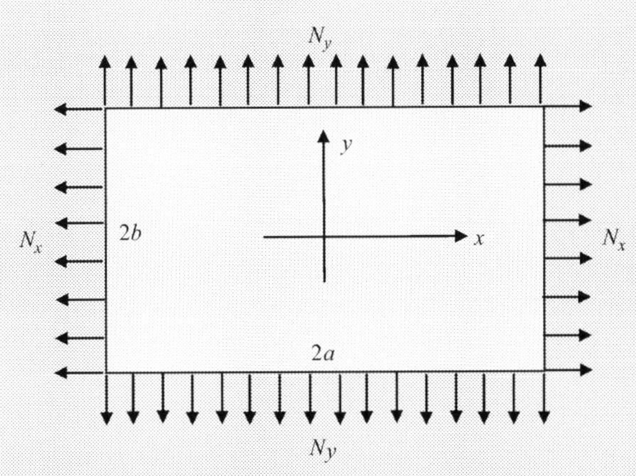

FIGURE 6.11

Rectangular plate under uniform biaxial loading.

The strains follow from Hooke's law $(6.2.31)_2$, for the plane stress case

$$\begin{aligned}
\varepsilon_x &= \frac{1}{E}(T_{xx} - \nu T_{yy}) = \frac{1}{E}(N_x - \nu N_y) \\
\varepsilon_y &= \frac{1}{E}(T_{yy} - \nu T_{xx}) = \frac{1}{E}(N_y - \nu N_x) \\
\varepsilon_{xy} &= \frac{1+\nu}{E} T_{xy} = 0
\end{aligned} \tag{6.2.54}$$

Finally, using the strain displacement relations (6.2.28)

$$\begin{aligned}
&\frac{\partial u}{\partial x} = \varepsilon_x = \frac{1}{E}(N_x - \nu N_y) \Rightarrow u = \frac{x}{E}(N_x - \nu N_y) + f(y) \\
&\frac{\partial v}{\partial y} = \varepsilon_y = \frac{1}{E}(N_y - \nu N_x) \Rightarrow v = \frac{y}{E}(N_y - \nu N_x) + g(x) \\
&\frac{\partial u}{\partial y} + \frac{\partial v}{\partial x} = 2\varepsilon_{xy} = 0 \Rightarrow f'(y) + g'(x) = 0
\end{aligned} \tag{6.2.55}$$

Following the arguments and steps in Example 3.11.2, we find that $g'(x) = -f'(y) = \text{constant}$, and these results can be easily integrated to get $f(y) = -\omega_o y + u_o$, and $g(x) = \omega_o x + v_o$. Thus, the displacement field becomes

$$\begin{aligned}
u &= \frac{x}{E}(N_x - \nu N_y) + u_o - \omega_o y \\
v &= \frac{y}{E}(N_y - \nu N_x) + v_o + \omega_o x
\end{aligned} \tag{6.2.56}$$

where ω_o, u_o, v_o are arbitrary constants of integration. The expressions $f(y)$ and $g(x)$ represent *rigid-body motion* terms where ω_o is the rotation about the

z-axis, and u_o and v_o are the translations in the x- and y-directions. Such terms will always result from the integration of the strain–displacement relations, and it is noted that they do not contribute to the strain or stress field. Thus *the displacements are determined from the strain field only up to an arbitrary rigid-body motion*. Additional boundary conditions on the displacements are needed to explicitly determine these terms. For example, if we agree that the center of the plate does not move, then $u_o = v_o = 0$; and if the x-axis does not rotate, then $\omega_o = 0$; and then all rigid-body terms will vanish and $f = g = 0$.

Notice for this very simple problem that the stress and strain field was uniform (no spatial variation) and the displacements varied linearly with position. The next two examples will explore elasticity solutions with significant variation in stress distribution including concentration and singularity behaviors.

EXAMPLE 6.2.4 SOLUTION TO A STRESS CONCENTRATION PROBLEM OF A STRESS-FREE HOLE IN AN INFINITE MEDIUM UNDER UNIAXIAL FAR-FIELD LOADING

Determine the elastic stress distribution in an infinite medium with a circular stress-free hole subjected to a uniform far-field tension T in a single direction as shown in Fig. 6.12. For ease of solution, we choose the loading direction to coincide with the x-axis and the hole radius is a.

Solution: For this geometry, it is best to formulate the problem in polar coordinates as shown. The boundary conditions on the problem are zero stresses on the hole and uniform axial stress at infinity:

$$\begin{aligned} &T_{rr}(a,\theta) = T_{r\theta}(a,\theta) = 0 \\ &T_{rr}(\infty,\theta) = \frac{T}{2}(1+\cos 2\theta) \\ &T_{\theta\theta}(\infty,\theta) = \frac{T}{2}(1-\cos 2\theta) \\ &T_{r\theta}(\infty,\theta) = -\frac{T}{2}\sin 2\theta \end{aligned} \tag{6.2.57}$$

Note that the far-field conditions have been determined using the transformation laws established in Exercise 4.4.

We first consider the state of stress in the medium if there was no hole. This stress field is simply $T_{xx} = T$, $T_{yy} = T_{xy} = 0$ and can be derived from the Airy stress function

$$\phi = \frac{1}{2}Ty^2 = \frac{T}{2}r^2\sin^2\theta = \frac{T}{4}r^2(1-\cos 2\theta)$$

The presence of the hole acts to disturb this uniform field. We expect that this disturbance will be local in nature, and thus the disturbed field will disappear as we move far away from the hole. Based on this, we choose a trial

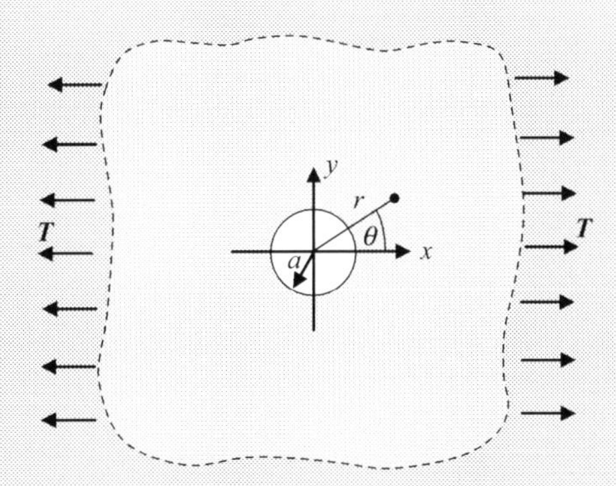

FIGURE 6.12

Stress-free hole in an infinite medium under uniform far-field loading.

solution that includes the axisymmetric and cos 2θ terms from the general Michell solution (6.2.50):

$$\begin{aligned}\phi &= a_0 + a_1 \log r + a_2 r^2 + a_3 r^2 \log r \\ &+ (a_{21} r^2 + a_{22} r^4 + a_{23} r^{-2} + a_{24}) \cos 2\theta\end{aligned} \tag{6.2.58}$$

Using the polar coordinate forms, the stresses corresponding to this Airy function are

$$\begin{aligned}T_{rr} &= \frac{1}{r}\frac{\partial \phi}{\partial r} + \frac{1}{r^2}\frac{\partial^2 \phi}{\partial \theta^2} = a_3(1 + 2\log r) + 2a_2 + \frac{a_1}{r^2} - (2a_{21} + \frac{6a_{23}}{r^4} + \frac{4a_{24}}{r^2}) \cos 2\theta \\ T_{\theta\theta} &= \frac{\partial^2 \phi}{\partial r^2} = a_3(3 + 2\log r) + 2a_2 - \frac{a_1}{r^2} + (2a_{21} + 12a_{22} r^4 + \frac{6a_{23}}{r^4}) \cos 2\theta \\ T_{r\theta} &= -\frac{\partial}{\partial r}\left(\frac{1}{r}\frac{\partial \phi}{\partial \theta}\right) = (2a_{21} + 6a_{22} r^2 - \frac{6a_{23}}{r^4} - \frac{2a_{24}}{r^2}) \sin 2\theta\end{aligned} \tag{6.2.59}$$

For finite stresses at infinity, we must take $a_3 = a_{22} = 0$. Applying the five boundary conditions in (6.2.57) generates five simple algebraic equations to solve for the remaining five unknown constants giving

$$a_1 = -\frac{a^2 T}{2}, \quad a_2 = \frac{T}{4}, \quad a_{21} = -\frac{T}{4}, \quad a_{23} = -\frac{a^4 T}{4}, \quad a_{24} = \frac{a^2 T}{2}$$

Substituting these values back into (6.2.59) gives the stress field

$$\begin{aligned}T_{rr} &= \frac{T}{2}\left(1 - \frac{a^2}{r^2}\right) + \frac{T}{2}\left(1 + \frac{3a^4}{r^4} - \frac{4a^2}{r^2}\right) \cos 2\theta \\ T_{\theta\theta} &= \frac{T}{2}\left(1 + \frac{a^2}{r^2}\right) - \frac{T}{2}\left(1 + \frac{3a^4}{r^4}\right) \cos 2\theta \\ T_{r\theta} &= -\frac{T}{2}\left(1 - \frac{3a^4}{r^4} + \frac{2a^2}{r^2}\right) \sin 2\theta\end{aligned} \tag{6.2.60}$$

The strain and displacement fields could then be determined using the standard procedures used previously.

The hoop stress variation around the boundary of the hole is given by

$$T_{\theta\theta}(a,\theta)=T(1-2\cos 2\theta) \tag{6.2.61}$$

and this is shown in the polar plot in Fig. 6.13. This distribution indicates that the stress is negative T at $\theta=0$, actually vanishes at $\theta=\pm 30°$ and leads to a maximum value at $\theta=\pm 90°$:

$$T_{\theta\theta\,\max}=T_{\theta\theta}(a,\pm\pi/2)=3T \tag{6.2.62}$$

Therefore, there is a stress concentration factor of 3 for this problem. The effects of the hole in perturbing the uniform stress field can be shown by plotting the stress variation with radial distance. Considering the case of the hoop stress $T_{\theta\theta}$ at an angle $\pi/2$, Fig. 6.13 also illustrates the distribution of $T_{\theta\theta}(r,\pi/2)/T$ vs nondimensional radial distance r/a. It is seen that the stress concentration around the hole is highly localized and decays very rapidly, essentially disappearing when $r > 5a$. These results come from MATLAB Code C-4.

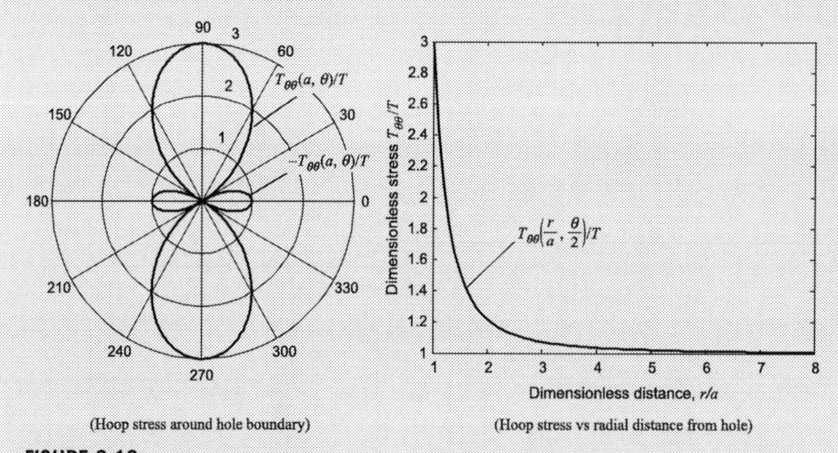

FIGURE 6.13

$T_{\theta\theta}$ stress behavior for stress concentration around hole.

EXAMPLE 6.2.5 SOLUTION TO A HALF-SPACE UNDER A CONCENTRATED NORMAL FORCE

Determine the stress distribution in an elastic half-space that carries a concentrated normal force acting on the free surface as illustrated in Fig. 6.14. This problem is commonly known as the *Flamant problem*.

Solution: Specifying boundary conditions for such problems with concentrated loadings requires a little modification of our previous examples. For this case, the tractions on any semicircular arc C enclosing the origin must balance the applied concentrated loadings. Since the area of such an arc is proportional to the radius r, the stresses must be of the order $1/r$ to allow such an equilibrium statement to hold on any radius. The appropriate terms in the general Michell solution (6.2.50) that will give stresses of the order $1/r$ are specified by

$$\phi = (a_{12} r \log r + a_{15} r\theta)\cos\theta + (b_{12} r \log r + b_{15} r\theta)\sin\theta \tag{6.2.63}$$

The stresses resulting from this stress function are

$$\begin{aligned} T_{rr} &= \frac{1}{r}[(a_{12} + 2b_{15})\cos\theta + (b_{12} - 2a_{15})\sin\theta] \\ T_{\theta\theta} &= \frac{1}{r}[a_{12}\cos\theta + b_{12}\sin\theta] \\ T_{r\theta} &= \frac{1}{r}[a_{12}\sin\theta - b_{12}\cos\theta] \end{aligned} \tag{6.2.64}$$

With zero normal and shear stresses on $\theta = 0$ and π, $a_{12} = b_{12} = 0$ and thus $T_{\theta\theta} = T_{r\theta} = 0$ everywhere. Therefore, this state of stress is sometimes called

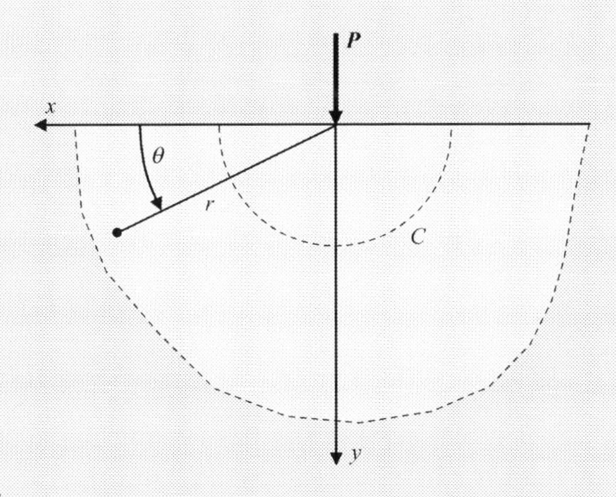

FIGURE 6.14

Half-space under concentrated normal force.

a *radial distribution*. To determine the remaining constants a_{15} and b_{15}, we apply the equilibrium statement that the summation of the tractions over the semicircular arc C of radius a must balance the applied loadings,

$$\begin{aligned} 0 &= -\int_0^\pi T_{rr}(a,\theta)a\cos\theta\, d\theta = -\pi b_{15} \\ P &= -\int_0^\pi T_{rr}(a,\theta)a\sin\theta\, d\theta = \pi a_{15} \end{aligned} \tag{6.2.65}$$

Thus, the constants are determined as $a_{15} = P/\pi$ and $b_{15} = 0$, and the stress field is now given by

$$\begin{aligned} T_{rr} &= -\frac{2P}{\pi r}\sin\theta \\ T_{\theta\theta} &= T_{r\theta} = 0 \end{aligned} \tag{6.2.66}$$

As expected, the stress field is singular at the origin directly under the point loading.

The Cartesian components corresponding to this stress field are determined using the transformation relations given in Exercise 4.5. The results are found to be

$$\begin{aligned} T_{xx} &= T_{rr}\cos^2\theta = -\frac{2Px^2y}{\pi(x^2+y^2)^2} \\ T_{yy} &= T_{rr}\sin^2\theta = -\frac{2Py^3}{\pi(x^2+y^2)^2} \\ T_{xy} &= T_{rr}\sin\theta\cos\theta = -\frac{2Pxy^2}{\pi(x^2+y^2)^2} \end{aligned} \tag{6.2.67}$$

The distribution of the normal and shearing stresses on a horizontal line located a distance a below the free surface of the half-space is shown in Fig. 6.15. The maximum normal stress directly under the load is given

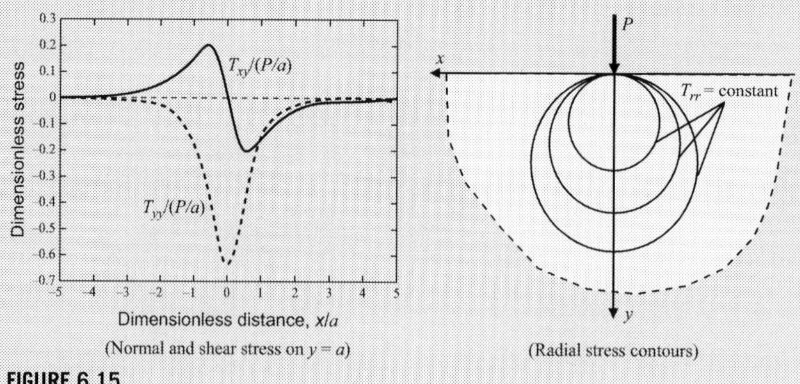

FIGURE 6.15

Stress distribution plots for Flamant problem.

by $|T_{yy}| = 2P/\pi a$. It is observed that the effects of the concentrated loading are highly localized, and the stresses are vanishingly small for distances where $x > 5a$. Stress contours of T_{rr} are also shown in Fig. 6.15. From solution (6.2.66), lines of constant radial stress are circles tangent to the half-space surface at the loading point. These results come from MATLAB Code C-5.

Many additional analytical elasticity solutions exist in the literature for a large variety of both two- and three-dimensional problem types. The interested reader is directed to the general references given at the beginning of this section. Dynamic problems in linear elasticity, sometimes referred to as *elastodynamics*, forms yet another field of study that contains hyperbolic partial differential field equations yielding wave propagation phenomena. This study will not be discussed here, but details can be found in Achenbach (1976) and Graff (1991).

6.3 IDEAL NONVISCOUS FLUIDS

We now begin our study of fluid mechanics and will first explore the simplest case of an ideal nonviscous fluid. Experiments indicate that some media cannot sustain a shear stress and be in static equilibrium. A deformation or *flow* will continue as long as any shear stress is applied. This type of behavior is normally associated with what we call a *fluid*. Fluids are commonly separated into liquids and gases. A liquid is generally difficult to compress, has an approximate fixed volume, and takes the shape the container that it is placed in. On the other hand, a gas can easily be compressed and will tend to expand unless it is confined in a container. Liquids will form a free surface (under normal gravity conditions), whereas a gas will not. Unlike elastic solids, the constitutive response for fluids indicates that the magnitude of applied shear stress is not a function of the strain, but rather is related to the *rate of strain*. However, in some cases, this functional relationship is very small, and so we can model this special type of material by concluding that the fluid is *frictionless* or *inviscid* and that no shearing stresses will be present during deformation. We will now briefly explore this type of fluid behavior and the viscous fluid case will be developed later in Section 6.4. Further, more detailed information on fluid mechanics can be found in Batchelor (2010), White (2016), Lamb (1993), Milne-Thomson (1974), and Schlichting (2017).

6.3.1 CONSTITUTIVE LAW

For the ideal nonviscous fluid, no shear stress will be present even under flow, and thus only a hydrostatic pressure p will exist in such a continuum. Therefore, the constitutive law reduces to the simple form

$$\boldsymbol{T} = -p\boldsymbol{I} \quad or \quad T_{ij} = -p\delta_{ij} \tag{6.3.1}$$

This type of material is commonly referred to as an *elastic fluid*. Often there exists an *equation of state* relating the pressure to the density

$$p = p(\rho) \tag{6.3.2}$$

and for such a case we say that the fluid is *barotropic*. Note that for a thermomechanical theory, the equation of state would include the temperature as well, that is, $p = p(\rho, \theta)$.

6.3.2 GENERAL FORMULATION

In order to provide a general formulation for inviscid flow problems, we start by using the constitutive form (6.3.1) in the equations of motion (5.36), and this gives the result

$$-\frac{1}{\rho}\nabla p + \boldsymbol{b} = \frac{\partial \boldsymbol{v}}{\partial t} + \boldsymbol{v} \cdot \nabla \boldsymbol{v} \tag{6.3.3}$$

which are called *Euler's equations of motion for inviscid fluids.* Notice that because of the convection term $\boldsymbol{v} \cdot \nabla \boldsymbol{v}$ in the acceleration, these equations will be nonlinear.

If we further restrict our study and consider only incompressible inviscid fluids, the equation of state reduces to

$$\rho = \text{constant} \tag{6.3.4}$$

and the conservation of mass or continuity equation as per relation (5.2.5) then becomes

$$\nabla \cdot \boldsymbol{v} = 0 \tag{6.3.5}$$

Consider next the case of *irrotational flows* where the vorticity or angular velocity vectors are zero. Using results (3.13.5),

$$\boldsymbol{w} = \frac{1}{2}\nabla \times \boldsymbol{v} = 0 \tag{6.3.6}$$

Next employing Stoke's Theorem (2.17.14), result (6.3.6) then implies that

$$\oint_C \boldsymbol{v} \cdot d\boldsymbol{s} = 0 \tag{6.3.7}$$

where *ds* is the tangential differential element along the closed contour C. In fluid mechanics, the integral appearing in relation (6.3.7) in commonly denoted as the *circulation* Γ around contour C. It can be shown (see Malvern, 1969 for details) that *for a barotropic fluid with conservative body forces, the circulation around any material contour will be zero*, a result known as *Kelvin's Theorem*. It can be shown that this theorem implies that if an inviscid fluid flow starts as irrotational it will remain that way, and this is one reason why this type of flow kinematics is assumed.

In light of relation (6.3.7), we conclude that the integrand $\boldsymbol{v} \cdot d\boldsymbol{s}$ must be an exact differential, and thus we can write

$$\boldsymbol{v} = \nabla \phi \tag{6.3.8}$$

where ϕ is a scalar *potential function*. Using this result in the continuity relation (6.3.5) gives

$$\nabla^2\phi = 0 \tag{6.3.9}$$

and thus the potential function satisfies the classical Laplace equation. This relation is then the governing equation for irrotational flows of incompressible inviscid fluids. This is often called *potential flow theory*.

For the potential flow problem, let us assume that the body force is derivable from a potential function such that $\boldsymbol{b} = -\nabla\Omega$. Under these conditions, Euler's equation (6.3.3) becomes

$$-\nabla\left(\frac{p}{\rho} + \Omega\right) = \frac{\partial \boldsymbol{v}}{\partial t} + \boldsymbol{v} \cdot \nabla \boldsymbol{v} \tag{6.3.10}$$

Using (6.3.8), this can be rewritten as

$$\nabla\left(\frac{\partial \phi}{\partial t} + \frac{v^2}{2} + \frac{p}{\rho} + \Omega\right) = 0 \tag{6.3.11}$$

where $v^2 = \boldsymbol{v} \cdot \boldsymbol{v}$. Relation (6.3.11) then implies that

$$\frac{\partial \phi}{\partial t} + \frac{v^2}{2} + \frac{p}{\rho} + \Omega = f(t) \tag{6.3.12}$$

where $f(t)$ is an arbitrary function of time. If the flow is *steady*, then (6.3.12) reduces to

$$\frac{v^2}{2} + \frac{p}{\rho} + \Omega = 0 \tag{6.3.13}$$

Relations (6.3.12) and (6.3.13) are known as *Bernoulli's equations*. This derivation also shows that irrotational flows are always dynamically possible under the stated conditions, since the equations of motion will always be integrable to yield (6.3.12) and (6.3.13).

We will focus our problem solution formulation for only two-dimensional flow problems in the x,y-plane, and thus the velocity vector will have two components v_x and v_y. The distributions of these velocities are the unknowns to be found. In addition to the potential formulation given by relations (6.3.8) and (6.3.9), another common solution approach is to introduce a stream function $\psi(x,y)$. Lines of ψ = constant are called *streamlines* and the fluid velocity vector is always *tangent* to streamlines. The potential function and the stream function are related through the following equations

$$\begin{aligned} v_x &= \frac{\partial \phi}{\partial x} = \frac{\partial \psi}{\partial y} \\ v_y &= \frac{\partial \phi}{\partial y} = -\frac{\partial \psi}{\partial x} \end{aligned} \tag{6.3.14}$$

Using relations (6.3.14), it can be shown that contours of the potential function and stream function are orthogonal to each other and that the stream function also satisfies Laplace's equation

$$\nabla^2\psi = \frac{\partial^2\psi}{\partial x^2} + \frac{\partial^2\psi}{\partial y^2} = 0 \tag{6.3.15}$$

Notice that the governing equations in terms of the stream or potential function reduce to a single relation, thus making the flow solution easier to be found.

The boundary conditions for such inviscid flow problems are quite simple. For fluid contact with a fixed surface, the normal velocity component would vanish and the tangential component would not be specified. Actually the tangential component will be determined as part of the flow solution. Thus, based on our previous definitions of the potential and stream function, we can either specify the normal derivative $d\phi/dn = 0$ or the condition $\psi =$ constant on a fixed surface in contact with the inviscid fluid flow. As we shall see in Section 6.4, these boundary conditions will be different for viscous fluids.

6.3.3 PROBLEM SOLUTIONS

We now explore the solution to a few selected two-dimensional potential flow problems. This involves solving the governing Laplace equation (6.3.9) or (6.3.15) for the specific problem geometry. Once the potential or stream function is determined, relation (6.3.14) can then be used to find the velocity distribution. We will use MATLAB software to plot velocity vector distributions. As before, analytical solutions are most likely to be found for two-dimensional problems of limited geometric complexity.

EXAMPLE 6.3.1 TWO-DIMENSIONAL UNIFORM FLOW

Consider uniform two-dimensional flow with velocity V in the x-direction. Determine the potential and stream functions and plot their contours in the x,y-plane.

Solution: For this simple problem, the velocity distribution is already given as $v_x = V$ and $v_y = 0$. Using relations (6.3.14),

$$\begin{aligned} v_x &= \frac{\partial\phi}{\partial x} = \frac{\partial\psi}{\partial y} = V \\ v_y &= \frac{\partial\phi}{\partial y} = -\frac{\partial\psi}{\partial x} = 0 \end{aligned} \Rightarrow \begin{aligned} \phi &= Vx + \text{constant} \\ \psi &= Vy + \text{constant} \end{aligned} \tag{6.3.16}$$

The velocity distribution and contours of the stream and potential functions are shown in Fig. 6.16 for the region $0 \le (x, y) \le 1$. The flow is uniform in the x-direction and the contours form an orthogonal vertical and horizontal mesh. Note these results come from MATLAB Code C-6.

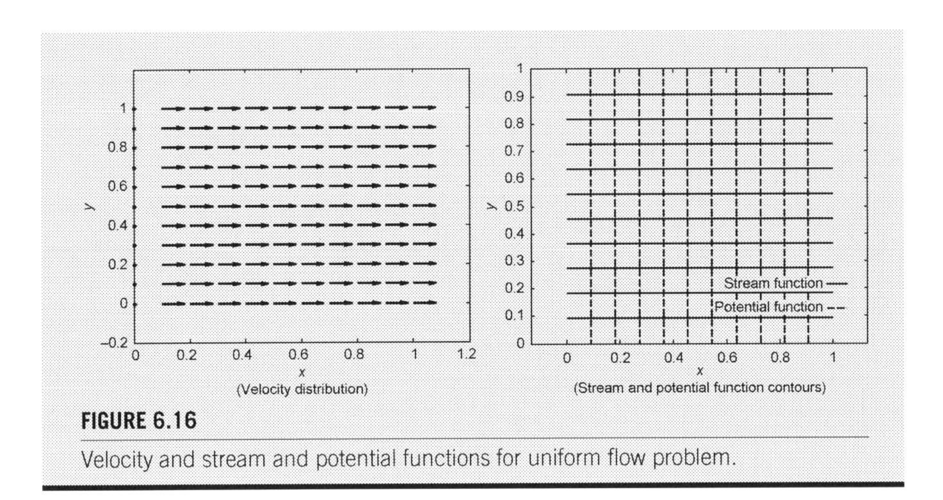

FIGURE 6.16

Velocity and stream and potential functions for uniform flow problem.

EXAMPLE 6.3.2 INVISCID FLOW IN A 90° CORNER

Determine the potential fluid flow in a 90° corner as shown in Fig. 6.17. Assume that the flow occurs down in the vertical and off to the right in the horizontal directions.

Solution: From our previous definitions, the boundaries $x = 0$ and $y = 0$ will be a streamline. Further, the origin $x = y = 0$ will be a stagnation point with zero fluid velocity. Since these boundary streamlines pass through the stagnation point, they must have a value of zero, and so $\psi(0, y) = \psi(x, 0) = 0$. Based on these boundary conditions, we look for the stream function solution to Laplace's equation (6.3.15) in the form

$$\psi(x, y) = Axy \tag{6.3.17}$$

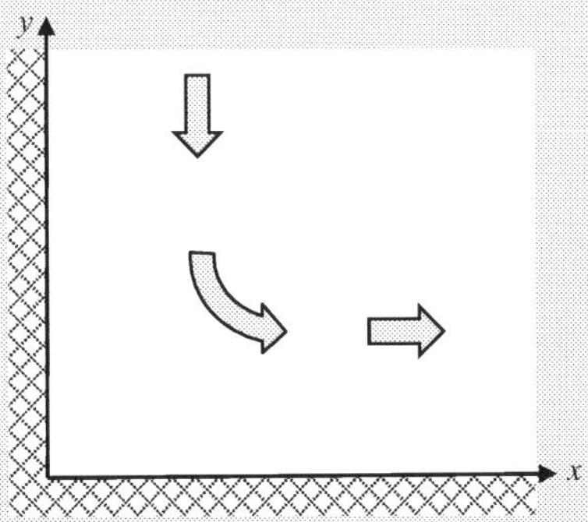

FIGURE 6.17

Flow in a corner.

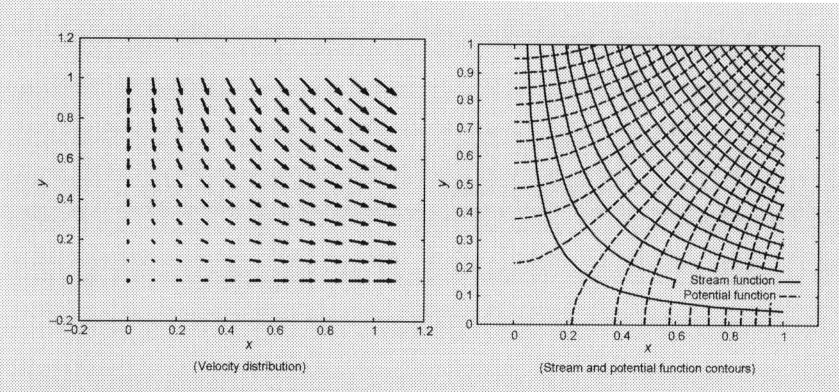

FIGURE 6.18

Velocity and stream and potential functions for corner flow problem.

where A is an arbitrary constant related to the far-field velocity. Using this form in (6.3.14), the velocity components are

$$v_x = \frac{\partial \psi}{\partial y} = Ax, \quad v_y = -\frac{\partial \psi}{\partial x} = -Ay \tag{6.3.18}$$

The potential function can then be obtained from integrating relations (6.3.14) to get

$$\phi(x,y) = \frac{1}{2}A(x^2 - y^2) \tag{6.3.19}$$

The velocity distribution and contours of the stream and potential functions are shown in Fig. 6.18 for the region $0 \le (x,y) \le 1$. As in the previous example, the contours form an orthogonal mesh. These plots come from MATLAB Code C-7.

EXAMPLE 6.3.3 INVISCID FLOW AROUND A CYLINDER

As a final example, consider the two-dimensional potential flow around a fixed cylinder of radius a as shown in Fig. 6.19. Far away from the cylinder, the flow is uniform with $v_x = U$ and $v_y = 0$. Determine the stream function solution and velocity distribution.

Solution: This problem is best formulated and solved in polar coordinates (r, θ) as shown the figure. For this case, the del operator becomes $\nabla = \boldsymbol{e}_r \frac{\partial}{\partial r} + \boldsymbol{e}_\theta \frac{1}{r}\frac{\partial}{\partial \theta}$ and the governing Laplace equation (6.3.15) can thus be written as

$$\nabla^2 \psi = \frac{\partial^2 \psi}{\partial r^2} + \frac{1}{r}\frac{\partial \psi}{\partial r} + \frac{1}{r^2}\frac{\partial^2 \psi}{\partial \theta^2} = 0 \tag{6.3.20}$$

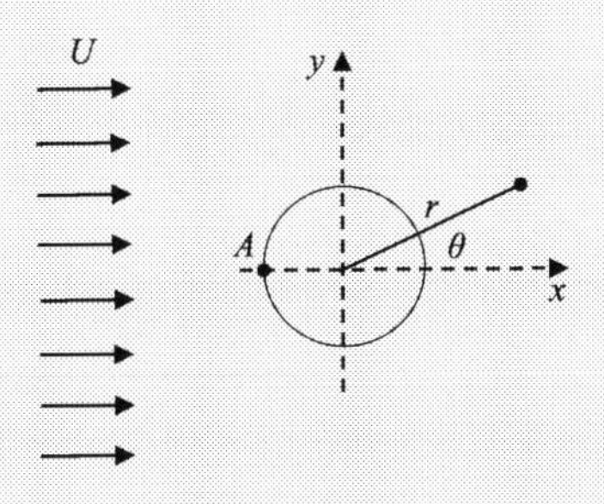

FIGURE 6.19

Two-dimensional flow around a fixed cylinder.

and the polar velocity components follow from the transformation relations in Example 2.18.1 giving the result

$$v_r = \frac{1}{r}\frac{\partial \psi}{\partial \theta}, \quad v_\theta = -\frac{\partial \psi}{\partial r} \tag{6.3.21}$$

Note that the relationship between the velocity components in Cartesian and polar coordinates can be expressed by (see Appendix B)

$$\begin{aligned} v_x &= v_r \cos\theta - v_\theta \sin\theta \\ v_y &= v_r \sin\theta + v_\theta \cos\theta \end{aligned} \tag{6.3.22}$$

Using the standard separation of variables approach, we choose $\psi = \psi(r,\theta) = R(r)T(\theta)$ and can further argue that the problem solution should be an odd function of θ. We thus try $T(\theta) = \sin\theta$, and substituting the form $\psi = R(r)\sin\theta$ into Eq. (6.3.20) gives

$$r^2\frac{d^2R}{dr^2} + r\frac{dR}{dr} - R = 0 \tag{6.3.23}$$

which is a Euler–Cauchy differential equation. The solution follows from standard methods giving $R = C_1 r + C_2 / r$, where C_1 and C_2 are arbitrary constants to be determined from the boundary conditions. Thus, the solution for the stream function takes the form

$$\psi = \left(C_1 r + \frac{C_2}{r}\right)\sin\theta \tag{6.3.24}$$

The boundary conditions are argued in the following manner. First, we note that point A on the cylinder is a *stagnation point* in that no flow velocity can occur at this location. Since the streamline that passes through a stagnation point has a value of zero, the stream function on the entire surface of the cylinder of radius a should vanish, leading to the boundary condition $\psi(a) = 0$. Next, far away from the cylinder, the flow is supposed to be uniform and only in the x-direction so that as $r \to \infty$, $v_x = v_r \cos\theta - v_\theta \sin\theta = U$, and

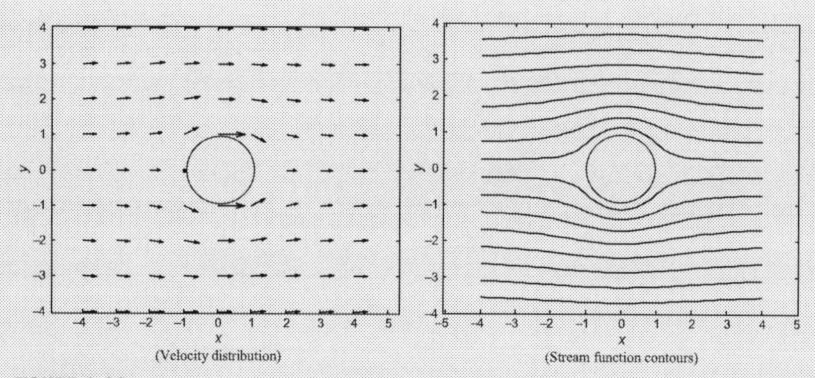

FIGURE 6.20

Velocity and stream function for flow around a cylinder.

$v_y = v_r \sin\theta + v_\theta \cos\theta = 0$. Note, however, that this second condition will be identically or automatically satisfied and thus will not yield a useful result.

Thus, the two useful boundary conditions become

$$\psi(a) = 0$$
$$\lim_{r\to\infty}\left(\frac{1}{r}\frac{\partial\psi}{\partial\theta}\cos\theta + \frac{\partial\psi}{\partial r}\sin\theta\right) = U \tag{6.3.25}$$

Using these two conditions gives the following values for the constants $C_1 = U,\ C_2 = -Ua^2$, and thus the solution to the problem becomes

$$\psi = Ur\left(1 - \frac{a^2}{r^2}\right)\sin\theta \tag{6.3.26}$$

Finally, the velocity solution follows from (6.3.21)

$$\begin{aligned} v_r &= \frac{1}{r}\frac{\partial\psi}{\partial\theta} = U\left(1 - \frac{a^2}{r^2}\right)\cos\theta \\ v_\theta &= -\frac{\partial\psi}{\partial r} = -U\left(1 + \frac{a^2}{r^2}\right)\sin\theta \end{aligned} \tag{6.3.27}$$

The velocity distribution and contours of the stream function are shown in Fig. 6.20. Notice the elevation of the fluid velocity near the cylinder. This behavior is analogous to the stress concentration problem shown in Example 6.2.4. These plots come from MATLAB Code C-8.

6.4 LINEAR VISCOUS FLUIDS

The previous section dealing with ideal inviscid fluids neglected the viscous effects between shear stress and rate of deformation. However, for many fluids and for many types of flows, viscous effects are significant and cannot be neglected. In light of this, the ideal fluid constitutive law (6.3.1) must be modified to include shear stresses in the fluid continuum. Experiments have shown that these shear stresses are related to the rate of deformation of the material. Over the years, several fluid constitutive relationships have been established, and in this section we will restrict ourselves to only the *linear viscous* case. More complicated nonlinear constitutive relations for fluids will be discussed later in Chapter 8. There is a wealth of literature dealing with linear viscous flows including Serrin (1959), Batchelor (2010),White (2016), and Schlichting (2017).

6.4.1 CONSTITUTIVE LAW

We now wish to determine the linear constitutive law that describes the relationship between the stress and the rate of deformation in fluids. Numerous experiments using instruments called *viscometers* have collected data on shear stress vs shear strain rate for special flow geometries in laboratory settings. Generally, such studies involve creating a local shearing flow in a thin gap and then measure the shear stress and rate of shear strain. Fig. 6.21 illustrates a common flow geometry and shows typical qualitative results for several material types such as oils, paints, pastes, gels, polymeric fluids, cornstarch liquids, etc. It is seen that various materials have flow characteristics that are quite different. In this section, however, we will focus our studies for the case originally proposed by Newton where the functional relationship is linear, that is, $T_{12} = \mu\dot{\gamma}_{12}$, where μ is a material constant known as the *viscosity*. Furthermore, experiments indicate that a fluid at rest (i.e. zero rate of deformation) will be in a state of hydrostatic pressure. Hence, similar to the previous elastic case, we can write a general constitutive equation for a viscous fluid as

$$\begin{aligned} \boldsymbol{T} &= -p\boldsymbol{I} + \boldsymbol{f}(\boldsymbol{D}) \\ T_{ij} &= -p\delta_{ij} + f_{ij}(D_{kl}) \end{aligned} \tag{6.4.1}$$

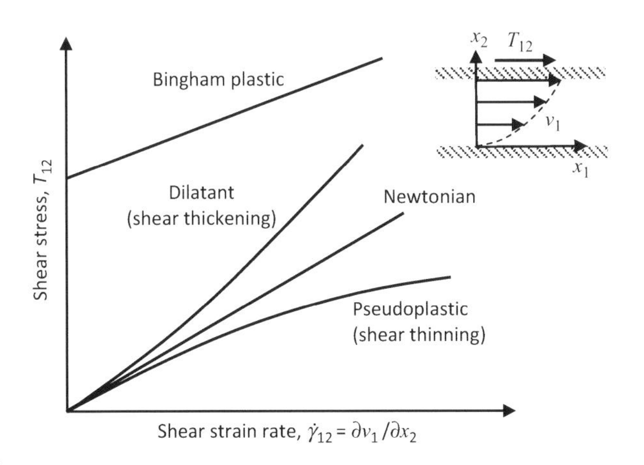

FIGURE 6.21

Typical shear stress vs shear strain rate behaviors.

where $\boldsymbol{D}$ is the rate of deformation tensor define by $(3.13.4)_1$ as $\boldsymbol{D}=\left((\nabla \boldsymbol{v})+(\nabla \boldsymbol{v})^T\right)/2$. Such a constitutive relation commonly defines a *Stokesian fluid*. It is further assumed that $\boldsymbol{f}(\boldsymbol{0})=\boldsymbol{0}$, so that under zero deformation rate, only hydrostatic pressure will be present.

Now the fluid constitutive form is to be linear and thus (6.4.1) must reduce to

$$T_{ij}=-p\delta_{ij}+M_{ijkl}D_{kl} \tag{6.4.2}$$

Normally, fluids are not nonhomogeneous or anisotropic, and therefore the fourth-order tensor M_{ijkl} is a constant tensor with the isotropic form

$$M_{ijkl}=\alpha\delta_{ij}\delta_{kl}+\beta\delta_{ik}\delta_{jl}+\gamma\delta_{il}\delta_{jk} \tag{6.4.3}$$

where α, β, and γ are arbitrary constants. Substituting (6.4.3) into (6.4.2) yields

$$T_{ij}=-p\delta_{ij}+\lambda D_{kk}\delta_{ij}+2\mu D_{ij} \tag{6.4.4}$$

where we have let $\lambda=\alpha$ and $2\mu=\beta+\gamma$. Constitutive relation (6.4.4) represents the *Navier–Poisson law of linear viscous or Newtonian fluids*. Note the similarity with the isotropic linear elastic constitutive development given by relation (6.2.20). Here, the two material constants λ and μ are referred to as the *dilatational and shear viscosity coefficients*, respectively. For the case of an incompressible fluid, $D_{kk}=0$, and (6.4.4) reduces to

$$T_{ij}=-p\delta_{ij}+2\mu D_{ij} \tag{6.4.5}$$

It should be pointed out that the previous constitutive relations were all developed for what is called *laminar or streamline flow*. Such a flow is generally described as orderly motion in which fluid particles move smoothly in adjacent layers sliding over each other with no mixing between layers. Motions of this type are common at low fluid velocities. However, at high speeds, fluid motions often become more random resulting in significant motions normal to these laminae directions. This type of random intermingling fluid motion is referred to a *turbulent flow*. For many particular flow geometries, the transition from laminar to turbulent flow can be determined by the value or range of a particular dimensionless parameter called the *Reynold's number*. This number is the ratio of the fluid's inertial force to the shearing force, that is, how fast the fluid is moving relative to how large is its viscosity. Thus, it is directly proportional to the fluid velocity and inversely proportional to the viscosity. For turbulent flow, the shear stress must be modified by momentum transfer between flow layers due to turbulent eddies and mixing. Thus, the original Newtonian linear relation $T_{12}=\mu\dot{\gamma}_{12}$ would require some modification with an additional term to account for this behavior. Such turbulent flow theories are well established (see Batchelor, 2010; White, 2016; Schlichting, 2017); however, we will not pursue them here.

Fig. 6.22 illustrates a special flow geometry called a *shear flow* velocity field which is specified by

$$\boldsymbol{v}=\{v_1(x_2),0,0\} \tag{6.4.6}$$

For this flow, the rate of deformation tensor becomes

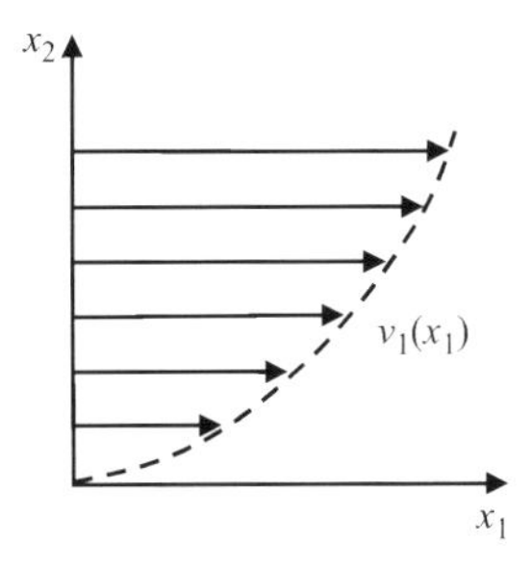

FIGURE 6.22

Shear flow geometry.

$$\boldsymbol{D}=\frac{1}{2}\begin{bmatrix} 0 & \dfrac{\partial v_1}{\partial x_2} & 0 \\ \dfrac{\partial v_1}{\partial x_2} & 0 & 0 \\ 0 & 0 & 0 \end{bmatrix} \tag{6.4.7}$$

Using the fluid constitutive law (6.4.4) then yields the simplified result for the shear stress

$$T_{12}=\mu\frac{\partial v_1}{\partial x_2}\Rightarrow\mu=\frac{T_{12}}{\partial v_1/\partial x_2} \tag{6.4.8}$$

Thus, the viscosity coefficient μ is the slope of the T_{12} vs $\partial v_1/\partial x_2$ curve as shown previously in Fig. 6.21.

Going back to the general flow case, the Newtonian fluid constitutive law (6.4.4) gives

$$\frac{1}{3}T_{kk}=-p+\left(\lambda+\frac{2}{3}\mu\right)D_{kk} \tag{6.4.9}$$

and thus

$$\begin{aligned}-\frac{1}{3}T_{kk}-p&=-\left(\lambda+\frac{2}{3}\mu\right)D_{kk}\\&=-\left(\lambda+\frac{2}{3}\mu\right)v_{k,k}\\&=\left(\lambda+\frac{2}{3}\mu\right)\frac{1}{\rho}\frac{D\rho}{Dt}\end{aligned} \tag{6.4.10}$$

where we have used the continuity equation (5.2.4). The quantity $(\lambda + 2/3\mu)$ is normally called the *bulk viscosity*. Stokes assumed that $p = -(1/3)T_{kk}$, and thus claimed that for general flows

$$\lambda+\frac{2}{3}\mu=0 \tag{6.4.11}$$

which is the somewhat controversial *Stokes condition* used throughout classical fluid mechanics. Truesdell (1966) gives an extended discussion on the history and some experimental contradictions associated with this assumption. Nevertheless, most fluid dynamics studies incorporate Stokes condition $\lambda = -(2/3)\mu$, which eliminates the dilatational viscosity and thus constitutive relation (6.4.4) becomes

$$T_{ij} = -p\delta_{ij} - \frac{2}{3}\mu D_{kk}\delta_{ij} + 2\mu D_{ij} \tag{6.4.12}$$

6.4.2 GENERAL FORMULATION

We now wish to formulate the basic governing equations of linear viscous fluids. Problems in fluid mechanics are commonly formulated in terms of the velocity components rather than stresses. This is similar to the displacement formulation previously discussed in the linear elasticity section. Using the rate of deformation relation $(3.13.4)_1$ and constitutive relation (6.4.12) in the equations of motion (5.3.5) gives

$$\begin{gathered} -p_{,i} + \frac{\mu}{3}v_{k,ki} + \mu v_{i,kk} + \rho b_i = \rho\left(\frac{\partial v_i}{\partial t} + v_j v_{i,j}\right) \\ -\nabla p + \frac{\mu}{3}\nabla(\nabla\cdot\boldsymbol{v}) + \mu\nabla^2\boldsymbol{v} + \rho\boldsymbol{b} = \rho\left(\frac{\partial\boldsymbol{v}}{\partial t} + \boldsymbol{v}\cdot\nabla\boldsymbol{v}\right) \end{gathered} \tag{6.4.13}$$

These relations are called the *Navier–Stokes equations* of classical fluid mechanics. For the compressible case, Eqs. (6.4.13) must be coupled with the continuity relations and an equation of state

$$\begin{gathered} \frac{\partial\rho}{\partial t} + (\rho v_k)_{,k} = 0, \quad \frac{\partial\rho}{\partial t} + \nabla\cdot(\rho\boldsymbol{v}) = 0 \\ p = p(\rho) \end{gathered} \tag{6.4.14}$$

This will then yield an isothermal mechanical theory with five equations for the five unknowns p, ρ, v_i.

For the incompressible case, the Navier–Stokes equations reduce to

$$\begin{gathered} -p_{,i} + \mu v_{i,kk} + \rho b_i = \rho\left(\frac{\partial v_i}{\partial t} + v_j v_{i,j}\right) \\ -\nabla p + \mu\nabla^2\boldsymbol{v} + \rho\boldsymbol{b} = \rho\left(\frac{\partial\boldsymbol{v}}{\partial t} + \boldsymbol{v}\cdot\nabla\boldsymbol{v}\right) \end{gathered} \tag{6.4.15}$$

and these must be coupled with the incompressible form of the continuity relation

$$v_{k,k} = 0, \quad \nabla\cdot\boldsymbol{v} = 0 \tag{6.4.16}$$

This will then yield an isothermal mechanical theory with four equations for the four unknowns p, v_i. Note that because of the convective term in the acceleration, either previous form of the Navier–Stokes equations is *nonlinear*. Thus, even though we incorporated a linear constitutive law, the general problem in Newtonian fluid dynamics is nonlinear.

Boundary conditions for this fluid mechanics theory would normally be applied at a solid or free surface. On a solid surface, the fluid velocity is assumed to match the velocity of the surface. This is the so-called *no-slip condition* in which fluid particles in contact with an external boundary move with the boundary's motion. For the case of a free surface (not in contact with another body), the fluid traction would have to match those applied from the surroundings. For example, if the free surface is simply exposed to atmospheric conditions, then neglecting surface tension effects, the normal traction would be equal to the ambient pressure while the shear tractions would have to vanish. For time dependent nonsteady flows, particular initial conditions on the velocity components would be required.

6.4.3 PROBLEM SOLUTIONS

We now explore the solution to a few selected two-dimensional linear viscous flow problems. This involves solving the governing Navier–Stokes equations (6.4.13) or (6.4.15) and the associated continuity equations (6.4.14) and (6.4.16) for some specific geometries. We will use MATLAB software to plot velocity distributions. As before, analytical solutions are most likely to be found for two-dimensional problems of limited geometric complexity. We will focus on incompressible flows and will avoid the nonlinear acceleration terms by considering only steady fluid motions.

EXAMPLE 6.4.1 PLANE POISEUILLE FLOW

Consider the two-dimensional steady flow of an incompressible linearly viscous fluid between two stationary plane boundaries as shown in Fig. 6.23. We assume that the boundaries are parallel to each other and are of infinite extent. The flow occurs only in the x_1-direction and is solely a function of the x_2-coordinate. This type of flow is commonly called *plane Poiseuille flow*. Determine the velocity field for this case.

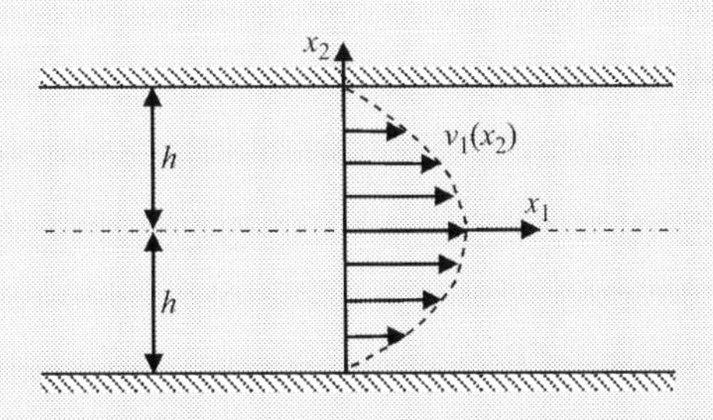

FIGURE 6.23

Plane Poiseuille flow geometry.

Solution: For this case, we assume no body forces, zero acceleration terms, and a velocity field of the form $v_1(x_2)$, $v_2 = v_3 = 0$. Under these conditions, the governing Navier–Stokes equations (6.4.13) reduce to

$$\begin{aligned} \frac{\partial p}{\partial x_1} &= \mu \frac{\partial^2 v_1}{\partial x_2^2} \\ \frac{\partial p}{\partial x_2} &= \frac{\partial p}{\partial x_3} = 0 \end{aligned} \tag{6.4.17}$$

and the continuity equation (6.4.16) is identically satisfied.

From the second equation in (6.4.17), we conclude that the pressure will depend only on x_1, $p = p(x_1)$. Using this result in relation $(6.4.17)_1$ allows us to argue that its left-hand side depends only on x_1, whereas the right-hand side is a function only of x_2, and therefore both sides must equal and common constant. Thus, we can write

$$\frac{dp}{dx_1} = \mu \frac{d^2 v_1}{dx_2^2} = -k \tag{6.4.18}$$

where $-k$ is the constant pressure gradient in the x_1-direction, and we have chosen the constant to be negative, thus indicating that the pressure is decreasing in the flow direction. Using the second relation in (6.4.18),

$$\frac{d^2 v_1}{dx_2^2} = -\frac{k}{\mu} \Rightarrow v_1 = -\frac{k}{\mu} x_2^2 + C_1 x_2 + C_2 \tag{6.4.19}$$

where C_1 and C_2 are arbitrary constants of integration. The boundary conditions on the problem are that the velocity must vanish at the channel's top and bottom, and thus $v_1(\pm h) = 0$. This statement then gives two simple equations that allow the solution for the two unknowns C_1 and C_2. Putting these results together then gives the final form for the velocity distribution

$$v_1 = \frac{k}{2\mu}(h^2 - x_2^2) \tag{6.4.20}$$

Result (6.4.20) indicates that the velocity distribution is parabolic with a maximum value occurring at mid-channel ($x_2 = 0$) given by

$$v_{1\max} = \frac{kh^2}{2\mu} \tag{6.4.21}$$

Fig. 6.24 illustrates a MATLAB (Code C-9) plot of this velocity distribution normalized by $kh^2/2\mu$. This distribution is the same at all sections (x_1 locations) along the channel. Another quantity that is sometimes of interest is

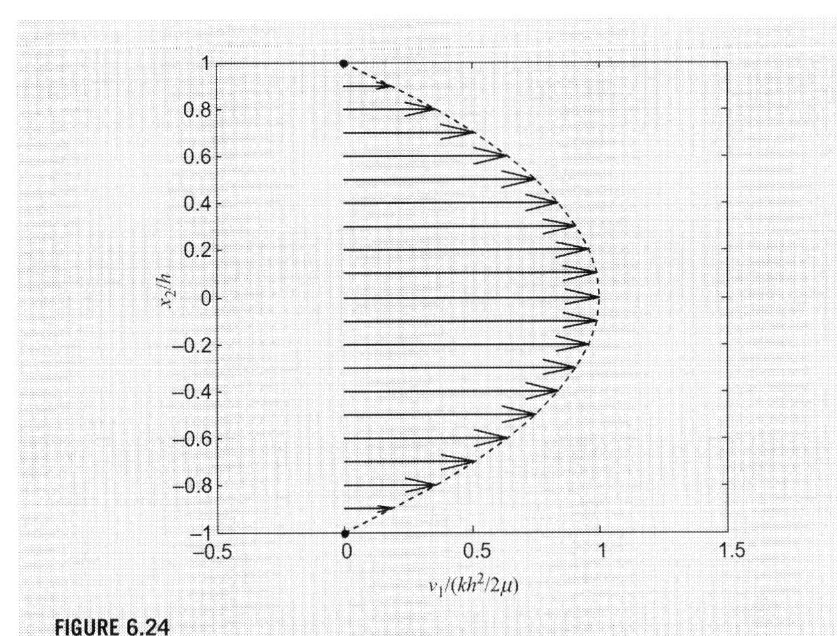

FIGURE 6.24

Plane Poiseuille flow field.

the *volume flow rate*, and for this problem the volume flow per unit time per unit x_3-width is given by integrating (6.4.18)

$$Q = \int_{-h}^{h} v_1 \, dx_2 = \frac{k}{\mu}\left(\frac{2h^3}{3}\right) \tag{6.4.22}$$

EXAMPLE 6.4.2 PLANE COUETTE FLOW

Consider the two-dimensional steady flow of an incompressible linearly viscous fluid between two plane boundaries. One boundary at $x_2 = 0$ is to remain fixed, whereas the other boundary at $x_2 = h$ moves with a constant speed V as shown in Fig. 6.25. We assume that the boundaries are parallel to each other and are of infinite extent. The flow occurs only in the x_1-direction and is solely a function of the x_2-coordinate. This type of flow is generally referred to as *plane Couette flow*. Determine the velocity field for this case.

Solution: Similar to the previous example, we assume no body forces, zero acceleration terms, and a velocity field of the form $v_1(x_2)$, $v_2 = v_3 = 0$.

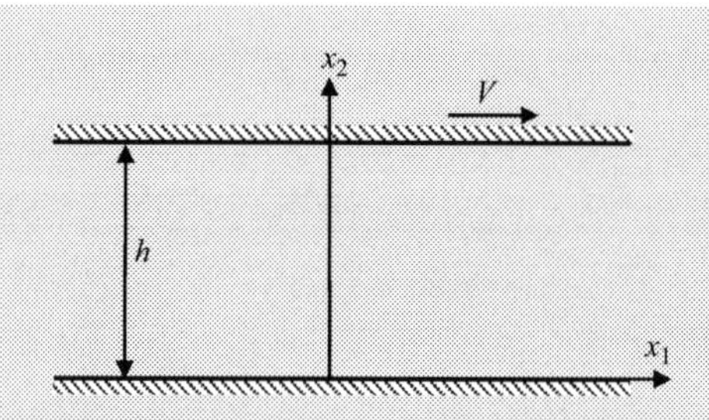

FIGURE 6.25

Plane Couette flow geometry.

Under these conditions, the continuity equation is identically satisfied and the Navier–Stokes equations again reduce to

$$\begin{aligned}\frac{\partial p}{\partial x_1} &= \mu \frac{\partial^2 v_1}{\partial x_2^2} \\ \frac{\partial p}{\partial x_2} &= \frac{\partial p}{\partial x_3} = 0\end{aligned} \tag{6.4.23}$$

Applying the same arguments used in the previous example, we conclude that the pressure must be of the form $p = p(x_1)$, and again find

$$\frac{dp}{dx_1} = \mu \frac{d^2 v_1}{dx_2^2} = -k \tag{6.4.24}$$

where $-k$ is the constant pressure gradient in the x_1-direction. This again allows direct integration of relation (6.4.24) to determine the unknown velocity component

$$v_1 = -\frac{k}{\mu} x_2^2 + C_1 x_2 + C_2 \tag{6.4.25}$$

where C_1 and C_2 are arbitrary constants of integration. The boundary conditions on this problem are that the velocity must match with the speed of the top of the channel and vanish at the bottom. These two conditions then yield values for C_1 and C_2:

$$\begin{aligned}&v_1(0) = 0 \Rightarrow C_2 = 0 \\ &v_1(h) = V \Rightarrow C_1 = \frac{V}{h} + \frac{k}{\mu} h\end{aligned} \tag{6.4.26}$$

Putting these results together then gives the final form for the velocity distribution

$$v_1 = \frac{V}{h} x_2 + \frac{k}{\mu}(h - x_2) x_2 \tag{6.4.27}$$

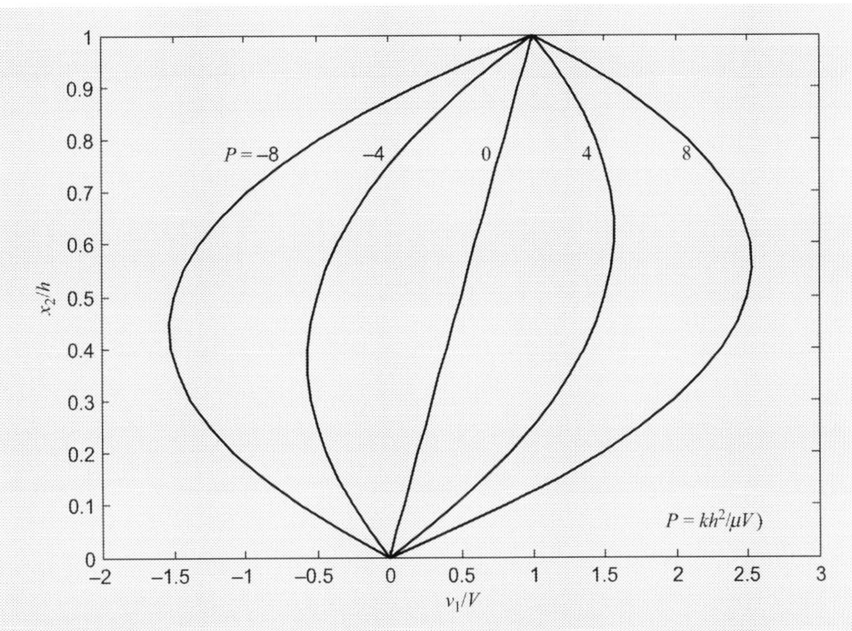

FIGURE 6.26

Velocity profiles for plane Couette flows.

For this example, the velocity distribution depends on V and will not in general be parabolic as found in Example 6.4.1 (unless $V = 0$). Fig. 6.26 shows several velocity profiles for different values of a dimensionless parameter $P = kh^2 / \mu V$. Plots are generated from MATLAB Code C-9. As $V \rightarrow 0$, $P \rightarrow \infty$ and the velocity profile becomes parabolic. Note that for the special case with no pressure gradient k, the velocity becomes simply $v_1 = \frac{V}{h} x_2$, which is just a linear distribution as shown in Fig. 6.26 for $P = 0$. It is also interesting to notice that for cases with $P < 0$, the flow exhibits reversal behavior with zones of both positive and negative flow directions.

EXAMPLE 6.4.3 HAGEN–POISUEILLE FLOW

We now explore the steady incompressible flow in a circular cylindrical tube and assume the velocity field has only one axisymmetric component along the tube's axis. Thus, we will use cylindrical coordinates to formulate and solve the problem which is shown in Fig. 6.27. For this flow case, the velocity field may be taken as $v_r = v_\theta = 0, \quad v_z = v_z(r)$. This type of flow is com-

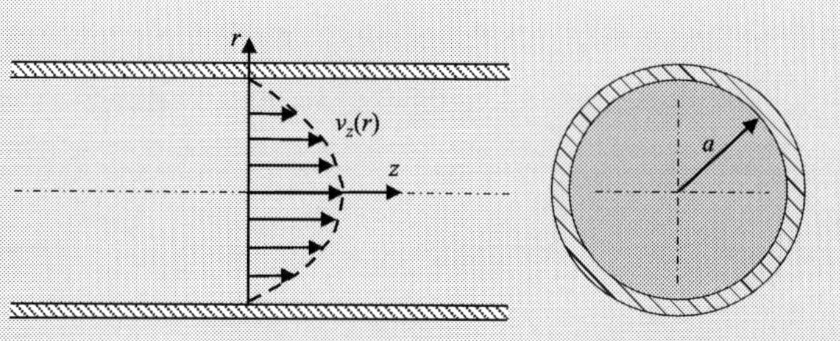

FIGURE 6.27

Hagen–Poiseuille flow geometry.

monly called *Hagen–Poiseuille flow*. We now wish to solve the viscous flow to determine the velocity $v_z = v_z(r)$.

Solution: We first explore the continuity equation (6.4.14) in cylindrical coordinates using some help from relation (2.18.20):

$$\nabla \cdot \boldsymbol{v} = \frac{1}{r}\frac{\partial}{\partial r}(rv_r) + \frac{1}{r}\frac{\partial v_\theta}{\partial \theta} + \frac{\partial v_z}{\partial z} = 0 \tag{6.4.28}$$

It is therefore observed that the given velocity field satisfies this relation identically. Next moving on to the Navier–Stokes equations in cylindrical coordinates, we find that the restricted velocity field reduces the system to

$$\begin{aligned} &\frac{\partial p}{\partial r} = \frac{\partial p}{\partial \theta} = 0 \\ &\frac{\partial p}{\partial z} = \mu\left[\frac{1}{r}\frac{d}{dr}\left(r\frac{\partial v_z}{\partial z}\right)\right] \end{aligned} \tag{6.4.29}$$

Thus, we can conclude that the pressure p depends only on z and can apply similar arguments as used in the previous examples to conclude that

$$\frac{dp}{dz} = \mu\left[\frac{1}{r}\frac{d}{dr}\left(r\frac{dv_z}{dz}\right)\right] = \text{constant} = -k \tag{6.4.30}$$

and so

$$\frac{1}{r}\frac{d}{dr}\left(r\frac{dv_z}{dz}\right) = \frac{-k}{\mu} \tag{6.4.31}$$

Relation (6.4.31) can be integrated to get

$$v_z = -\frac{k}{4\mu}r^2 + C_1 \ln r + C_2 \tag{6.4.32}$$

where C_1 and C_2 are arbitrary constants of integration. The boundary conditions on the problem are that the velocity must vanish at $r = a$, and the solution must remain bounded for $0 \le r \le a$. The boundedness condition implies that C_1 must vanish, while the condition at $r = a$ gives $C_2 = ka^2 / 4\mu$. Putting these results together provides the final form for the velocity distribution

$$v_z = \frac{k}{4\mu}(a^2 - r^2) \tag{6.4.33}$$

and thus the axial velocity component is distributed as a paraboloid of revolution, which could be generated by a 2π rotation of the plane distribution shown in Fig. 6.24. This plot comes from MATALB Code C-9. The three-dimensional velocity vector distribution is shown in Fig. 6.28. The maximum velocity occurs at $r = 0$ and is given by

$$v_{z\,\max} = \frac{ka^2}{4\mu} \tag{6.4.34}$$

The volume flow rate is again found by integrating the velocity distribution (6.4.31):

$$Q = \int_0^a v_z 2\pi r\, dr = \frac{k\pi a^4}{8\mu} \tag{6.4.35}$$

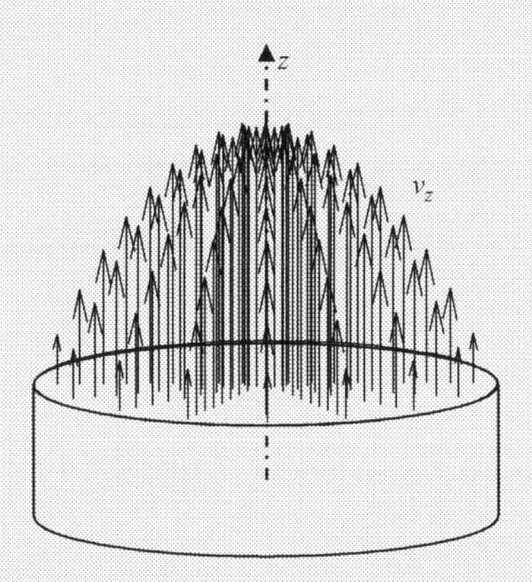

FIGURE 6.28

Hagen–Poiseuille velocity distribution.

Many additional analytical solutions to the Navier–Stokes equations exist in the literature for more general fluid conditions and more complicated flow geometries. The interested reader is directed to the general references given at the beginning of this section.

6.5 LINEAR VISCOELASTIC MATERIALS

Up to this point, we have explored the basics of linear elastic and linear viscous material models. We found that for the linear elastic case, the stress was proportional to strain and the theory was applicable to many solid materials under small deformations. On the other hand, for the linear viscous model, stress was proportional to the rate of strain and this theory had applications to materials that flow, that is, fluid mechanics. For many materials, both elastic and viscous behaviors are observed together, and such materials are generally called *viscoelastic*. For such media, the distinction between solid and fluid behavior can become somewhat difficult to make. Typical behaviors commonly include a time-dependent response with permanent deformation, and these phenomena manifest themselves as *creep*, *stress relaxation*, *hysteresis*, and *history-dependent response*. Such behaviors are common in high polymers (see amorphous material illustration in Fig. 6.1), many metals and plastics at elevated temperatures, bitumens, and some biological materials. In this section, we will limit discussion to linear viscoelasticity where the strains and strain rates are small and the constitutive relations will be linear. Chapter 8 will consider the more general case of nonlinear viscoelastic modeling. As with the previous linear continuum theories, our presentation here will be only introductory. Entire texts have been devoted to linear viscoelasticity, and the interested reader is directed to Christensen (2010), Flügge (1975), Bland (2016), Golden and Graham (1988), and Gutierrez-Lemini (2014).

6.5.1 CONSTITUTIVE LAWS

Often viscoelastic behaviors can be thought of as a range of response with elastic behavior, at one extreme, and viscous behavior, at the other end. In between these two ends, the material response is a combination of both elastic and viscous behaviors. As we will see, our developed constitutive models will often exhibit this same structure. To demonstrate this point, it is instructive to explore the behavior of an old toy from the 1950s called *Silly Putty.* This pliable material (silicone polymer) is a classic viscoelastic as it behaves as an elastic solid under rapid deformation, but completely changes to a flowing creeping liquid under slow loadings. The author has used this material as a classroom demonstration countless times. Fig. 6.29 illustrates two of these behaviors using a sample initially rolled into an approximate spherical ball shape. Fig. 6.29A shows a time sequence of the ball dropped onto a rigid horizontal surface. Under such dynamic impact loading, the sample bounces off of the surface with a high degree of elastic response. Fig. 6.29B illustrates the same ball sample gently placed on the rigid horizontal surface at time $t = 0$. Under just static self-weight loading, the ball slowly begins to change shape through a flowing creep deformation. After about 12 h (at room temperatures), the shape has significantly changed as shown. For larger static loadings and/or higher temperatures, this material would show even more flow behavior.

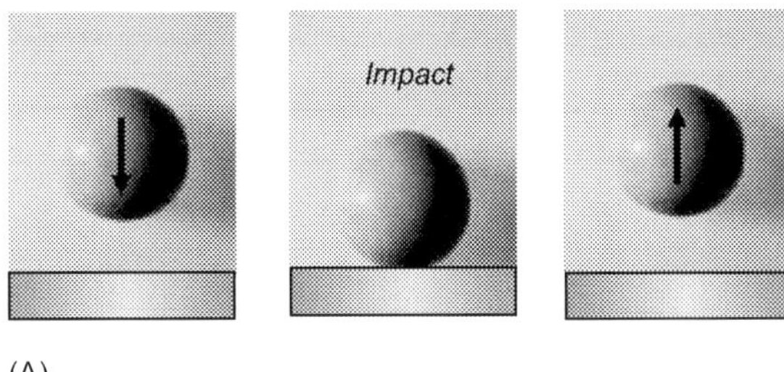

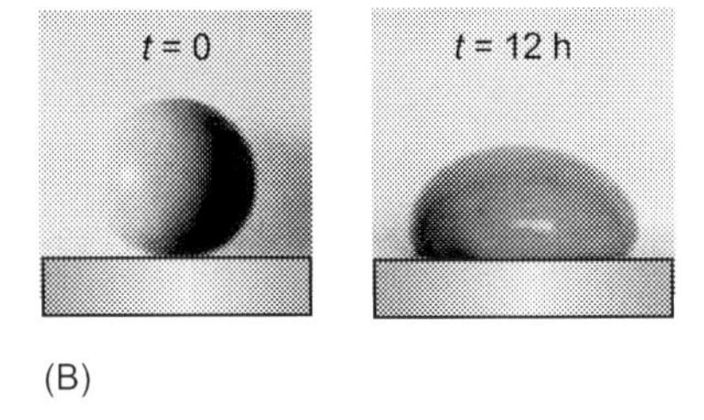

(B)

FIGURE 6.29

Viscoelastic behaviors of silly putty: (A) silly putty ball bounces off rigid horizontal surface behaving like elastic solid under rapid impact loading; (B) silly putty ball exhibiting flowing creep deformation under static loading of self-weight.

We again start our constitutive equation development with some comments on experimental observations. Extensive data on many materials indicate that if we subject a one-dimensional sample to a step change in strain

$$\varepsilon = \varepsilon_o H(t) \tag{6.5.1}$$

where $H(t)$ is the Heaviside step function defined by

$$H(t) = \begin{cases} 1, & t > 0 \\ 0, & t < 0 \end{cases} \tag{6.5.2}$$

then the stress response will be time dependent of the form

$$T = G(t)\varepsilon_o \tag{6.5.3}$$

This type of experiment is called the *stress relaxation test*, and the time function $G(t)$ is referred to as the *stress relaxation function*. Typical time plots of this behavior are shown in Fig. 6.30. It should be noted that for linear viscoelastic behavior, the relaxation function is independent of the applied strain ε_o. If $G(t) \to 0$ as $t \to \infty$, we say the material is *fluid-like*, whereas if $G(t) \to G_o > 0$ as $t \to \infty$, the material is *solid-like*. Each of these cases are illustrated in the figure.

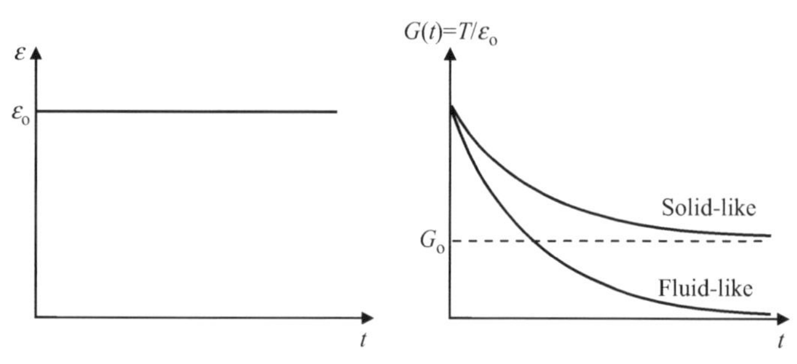

FIGURE 6.30

Stress relaxation behaviors for viscoelastic materials.

Next, consider another one-dimensional standard experiment where the sample is subjected to a step function in stress

$$T = T_o H(t) \tag{6.5.4}$$

Under this type of loading, many materials will exhibit the following time dependent strain response

$$\varepsilon = J(t) T_o \tag{6.5.5}$$

where $J(t)$ is called the *creep function* or *creep compliance*. Again note that for linear viscoelastic behavior, the creep function is independent of the applied stress T_o. If $J(t) \to \infty$ as $t \to \infty$, we say the material is *fluid-like*, whereas if $J(t) \to J_o > 0$ as $t \to \infty$, the material is *solid-like*. These cases are illustrated in Fig. 6.31.

6.5.1.1 Analog or mechanical viscoelastic constitutive models

Past studies of viscoelastic materials have often employed one-dimensional analog or mechanical models to develop simplified constitutive relations between stress, stress rates, strain, and strain rates. While this approach is simplistic and nontensorial, it does allow one to combine elastic and viscous behaviors in various ways. The

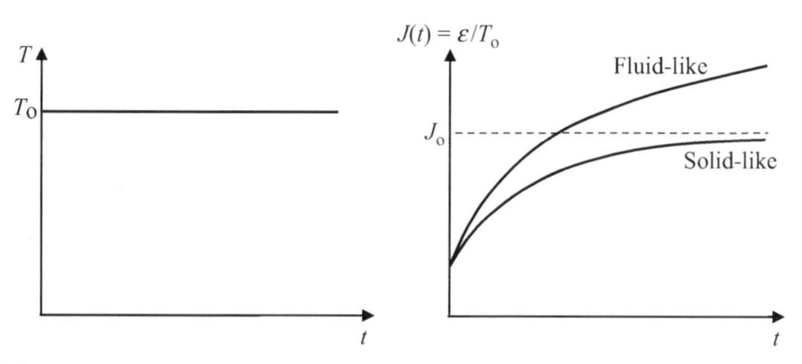

FIGURE 6.31

Creep compliance behaviors for viscoelastic materials.

scheme also brings out fundamental properties and definitions and provides a means to extend these fundamentals to multidimensional relations.

We start this method by defining the basic elastic and viscous elements. Within these models, axial force will represent the continuum stress, and axial deformation and velocity will represent the strain and strain rate. The particular stress or strain component to be modeled could be normal, shear, or volumetric; however, further discussion will be needed to extend things to multi-dimensional relations. The fundamental elastic element is a *spring* with a linear force–deformation stiffness E, as shown in Fig. 6.32. The element's constitutive law is then simply

$$T = E\varepsilon \tag{6.5.6}$$

(a) Elastic element, $T = E\varepsilon$

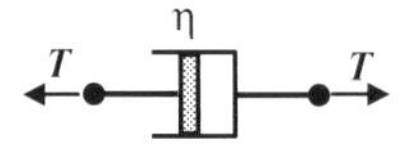

(b) Viscous element, $T = \eta\dot{\varepsilon}$

FIGURE 6.32

Basic elastic and viscous elements.

The relaxation and creep functions for just the spring element follow from the previous definitions

$$\begin{aligned} G(t) &= E \\ J(t) &= 1/E \end{aligned} \tag{6.5.7}$$

which are all constants.

The fundamental viscous element is a *dashpot* with a linear force–deformation rate viscosity η, as illustrated in Fig. 6.32, and thus the constitutive law is

$$T = \eta\dot{\varepsilon} \tag{6.5.8}$$

Note that superimposed dots used here simply indicate ordinary time derivatives. The relaxation and creep functions for the dashpot are

$$\begin{aligned} G(t) &= \eta\delta(t) \\ J(t) &= \frac{1}{\eta}t \end{aligned} \tag{6.5.9}$$

where $\delta(t)$ is the *Dirac delta function* defined by the following properties:

$$\delta(t) = \begin{cases} 0, & t < 0, \\ \infty, & t = 0, \\ 0, & t > 0, \end{cases} \quad \int_{-\infty}^{\infty} \delta(t)\,dt = 1, \quad \delta(t) = \frac{dH(t)}{dt} \tag{6.5.10}$$

6.5.1.2 Maxwell model

Now the primary idea in this modeling approach is to connect spring and dashpot models in various ways to establish particular constitutive equations that may be useful to simulate real material behavior. We start with a very simple fundamental combination

FIGURE 6.33

Maxwell fluid model.

of connecting a spring and dashpot in a series arrangement as shown in Fig. 6.33, and this is referred to as a *Maxwell model* that has characteristics of a fluid-like material. For this configuration, the stress in each element is identical, while the sum of the strains in the spring and dashpot equals the total strain in the combined model. This information yields the following relations and produces the constitutive law:

$$\left.\begin{array}{l} T_s = T_d = T \\ \varepsilon_s + \varepsilon_d = \varepsilon \end{array}\right\} \Rightarrow \dot{T} + \frac{E}{\eta}T = E\dot{\varepsilon} \tag{6.5.11}$$

We can see that even in this most simple viscoelastic model, the constitutive law contains rate-dependent terms. Thus, if the strain is specified, (6.5.11) is an ordinary differential equation with constant coefficients whose solution would give the stress response. Note that an often used parameter $\tau = \eta/E$ is commonly referred to as the *relaxation or retardation time constant.* Since this constitutive model only includes two material constants E and η, it will likely be limited in its ability to quantitatively predict real material behavior. Nevertheless, it does provide a simple starting point to understand viscoelastic modeling.

EXAMPLE 6.5.1 RELAXATION AND CREEP FUNCTIONS FOR MAXWELL MODEL

Determine the relaxation and creep functions for the Maxwell model and plot the results for different values of relaxation times.

Solution: For the stress relaxation behavior, the strain is specified by (6.5.1), $\varepsilon = \varepsilon_o H(t)$. Substituting this form into constitutive relation (6.5.11) gives

$$\dot{T} + \frac{E}{\eta}T = 0, \quad t > 0$$

The solution to this standard first-order ordinary differential equation is given by

$$T = Ke^{-(E/\eta)t}$$

where K is an arbitrary constant. We assume an initial condition on this case to be $T(0) = E\varepsilon_o$, and thus the constant is determined to be $K = E\varepsilon_o$. So, the stress response is

$$T = E\varepsilon_o e^{-(E/\eta)t}$$

and thus using (6.5.3) the relaxation function follows to be

$$G(t)=Ee^{-(E/\eta)t}=Ee^{-t/\tau} \tag{6.5.12}$$

For the creep function, the stress is specified by $T=T_oH(t)$, and using this in constitutive relation (6.5.11) gives

$$\dot{\varepsilon}=T_o/\eta,\quad t>0$$

This differential equation can easily be integrated to get

$$\varepsilon=T_ot/\eta+C$$

where C is an arbitrary constant. Again applying the expected initial condition $\varepsilon(0)=T_o/E$ determines the constant C and gives the final form for the strain response

$$\varepsilon=T_o\left(\frac{t}{\eta}+\frac{1}{E}\right)$$

Using (6.5.5) the creep function follows is then given by

$$J(t)=\frac{t}{\eta}+\frac{1}{E} \tag{6.5.13}$$

Fig. 6.34 generated by MATLAB Code C-10 illustrates the relaxation and creep behaviors of the Maxwell model for different relaxation times τ. It can be seen that for a given value of time, as the relaxation time parameter increases, less model relaxation is predicted. Likewise, the amount of creep predicted will be less for larger values of τ. This is typical relaxation time behavior for viscoelastic models; however, the linear creep prediction with time is not likely to match with real material behavior.

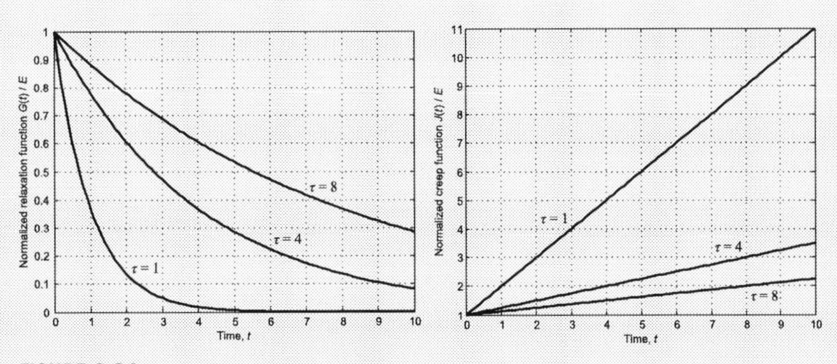

FIGURE 6.34

Relaxation and creep functions for Maxwell model.

6.5.1.3 Kelvin–voigt model

Next, consider the case of connecting a spring and dashpot in a parallel combination as shown in Fig. 6.35. This produces the *Kelvin–Voigt model* that has characteristics of a solid-like material. For this configuration, the strain in each element is identical, while the sum of the stresses in the spring and dashpot equals the total stress in the combined model. This information yields the following relations and produces the constitutive law:

$$\left.\begin{array}{l} T_s + T_d = T \\ \varepsilon_s = \varepsilon_d = \varepsilon \end{array}\right\} \Rightarrow T = E\varepsilon + \eta\dot{\varepsilon} \tag{6.5.14}$$

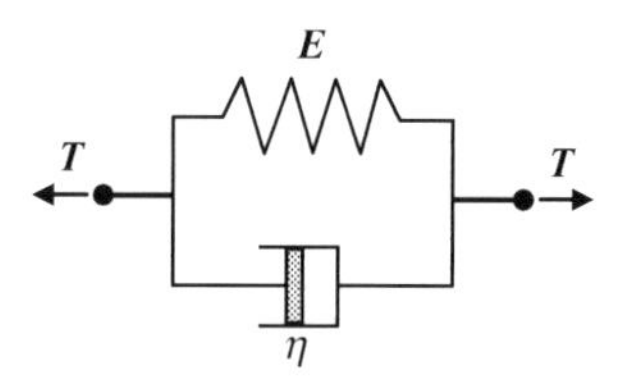

FIGURE 6.35

Kelvin–Voigt model.

As in the previous Maxwell model, we observe a constitutive law that contains rate-dependent terms. Thus, if the stress is specified, (6.5.14) is an ordinary differential equation whose solution would give the strain response. As before, since this constitutive model only includes two material constants E and η, it will likely be limited in its ability to predict real material behavior. However, like the Maxwell model, it does provide a simple starting point to understand viscoelastic modeling.

EXAMPLE 6.5.2 RELAXATION AND CREEP FUNCTIONS FOR KELVIN–VOIGT MODEL

Determine the relaxation and creep functions for the Kelvin–Voigt model and plot the results for different values of relaxation times.

Solution: For the stress relaxation behavior, the strain is specified by (6.5.1), $\varepsilon = \varepsilon_o H(t)$. Substituting this form into constitutive relation (6.5.14) gives

$$T = E\varepsilon_o, \quad t > 0$$

and using (6.5.3) the relaxation function follows to be

$$G(t) = E \tag{6.5.15}$$

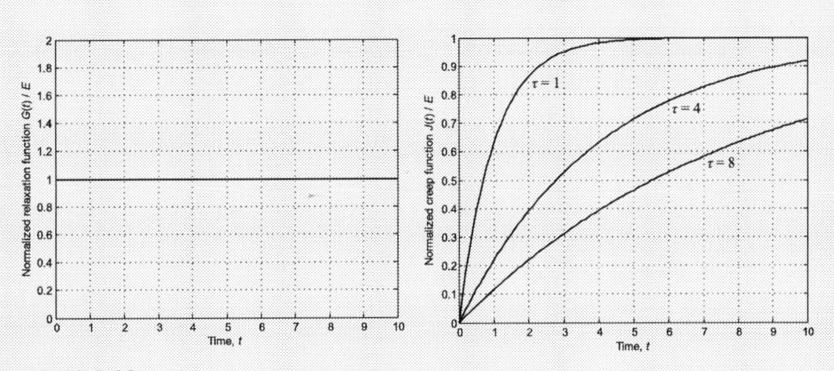

FIGURE 6.36

Relaxation and creep functions for Kelvin model.

For the creep function, the stress is specified by $T = T_o H(t)$, and using this in constitutive relation (6.5.14) gives

$$T_o = E\varepsilon + \eta\dot{\varepsilon}$$

The solution to this nonhomogeneous first-order differential equation follows from standard methods to be

$$\varepsilon = Ce^{-(E/\eta)t} + \frac{T_o}{E}$$

where C is an arbitrary constant. Again applying the expected initial condition $\varepsilon(0) = 0$ determines the constant C and gives the final form for the strain response

$$\varepsilon = \frac{T_o}{E}\left(1 - e^{-(E/\eta)t}\right)$$

Using (6.5.5), the creep function is then given by

$$J(t) = \frac{1}{E}\left(1 - e^{-(E/\eta)t}\right) \tag{6.5.16}$$

Fig. 6.36 generated by MATLAB Code C-11 illustrates the relaxation and creep behaviors of the Kelvin model for different relaxation times τ. It can be seen that the relaxation predictions are independent of the relaxation time parameter as shown in relation (6.5.15). This situation would indicate that the Kelvin model is not likely to be able to predict real material relaxation behavior. Similar to the Maxwell case, the amount of creep predicted for the Kelvin model will be less for larger values of τ.

EXAMPLE 6.5.3 STRAIN RECOVERY AFTER UNLOADING FOR MAXWELL AND KELVIN–VOIGT MODELS

Using the following loading history

$$T(t)=\begin{cases} T_o, & 0\le t\le t_1 \\ 0, & t>t_1 \end{cases}$$

determine the strain behavior or *recovery after unloading* for the Maxwell and Kelvin models and plot the strain responses.

Solution: Denote the strain responses in the Maxwell and Kelvin models by ε_M and ε_K. Over the time range $0\le t\le t_1$, this stress history simply creates the standard creep strain responses given previously in Examples 6.5.1 and 6.5.2:

$$\varepsilon_M=T_o\left(\frac{t}{\eta}+\frac{1}{E}\right),\quad \varepsilon_K=\frac{T_o}{E}\left(1-e^{-(E/\eta)t}\right),\quad 0\le t\le t_1 \tag{6.5.17}$$

For time $t>t_1$, the stress is now zero. For the Maxwell model, the instantaneous drop in stress T_o at $t=t_1$ would result in an immediate elastic recovery of T_o/E from just the spring. Since for $t>t_1$, both the stress and stress rate are zero, this implies that $\dot{\varepsilon}_M=0$. This results in a constant strain over this time interval

$$\varepsilon_M=T_o\left(\frac{t_1}{\eta}+\frac{1}{E}\right)-\frac{T_o}{E}=\frac{t_1}{\eta}T_o,\quad t>t_1 \tag{6.5.18}$$

For time $t>t_1$, the Kelvin model would predict an exponentially decreasing strain response of the form

$$\varepsilon=Ke^{-(E/\eta)(t-t_1)}$$

where K is a constant. For this model, there can be no strain discontinuity at $t=t_1$, and enforcing this condition gives an equation to determine the constant K:

$$K=\frac{T_o}{E}\left(1-e^{-(E/\eta)t_1}\right)$$

and so the Kelvin strain response for $t>t_1$ is then given by

$$\varepsilon_K=\frac{T_o}{E}\left(1-e^{-(E/\eta)t_1}\right)e^{-(E/\eta)(t-t_1)},\quad t>t_1 \tag{6.5.19}$$

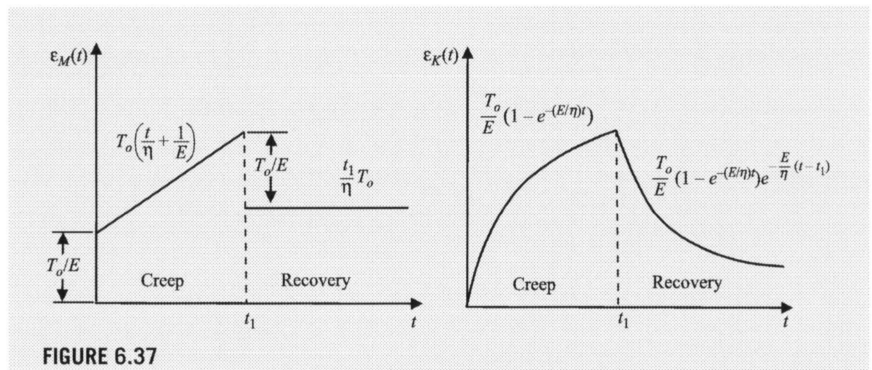

FIGURE 6.37

Recovery behavior for Maxwell and Kelvin models.

Thus, we have determined two different strain–time responses for each model and these are put together and shown in Fig. 6.37. The Kelvin predictions are qualitatively correct with a nonlinear increasing creep phase followed by a decreasing recovery response. However, as mentioned previously, these simple model predictions would likely not match with experimental data for real materials. For the Maxwell model, no inelastic recovery is found, but a permanent strain is present for large time.

EXAMPLE 6.5.4 CONSTANT LOADING RATE RESPONSE FOR KELVIN–VOIGT MODEL USING MATLAB TO SOLVE ODE CONSTITUTIVE LAW

Determine the stress-–time, strain–time, and stress–strain response of the Kelvin model for a loading rate specified by $T = Rt$, where R is the constant loading rate.

Solution: Although the Kelvin model under these conditions will yield an ordinary differential equation that has a known analytical solution, we will solve and plot the required results using the numerical tools in MATLAB. Code C-12 in Appendix C lists the fairly simple code using the built-in ODE45 solver that handles differential equations of first order. The results are shown in Fig. 6.38 as three separate plots of the stress–time, strain–time, and stress–strain behaviors for a set of three loading rates $R = 10$, 20, and 40. Using suitable units, the model parameters were chosen to be $E = 50$ and $\eta = 100$. The figure illustrates the loading rate effect on viscoelastic material behavior.

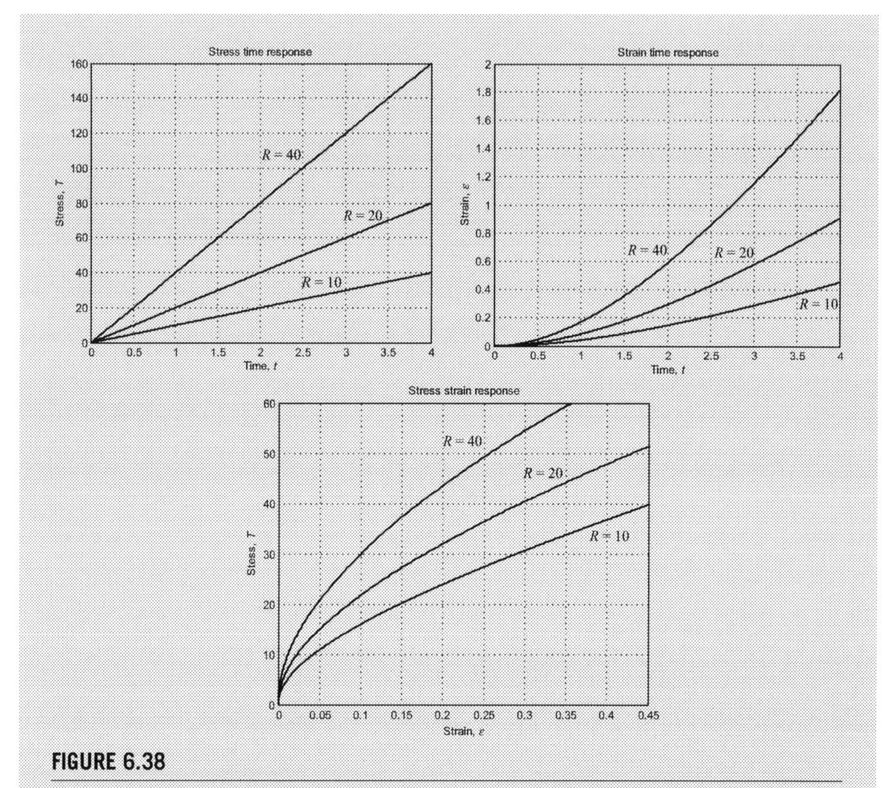

FIGURE 6.38

Stress–time, strain–time, and stress–strain response of Kelvin model under constant loading rate $T = Rt$. $R = 10, 20, 40$, and $E = 50$ and $\eta = 100$.

The stress–strain curves show the usual rate dependency with higher values of stress occurring at higher loading rates.

EXAMPLE 6.5.5 LOADING AND UNLOADING RESPONSE FOR KELVIN–VOIGT MODEL USING MATLAB TO SOLVE ODE CONSTITUTIVE LAW

Determine the stress–strain response of the Kelvin model for a loading history shown in Fig. 6.39. This history corresponds to a constant loading rate followed by a constant unloading rate.

Solution: Although the Kelvin model under these conditions will again yield an ordinary differential equation that has a known analytical solution, as in the previous example we will solve and plot the required results using the numerical tools in MATLAB. Code C-13 in Appendix C lists the code using the built-in ODE45 solver that handles differential equations of first order.

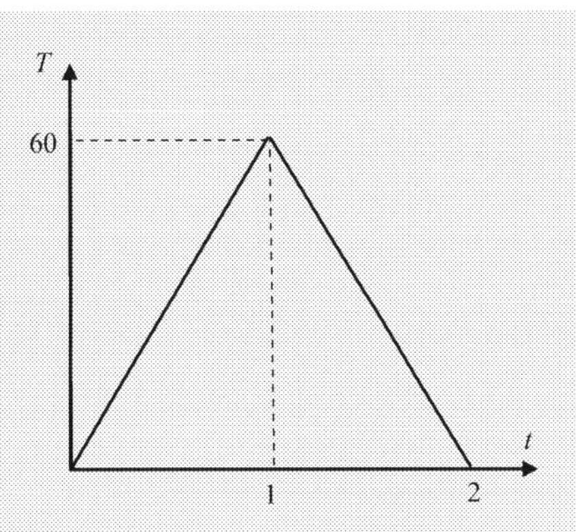

FIGURE 6.39

Loading history for Example 6.5.5.

The stress–strain results are shown in Fig. 6.40 for the case with model parameters $E = 200$ and $\eta = 100$. It is noted that the behavior is of course inelastic with no complete strain recovery even though the stress was completely removed. Permanent deformation is thus predicted, and an energy loss will result over this loading cycle.

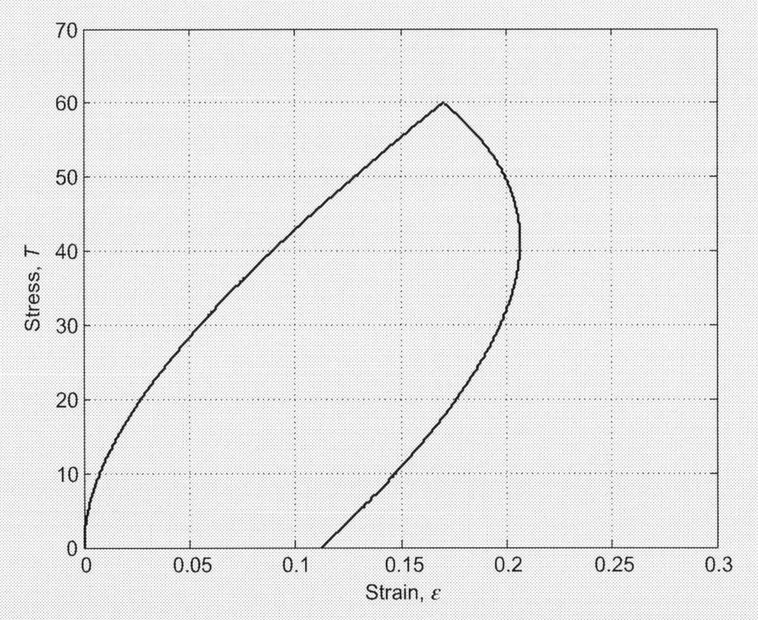

FIGURE 6.40

Stress–strain response for Kelvin model in Example 6.5.5 with parameters $E = 200$ and $\eta = 100$.

The previous five examples illustrate many of the unique behaviors associated with viscoelastic materials. Using simple Maxwell and Kelvin models, qualitative behaviors of stress relaxation, creep, strain recovery, and inelastic loading–unloading stress–strain responses were demonstrated. Next, we explore more sophisticated analog models that will have improved chances of quantitatively modeling real material behaviors.

6.5.1.4 More general analog models

As previously mentioned, while the simple Maxwell and Kelvin models provide many qualitatively correct viscoelastic behaviors, they will not in general simulate actual data of real materials. With this in mind, efforts to improve analog modeling have combined additional spring and dashpot elements in more complex arrangements. Some of these combined models are shown in Fig. 6.41. Note that models

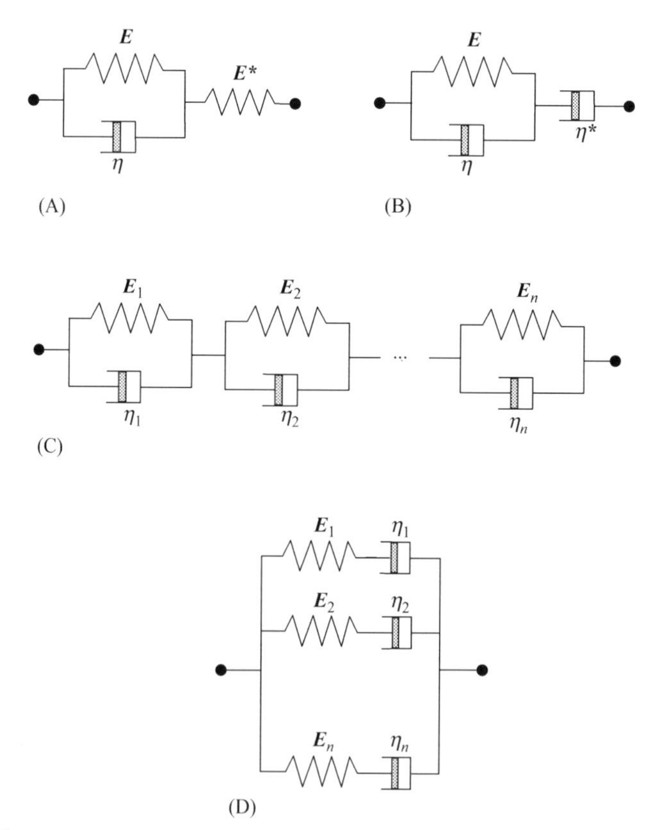

FIGURE 6.41

More general analog viscoelastic models: (A) Three-parameter solid; (B) three-parameter fluid; (C) generalized Kelvin chain; (D) generalized Maxwell chain.

with a free spring as shown in Fig. 6.41A will generally produce solid-like behaviors, whereas a free dashpot as shown in Fig. 6.41B will result in a fluid-like long-term response.

The three-parameter solid shown in Fig. 6.41A represents a Kelvin model in series with an elastic element. To arrive at the constitutive model for the entire combination, we note that the sum of the strains in Kelvin and elastic components add to give the total overall strain, and the stress in each of these components are the same and equal the overall stress

$$\left.\begin{array}{l} T_K = T_s = T \\ \varepsilon_K + \varepsilon_s = \varepsilon \end{array}\right\} \Rightarrow T + p_1\dot{T} = q_o\varepsilon + q_1\dot{\varepsilon} \tag{6.5.20}$$

where the coefficients are given by

$$p_1 = \frac{\eta}{E+E^*}, \quad q_o = \frac{EE^*}{E+E^*}, \quad q_1 = \frac{\eta E^*}{E+E^*} \tag{6.5.21}$$

The relaxation and creep functions for this model are found to be

$$\begin{aligned} G(t) &= \frac{q_1}{p_1}e^{-t/p_1} + q_o(1-e^{-t/p_1}) \\ J(t) &= \frac{p_1}{q_1}e^{-q_o t/q_1} + \frac{1}{q_o}(1-e^{-q_o t/q_1}) \end{aligned} \tag{6.5.22}$$

Note that this model carries the restriction $q_1 > p_1 q_o$ to guarantee that the relaxation/creep functions remain positive (see Flügge, 1975) for more details on this requirement.

Following a similar analysis, the constitutive relation for the three-parameter fluid shown in Fig. 6.41B can be developed as the form

$$T + p_1\dot{T} = q_1\dot{\varepsilon} + q_2\ddot{\varepsilon} \tag{6.5.23}$$

where the coefficients are given by

$$p_1 = \frac{\eta+\eta^*}{E}, \quad q_1 = \eta^*, \quad q_2 = \frac{\eta\eta^*}{E} \tag{6.5.24}$$

The relaxation and creep functions for this model are given by

$$\begin{aligned} G(t) &= \frac{q_1^2}{p_1 q_1 - q_2}e^{-t/p_1} \\ J(t) &= \frac{t}{q_1} + \frac{p_1 q_1 - q_2}{q_1^2}(1-e^{-q_1 t/q_2}) \end{aligned} \tag{6.5.25}$$

Similar to the three-parameter solid, this model also carries a restriction $p_1 q_1 > q_2$ to guarantee that relaxation/creep functions remain positive (see Flügge, 1975).

Clearly, these three-parameter models offer additional model constants to fit with real material relaxation and/or creep behavior. This concept can be extended to even larger numbers of Maxwell or Kelvin chains as shown in Fig. 6.41C and D, thus producing a general differential constitutive relations of the form

$$p_o T + p_1 \dot{T} + p_2 \ddot{T} + \cdots = q_o \varepsilon + q_1 \dot{\varepsilon} + q_2 \ddot{\varepsilon} + \cdots$$
$$\sum_{k=0}^{m} p_k \frac{d^k T}{dt^k} = \sum_{k=0}^{m} q_k \frac{d^k \varepsilon}{dt^k} \tag{6.5.26}$$

or in operator form

$$\boldsymbol{P}T = \boldsymbol{Q}\varepsilon$$
$$\text{where} \quad \boldsymbol{P} = \sum_{k=0}^{m} p_k \frac{d^k}{dt^k}, \quad \text{and} \quad \boldsymbol{Q} = \sum_{k=0}^{m} q_k \frac{d^k}{dt^k} \tag{6.5.27}$$

Of course we notice that including more analog elements raises the order of the differential equation constitutive law. This complexity can be offset by using Laplace transforms to solve most of these higher-order ODEs. However, we will not pursue this approach to solve specific problems, but will employ Laplace transforms later in a more general way. The MATLAB ODE solver used in Examples 6.5.4 and 6.5.5 can also easily handle such higher-order equations (see Exercise 6.32). At this stage, we could rethink the analog constitutive concepts and eliminate the idea of spring–dashpot models and simply view (6.5.26) or (6.5.27) as just a general rate-dependent constitutive relation. In this regard, we could think of the general relaxation solution coming from a series form $T(t) = \varepsilon_o \sum_k G_k e^{-t/\tau_k}$, thus yielding a relaxation function $G(t) = \sum_k G_k e^{-t/\tau_k}$ with a discrete set of relaxation time τ_k.

Next, we wish to pursue the issue of how to apply these analog models to multi-dimensional continuum problems. Starting with either relation (6.5.26) or (6.5.27), it would be tempting to simply substitute the Cauchy stress tensor T_{ij} for the scalar stress T, and likewise substitute the small strain tensor ε_{ij} for the scalar variable ε. However, many experiments have shown that most isotropic engineering materials behave elastically in volumetric deformation, and this would imply that isotropic viscoelastic constitutive laws should only be applied to the deviatoric stress and strain tensors. With this in mind, let us review the decomposition of stress and strain into spherical and deviatoric components. Sections 3.10 and 4.5 provide the appropriate decomposition relations for the small strain and Cauchy stress tensors and are repeated here

$$\varepsilon_{ij} = \tilde{\varepsilon}_{ij} + \hat{\varepsilon}_{ij}$$
$$\tilde{\varepsilon}_{ij} = \frac{1}{3}\varepsilon_{kk}\delta_{ij}, \quad \hat{\varepsilon}_{ij} = \varepsilon_{ij} - \frac{1}{3}\varepsilon_{kk}\delta_{ij} \tag{6.5.28}$$

$$T_{ij} = \tilde{T}_{ij} + \hat{T}_{ij}$$
$$\tilde{T}_{ij} = \frac{1}{3}T_{kk}\delta_{ij}, \hat{T}_{ij} = T_{ij} - \frac{1}{3}T_{kk}\delta_{ij} \tag{6.5.29}$$

It can easily be shown that the isotropic form of Hooke's law (6.2.20) can be decomposed into spherical and deviatoric expressions (see Exercise 6.4):

$$\begin{aligned} \tilde{T}_{kk} &= 3k\tilde{\varepsilon}_{kk} \quad \text{or} \quad T_{kk} = 3k\varepsilon_{kk} \\ \hat{T}_{ij} &= 2\mu\hat{\varepsilon}_{ij} \end{aligned} \tag{6.5.30}$$

Likewise, a similar decomposition can be made for the linear viscous constitutive law (6.4.4):

$$T_{kk} = -3p + (3\lambda + 2\mu)D_{kk}, \quad \hat{T}_{ij} = 2\mu\hat{D}_{ij} \tag{6.5.31}$$

where for (6.5.31), μ would now correspond to the fluid viscosity. Noting that $D_{ij} \approx \dot{\varepsilon}_{ij}q$, the second relation in (6.5.31) would imply that the constitutive relation for the viscous element would $\hat{T}_{ij} = 2\mu\dot{\hat{\varepsilon}}_{ij}$. All of this then would allow elastic and viscous components to be put together in a tensor format in terms of deviatoric components. Thus, say, for a Kelvin model, the one-dimensional relation (6.5.14) could be written in tensor form as

$$\hat{T}_{ij} = 2\mu\hat{\varepsilon}_{ij} + 2\eta\dot{\hat{\varepsilon}}_{ij} \tag{6.5.32}$$

where μ is the elastic shear modulus and η is the viscosity. Likewise, the general form (6.5.27) could be expressed as

$$\boldsymbol{P}\hat{T}_{ij} = 2\boldsymbol{Q}\hat{\varepsilon}_{ij} \tag{6.5.33}$$

The spherical or volumetric deformation would then be governed by just the elastic constitutive relation.

6.5.1.5 Linear integral constitutive relations

Although the previous viscoelastic analog modeling provided an efficient way to combine elastic and viscous material response, the resulting general relation (6.5.27) becomes cumbersome if many terms are needed to accurately model real materials. To avoid this, a linear integral constitutive form can be constructed that will provide a more convenient means to model viscoelastic behavior. Using the *Boltzmann principle of superposition* for our linear theory, we can state that the total strain is equal to the sum of a sequence of strains caused by individual stresses. This concept is illustrated in Fig. 6.42 where the stresses are changing in incremental steps, and the resulting strains come from using the creep function definition over each step. The superposition concept can thus be written as

$$\begin{aligned} \varepsilon(t) &= J(t)T_o + J(t-t_1)\Delta T_1 + J(t-t_2)\Delta T_2 + \cdots \\ &= J(t)T_o + \sum_k J(t-t_k)\Delta T_k \end{aligned} \tag{6.5.34}$$

where J is the material's creep function. In the limit as the step size becomes infinitesimal and the number of steps goes to infinity, the summation in (6.5.34) becomes an integral

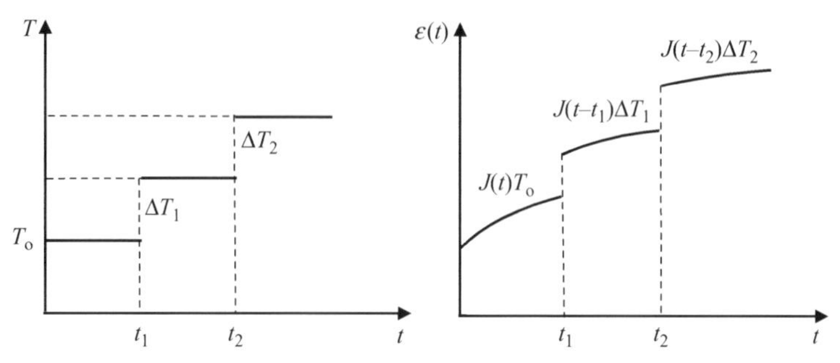

FIGURE 6.42

Stepwise stress loading and resulting strain response.

$$\begin{aligned}
\varepsilon(t) &= J(t)T_o + \int_0^t J(t-t')dT(t') \\
&= J(t)T_o + \int_0^t J(t-t')\frac{dT(t')}{dt'}dt' \\
&= \int_{-\infty}^t J(t-t')\frac{dT(t')}{dt'}dt'
\end{aligned} \tag{6.5.35}$$

where the last line comes from the fact that $J(t) = 0$ for $t < 0$. Note that the integrand $J(t-t'$ could also be thought of as an *influence function*.

Relations (6.5.35) indicate that the strain at time t depends on the entire *stress history* $T(t')$, where $0 \le t' \le$ or $-\infty \le t' \le$. This fact is quite different than what occurs in elastic models where the strain at time t depends only on the stress at that particular time. The integrals in (6.5.35) are commonly referred to as *hereditary integrals*, and the particular form in $(6.5.35)_1$ is often called a *Stieltjes integral*. This history-dependent behavior brings up the interesting issue of how the influence or creep function $J(t-t')$ will vary over the integration range. We would normally expect that the recent past history would have a more significant effect on the behavior at time t than the effects that occurred in the distant history. Based on this, we would claim that the material's response has *fading memory*, and thus $J(t-t')$ would be a decreasing function of its argument. More significant use of the fading memory postulate is made in the nonlinear theories of material behavior.

Several alternative forms of relation (6.5.35) can be developed. Integration by parts will yield

$$\begin{aligned}
\varepsilon(t) &= J(0)T(t) + \int_0^t \frac{dJ(t-t')}{d(t-t')}T(t')dt' \\
&= J(0)T(t) + \int_0^t \frac{dJ(t')}{d(t')}T(t-t')dt'
\end{aligned} \tag{6.5.36}$$

It should be obvious that in the previous discussion leading to relations (6.5.35) and (6.5.36), the variables of strain and stress could be switched around by using the relaxation function $G(t)$ instead of the creep function $J(t)$. Thus, we can express the stress $T(t)$ in terms of the *strain history* in any of the following hereditary integral forms:

$$\begin{aligned} T(t) &= G(t)\varepsilon_o + \int_0^t G(t-t')\,d\varepsilon(t') \\ &= G(t)\varepsilon_o + \int_0^t G(t-t')\frac{d\varepsilon(t')}{dt'}dt' \\ &= \int_{-\infty}^t G(t-t')\frac{d\varepsilon(t')}{dt'}dt' \\ &= G(0)\varepsilon(t) + \int_0^t \frac{dG(t-t')}{d(t-t')}\varepsilon(t')\,dt' \\ &= G(0)\varepsilon(t) + \int_0^t \frac{dG(t')}{d(t')}\varepsilon(t-t')\,dt' \end{aligned} \tag{6.5.37}$$

EXAMPLE 6.5.6 CORRESPONDENCE BETWEEN INTEGRAL AND ANALOG CONSTITUTIVE FORMS FOR MAXWELL MODEL CASE

Consider the hereditary integral form from $(6.5.37)_3$, with the following form for the relaxation function $G(t)=Ee^{-t/\tau}$, where E is the elastic stiffness and $\tau = \eta/E$ is the relaxation time. This form corresponds to a single relaxation time function and is actually the Maxwell model prediction for the relaxation function (see Eq. (6.5.12)). Show by differentiation of the integral form, the resulting expression will be the usual Maxwell constitutive law (6.5.11).

Solution: Using Leibnitz rule, we formally take the time derivative of the integral form with the given relaxation function

$$\begin{aligned} \dot{T} &= \frac{d}{dt}\int_{-\infty}^t Ee^{-(t-t')/\tau}\frac{d\varepsilon(t')}{dt'}dt' \\ &= E\dot{\varepsilon} + \int_{-\infty}^t E\frac{d}{dt}e^{-(t-t')/\tau}\frac{d\varepsilon(t')}{dt'}dt' \\ &= E\dot{\varepsilon} - \frac{1}{\tau}\int_{-\infty}^t Ee^{-(t-t')/\tau}\frac{d\varepsilon(t')}{dt'}dt' \\ &= E\dot{\varepsilon} - \frac{1}{\tau}T \\ &\Rightarrow \dot{T} + \frac{E}{\eta}T = E\dot{\varepsilon} \end{aligned}$$

Thus, we arrive at relation (6.5.11), the usual Maxwell differential constitutive law.

Next we move on to incorporating the integral constitutive form for general three-dimensional behaviors. This can be done by changing the scalar variables to tensors of appropriate orders. If $\varepsilon_{ij}(t')$ represents the strain tensor history $(-\infty \le t' \le t)$, we can then express the stress response $T_{ij}(t)$ in any of the following forms:

$$\begin{aligned} T_{ij}(t) &= \int_{-\infty}^{t} G_{ijkl}(t-t')\frac{d\varepsilon_{kl}(t')}{dt'}dt' \\ &= G_{ijkl}(0)\varepsilon_{kl}(t) + \int_{0}^{t}\frac{dG_{ijkl}(t-t')}{d(t-t')}\varepsilon_{kl}(t')\,dt' \\ &= G_{ijkl}(0)\varepsilon_{kl}(t) + \int_{0}^{t}\frac{dG_{ijkl}(t')}{d(t')}\varepsilon_{kl}(t-t')\,dt' \end{aligned} \tag{6.5.38}$$

where G_{ijkl} is the fourth-order tensor relaxation function. As before, the roles of stress and strain can be interchanged leading to similar expressions

$$\begin{aligned} \varepsilon_{ij}(t) &= \int_{-\infty}^{t} J_{ijkl}(t-t')\frac{dT_{kl}(t')}{dt'}dt' \\ &= J_{ijkl}(0)T_{kl}(t) + \int_{0}^{t}\frac{dJ_{ijkl}(t-t')}{d(t-t')}T_{kl}(t')\,dt' \\ &= J_{ijkl}(0)T_{kl}(t) + \int_{0}^{t}\frac{dJ_{ijkl}(t')}{d(t')}T_{kl}(t-t')\,dt' \end{aligned} \tag{6.5.39}$$

where J_{ijkl} is the fourth-order tensor creep function. Based on similar arguments used in the elasticity theory, the tensors $\boldsymbol{G}$ and $\boldsymbol{J}$ must satisfy

$$\begin{aligned} G_{ijkl} &= G_{jikl} = G_{ijlk} \\ J_{ijkl} &= J_{jikl} = J_{ijlk} \end{aligned} \tag{6.5.40}$$

If the material is isotropic, then following our previous discussion for elastic materials, the fourth-order tensor $\boldsymbol{G}$ can be expressed as

$$G_{ijkl}(t) = \lambda(t)\delta_{ij}\delta_{kl} + \mu(t)(\delta_{ik}\delta_{jl} + \delta_{il}\delta_{jk}) \tag{6.5.41}$$

where $\lambda(t)$ and $\mu(t)$ are additional material relaxation functions. Using this in constitutive relation $(6.5.38)_1$ yields

$$T_{ij}(t) = \delta_{ij}\int_{-\infty}^{t}\lambda(t-t')\dot{\varepsilon}_{kk}(t')\,dt' + \int_{-\infty}^{t}2\mu(t-t')\dot{\varepsilon}_{ij}(t')\,dt' \tag{6.5.42}$$

As before we can separate (6.5.42) into spherical and deviatoric parts and write

$$\begin{aligned} \hat{T}_{ij}(t) &= \int_{-\infty}^{t}2\mu(t-t')\dot{\hat{\varepsilon}}_{ij}(t')\,dt' \\ T_{kk}(t) &= \int_{-\infty}^{t}3k(t-t')\dot{\varepsilon}_{kk}(t')\,dt' \end{aligned} \tag{6.5.43}$$

where $k(t) = [3\lambda(t) + 2\mu(t)]/3$. When viewed in this scheme, $\mu(t)$ would be referred to as the *shear relaxation function*, while $k(t)$ would be called the *bulk relaxation function*. Note that as mentioned previously, we could drop the viscoelastic spherical or volumetric response relation $(6.5.43)_2$ and assume this behavior is governed by the elastic relation $T_{kk} = 3k\varepsilon_{kk}$.

6.5.2 GENERAL FORMULATION

We now continue with the linear viscoelastic formulation by coupling the previous constitutive law(s) with other governing equations. Some further manipulations are necessary and we introduce the *correspondence principle* which is a very convenient strategy to find solutions to many problems. As with the previous elasticity case, we will limit our study to only quasi-static problems and will thus drop the acceleration term in the equations of motion. It should be noted that even with this assumption, viscoelastic deformation will produce time-dependent stresses, strains, and displacements. Since we are again setting up a mathematical boundary value problem, we must also establish appropriate boundary conditions applicable for the theory.

The general formulation of linear viscoelasticity is quite similar to that of linear elasticity previously presented in Section 6.2. The governing equations include

$$\text{Strain-Displacement Relations} \quad \varepsilon_{ij} = \frac{1}{2}(u_{i,j} + u_{j,i}) \tag{6.5.44}$$

$$\text{Compatibility Relations} \quad \varepsilon_{ij,kl} + \varepsilon_{kl,ij} - \varepsilon_{ik,jl} - \varepsilon_{jl,ik} = 0 \tag{6.5.45}$$

$$\text{Equilibrium Equations} \quad T_{ij,j} + \rho b_i = 0 \tag{6.5.46}$$

$$\text{Constitutive Law} \quad T_{ij}(t) = \delta_{ij}\int_{-\infty}^{t} \lambda(t-t')\,\dot{\varepsilon}_{kk}(t')\,dt' + \int_{-\infty}^{t} 2\mu(t-t')\,\dot{\varepsilon}_{ij}(t')\,dt' \tag{6.5.47}$$

With time-dependent deformations, the conservation of mass relation could also be included in this set to determine mass density changes.

As previously mentioned, the compatibility relations ensure that the displacements are continuous and single-valued and are necessary only when the strains are arbitrarily specified. If the displacements are included in the original problem formulation, the solution will normally generate single-valued displacements and strain compatibility will automatically be satisfied. Therefore, similar to the elasticity case, the compatibility relations are normally set aside, to be used only with the stress formulation. This system involves 15 unknowns including three displacements u_i, six strains ε_{ij}, and six stresses T_{ij}. For isotropic materials, the equation system also includes two viscoelastic material functions $\lambda(t)$ and $\mu(t)$ and a body force density b_i, and these are to be given *a priori* with the problem formulation. As with the elasticity case, this general system of equations is of such complexity that solutions via analytical methods are very difficult and further simplification is required to solve even simple problems.

The common types of boundary conditions for linear viscoelasticity are quite similar to those of elasticity and normally include specification of how the body is being *supported* or *loaded*. This concept is mathematically formulated by specifying either the *displacements* or *tractions* at boundary points, and for the viscoelastic case we could have time-dependent values for each of these quantities. Referring back to the elastic case in Fig. 6.8, this graphic illustrates the three common cases including tractions, displacements, and a mixed case where tractions are specified on boundary S_t and displacements are given on the remaining portion S_u such that the total boundary is given by $S = S_t + S_u$.

6.5.2.1 Correspondence principle

It has been repeatedly mentioned in this section that there exists considerable similarity between the elastic and viscoelastic field equations. The correspondence principle establishes a direct connection between viscoelastic and elastic problem solutions. Since there is a large body of elasticity solutions that exist in the open literature, this scheme provides a very useful method to solve viscoelastic problems.

The technique uses *Laplace transforms*, and so we will briefly review this mathematical method. Laplace transformation is one of several linear integral transform methods that are useful for the solution to both ordinary and partial differential equations. The common useful property is that the transform changes particular governing differential and integral forms into algebraic relations, thus making the governing equation easier to solve. For a general function $f(t),\ (t>0)$, the Laplace transform is denoted by $\bar{f}(s)$ and defined by

$$\bar{f}(s) = \mathfrak{L}\{f(t)\} = \int_0^\infty f(t)e^{st}\,dt \tag{6.5.48}$$

where s is the transformed time variable. Thus functions of t are transformed into functions of s. Particular special properties of this transform useful in this context are

$$\begin{aligned} &\mathfrak{L}\{\dot{f}(t)\} = s\bar{f}(s) - f(0) \\ &\mathfrak{L}\{\ddot{f}(t)\} = s^2\bar{f}(s) - sf(0) - \dot{f}(0) \\ &\mathfrak{L}\left\{\int_0^t f(t-\tau)g(\tau)\,d\tau\right\} = \bar{f}(s)\bar{g}(s) \end{aligned} \tag{6.5.49}$$

The inverse Laplace transform may be written in operator form as $f(t) = \mathfrak{L}^{-1}\{\bar{f}(s)\}$. Taking the transform of various functions and doing the inverse transformation to get back to forms involving the real variable t is commonly done using Laplace transform tables found in many texts and on-line. Sometimes the inverse transform is done using contour integration in the complex plane.

So if we relist our viscoelastic field equations assuming the case of deformations starting at $t = 0$ and take the Laplace transform of the entire system, we get

$$\mathfrak{L}\left\{\begin{gathered} \varepsilon_{ij} = \frac{1}{2}(u_{i,j} + u_{j,i}) \\ T_{ij,j} + \rho b_i = 0 \\ T_{ij} = \delta_{ij}\int_0^t \lambda(t-t')\dot{\varepsilon}_{kk}(t')\,dt' + \int_0^t 2\mu(t-t')\dot{\varepsilon}_{ij}(t')\,dt' \end{gathered}\right\} = \left\{\begin{gathered} \bar{\varepsilon}_{ij} = \frac{1}{2}(\bar{u}_{i,j} + \bar{u}_{j,i}) \\ \bar{T}_{ij,j} + \rho\bar{b}_i = 0 \\ \bar{T}_{ij} = s\bar{\lambda}(s)\bar{\varepsilon}_{kk}\delta_{ij} + 2s\bar{\mu}(s)\bar{\varepsilon}_{ij} \end{gathered}\right\} \tag{6.5.50}$$

where we have assumed zero initial conditions on all transformed variables. The resulting transformed system (6.5.50) is identical to linear elasticity field equations in terms of the transformed variables $\{\bar{T}_{ij}, \bar{\varepsilon}_{ij}, \bar{u}_i, \bar{b}_i\}$ providing that $s\bar{\lambda}(s)$ and $s\bar{\mu}(s)$ correspond to the elastic moduli λ and μ. It is then noted that for problems of this type, the Laplace transform viscoelastic solution can be obtained directly from the corresponding elastic solution by replacing λ and μ with $s\bar{\lambda}(s)$ and $s\bar{\mu}(s)$, respectively. All time-dependent loadings must also be replaced by their corresponding Laplace transforms. The solution can then be completed by inverting the transformed solution back to the time t domain. This association between viscoelastic and elastic problems is called the *correspondence principle*. Note that many elasticity solutions contain parts (either the stresses or the displacements) that do not contain any elastic constants. For such a case, the viscoelastic solution then coincides with that portion of the elastic solution. We will see this case in subsequent examples.

Laplace transforms can also be used to demonstrate a correspondence between analog and hereditary integral models. Neglecting history effects before $t = 0$, the one-dimensional hereditary integral relation (6.5.37) can be written as

$$T(t) = G(0)\varepsilon(t) + \int_0^t \frac{dG(t-t')}{d(t-t')}\varepsilon(t')\,dt' \tag{6.5.51}$$

Taking the Laplace transform of this relation yields

$$\bar{T}(s) = G(0)\bar{\varepsilon}(s) + \bar{\dot{G}}(s)\bar{\varepsilon}(s) = s\bar{G}(s)\bar{\varepsilon}(s) \tag{6.5.52}$$

Now the most general analog model was given by relation (6.5.26), and taking the Laplace transform of this equation for the case of zero initial conditions gives

$$(p_o + p_1 s + p_2 s^2 + \cdots)\bar{T}(s) = (q_o + q_1 s + q_2 s^2 + \cdots)\bar{\varepsilon}(s) \tag{6.5.53}$$

and solving for the stress gives

$$\bar{T}(s) = \frac{(q_o + q_1 s + q_2 s^2 + \cdots)}{(p_o + p_1 s + p_2 s^2 + \cdots)}\bar{\varepsilon}(s) \tag{6.5.54}$$

Comparing Eqs. (6.5.52) and (6.5.54), we find that the continuous relaxation function from the hereditary integral form is related to a combination of discrete analog models through the expression

$$\bar{G}(s) = \frac{(q_o + q_1 s + q_2 s^2 + \cdots)}{s(p_o + p_1 s + p_2 s^2 + \cdots)} \tag{6.5.55}$$

6.5.3 PROBLEM SOLUTIONS

We now wish to use the previous constitutive relations and formulation equations to explore a few analytical solutions to some viscoelastic problems. These specific examples will further demonstrate important features of the linear viscoelastic model.

EXAMPLE 6.5.7 HARMONIC RESPONSE OF LINEAR VISCOELASTIC MODELS—COMPLEX MODULUS AND COMPLIANCE

Consider the response of our linear viscoelastic constitutive models under harmonic loading or deformation. If we were to subject a viscoelastic sample to a harmonic input strain, we would expect a harmonic output stress response of the same frequency. Likewise, we would also expect the same cause and effect if the stress was the input and the strain was the output. In this example, we wish to explore this type of behavior for one-dimensional Maxwell, Kelvin, and integral models.

Solution: We first start with a harmonic input strain using the more convenient complex exponential form $\varepsilon(t) = \varepsilon_o e^{i\omega t}$, where ω is the frequency and ε_o is the amplitude. The expected harmonic output stress would then be $T(t) = T_o e^{i\omega t}$. Substituting these forms into the Maxwell and Kelvin models and cancelling the common $e^{i\omega t}$ term yields

$$\text{Maxwell model: } \dot{T} + \frac{1}{\tau}T = E\dot{\varepsilon} \Rightarrow T_o i\omega + \frac{1}{\tau}T_o = E\varepsilon_o i\omega \Rightarrow T_o = \frac{Ei\omega}{i\omega + (1/\tau)}\varepsilon_o \Rightarrow$$

$$T_o = (G_1 + iG_2)\varepsilon_o, \quad G_1 = \frac{E\tau^2\omega^2}{1+\tau^2\omega^2}, \quad G_2 = \frac{E\tau\omega}{1+\tau^2\omega^2} \tag{6.5.56}$$

$$\varepsilon_o = (J_1 + iJ_2)T_o, \quad J_1 = -\frac{1}{E}, \quad J_2 = \frac{1}{\eta\omega}$$

$$\text{Kelvin model: } T = E\varepsilon + \eta\dot{\varepsilon} \Rightarrow T_o = E\varepsilon_o + \eta i\omega\varepsilon_o \Rightarrow T_o = (E + i\eta\omega)\varepsilon_o \Rightarrow$$

$$T_o = (G_1 + iG_2)\varepsilon_o, \quad G_1 = E, \quad G_2 = \eta\omega$$

$$\varepsilon_o = (J_1 + iJ_2)T_o, \quad J_1 = \frac{E}{E^2 + \eta^2\omega^2}, \quad J_2 = \frac{-\eta\omega}{E^2 + \eta^2\omega^2} \tag{6.5.57}$$

G_1 and G_2 are called the *real and imaginary parts of the complex modulus* and J_1 and J_2 are called the *real and imaginary parts of the complex compliance*. It can be shown that G_1 is related to the energy stored in a cycle and thus is often referred to as the *storage modulus*, while G_2 can be related to the energy dissipated over a cycle and is appropriately called the *loss modulus*.

For the general differential constitutive form

$$\sum_{k=0}^{m} p_k \frac{d^k T}{dt^k} = \sum_{k=0}^{m} q_k \frac{d^k \varepsilon}{dt^k} \Rightarrow T_o \sum_{k=0}^{m} p_k (i\omega)^k = \varepsilon_o \sum_{k=0}^{m} q_k (i\omega)^k \Rightarrow T_o = \frac{\sum_{k=0}^{m} q_k (i\omega)^k}{\sum_{k=0}^{m} p_k (i\omega)^k}\varepsilon_o \Rightarrow$$

$$G_1 = Re\left\{\frac{\sum_{k=0}^{m} q_k (i\omega)^k}{\sum_{k=0}^{m} p_k (i\omega)^k}\right\} \quad \text{and} \quad G_2 = Im\left\{\frac{\sum_{k=0}^{m} q_k (i\omega)^k}{\sum_{k=0}^{m} p_k (i\omega)^k}\right\} \tag{6.5.58}$$

Finally exploring the one-dimensional integral constitutive form

$$T(t) = \int_0^t G(t-t')\frac{d\varepsilon(t')}{dt'}dt' \Rightarrow T_o = i\omega\,\varepsilon_o \int_0^t G(t-t')e^{-i\omega(t-t')}dt' \Rightarrow$$

$$G_1 = Re\left\{i\omega\int_0^t G(t-t')e^{-i\omega(t-t')}dt'\right\} \text{ and } G_2 = Im\left\{i\omega\int_0^t G(t-t')e^{-i\omega(t-t')}dt'\right\} \tag{6.5.59}$$

EXAMPLE 6.5.8 ONE-DIMENSIONAL RELAXATION FUNCTIONS FROM THREE-DIMENSIONAL VISCOELASTIC CONSTITUTIVE RELATIONS

Determine the tensile relaxation function and Poisson's ratio function from the three-dimensional integral constitutive relations (6.5.47). Also explore the constitutive case where volumetric and deviatoric responses are separated as in (6.5.43) and the volumetric behavior is taken to be elastic.

Solution: For a uniaxial relaxation deformation, we consider the stress and strain matrices as

$$T_{ij}(t)=\begin{bmatrix} T_o(t) & 0 & 0 \\ 0 & 0 & 0 \\ 0 & 0 & 0 \end{bmatrix},\quad \varepsilon_{ij}(t)=\begin{bmatrix} \varepsilon_o(t) & 0 & 0 \\ 0 & \varepsilon_{22}(t) & 0 \\ 0 & 0 & \varepsilon_{33}(t) \end{bmatrix}H(t) \quad (6.5.60)$$

with $\varepsilon_{22}(t)=\varepsilon_{33}(t)=\varepsilon_t(t)$. The Laplace transformed relation (6.5.50) gives

$$\begin{aligned} \bar{T}_{ij} &= s\bar{\lambda}(s)\bar{\varepsilon}_{kk}\delta_{ij}+2s\bar{\mu}(s)\bar{\varepsilon}_{ij} \Rightarrow \\ \bar{T}_o &= s\bar{\lambda}(s)(\bar{\varepsilon}_o+2\bar{\varepsilon}_t)+2s\bar{\mu}(s)\bar{\varepsilon}_o \\ \bar{T}_{22} &= \bar{T}_{33}=0=s\bar{\lambda}(s)(\bar{\varepsilon}_o+2\bar{\varepsilon}_t)+2s\bar{\mu}(s)\bar{\varepsilon}_t \end{aligned} \quad (6.5.61)$$

Solving $(6.5.61)_3$ for the transverse strain gives $\bar{\varepsilon}_t=\dfrac{-\bar{\lambda}(s)}{2[\bar{\lambda}(s)+\bar{\mu}(s)]}\bar{\varepsilon}_o$. Then using this in $(6.5.61)_2$ gives

$$\bar{T}_o=\frac{s\bar{\mu}(s)[3\bar{\lambda}(s)+2\bar{\mu}(s)]}{\bar{\lambda}(s)+\bar{\mu}(s)}\bar{\varepsilon}_o=s\bar{E}(s)\bar{\varepsilon}_o \quad (6.5.62)$$

and thus taking the inverse Laplace transform

$$T_o(t)=\int_0^t E(t-t')\frac{d\varepsilon_o(t')}{dt'}dt',\quad \text{where } E(t)=\mathcal{L}^{-1}\left\{\frac{\bar{\mu}(s)[3\bar{\lambda}(s)+2\bar{\mu}(s)]}{\bar{\lambda}(s)+\bar{\mu}(s)}\right\} \quad (6.5.63)$$

$E(t)$ is called the *tensile relaxation modulus*.

Reusing the previous relations between transverse and axial strains gives the ratio:

$$\frac{\bar{\varepsilon}_t}{\bar{\varepsilon}_o}=\frac{-\bar{\lambda}(s)}{2[\bar{\lambda}(s)+\bar{\mu}(s)]}$$

and the inverse Laplace transform gives the *Poisson's ratio modulus* ν:

$$\nu(t)=\left|\frac{\bar{\varepsilon}_t}{\bar{\varepsilon}_o}\right|=\mathcal{L}^{-1}\left\{\frac{\bar{\lambda}(s)}{2[\bar{\lambda}(s)+\bar{\mu}(s)]}\right\} \quad (6.5.64)$$

For the case where we use the separation of volumetric and deviatoric behaviors based on relation (6.5.43) with the volumetric response governed by the elastic relation, we have

$$T_{kk} = 3k\varepsilon_{kk},\ \hat{T}_{ij}(t) = \int_0^t 2\mu(t-t')\dot{\hat{\varepsilon}}_{ij}(t')dt' \tag{6.5.65}$$

Taking the Laplace transform of these relations gives

$$\begin{aligned}
&\overline{T}_{kk} = 3k\overline{\varepsilon}_{kk},\quad \overline{\hat{T}}_{ij} = 2s\overline{\mu}(s)\overline{\hat{\varepsilon}}_{ij} \Rightarrow \\
&\overline{T}_{ij} - \frac{1}{3}\overline{T}_{kk}\delta_{ij} = 2s\overline{\mu}(s)\left(\overline{\varepsilon}_{ij} - \frac{1}{3}\overline{\varepsilon}_{kk}\delta_{ij}\right) \Rightarrow \\
&\overline{T}_{ij} - \frac{1}{3}\overline{T}_{kk}\delta_{ij} = 2s\overline{\mu}(s)\overline{\varepsilon}_{ij} - \frac{2}{9k}s\overline{\mu}(s)\overline{T}_{kk}\delta_{ij}
\end{aligned} \tag{6.5.66}$$

For our one-dimensional loading case

$$\begin{aligned}
&\overline{T}_{11} - \frac{1}{3}\overline{T}_{11} = 2s\overline{\mu}(s)\overline{\varepsilon}_o - \frac{2}{9k}s\overline{\mu}(s)\overline{T}_{11} \Rightarrow \\
&\overline{T}_{11} = \frac{9ks\overline{\mu}(s)}{3k + s\overline{\mu}(s)}\overline{\varepsilon}_o = s\overline{E}(s)\overline{\varepsilon}_o
\end{aligned} \tag{6.5.67}$$

So for this case

$$T_o(t) = \int_0^t E(t-t')\frac{d\varepsilon_o(t')}{dt'}dt', \quad \text{where} \quad E(t) = \mathcal{L}^{-1}\left\{\frac{9k\overline{\mu}(s)}{3k + s\overline{\mu}(s)}\right\} \tag{6.5.68}$$

and the Poisson's ratio function becomes

$$\nu(t) = \mathcal{L}^{-1}\left|\frac{\overline{\varepsilon}_t}{\overline{\varepsilon}_o}\right| = \mathcal{L}^{-1}\left\{\frac{3k - 2\overline{\mu}(s)}{6k + 2\overline{\mu}(s)}\right\} \tag{6.5.69}$$

Some of these results will be useful in problem solutions to come.

EXAMPLE 6.5.9 VISCOELASTIC STRESSES AND DISPLACEMENTS IN A THICK-WALLED TUBE UNDER INTERNAL PRESSURE

Determine the viscoelastic stress and displacement solution for a thick-walled cylindrical tube under internal pressure $p(t) = pH(t)$ as shown in Fig. 6.43.

Solution: We wish to use the correspondence principle, so we first seek the elasticity solution to this problem. This solution is normally developed under plane strain conditions in polar coordinates. Sadd (2014) provides the necessary solution

$$\begin{aligned}
&T_{rr} = \frac{r_1^2 p}{r_2^2 - r_1^2}\left(1 - \frac{r_2^2}{r^2}\right),\quad T_{\theta\theta} = \frac{r_1^2 p}{r_2^2 - r_1^2}\left(1 + \frac{r_2^2}{r^2}\right) \\
&u_r = \frac{1}{2\mu}\frac{r_1^2 p}{r_2^2 - r_1^2}\left(\frac{r_2^2}{r} + \frac{\mu}{\lambda + \mu}r\right)
\end{aligned} \tag{6.5.70}$$

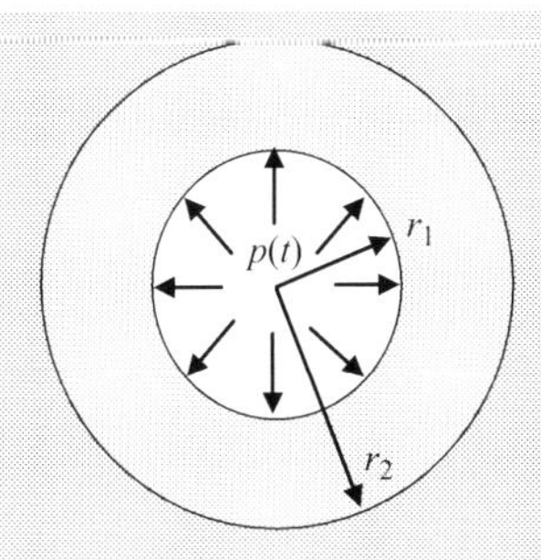

FIGURE 6.43

Pressurized thick-walled cylindrical tube.

Thus, according to the correspondence principle, the Laplace transformed viscoelastic solution will be

$$\begin{aligned} \bar{T}_{rr} &= \frac{r_1^2 \bar{p}}{r_2^2 - r_1^2}\left(1 - \frac{r_2^2}{r^2}\right), \quad \bar{T}_{\theta\theta} = \frac{r_1^2 \bar{p}}{r_2^2 - r_1^2}\left(1 + \frac{r_2^2}{r^2}\right) \\ \bar{u}_r &= \frac{1}{2s\bar{\mu}(s)} \frac{r_1^2 \bar{p}}{r_2^2 - r_1^2}\left(\frac{r_2^2}{r} + \frac{s\bar{\mu}(s)}{s\bar{\lambda}(s) + s\bar{\mu}(s)} r\right) \end{aligned} \tag{6.5.71}$$

Using the fact that $\bar{p} = p/s$, and taking the inverse Laplace transform of this set gives the viscoelastic solution

$$\begin{aligned} T_{rr} &= \frac{r_1^2 p}{r_2^2 - r_1^2}\left(1 - \frac{r_2^2}{r^2}\right), \quad T_{\theta\theta} = \frac{r_1^2 p}{r_2^2 - r_1^2}\left(1 + \frac{r_2^2}{r^2}\right) \\ u_r &= \frac{r_1^2 r_2^2 p}{2(r_2^2 - r_1^2) r} \mathfrak{L}^{-1}\left(\frac{1}{s^2 \bar{\mu}(s)}\right) + \frac{r_1^2 r p}{2(r_2^2 - r_1^2)} \mathfrak{L}^{-1}\left(\frac{1}{s^2[\bar{\lambda}(s) + \bar{\mu}(s)]}\right) \end{aligned} \tag{6.5.72}$$

To complete the solution, we need to decide on the forms of the relaxation functions $\lambda(t)$ and $\mu(t)$. For convenience, choose the simple forms

$$\lambda(t) = \lambda_o e^{-t/\tau}, \quad \mu(t) = \mu_o e^{-t/\tau} \tag{6.5.73}$$

With this choice $\bar{\lambda}(s) = \dfrac{\lambda_o}{s + (1/\tau)}$, $\bar{\mu}(s) = \dfrac{\mu_o}{s + (1/\tau)}$, and thus

$$\frac{1}{s^2 \bar{\mu}(s)} = \frac{s + (1/\tau)}{s^2 \mu_o}, \quad \frac{1}{s^2[\bar{\lambda}(s) + \bar{\mu}(s)]} = \frac{s + (1/\tau)}{(\lambda_o + \mu_o) s^2} \tag{6.5.74}$$

Taking the inverse Laplace transforms from standard tables gives

$$\begin{aligned} \mathfrak{L}^{-1}\left(\frac{1}{s^2 \bar{\mu}(s)}\right) &= \frac{1}{\mu_o} \mathfrak{L}^{-1}\left(\frac{1}{s} + \frac{1}{\tau s^2}\right) = \frac{1}{\mu_o}(H(t) + t) \\ \mathfrak{L}^{-1}\left(\frac{1}{s^2[\bar{\lambda}(s) + \bar{\mu}(s)]}\right) &= \frac{1}{(\lambda_o + \mu_o)} \mathfrak{L}^{-1}\left(\frac{1}{s} + \frac{1}{\tau s^2}\right) = \frac{1}{(\lambda_o + \mu_o)}(H(t) + t) \end{aligned} \tag{6.5.75}$$

Putting all these results together gives the final form for the viscoelastic solution

$$T_{rr} = \frac{r_1^2 p}{r_2^2 - r_1^2}\left(1 - \frac{r_2^2}{r^2}\right),\ T_{\theta\theta} = \frac{r_1^2 p}{r_2^2 - r_1^2}\left(1 + \frac{r_2^2}{r^2}\right)$$
$$u_r = \frac{r_1^2 p(1+t)}{2(r_2^2 - r_1^2)}\left(\frac{1}{\mu_o}\frac{r_2^2}{r} + \frac{r}{(\lambda_o + \mu_o)}\right) \tag{6.5.76}$$

Notice that for this problem the viscoelastic stresses are the same as those from elasticity since these fields did not contain any elastic constants. Furthermore, our simple relaxation models specified in (6.5.73) coincide with a Maxwell type model, and thus they predict fluid-like behavior in the radial displacement growing linearly with time t. This would of course produce unbounded displacements and eventually negate our small deformation assumption. For the case $t = 0$, the radial displacement will match with the elasticity solution with elastic moduli λ_o and μ_o. Different relaxation functions for $\lambda(t)$ and $\mu(t)$ would produce more reasonable temporal radial displacement predictions for solid-like behavior.

EXAMPLE 6.5.10 VISCOELASTIC STRESSES IN A HALF-SPACE UNDER CONCENTRATED NORMAL LOAD—BOUSSINESQ'S PROBLEM

Determine the viscoelastic stresses in a half-space under a concentrated load $P(t) = PH(t)$ acting normal to the free surface as shown in Fig. 6.44. This is a classic axisymmetric problem in elasticity theory and is called the *Boussinesq problem*.

Solution: We again wish to use the correspondence principle, so we first seek the elasticity solution to this problem. This axisymmetric solution is normally developed using cylindrical coordinates, and the complete stress field solution is given by Sadd (2014):

$$T_{rr} = \frac{P}{2\pi R^2}\left[-\frac{3r^2 z}{R^3} + \frac{(1-2\nu)R}{R+z}\right]$$
$$T_{\theta\theta} = \frac{(1-2\nu)P}{2\pi R^2}\left[\frac{z}{R} - \frac{R}{R+z}\right] \tag{6.5.77}$$
$$T_{zz} = -\frac{3Pz^3}{2\pi R^5},\ T_{rz} = -\frac{3Prz^2}{2\pi R^5},\ T_{r\theta} = T_{z\theta} = 0$$

where $r = \sqrt{x^2 + y^2}$ and $R = \sqrt{r^2 + z^2}$. Note that the stresses T_{zz} and T_{rz} do not contain any elastic moduli and so they would also be the solution to the corresponding viscoelastic problem. Stress components T_{rr} and $T_{\theta\theta}$ contain the common moduli term $(1-2\nu)$, and so we will only work on the T_{rr} component in detail.

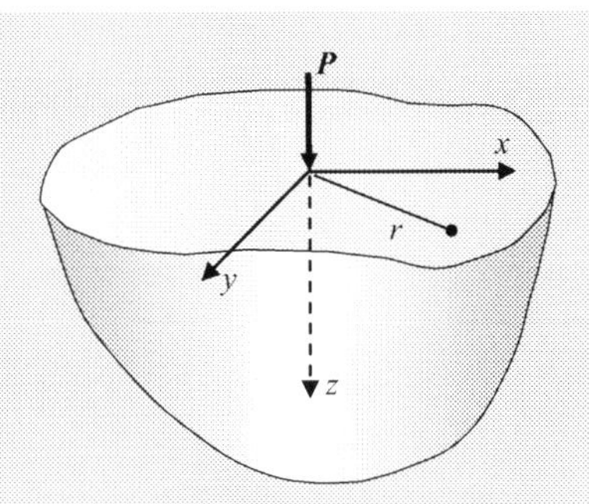

FIGURE 6.44

Boussinesq problem: normal force on the surface of a half-space.

Thus, according to the correspondence principle, the Laplace transformed viscoelastic solution will be

$$\bar{T}_{rr} = \frac{P}{2\pi R^2 s}\left[-\frac{3r^2 z}{R^3} + (1-2\bar{\nu}(s))\frac{R}{R+z}\right] \tag{6.5.78}$$

Taking the inverse Laplace transform gives the viscoelastic solution

$$T_{rr} = \frac{P}{2\pi R^2}\left[-\frac{3r^2 z}{R^3} + \mathfrak{L}^{-1}\left(\frac{1-2\bar{\nu}(s)}{s}\right)\frac{R}{R+z}\right] \tag{6.5.79}$$

To complete the solution, we need to decide on the forms of the material Poisson ratio function $\nu(t)$. For this example, we will choose the constitutive scheme based on the separation of volumetric and deviatoric behaviors with the volumetric response governed by the elastic relation. Using the results from Example 6.5.8, the Poisson's ratio was given by

$$\bar{\nu}(s) = \frac{3k - 2\bar{\mu}(s)}{6k + 2\bar{\mu}(s)} \Rightarrow \frac{1-2\bar{\nu}(s)}{s} = \frac{3\bar{\mu}(s)}{s[3k + \bar{\mu}(s)]} \tag{6.5.80}$$

Next choosing a Kelvin-type shear relaxation function

$$\bar{\mu}(s) = \mu_o + \eta s \tag{6.5.81}$$

$$\begin{aligned}
\mathfrak{L}^{-1}\left(\frac{1-2\bar{\nu}(s)}{s}\right) &= \mathfrak{L}^{-1}\left(\frac{3\bar{\mu}(s)}{s[3k + \bar{\mu}(s)]}\right) = \mathfrak{L}^{-1}\left(\frac{3(\mu_o + \eta s)}{s[3k + \mu_o + \eta s\ ]}\right) \\
&= \mathfrak{L}^{-1}\left(\frac{3\mu_o}{s[3k + \mu_o + \eta s\ ]}\right) + \mathfrak{L}^{-1}\left(\frac{3\eta}{[3k + \mu_o + \eta s\ ]}\right) \\
&= \frac{3\mu_o}{3k + \mu_o}\left(1 - e^{-\frac{3k+\mu_o}{\eta}t}\right) + 3e^{-\frac{3k+\mu_o}{\eta}t} \\
&= \frac{3\mu_o}{3k + \mu_o} + \frac{9k}{3k + \mu_o}e^{-\frac{3k+\mu_o}{\eta}t}
\end{aligned}$$

Putting these inverse transform results back into (6.5.79) gives the viscoelastic solution for the T_{rr} stress component

$$T_{rr}(r,z,t)=\frac{P}{2\pi R^2}\left[-\frac{3r^2z}{R^3}+\left(\frac{3\mu_o}{3k+\mu_o}+\frac{9k}{3k+\mu_o}e^{-\frac{3k+\mu_o}{\eta}t}\right)\frac{R}{R+z}\right] \tag{6.5.82}$$

The other component $T_{\theta\theta}$ would follow by similar analysis. It is interesting to note that as $t\rightarrow\infty$,

$$T_{rr}(r,z,t)=\frac{P}{2\pi R^2}\left[-\frac{3r^2z}{R^3}+\frac{3\mu_o}{3k+\mu_o}\frac{R}{R+z}\right] \tag{6.5.83}$$

which coincides with the original elastic solution since $1-2\nu=\dfrac{3\mu}{3k+\mu}$.

6.6 CLASSICAL PLASTIC MATERIALS

We now extend our study to explore another type of inelastic behavior of solids called *plasticity*. This behavior has some fundamental differences from our previous discussion on viscoelasticity. As we have seen, many solids have linear elastic behavior followed by a change in the stress–strain response to a more flow-type behavior that will result in permanent material deformation when the loadings are removed. Plasticity theory can be a very complex study since it can include large deformations, rate-dependent effects, and various yield and strain hardening principles. Here we will only present a brief look at the classical small deformation, rate-independent models. More detailed presentations on plasticity are given in Hill (1950), Chakrabarty (1987), Lubliner (1990), Wu (2005), and Bower (2010). Applications of plasticity theory provide many important behavior models used in metal forming and machining, crash resistant structures, and inelastic structural analysis and design.

6.6.1 YIELD CRITERIA AND CONSTITUTIVE LAW

Consider again in more detail typical uniaxial tensile behavior of a ductile material as shown in Fig. 6.45A. We will describe the behavior in general approximate terms realizing that particular materials may deviate somewhat from our simplified descriptions. As loading begins, the material initially behaves linearly elastic under small deformations. In this region, the loading and unloading paths are the same and remain on line OA. As mentioned in Section 6.2.1, as loading continues a point will be reached where the material response will fundamentally change. At this location, linear elastic behavior will end, and a new and different constitutive response will begin. Again as per our previous discussion, several different descriptors could be used to describe these changes including the *proportional limit*, *elastic limit*, and the *yield point*. It is common that the distinction between these different events and locations on the stress–strain curve are normally dropped, and we simply mark a single location labeled A, and refer to it as the *yield point*. The primary descriptor of this point is normally the stress value called the *yield stress*, designated by Y. Loading beyond point A invokes a new inelastic constitutive behavior with flow characteristics in the plastic region. Note that for many materials the stress will still exhibit an increasing function with strain during plastic flow, and this is called *work* or *strain hardening*. If the loading continues to point B, and then unloading is done, the downward unloading path will be approximately parallel to line OA. Thus, unloading is normally done elastically. If unloading is taken to zero stress, there will be permanent irreversible deformation, ε_p. Thus, at any generic point B, we can separate the total strain through the additive relation $\varepsilon = \varepsilon_e + \varepsilon_p$. If the unloading continues into the compression zone, a new compressive yield point will be reached. It is generally found that this compressive yield point (in absolute value) is smaller than the initial value in tension, and this phenomenon is called the *Bauschinger effect*. Finally, if the material is reloaded, the new loading path will not exactly coincide with path CB. This lack of coincidence of loading and unloading paths implies that the stress is a *history-dependent function of the strain* for plastic deformations.

Fig. 6.45B illustrates typical stress–strain behavior for different loading or strain rates. The elastic modulus and yield point normally increase with loading/strain rate. Thus, the yield behavior is commonly *rate dependent*, and this is normally referred to as *viscoplastic* behavior. We will not explore this type of response as our discussion will be limited to classical rate-independent plasticity theories. Furthermore, we assume isotropic behavior, incompressible plastic deformation, and will generally drop any Bauschinger effects.

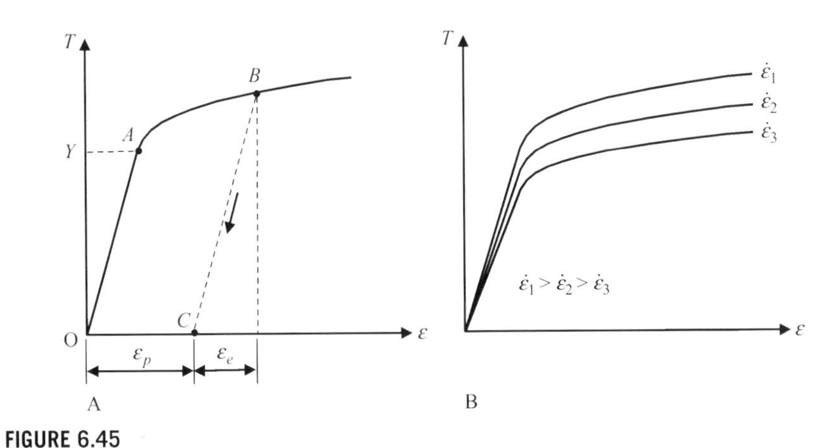

FIGURE 6.45

Typical uniaxial stress–strain behavior for ductile materials: (A) inelastic behavior; (B) rate-dependent behavior.

Summarizing our previous comments, experiments indicate that for many materials such as ductile metals, plastics, polymers, soils, rock, etc., the stress–strain constitutive behavior contains three features:

1. *initial elastic behavior*, followed by,
2. a combined stress or strain condition producing a change in the stress–strain behavior, that is, a *yield condition*,
3. a *plastic or flow response* governed by a different constitutive law.

We generally call this material behavior plasticity, and the essential characteristics are the yield condition and the plastic constitutive relation which is sometimes referred to as the *flow law*.

Before going into three-dimensional constitutive details, we first consider some simple one-dimensional models of idealized uniaxial plastic behaviors that have been developed in the literature. Fig. 6.46 illustrates four such models including *rigid perfectly plastic*, *elastic perfectly plastic*, *rigid linear strain hardening*, and *elastic linear strain hardening*. These models are somewhat similar to our previous analog spring–dashpot

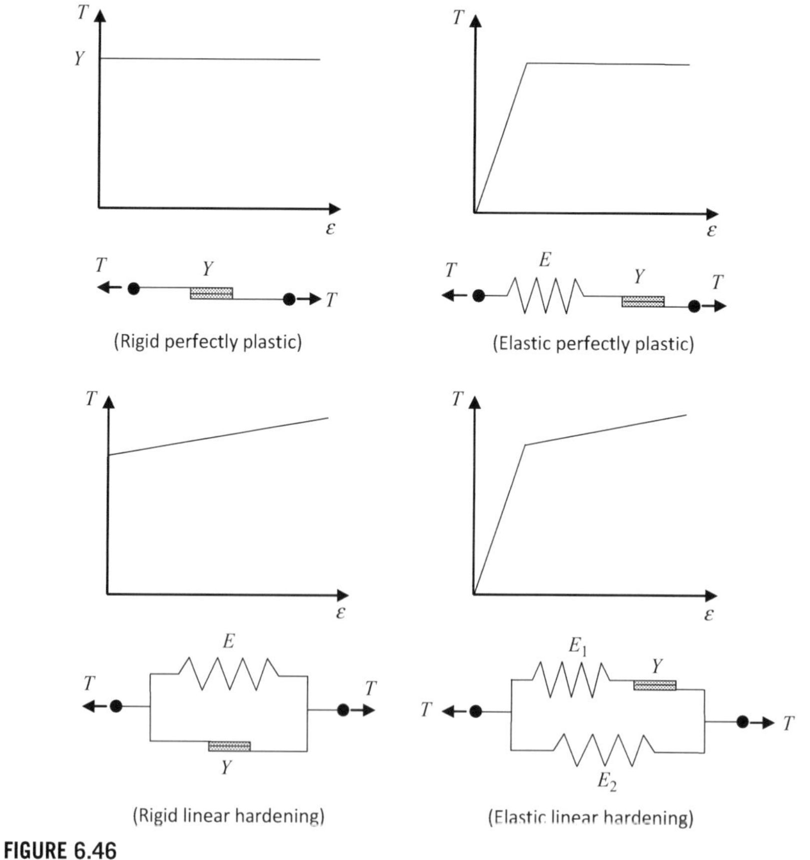

FIGURE 6.46

Idealized uniaxial plasticity models.

models used in linear viscoelasticity. The rigid perfectly plastic case can be generated by a simple *frictional element* with yield value Y. This element gives no deformation until the stress level reaches Y, and then unrestrained deformation occurs thereafter. The other cases shown are created through various combinations of frictional and elastic elements. Stress–stain curves for each of these models are illustrated, and these can be compared with more realistic behaviors shown in Fig. 6.45. These cases can be further generalized into viscoplastic models by including viscous dashpots previously presented.

6.6.1.1 Yield function

Development of yield functions goes back a century or more, and early work was related to various failure theories of solids. Such theories were often described by maximum principal stresses, or maximum shear stresses or maximum distortion energy. When particular points in the solid reached such a maximum strength value, it was assumed that material failure had initiated. Initially, we will explore a more general approach and then will focus attention on two particular theories.

Classical plasticity uses small deformation theory, and thus the total strain ε_{ij} can be decomposed into two parts, elastic strains ε_{ij}^{e}, and plastic strains ε_{ij}^{p} in the simple additive form

$$\varepsilon_{ij} = \varepsilon_{ij}^{e} + \varepsilon_{ij}^{p} \tag{6.6.1}$$

For the general three-dimensional case, the basic assumption of classical plasticity theory is that there exists a scalar *yield function* (or *loading function*), which depends on the *stress, plastic strain,* and the *history of loading*, in such a way as to characterize the material's yield behavior. This idea may be written as

$$f(T_{ij}, \varepsilon_{ij}^{p}, \kappa) = 0 \tag{6.6.2}$$

where κ is known as the *work-hardening parameter*. No change in plastic deformation occurs when $f < 0$. When $f = 0$, changes in plastic deformation occur, and no meaning is associated with $f > 0$. The work-hardening parameter κ is normally assumed to depend on the plastic deformation history of the material.

To clarify what is meant by loading and unloading, consider the time rate of change of f:

$$\dot{f} = \frac{\partial f}{\partial T_{ij}}\dot{T}_{ij} + \frac{\partial f}{\partial \varepsilon_{ij}^{p}}\dot{\varepsilon}_{ij}^{p} + \frac{\partial f}{\partial \kappa}\dot{\kappa} \tag{6.6.3}$$

The condition $f = 0$ and $\dot{f} < 0$ would imply that $f < 0$ at the next instant of time. This gives *unloading*. However, we also require that during the unloading, no plastic strain occurs, so $\varepsilon_{ij}^{p} = 0$, and that the rate of change of the strain hardening parameter κ must also vanish. Hence for *unloading and loading we may write*

$$\begin{aligned} f = 0, \quad & \frac{\partial f}{\partial T_{ij}}\dot{T}_{ij} < 0, \quad \textit{unloading} \\ f = 0, \quad & \frac{\partial f}{\partial T_{ij}}\dot{T}_{ij} > 0, \quad \textit{loading} \end{aligned} \tag{6.6.4}$$

Because of these definitions, f is sometimes referred to as the loading function.

Classical theories of plasticity often consider simplified yield functions which only depend on the stress and neglect the effects of the plastic strain and work hardening parameters. These are referred to as perfect plasticity, and a couple of simple examples were shown in the one-dimensional models in Fig. 6.46. Considering now only this case, we can express the yield function as

$$f = f(T_{ij}) \tag{6.6.5}$$

If the material is *isotropic*, then f must be invariant with respect to coordinate rotations, and hence must be only a function of the stress invariants. This is equivalent to employing the scalar-valued representation theorem (2.14.1). Thus, we can write

$$f = f(I_T, II_T, III_T) \tag{6.6.6}$$

or equivalently in terms of the principal stresses

$$f = f(T_1, T_2, T_3) \tag{6.6.7}$$

Relation (6.6.7) may be conveniently visualized as surface in a three-dimensional principal stress space as shown in Fig. 6.47. Stress states lying inside of the $f = 0$ surface will be elastic, while points on the $f = 0$ will have reached yield. In visualizing this concept for a loading situation, we start with zero stress at the origin. As loading increases, a loading path will be generated in the stress space. This path will depend on the multidimensional stress system that is being applied. As long as the loading path remains inside of the $f = 0$ surface, the material response will be elastic. Once any portion of the loading path reaches the yield surface, plastic deformation is initiated. What happens after this point will depend on the hardening characteristics and will be briefly discussed later.

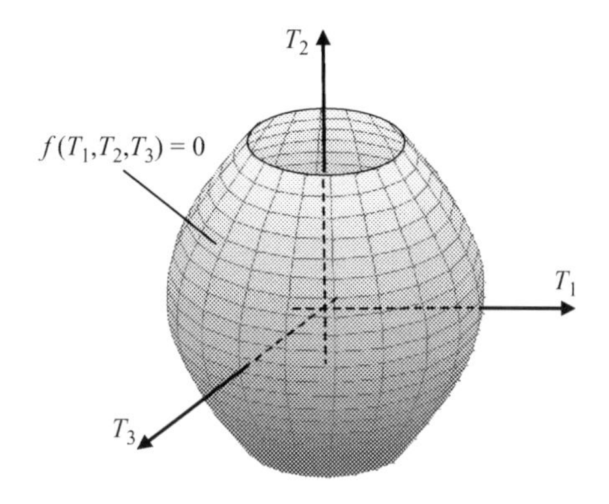

FIGURE 6.47

Yield surface in principal stress space.

Although not true for all materials, considerable experimental evidence for most metals indicates that the plastic yielding is independent of the hydrostatic pressure. Using this fact, relation (6.6.6) reduces to

$$f = f(II_T, III_T) \tag{6.6.8}$$

We can conclude that if the yield function is independent of $I_T = T_{kk}$, it is also independent of the spherical part of the stress tensor and must then depend only on the deviatoric stress $\hat{T}_{ij}$. Thus, the yield function further reduces to

$$f = f(II_{\hat{T}}, III_{\hat{T}}) \quad \text{or} \quad f = f(J_2, J_3) \tag{6.6.9}$$

It is common to denote the second and third invariants of the deviatoric stress tensor as J_2 and J_3, in the following fashion:

$$II_{\hat{T}} = -J_2 = -\frac{1}{2}\hat{T}_{ij}\hat{T}_{ij}\ ,\ III_{\hat{T}} = J_3 = \det(\hat{T}_{ij}) \tag{6.6.10}$$

We now explore in detail two classical yield functions of this type called *Mises* and *Tresca* yield criteria.

6.6.1.2 Mises yield condition

The *Mises yield condition* is defined from elastic strain energy concepts. From previous relation (6.2.41), the total strain energy was expressed in terms of the stress for isotropic materials. We can decompose this energy into two parts, one associated with *volumetric* change U_v and the other due to *distortional* (change in shape) deformation U_d:

$$U = U_v + U_d \tag{6.6.11}$$

where

$$\begin{aligned} U_v &= \frac{1-2\nu}{6E} I_T^2 = \frac{1-2\nu}{6E}\left(T_1 + T_2 + T_3\right)^2 \\ U_d &= -\frac{1}{6\mu}\left(I_T^2 + 3II_T\right) \\ &= \frac{1}{12\mu}\left[\left(T_1 - T_2\right)^2 + \left(T_2 - T_3\right)^2 + \left(T_3 - T_1\right)^2\right] = \frac{1}{4\mu}\hat{T}_{ij}\hat{T}_{ij} = \frac{1}{2\mu}J_2 \end{aligned} \tag{6.6.12}$$

It has been previously argued that the yield function will not depend on the hydrostatic pressure and thus cannot depend on the first invariant of stress. Thus, from (6.6.12) only U_d can be included in such an energy condition, and this distortional strain energy is linearly related to J_2. Therefore, the Mises yield condition related to maximum distortional strain energy is defined by

$$\begin{aligned} & f(T_{ij}) = J_2 - k^2 \Rightarrow \\ & \left(T_1 - T_2\right)^2 + \left(T_2 - T_3\right)^2 + \left(T_3 - T_1\right)^2 = 6k^2 \\ & T_1^2 + T_2^2 + T_3^2 - T_1T_2 - T_2T_3 - T_3T_1 = 3k^2 \end{aligned} \tag{6.6.13}$$

where k is a constant normally independent of the strain history. Relation (6.6.13) is sometimes called J_2 *flow theory*. Hence we have $J_2 \leq k^2$, with plastic flow for the equality case. For a work hardening material k can be allowed to change with strain history.

Considering the simple shear state of stress $\boldsymbol{T} = T_{12}\boldsymbol{e}_1\boldsymbol{e}_2 + T_{12}\boldsymbol{e}_2\boldsymbol{e}_1$, for this case $J_2 = \hat{T}_{ij}\hat{T}_{ij}/2 = T_{12}^2$ and thus $f = 0 \Rightarrow k = (T_{12})_y$. Hence k corresponds to the yield stress in pure simple shear. Furthermore, if we consider the uniaxial tension state of stress $\boldsymbol{T} = T_{11}\boldsymbol{e}_1\boldsymbol{e}_1$, we find that $J_2 = \hat{T}_{ij}\hat{T}_{ij}/2 = T_{11}^2/3$, and for $f = 0 \Rightarrow k = (T_{11})_y/\sqrt{3} = Y/\sqrt{3}$. So k can also be related to the yield tensile stress.

6.6.1.3 Tresca yield condition

The *Tresca yield condition* proposes that the primary factor for yielding is the maximum shear stress in the material. The criterion then stipulates that the maximum shear stress must be equal to a constant material value of k during plastic flow.

To express this idea analytically, it is easier to use the principal stresses T_1, T_2, T_3. From our theory in Section 4.4 and Fig. 4.7, we can write the Tresca yield condition as

$$\begin{aligned}
f &= T_1 - T_3 - 2k, \quad && T_1 \geq T_2 \geq T_3 \\
f &= T_3 - T_1 - 2k, && T_3 \geq T_2 \geq T_1 \\
f &= T_2 - T_1 - 2k, && T_2 \geq T_3 \geq T_1 \\
f &= T_1 - T_2 - 2k, && T_1 \geq T_3 \geq T_2 \\
f &= T_3 - T_2 - 2k, && T_3 \geq T_1 \geq T_2 \\
f &= T_2 - T_3 - 2k, && T_2 \geq T_1 \geq T_3
\end{aligned} \tag{6.6.14}$$

We can collect all these separate conditions together by writing a product form

$$f = [(T_1 - T_2)^2 - 4k^2][(T_2 - T_3)^2 - 4k^2][(T_3 - T_1)^2 - 4k^2] \tag{6.6.15}$$

Finally, condition (6.6.15) can be written in an invariant form by

$$f = 4J_2^3 - 27J_3^2 - 36k^2J_2^2 + 96k^4J_2 - 64k^6 \tag{6.6.16}$$

For the case of uniaxial tension, $\boldsymbol{T} = T_{11}\boldsymbol{e}_1\boldsymbol{e}_1$, the Tresca condition $f = 0 \Rightarrow k = (T_{11})_y/2 = Y/2$.

These two yield functions may be plotted in principal stress space as shown in Fig. 6.48. Note that both Mises and Tresca conditions fall into the general form given by relation (6.6.9), and because they are independent of the hydrostatic stress both will be an open cylindrical shape with axis along the line OH defined by $T_1 = T_2 = T_3$ which makes equal angles with the T_1, T_2, T_3 axes. Thus, in principle no matter how large the stress, a loading path along line OH will never reach yield. The Mises function is a circular cylinder, whereas the Tresca condition is a hexagonal cylinder, both having the same axis along OH. The plane normal to line OH is referred to as the *deviatoric plane*. Stress states lying on this plane will be deviatoric, while states normal to the plane will be hydrostatic. Based on these observations, it is understandable why these two yield surfaces will be an open cylinder oriented with axes along OH.

The projection of the Mises and Tresca yield surface on the deviatoric plane is shown in Fig. 6.49. The Mises locus is a circle and the Tresca shape is a regular hexagon. Each

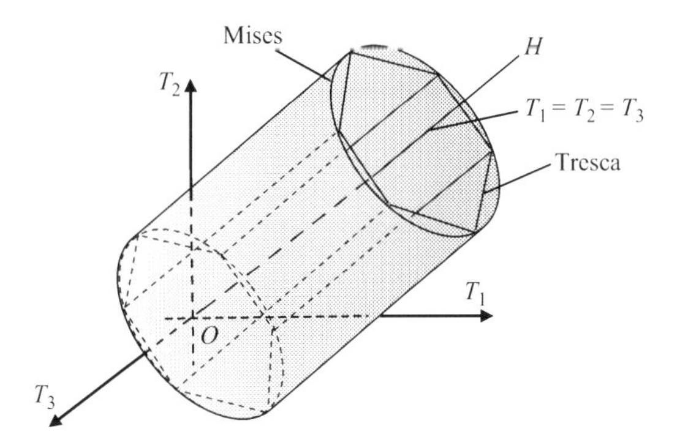

FIGURE 6.48

Mises and Tresca yield surfaces in principal stress space.

side of the Tresca hexagon represents one of the six equations listed in relation (6.6.14). We now make a few general comments about these deviatoric yield loci. For our isotropic case, the order of the principal stresses is immaterial, and this implies that the yield locus is symmetric about the projection of each principal axis. Furthermore, if we neglect any difference between tension and compression yield (no Bauschinger effect), then we have additional symmetry about directions orthogonal to each project principal axes. Collecting these ideas together provides the conclusion that a general yield locus must repeat over twelve 30° angular sections as verified in Fig. 6.49.

Since many experiments are often two-dimensional, it is more convenient to consider a plane stress case where one of the principal stresses is dropped. Thus, if we choose $T_3 = 0$, the Mises and Tresca yield conditions reduce to

$$\begin{aligned} &T_1^2 - T_1T_2 + T_2^2 = Y \ldots \text{Mises} \\ &T_1 = \pm Y, T_2 = \pm Y, T_1 - T_2 = \pm Y \ldots \text{Tresca} \end{aligned} \quad (6.6.17)$$

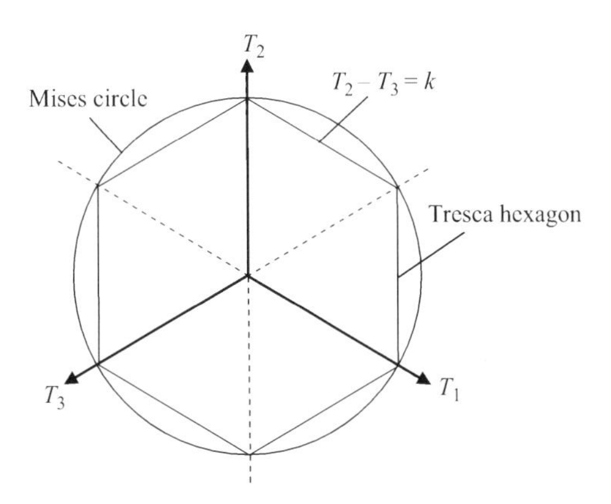

FIGURE 6.49

Mises and Tresca yield loci in deviatoric plane.

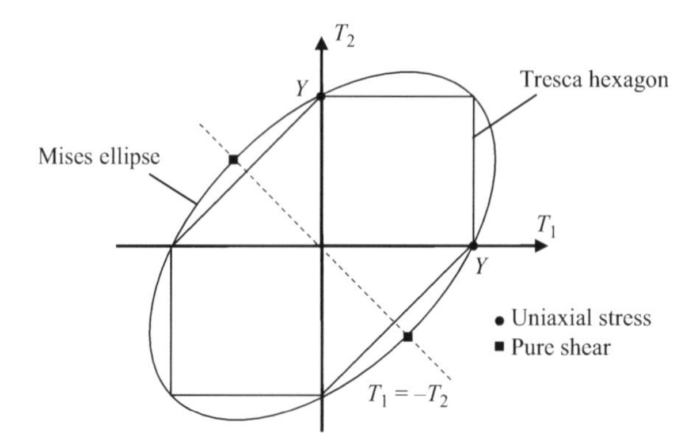

FIGURE 6.50

Mises and Tresca yield loci in T_1, T_2-plane.

Fig. 6.50 illustrates these yield loci in a T_1,T_2-system. For this case, the Mises locus is an ellipse oriented at a 45° angle as shown, whereas the Tresca criterion takes a skewed hexagonal shape within the Mises ellipse. Particular states of stress on the Mises ellipse can be identified, and the cases of uniaxial stress and pure shear are shown. Experimental data on several ductile metals commonly match better with the Mises yield model (Chakrabarty, 1987). However, differences between these two yield functions are generally small.

Several other more specialized yield criteria have been proposed. For many geomaterials such as sand and clay soils, rocks, and concretes, the inelastic deformation results from internal frictional sliding. This type of behavior then points out that normal stresses (including the hydrostatic stress) will have an effect on yielding. This reasoning has led to the development of *Mohr–Coulomb and Ducker–Prager yield criteria*; Asaro and Lubarda (2006) provide some specific details on these and a few other specialized yield functions.

Although we will not pursue strain hardening in detail, at this point we briefly present some fundamental qualitative aspects on how this would relate to the yield function. The basic idea is that once yield is reached, hardening will cause the yield surface to move in stress space. Fig. 6.51 shows two of the common model examples in T_1,T_2 stress space for the Mises criterion. Fig. 6.51A illustrates *isotropic hardening* where the yield surface undergoes uniform expansion. For this case, the material becomes stronger in all directions, thus implying no Bauschinger effect. The fundamental modification in the yield function could be expressed by $f = J_2 - \kappa^2$, where κ is no longer a constant but rather is a scalar variable dependent on the deformation history. For the Mises yield function in this stress space $T_1^2 - T_1T_2 + T_2^2 = Y$, we could simply add an addition term and write $T_1^2 - T_1T_2 + T_2^2 = Y + T_H$, where T_H would be a hardening stress increment that would result in uniform expansion of the Mises ellipse in all directions as shown. Another common model shown in Fig. 6.51B is *kinematic hardening* that takes into account the Bauschinger effect whereby yield in tension produces reduction in the yield value in compression. This can be accomplished if we

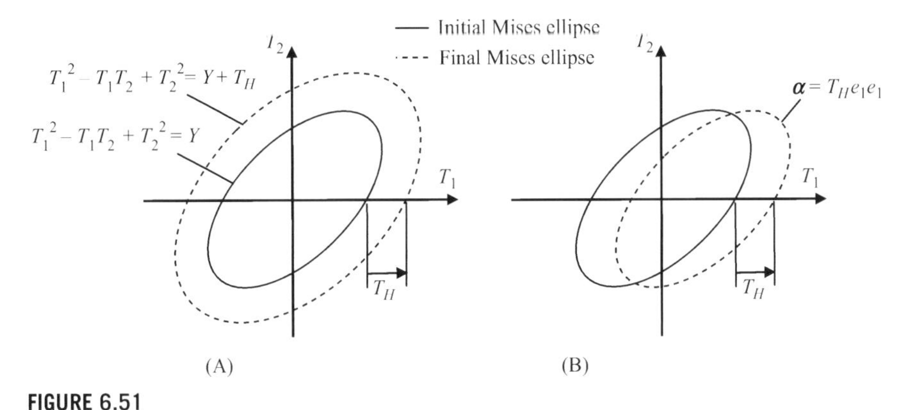

FIGURE 6.51

Isotropic and kinematic hardening in T_1, T_2-plane for Mises criterion: (A) isotropic hardening ($f = J_2 - \kappa^2$); (B) kinematic hardening ($f = f(T_{ij} - \alpha_{ij})$).

allow the yield locus to translate in the direction of the loading path. The fundamental modification in the yield function for this case could be expressed by $f = f(T_{ij} - \alpha_{ij})$, where α_{ij} is related to the translation of the yield surface center and is sometimes called the *back stress*. The situation shown in Fig. 6.51B would be for the case where $\alpha = T_H \boldsymbol{e}_1 \boldsymbol{e}_1$. This situation produces *stress-induced anisotropy* since the loading has modified the yield criterion in different directions in stress space. For some applications, it has been found necessary to use mixed hardening whereby both isotropic and kinematic hardening are included in particular combinations.

6.6.1.4 Plastic stress–strain relations

We now wish to explore plastic stress–strain constitutive laws. These are often referred to as *flow laws* since such deformations have flow-like time-dependent features. Generally the plastic strains depend on load history, and normally such constitutive relations will include strain and/or stress rates or use incremental forms of these variables. Commonly, the basic idea is that once the material has reached yield, we wish to determine the small change in the plastic strain $d\varepsilon_{ij}^p$ (or sometimes the total strain $d\varepsilon_{ij}$) due to a small change in applied stress dT_{ij}. Often numerical schemes like finite-element method codes are used to integrate such incremental relations. This study can be extensive and considerably complex; however, we will limit the presentation to only some of the more basic aspects of small deformation plasticity.

We start by neglecting elastic deformation and strain hardening and will thus focus on the rigid perfectly plastic response as shown in Fig. 6.46. We might expect such a constitutive equation for plastic deformation to be somewhat similar to that of a Newtonian fluid where the deviatoric stress would depend linearly on the rate of deformation. However, experimental observations do not support that type of relation. Over a century ago, Levy and later von Mises proposed that under plastic deformation, the principal axes of strain increment coincides with the principal axes of deviatoric stress and thus they suggested the equality of the following ratios:

$$\frac{d\varepsilon_{11}}{\hat{T}_{11}} = \frac{d\varepsilon_{22}}{\hat{T}_{22}} = \frac{d\varepsilon_{33}}{\hat{T}_{33}} = \frac{d\varepsilon_{12}}{\hat{T}_{12}} = \frac{d\varepsilon_{23}}{\hat{T}_{23}} = \frac{d\varepsilon_{31}}{\hat{T}_{31}} \tag{6.6.18}$$

or more compactly

$$d\varepsilon_{ij} = \hat{T}_{ij} d\lambda \tag{6.6.19}$$

where $d\lambda$ is a scalar proportionality factor. Within the context of small deformations, $\dot{\varepsilon}_{ij} \approx D_{ij}$, and thus relation (6.6.19) can be recast into the following flow law between the deviatoric stress and the rate of deformation:

$$\hat{T}_{ij} = \frac{k}{\sqrt{|II_D|}} D_{ij} \tag{6.6.20}$$

where $II_D = D_{ij}D_{ij}/2$ is the second invariant of the rate of deformation tensor, and k is a constant. Forms (6.6.18)–(6.6.20) are referred to as the *Levy–Mises plastic constitutive relations.*

Constitutive form (6.6.20) directly gives the result $D_{kk} = 0$, thus implying plastic incompressibility. This constitutive form also actually implies the Mises yield condition since

$$\frac{1}{2}\hat{T}_{ij}\hat{T}_{ij} = \frac{k^2}{2|II_D|} D_{ij}D_{ij} = k^2$$

Note that it was previously shown that $k = (T_{11})_y/\sqrt{3} = (T_{12})_y$ and that there will be no strain hardening for this case. Within the Levy–Mises law, there is no elastic strain and thus $d\varepsilon_{ij} = d\varepsilon_{ij}^p$.

Next consider a generalization of the Levy–Mises constitutive model that includes the elastic strains. This case would then represent the elastic perfectly plastic model shown in Fig. 6.46. Using the additive strain decomposition (6.6.1), we can write

$$\begin{aligned} d\varepsilon_{ij} &= d\varepsilon_{ij}^e + d\varepsilon_{ij}^p \\ d\varepsilon_{ij}^e &= \frac{1+\nu}{E} dT_{ij} - \frac{\nu}{E}\delta_{ij}T_{kk} \\ &= \frac{1+\nu}{E} d\hat{T}_{ij} + \frac{1-2\nu}{E}\delta_{ij}T_{kk} \\ d\varepsilon_{ij}^p &= \hat{T}_{ij} d\lambda \\ d\varepsilon_{kk}^p &= 0 \end{aligned} \tag{6.6.21}$$

and using the Mises yield criteria gives

$$d\lambda = \frac{1}{k}\sqrt{|II_{d\varepsilon^p}|} = \frac{1}{k}\sqrt{d\varepsilon_{ij}^p d\varepsilon_{ij}^p/2} = \frac{1}{Y}\sqrt{\frac{3}{2}d\varepsilon_{ij}^p d\varepsilon_{ij}^p} \tag{6.6.22}$$

Relations (6.6.21) and (6.6.22) are referred to as the *Prandtl–Reuss plastic constitutive laws.*

Another general flow rule concept is to represent the plastic strain increment as the derivative of a *potential function*

$$d\varepsilon_{ij}^{p} = \lambda \frac{\partial g}{\partial T_{ij}} \tag{6.6.23}$$

where g is a general plastic potential function that would be determined from experiments. If we set $g = f$, that is, make the potential function the same as the yield function, and choose the Mises yield criterion then

$$d\varepsilon_{ij}^{p} = \lambda \frac{\partial f}{\partial T_{ij}} = \lambda \hat{T}_{ij} \tag{6.6.24}$$

which is the same flow rule as (6.6.19). Furthermore, (6.6.24) indicates that the plastic strain increment is proportional to the gradient of the yield surface and is therefore *normal* to the surface. This fact is called the *normality condition*, and a flow rule that obeys this condition is called an *associated flow rule*. For the case where $g \neq f$, the plastic strain increment will not necessarily be normal to the yield surface and the flow rule is referred to as a *nonassociated flow rule*. Nonassociated flow rules are often used in geomechanics applications where the yield criterion depends on the hydrostatic pressure.

Much more could be said about additional features related to yield criteria and plastic flow rules, but because of space limitations we will end this discussion and move on to a few plasticity problem solutions.

6.6.2 PROBLEM SOLUTIONS

We now wish to use the previous yield criteria and constitutive relations to explore a few analytical solutions to some simple plasticity problems. The examples explore continuum elastic–plastic boundary-value problems starting with the elasticity solution, and then moving on to the case where yielding becomes present to develop plastic zones. These specific examples demonstrate some features on how the plasticity model is incorporated into problem solution.

EXAMPLE 6.6.1 ELASTOPLASTIC STRESSES IN A THICK-WALLED TUBE UNDER INTERNAL PRESSURE

Determine the elastic–plastic deformation behavior of a thick-walled cylindrical tube under internal pressure p that was previously shown in Fig. 6.43.

Solution: The elasticity solution is normally developed under plane strain conditions in polar coordinates ($\varepsilon_{zz} = 0$) and was previously provided in Example 6.5.9:

$$\begin{gathered} T_{rr} = \frac{r_1^2 p}{r_2^2 - r_1^2}\left(1 - \frac{r_2^2}{r^2}\right), T_{\theta\theta} = \frac{r_1^2 p}{r_2^2 - r_1^2}\left(1 + \frac{r_2^2}{r^2}\right) \\ T_{zz} = \nu(T_{rr} + T_{\theta\theta}), T_{r\theta} = T_{\theta z} = T_{rz} = 0 \\ u_r = \frac{1}{2\mu}\frac{r_1^2 p}{r_2^2 - r_1^2}\left(\frac{r_2^2}{r} + \frac{\mu}{\lambda + \mu} r\right) \end{gathered} \tag{6.6.25}$$

Since all shear stresses vanish, T_{rr}, $T_{\theta\theta}$, T_{zz} are the principal stresses, and we can verify that $T_{\theta\theta} > T_{zz} > T_{rr}$.

If we use the Tresca yield criteria, the onset of yielding is given by

$$T_{\theta\theta} - T_{rr} = 2k \Rightarrow T_{\theta\theta} - T_{rr} = \frac{2r_1^2 p}{r_2^2 - r_1^2}\left(\frac{r_2^2}{r^2}\right) \tag{6.6.26}$$

This expression takes on a maximum value at $r = r_1$, and thus yielding begins at the inner radius when the applied pressure becomes

$$p_o = k\left(1 - \frac{r_1^2}{r_2^2}\right) \tag{6.6.27}$$

where k is equal to one-half the yield stress in tension.

For the case where we use the Mises yield criteria, we can use relation (6.6.13) to write

$$(T_{rr} - T_{\theta\theta})^2 + (T_{\theta\theta} - T_{zz})^2 + (T_{zz} - T_{rr})^2 = 6k^2 \tag{6.6.28}$$

Using the given elastic solution (6.6.25), the yield relation reduces to

$$\left[\frac{1}{3}(1-2\nu)^2 + \frac{r_2^4}{r^4}\right]K^2 p^2 = k^2 \tag{6.6.29}$$

where $K = r_1^2/(r_2^2 - r_1^2)$. Similar to the Tresca case, it can be shown that Mises yielding will also begin at $r = r_1$, and relation (6.6.29) can then be solved for the initial yield pressure

$$p_o = \frac{k\left(1 - r_1^2/r_2^2\right)}{\sqrt{1 + \frac{1}{3}(1-2\nu)^2 r_1^4/r_2^4}} \tag{6.6.30}$$

where for the Mises case, $k = Y/\sqrt{3}$.

When the internal pressure increases beyond p_o, a *plastic zone* will spread from the inner boundary r_1 to some larger radius that we will denote by r_p as shown in Fig. 6.52. In the remaining elastic region $r_p < r < r_2$, the stresses will be given by relations similar to (6.6.25) but satisfying boundary conditions $T_{rr}(r_2) = 0$ and $T_{rr}(r_p) = -p_p$, where p_p is the pressure on the elastic–plastic boundary such that the yield criterion is satisfied.

$$\begin{aligned} T_{rr} &= \frac{r_p^2 p_p}{r_2^2 - r_p^2}\left(1 - \frac{r_2^2}{r^2}\right), \quad T_{\theta\theta} = \frac{r_p^2 p_p}{r_2^2 - r_p^2}\left(1 + \frac{r_2^2}{r^2}\right) \\ T_{zz} &= \nu\left(T_{rr} + T_{\theta\theta}\right) = \frac{2\nu r_p^2 p_p}{r_2^2 - r_p^2}, \quad T_{r\theta} = T_{\theta z} = T_{rz} = 0 \end{aligned} \tag{6.6.31}$$

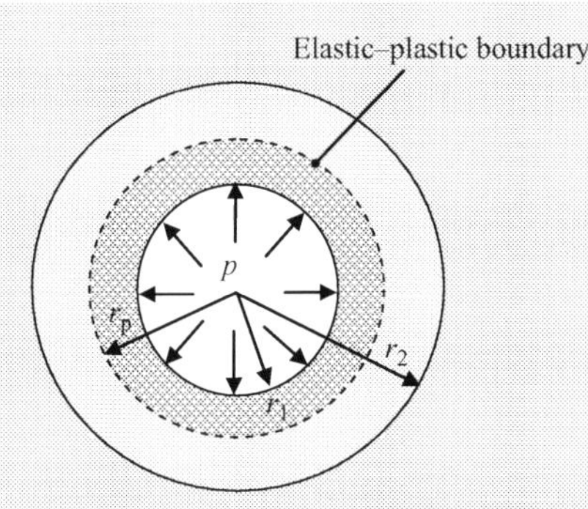

FIGURE 6.52

Elastic–plastic response of pressurized thick-walled cylindrical tube.

Using the Tresca yield function with $T_{\theta\theta} > T_{zz} > T_{rr}$, at $r = r_p$:

$$T_{\theta\theta} - T_{rr} = 2k \Rightarrow T_{\theta\theta} - T_{rr} = \frac{2r_2^2 p_p}{r_2^2 - r_p^2} = 2k \Rightarrow p_p = k\left(1 - \frac{r_p^2}{r_2^2}\right) \tag{6.6.32}$$

Within the plastic zone $r_1 < r < r_p$, the Tresca condition $T_{\theta\theta} - T_{rr} = 2k$ must be maintained. The stresses within this region can be found by going back to the basic equilibrium equation for this problem in polar coordinates (see Appendix A)

$$\frac{dT_{rr}}{dr} + \frac{T_{rr} - T_{\theta\theta}}{r} = 0 \Rightarrow \frac{dT_{rr}}{dr} = \frac{2k}{r} \Rightarrow T_{rr} = 2k \log r + C$$

Applying boundary condition $T_{rr}(r_1) = -p$, determines the value of the constant C and the form of the radial stress:

$$T_{rr} = -p + k \log\left(\frac{r^2}{r_1^2}\right) \tag{6.6.33}$$

Next, using the continuity relation across the elastic–plastic boundary $T_{rr}(r_p) = -p_p$ gives the internal pressure result

$$p = k\left[1 - \frac{r_p^2}{r_2^2} + \log\left(\frac{r_p^2}{r_1^2}\right)\right] \tag{6.6.34}$$

Collecting these results then determines the final results for the stresses in the plastic zone $r_1 < r < r_p$:

$$T_{rr} = -k\left[1 - \frac{r_p^2}{r_2^2} + \log\left(\frac{r_p^2}{r^2}\right)\right]$$
$$T_{\theta\theta} = k\left[1 + \frac{r_p^2}{r_2^2} - \log\left(\frac{r_p^2}{r^2}\right)\right] \qquad (6.6.35)$$
$$T_{zz} = 2\nu k\left[\frac{r_p^2}{r_2^2} - \log\left(\frac{r_p^2}{r^2}\right)\right]$$

and in the elastic zone $r_p < r < r_2$:

$$T_{rr} = -k\left(\frac{r_p^2}{r^2} - \frac{r_p^2}{r_2^2}\right)$$
$$T_{\theta\theta} = k\left(\frac{r_p^2}{r^2} + \frac{r_p^2}{r_2^2}\right) \qquad (6.6.36)$$
$$T_{zz} = 2k\nu\frac{r_p^2}{r_2^2}$$

The distribution of the in-plane stresses T_{rr} and $T_{\theta\theta}$ are shown in Fig. 6.53 for the case $r_1/r_2 = 0.5$ with several different r_p/r_2 ratios. MATLAB Code C-15 was used for the calculations and plotting. The case $r_p/r_2 = 0.5$

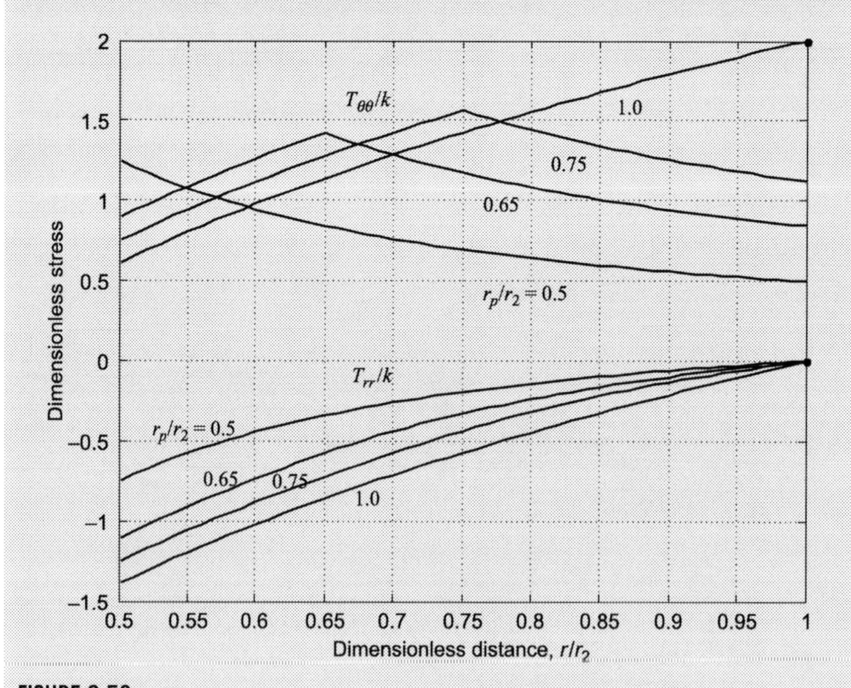

FIGURE 6.53

Elastic–plastic radial and hoop stresses in thick-walled tube problem.

corresponds to the elastic solution with no plastic zone, whereas the case $r_p / r_2 = 1.0$ represents the fully plastic situation with no elastic zone. The radial stresses T_{rr} are all negative, continuous, and approach zero at the outer boundary $r = r_2$. The hoop stresses $T_{\theta\theta}$ are all positive and continuous but the cases with $0.5 < r_p / r_2 < 1.0$ show a discontinuity in slope at elastic–plastic boundary $r = r_p$. It should be noted that there exists significant differences in hoop stress behavior between the fully elastic and fully plastic distributions.

The previous solution based on the Tresca yield condition was carried out without reference to the deformation field. Use was only made of the static equilibrium equation and the stresses were determined in terms of the elastic–plastic radius r_p. Attempting this same solution using the Mises yield function will result in a much more complicated analysis requiring the use of the Prandtl–Reuss relations (6.6.21). We will not pursue this analysis, and the interested reader is referred to Hill (1950), Chakrabarty (1987), or Lubliner (1990) for further details. Note that elastic–plastic expansion of a pressurized thick-walled spherical shell can also be handled in much the same way. Details on this can be found in the previous three references.

EXAMPLE 6.6.2 ELASTOPLASTIC STRESSES IN A CYLINDRICAL BAR UNDER TORSIONAL LOADING

Determine the elastic–plastic stress distribution in a cylindrical bar under pure torsional loading as shown in Fig. 6.54. The problem geometry is initially described in Cartesian coordinates (x,y,z) and involves the torsional loading of a prismatic cylinder with axis along the z-direction. The general cross-section is denoted by R but for simplicity, we will eventually only consider in detail the case of a circular cross-section.

Solution: As with the previous example, we start with the elasticity formulation and solution. The torsion problem is a classical one in elasticity and the complete formulation may be found in Sadd (2014). We outline here only a few of the necessary basic relations. Following the classical Saint Venant semi-inverse formulation, it is assumed that the character of the elastic stress field in most locations within the bar would depend only in a secondary way on the exact distribution of tractions on the ends. Thus, only the end resultant loading T is used as a condition on the problem. Following the semi-inverse scheme, the cross-section displacements are assumed to be rigid-body rotation about the z-axis, whereas the out-of-plane deformation is taken to be an unknown function of x and y. The section angle of twist is taken to vary linearly with the axial coordinate z. Under these assumptions, the displacements become

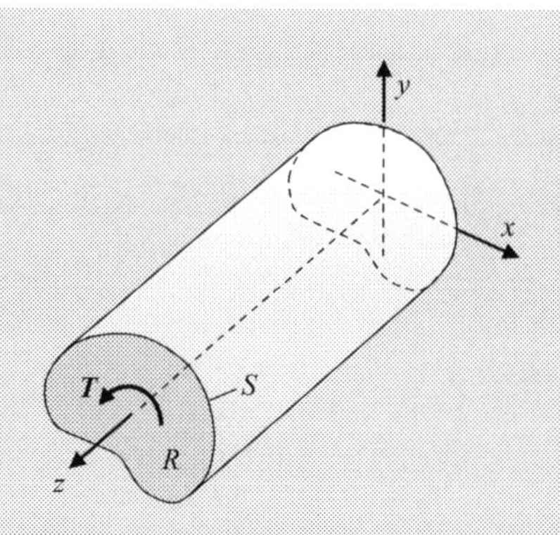

FIGURE 6.54

Torsion of a cylindrical bar.

$$u_x = -\alpha yz, u_y = \alpha xz, u_z = u_z(x, y) \tag{6.6.37}$$

where α is the *angle of twist per unit length*. This set of displacements produces a stress field with only two nonzero components of the form

$$T_{xx} = T_{yy} = T_{zz} = T_{xy} = 0$$
$$T_{xz} = T_{xz}(x, y),\ T_{yz} = T_{yz}(x, y)$$

The elasticity formulation commonly uses the *Prandtl stress function* $\phi = \phi(x, y)$ such that the two nonzero stresses can be expressed by

$$T_{xz} = \frac{\partial \phi}{\partial y}, T_{yz} = -\frac{\partial \phi}{\partial x} \tag{6.6.38}$$

Using this representation, the equilibrium equations will be identically satisfied and the compatibility condition gives a single relation

$$\nabla^2 \phi = \frac{\partial^2 \phi}{\partial x^2} + \frac{\partial^2 \phi}{\partial y^2} = -2\mu\alpha \tag{6.6.39}$$

which becomes the governing equation for the problem. Zero tractions on the lateral sides of the cylinder result in a boundary condition specifying that the stress function is a constant on section boundary S. For a simply connected cross-section, this constant may be chosen to be zero

$$\phi = 0 \quad \text{on } S \tag{6.6.40}$$

Finally, we can relate the applied torque T to the stress function by using

$$T = \iint_R (xT_{yz} - yT_{xz})\,dx\,dy = -\iint_R \left(x\frac{\partial \varphi}{\partial x} + y\frac{\partial \varphi}{\partial y}\right)dx\,dy = 2\iint_R \varphi\,dx\,dy \tag{6.6.41}$$

Relations (6.6.39)–(6.6.41) provide the basic governing stress formulation for the elasticity solution to the torsion problem.

We will now explore a specific cross-sectional shape and choose a solid circular section of radius a as shown in Fig. 6.55. Because of the symmetric section shape, the elastic–plastic solution will be straightforward with limited complexity. We begin by exploiting the problem symmetry and argue that all variables will be functions of only the radial coordinate r, and thus choose polar coordinates to formulate and solve the problem. For this case, governing equation (6.6.39) becomes

$$\nabla^2\phi = \frac{1}{r}\frac{d}{dr}\left(r\frac{d\phi}{dr}\right) = -2\mu\alpha \tag{6.6.42}$$

This ordinary differential equation can be easily integrated and applying the boundary condition (6.6.40) gives the result

$$\phi = \frac{\mu\alpha}{2}(a^2 - r^2) \tag{6.6.43}$$

In polar coordinates, the only nonzero stress component is given by

$$T_{\theta z} = \tau = -\frac{d\phi}{dr} = \mu\alpha r \tag{6.6.44}$$

The torque T can then be found using Eq. (6.6.41) giving the result

$$T = 2\iint_R \varphi\, dx\, dy = \frac{1}{2}\pi\mu\alpha a^4 \tag{6.6.45}$$

Combining the previous two equations determines the stress in terms of the torque loading

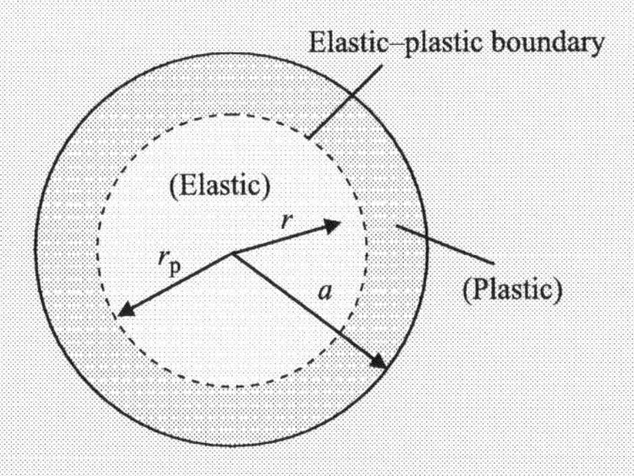

FIGURE 6.55

Elastic–plastic response for the torsion of bar with circular section.

$$\tau = \frac{2Tr}{\pi a^4} \tag{6.6.46}$$

These relations then provide the elastic stress distribution and load carrying capacity.

It is noted that the maximum stress occurs on the outer boundary $r = a$, and this is where yielding will begin as the applied torque or the angle of twist is increased. It can be shown that for any general cross-sectional shape, the maximum stress will always occur at the boundary. Both Tresca and Mises criteria predict yielding when the resultant shear stress $\tau = T_{\theta z} = \sqrt{T_{xz}^2 + T_{yz}^2} = k$, where for the Tresca case $k = Y/2$ and for the Mises criteria $k = Y/\sqrt{3}$. Therefore, the torque and angle of twist at yield initiation follows from (6.6.45) and (6.6.46):

$$T_p = \frac{\pi k a^3}{2} \quad \text{and} \quad \alpha_p = \frac{k}{\mu a} \tag{6.6.47}$$

As the torque or angle of twist increases beyond these initiation values, an annular plastic zone will expand uniformly from the outer boundary. The elastic region remains in the inner core of the bar, and the elastic–plastic boundary is defined by radius r_p as illustrated in Fig. 6.55. The stress distribution in the elastic zone is linear with the radial coordinate as per (6.6.46). Neglecting any hardening effects, the stress in the entire plastic region will be constant, equaling the yield value k. Invoking the continuity of the stress at r_p, (6.6.44) determines this location

$$r_p = \frac{k}{\mu \alpha} \tag{6.6.48}$$

Thus, the stress distribution in each zone is given by

$$\begin{aligned} \tau &= k\frac{r}{r_p}, \quad 0 \le r \le r_p \\ \tau &= k, \quad r_p \le r \le a \end{aligned} \tag{6.6.49}$$

Note again that we have been able to determine the stresses without knowing the deformation, and thus this problem is statically determined. For the elastic–plastic case, the torque can be computed as

$$\begin{aligned} T &= 2\pi \int_0^a \tau r^2\, dr = 2\pi \left[\int_0^{r_p} \frac{kr^3}{r_p} dr + \int_{r_p}^a kr^2\, dr \right] \\ &= \frac{2}{3}\pi k \left[a^3 - \frac{r_p^3}{4} \right] = \frac{1}{3} T_p \left[4 - \left(\frac{\alpha_p}{\alpha} \right)^3 \right] = T_U \left[1 - \frac{1}{4} \left(\frac{\alpha_p}{\alpha} \right)^3 \right] \end{aligned} \tag{6.6.50}$$

The fully plastic torque T_U occurs as $r_p \to 0$, and this gives $T_U = \frac{2}{3}\pi k a^3$. A dimensionless plot of the torque vs angle of twist is shown in Fig. 6.56. MATLAB Code C-16 was used for the calculations and plotting. It is observed that the torque rapidly approaches the fully plastic ultimate value.

While elastic torsion solutions for other cross-sectional shapes can be determined, finding such plastic solutions is difficult. As previously mentioned, for any general cross-sectional shape, the maximum stress will always occur on the boundary. However, finding the subsequent development of the elastic–plastic zones is a challenging task, and most such problems are handled numerically.

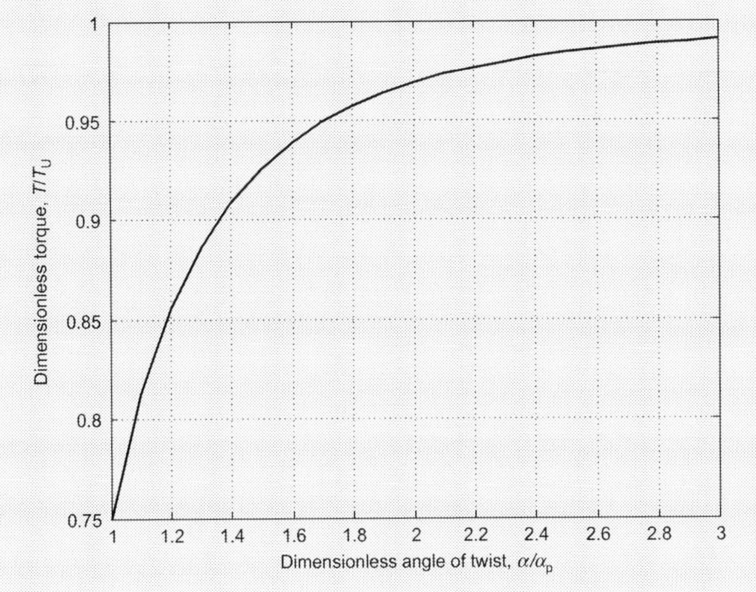

FIGURE 6.56

Torque vs angle of twist for circular section under elastic–plastic torsion.

REFERENCES

Achenbach, J.D., 1976. Wave Propagation in Elastic Solids. North Holland, Amsterdam.

Asaro, R.J., Lubarda, V.A., 2006. Mechanics of Solids and Materials. Cambridge University Press, Cambridge.

Barber, J.R., 2010. Elasticity. Springer, Dordrecht, the Netherlands.

Batchelor, G.K., 2010. An Introduction to Fluid Dynamics. Cambridge University Press, Cambridge.

Bland, D.R., 2016. The Linear Theory of Viscoelasticity. Dover, Mineola, NY.
Bower, A.F., 2010. Applied Mechanics of Solids. CRC Press, Boca Raton, FL.
Chakrabarty, J., 1987. Theory of Plasticity. McGraw Hill, New York.
Christensen, R.M., 2010. Theory of Viscoelasticity. Dover, New York.
Cowin, S.C., Mehrabadi, M.M., 1995. Anisotropic Symmetries of Linear Elasticity, Applied Mechanics Reviews. ASME 48, 247–285.
Flügge, W., 1975. Viscoelasticity. Springer, Berlin.
Golden, J.M., Graham, G.A.C., 1988. Boundary Value Problems in Linear Viscoelasticity. Springer, Berlin.
Graff, K.F., 1991. Wave Motion in Elastic Solids. Dover, New York.
Gutierrez-Lemini, D., 2014. Engineering Viscoelasticity. Springer, New York.
Hill, R., 1950. The Mathematical Theory of Plasticity. Oxford Press, Oxford.
Kachanov, M., Shafino, B., Tsukrov, I., 2003. Handbook of Elasticity Solutions. Kluwer Academic Press, Dordrecht, the Netherlands.
Lamb, S.H., 1993. Hydrodynamics. Cambridge University Press, Cambridge.
Lubliner, J., 1990. Plasticity Theory. Macmillan, New York.
Malvern, L.E., 1969. Introduction to the Mechanics of a Continuous Medium. Prentice-Hall, Englewood Cliffs, NJ.
Michell, J.H., 1899. On the direct determination of stress in an elastic solid with application of the theory of plates. Proc. London Math. Soc. 31, 100–124.
Milne-Thomson, L.M., 1974. Theoretical Hydrodynamics. Dover, New York.
Sadd, M.H., 2014. Elasticity: Theory, Applications and Numerics, Third ed. Elsevier, Waltham, MA.
Schlichting, H., 2017. Boundary-Layer Theory, Ninth ed. Springer, Berlin.
Sendeckyj, G.P., 1975. Some topics of anisotropic elasticity, composite materials. Chamis, C.C. (Ed.), Structural Design and Analysis Part I, 7, Academic Press, New York.
Serrin, J., 1959. Mathematical principles of classical fluid mechanics. Flugge, S. (Ed.), Handbuch der Physik, VIII/1, Springer, Berlin.
Timoshenko, S.P., Goodier, J.N., 1970. Theory of Elasticity. McGraw-Hill, New York.
Ting, T.C.T., 1996. Anisotropic Elasticity. Theory and Applications. Oxford University Press, New York.
Truesdell, C.A., 1966. Continuum Mechanics. I. The Mechanical Foundations of Elasticity and Fluid Dynamics. Gordon & Breach, New York.
White, F.M., 2016. Fluid Mechanics. McGraw-Hill, New York.
Wu, H.C., 2005. Continuum Mechanics and Plasticity. Boca Raton, FL, Chapman & Hall, CRC.
Zheng, Q.S., Spencer, A.J.M., 1993. Tensors which characterize anisotropies. Int. J. Eng. Sci. 31, 679–693.

EXERCISES

6.1 Using relations (6.2.2) and (6.2.7), explicitly show that $C_{ijkl} = C_{klij}$.

6.2 Starting with the general form (6.2.17), show that the isotropic fourth-order elasticity tensor can be expressed in the following forms:

$$C_{ijkl} = \lambda\delta_{ij}\delta_{kl} + \mu(\delta_{il}\delta_{jk} + \delta_{ik}\delta_{jl})$$
$$C_{ijkl} = \mu(\delta_{il}\delta_{jk} + \delta_{ik}\delta_{jl}) + (k - \frac{2}{3}\mu)\delta_{ij}\delta_{kl}$$
$$C_{ijkl} = \frac{E\nu}{(1+\nu)(1-2\nu)}\delta_{ij}\delta_{kl} + \frac{E}{2(1+\nu)}(\delta_{il}\delta_{jk} + \delta_{ik}\delta_{jl})$$

6.3 If the elastic constants E, k, and μ are required to be positive, show that Poisson's ratio must satisfy the inequality $-1 < \nu < \frac{1}{2}$. For most real materials, it has been found that $0 < \nu < \frac{1}{2}$. Show that this more restrictive inequality in this problem implies that $\lambda > 0$.

6.4 Show that Hooke's law for an isotropic material may be expressed in terms of spherical and deviatoric tensors by the two relations

$$\tilde{T}_{ij} = 3k\tilde{\varepsilon}_{ij}, \hat{T}_{ij} = 2\mu\hat{\varepsilon}_{ij}$$

6.5 For incompressible elastic materials, there will be a constraint on all deformations such that the change in volume must be zero, thus implying that $\varepsilon_{kk} = 0$. First show that under this constraint, Poisson's ratio will become 1/2 and the bulk modulus and Lamé's constant will become unbounded. Next show that the usual form of Hooke's law $T_{ij} = \lambda\varepsilon_{kk}\delta_{ij} + 2\mu\varepsilon_{ij}$ will now contain an indeterminate term. For such cases, Hooke's law is commonly rewritten in the form $T_{ij} = -p\delta_{ij} + 2\mu\varepsilon_{ij}$, where p is referred to as the *hydrostatic pressure* which cannot be determined directly from the strain field but is normally found by solving the boundary-value problem. Finally, justify that $p = T_{kk}/3$.

6.6 Go through the details and explicitly develop the Beltrami–Michell compatibility equations (6.2.35).

6.7 For the elasticity displacement formulation, use relations (6.2.36) in the equilibrium equations and develop the Navier equations (6.2.37).

6.8 Starting with the general strain energy expression (6.2.39), develop the strain and stress forms (6.2.40) and (6.2.41).

6.9 Since the elastic strain energy has physical meaning that is independent of the choice of coordinate axes, it must be invariant to all coordinate transformations. Because U is a *quadratic form* in the strains or stresses, it cannot depend on the third invariants III_ε or III_T, and so it must depend only on the first two invariants of the strain or stress tensors. Show that the strain energy can be written in the following two forms:

$$U = \left(\frac{1}{2}\lambda + \mu\right)I_\varepsilon^2 - 2\mu II_\varepsilon$$
$$= \frac{1}{2E}\left(I_T^2 - 2(1+\nu)II_T\right)$$

6.10 Using the stress–Airy stress function relations (6.2.47), show that this form does automatically satisfy the two-dimensional equilibrium equations with no body forces and that the compatibility relation will then lead to the biharmonic equation (6.2.48).

6.11 Verify that the Airy stress function

$$\phi = \frac{s}{4}\left(xy + \frac{ly^2}{c} + \frac{ly^3}{c^2} - \frac{xy^2}{c} + \frac{xy^3}{c^2} \right)$$

solves the problem of a cantilever beam loaded by uniform shear along its bottom edge as shown. Use pointwise boundary conditions on $y = \pm c$ and only resultant effects at ends $x = 0$ and l. Note, however, you should be able to show that T_{xx} vanishes at $x = l$.

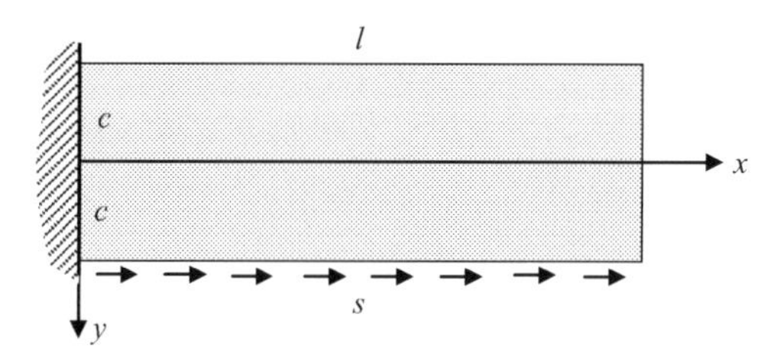

6.12 Consider the axisymmetric problem of an annular disk with a fixed inner radius and loaded with uniform shear stress τ over the outer radius. Using the Airy stress function term $a_4\theta$, show that stress solution for this problem is given by $T_{rr} = T_{\theta\theta} = 0, \quad T_{r\theta} = \tau r_2^2 / r^2$. Use the polar form of the stress–Airy stress function relations given in (6.2.59).

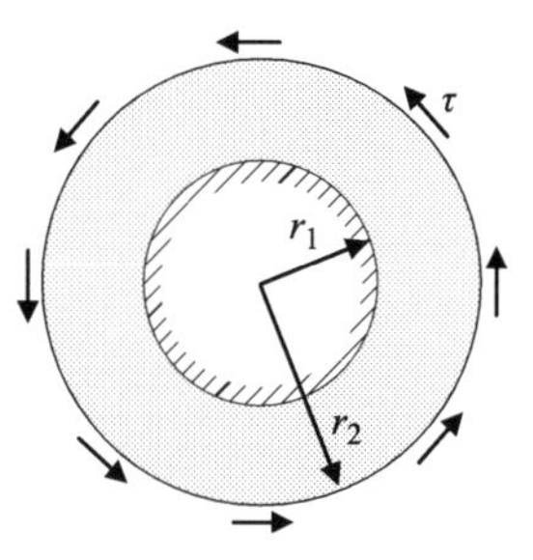

6.13 The starting point for the elastic torsion problem shown in Fig. 6.54 had displacements of the form $u_x = -\alpha yz, u_y = \alpha xz, u_z = u_z(x, y)$, where α is the constant angle of twist per unit length. Verify that this set of displacements produces a stress field with only two nonzero components of the form

$$T_{xx} = T_{yy} = T_{zz} = T_{xy} = 0$$
$$T_{xz} = T_{xz}(x, y),\ T_{yz} = T_{yz}(x, y)$$

6.14 Recall that for fluid flows, the velocity vector is always tangent to a streamline, and thus $\boldsymbol{v} \times d\boldsymbol{x} = 0$, where $d\boldsymbol{x}$ is the differential tangent vector to the streamline. Considering only a two-dimensional flow situation, show that this cross-product will yield a relation that confirms that the stream function ψ will be a constant along a streamline.

6.15 Prove that the streamlines are orthogonal to the potential lines. Hint: start with the relation $\phi = \text{constant} \Rightarrow d\phi = \frac{\partial \phi}{\partial x}dx + \frac{\partial \phi}{\partial y}dy = 0$ and a similar expression for the stream function. Then use relations (6.3.14).

6.16 Consider the following two-dimensional flow field $v_x = Ax$, $v_y = By$. Determine the relationship between the constants A and B for the flow to be incompressible (isochoric). Next show that the flow is irrotational. Finally determine the potential function ϕ defined by (6.3.8).

6.17 Using polar coordinates, explore the nature of a two-dimensional inviscid flow field for the case with stream function $\psi = K\theta$, where K is a constant. Make use of the polar coordinate relations in (6.3.21). Determine the velocity components and describe the nature of the flow field. Finally sketch the streamlines and potential lines.

6.18 Repeat Exercise 6.17 for the case where $\psi = K \log r$.

6.19 For the case with nonzero bulk viscosity, show that the Navier–Stokes equations become

$$-p_{,i} + (\lambda + \mu)v_{k,ki} + \mu v_{i,kk} + \rho b_i = \rho\left(\frac{\partial v_i}{\partial t} + v_j v_{i,j}\right)$$

$$-\nabla p + (\lambda + \mu)\nabla(\nabla \cdot v) + \mu \nabla^2 v + \rho b = \rho\left(\frac{\partial v}{\partial t} + v \cdot \nabla v\right)$$

6.20 Develop the differential form of the energy equation (5.6.12) using the linear viscous fluid constitutive law (6.4.12).

6.21 For very slow flows of an incompressible linearly viscous fluid, show that the Navier–Stokes equation with no body forces reduce to $\nabla p = \mu \nabla^2 \boldsymbol{v}$. Next show that the pressure field must be harmonic, that is, $\nabla^2 p = 0$. For the two-dimensional case, show that if the velocity is given by $\boldsymbol{v} = \nabla \times \boldsymbol{\psi}$, where $\boldsymbol{\psi} = (0, 0, \psi)$ is the stream function, then all equations are satisfied if the stream function is biharmonic, $\nabla^4 \psi = 0$.

6.22 The Navier–Stokes equation for an incompressible fluid without body forces can be written as

$$\frac{\partial v}{\partial t} + v \cdot \nabla v = -\frac{1}{\rho}\nabla p + \nu \nabla^2 v$$

where $\nu = \mu / \rho$ is called the ***kinematic viscosity***. If we define the vorticity vector by $\boldsymbol{\omega} = \nabla \times \boldsymbol{v}$, show that by taking the curl of the Navier–Stokes relation we can transform it into the ***vorticity equation***

$$\frac{D\boldsymbol{\omega}}{Dt} = \frac{\partial \boldsymbol{\omega}}{\partial t} + \boldsymbol{v} \cdot \nabla \boldsymbol{\omega} = (\boldsymbol{\omega} \cdot \nabla)\boldsymbol{v} + \nu \nabla^2 \boldsymbol{\omega}$$

6.23 Consider a Couette flow between two concentric cylinders with inner radius r_i and outer radius r_o. We will assume that the inner cylinder is fixed, whereas the outer cylinder is rotating with a constant angular velocity Ω. This will establish the following cylindrical shearing flow field:

$$v_r = 0, v_\theta = v_\theta(r), v_z = 0$$

Show that in cylindrical coordinates, the Navier–Stokes equations reduce to the single relation

$$\frac{d^2 v_\theta}{dr^2} + \frac{1}{r}\frac{dv_\theta}{dr} - \frac{v_\theta}{r^2} = 0$$

Subject to the boundary conditions $v_\theta(r_i) = 0, v_\theta(r_o) = \Omega r_o$, show that the solution to this equation is

$$v_\theta = \frac{\Omega r_i^2 r_o^2}{r_o^2 - r_i^2}\left(\frac{r}{r_i^2} - \frac{1}{r}\right)$$

6.24 For Exercise 6.23, show that the shearing stress on the outer cylindrical surface is given by

$$T_{r\theta}(r_o) = 2\mu D_{r\theta}(r_o) = \mu r_o \frac{d}{dr}\left(\frac{v_\theta}{r}\right)\bigg|_{r=r_o} = \frac{2\Omega \mu r_i^2}{r_o^2 - r_i^2}$$

Using this result show that the torque per unit length of cylinder necessary to maintain the angular velocity Ω is given by $T = \dfrac{4\pi\Omega\mu r_i^2 r_o^2}{r_o^2 - r_i^2}$. Note that this relation could be used to determine the fluid viscosity if we know the torque, angular velocity, and the cylinder radii.

6.25 Recall from the energy equation (5.6.12), the rate of working of the stresses or stress power was contained in the term $T_{ij}D_{ij}$. For the general linearly viscous fluid (6.4.4) show that this term is given by $\Phi = T_{ij}D_{ij} = -pD_{kk} + \lambda(D_{kk})^2 + 2\mu D_{ij}D_{ij}$. Reduce this form for the incompressible case. These expressions are commonly called the *dissipation functions.*

6.26 Consider the response of a *viscoelastic fluid-like material* subjected to *simple shearing deformation* $\varepsilon_{12} = H(t)\kappa_o t$. Using a *Maxwell constitutive model*, determine the shear stress history $T_{12}(t)$ and make a temporal plot of this response. Also develop and plot the corresponding linearly viscous fluid case. Assume an initial condition $T_{12}(0) = 0$.

6.27 For the viscoelastic three-parameter solid model shown in Fig. 6.41A, the sum of the strains in Kelvin and elastic components add to give the total overall strain, and the stress in each of these components are the same and equal the overall stress. Using these facts, develop the governing constitutive law (6.5.20) and relations (6.5.21).

6.28 For the viscoelastic three-parameter fluid model shown in Fig. 6.41B, following similar steps as suggested in Exercise 6.27, develop the governing constitutive law (6.5.23) and relations (6.5.24).

6.29 For the viscoelastic three-parameter solid model shown in Fig. 6.41A, verify the forms of the stress relaxation and creep functions given by relations (6.5.22).

6.30 For the viscoelastic three-parameter fluid model shown in Fig. 6.41B, verify the forms of the stress relaxation and creep functions given by relations (6.5.25).

6.31 Use the given MATLAB Code C-14 to make plots of the given relaxation and creep functions for the three-parameter solid (6.5.22) (with $p_1 = 5$, $q_o = 0.5$, $q_1 = 10$) and liquid (6.5.25) (with $p_1 = 5$, $q_1 = 10$, $q_2 = 10$). Next we wish to modify the three-parameter solid model parameters such that $E^* \to \infty \Rightarrow p_1 \to 0$, and for the three-parameter fluid $\eta \to 0 \Rightarrow q_2 \to 0$. Investigate this case numerically using the code with the parameter sets: three-parameter solid ($p_1 = 0.01$, $q_o = 0.5$, $q_1 = 10$) and liquid ($p_1 = 5$, $q_1 = 10$, $q_2 = 0.1$). Compare these modified numerical predictions with those from the previous Maxwell and Kelvin models.

6.32 Similar to Example 6.5.5, explore the loading and unloading behavior of the three-parameter solid model using the MATLAB ODE solver. Use parameters $p_1 = 0.3$, $q_o = 200$, $q_1 = 300$.

6.33 Using numerical methods, make plots of the Maxwell and Kelvin storage and loss moduli(given in Example 6.5.7) vs the frequency ω. For convenience, plot G_1/E and G_2/η for the case with $\tau = 1$.

6.34 Show that the storage and loss modulus for the three-parameter solid are given by

$$G_1 = \frac{q_o + p_1 q_1 \omega^2}{1 + p_1^2 \omega^2}, G_2 = \frac{(q_1 - q_o p_1)\omega}{1 + p_1^2 \omega^2}$$

6.35 Consider the one-dimensional integral relaxation and creep relations for a viscoelastic material with loading and deformation starting at $t = 0$:

$$T(t) = \int_0^t G(t-t') \frac{d\varepsilon(t')}{dt'} dt' \text{ , } \varepsilon(t) = \int_0^t J(t-t') \frac{dT(t')}{dt'} dt$$

Taking the Laplace transform of each of these relations, show that $\bar{G}(s)\bar{J}(s)=\frac{1}{s^2}$, and thus

$$\int_0^t G(t')J(t-t')\,dt'=t$$

Make use of property $(6.5.49)_3$ and the fact that $\mathfrak{L}^{-1}\{1/s^2\}=t$.

6.36 Determine the solution to the thick-walled tube Example 6.5.9 for the case where the viscoelastic response is governed by Kelvin-like properties such that $\bar{\mu}(s)=\mu_o+\eta s$ and $\bar{\lambda}(s)=\lambda_o+\eta s$.

6.37 Determine the solution to the Boussinesq's Problem in Example 6.5.10 for the case where the viscoelastic response is governed by a Maxwell type law such that $\mu(t)=\mu_o e^{-t/\tau}$.

6.38 The one-dimensional nonlinear stress–strain behavior of materials that do not exhibit a well-defined yield has often been modeled by the *Ramberg–Osgood relation* $\varepsilon=\frac{T}{E}\left(1+\alpha\left(\frac{T}{T_o}\right)^{m-1}\right)=\frac{T}{E}+\alpha\frac{T_o}{E}\left(\frac{T}{T_o}\right)^m$ where α and m are constants and T_o is a reference yield-like stress. Using a numerical routine (MATLAB) make a series of dimensionless stress–strain plots of T/T_o vs $E\varepsilon/T_o$ for values of $\alpha=1$ and $m=2, 3, 5, 10, 50$. Discuss the case $m\to\infty$. From your plots, what is the role of the α-parameter?

6.39 Using the fact that the volumetric strain energy may be determined by

$$U_v=\frac{1}{2}\tilde{T}_{ij}\tilde{\varepsilon}_{ij}=\frac{1}{6}T_{jj}\varepsilon_{kk}=\frac{1-2\nu}{6E}T_{jj}T_{kk}$$

explicitly justify relations (6.6.12).

6.40 Starting with relation (6.6.15) develop the invariant form of the Tresca yield condition (6.6.16).

6.41 As shown in Fig. 6.48, along line OH, $T_1=T_2=T_3$. Justify that along this line that according to both Mises and Tresca criteria, no yielding can occur.

6.42 Relation (6.6.13) expressed the Mises yield function in terms of the principal stresses. However, it is often more convenient to express this relation in terms of just the stresses. Show that this can be done with the expression $\sigma_e^2=3k^2$, where σ_e is the effective or von Mises stress defined previously in (4.5.6) as

$$\sigma_e=\sigma_{von\ Mises}=\sqrt{\frac{3}{2}\hat{T}_{ij}\hat{T}_{ij}}$$
$$=\frac{1}{\sqrt{2}}\left[(T_{11}-T_{22})^2+(T_{22}-T_{33})^2+(T_{33}-T_{11})^2+6\left(T_{12}^2+T_{23}^2+T_{31}^2\right)\right]^{1/2}$$

6.43 Consider the Levy–Mises plasticity model, for a *plane strain deformation* in the x_1,x_2-plane. For such a case, the deformation in the x_3-direction will vanish, and the only nonzero stress components are $T_{11}, T_{22}, T_{33}, T_{12}$. First show that for the case of material incompressibility, $T_{33} = (T_{11} + T_{22})/2$. Next demonstrate that both the Mises and Tresca yield conditions give the identical result $(T_{11} - T_{22})^2 + 4T_{12} = 4k^2$.

6.44 For the thick-walled tube problem in Example 6.6.1, using the elastic relations (6.6.25) justify that $T_{\theta\theta} > T_{zz} > T_{rr}$. Next verify the Tresca and Mises yield relations (6.6.26) and (6.6.29).

6.45 For the thick-walled tube problem in Example 6.6.1, use a numerical routine (MATLAB) to make a plot of the internal pressure versus plastic radius given by relation (6.6.34). Keep things dimensionless by plotting p/k vs r_p/r_2, and consider the cases of $r_1/r_2 = 0.1, 0.2, 0.3, 0.4, 0.5$. Comment on the significant differences in the pressure over the range of r_1/r_2.

6.46 For the torsion problem in Example 6.6.2, first verify the elastic solution details given by relations (6.6.43)–(6.6.46). Next show that the both Tresca and Mises criteria give the same result $\tau = T_{\theta z} = \sqrt{T_{xz}^2 + T_{yz}^2} = k$, where for the Tresca case $k = Y/2$ and for the Mises criteria $k = Y/\sqrt{3}$.

6.47 Consider the torsion of the elliptical shape. It can be shown (Sadd, 2014) that the elasticity solution to this problem results in the stress distribution

$$T_{xz} = -\frac{2a^2\mu\alpha}{a^2+b^2}y = -\frac{2Ty}{\pi ab^3}, \quad T_{yz} = \frac{2b^2\mu\alpha}{a^2+b^2}x = \frac{2Tx}{\pi ba^3}$$

This yields the resultant shear stress in the cross-section $\tau = \dfrac{2T}{\pi ab}\sqrt{\dfrac{x^2}{a^4} + \dfrac{y^2}{b^4}}$.

For the case $a > b$, the maximum value of τ occurs at $x = 0$ and $y = \pm b$. Determine the torque and angle of twist that will initiate yielding.

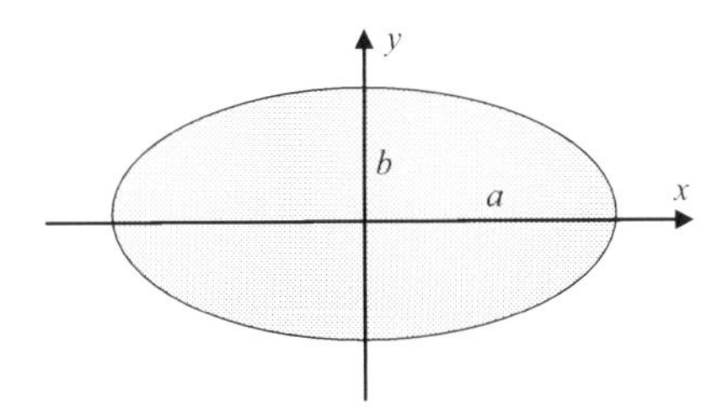

CHAPTER

Constitutive relations and formulation of theories involving multiple constitutive fields

7

7.1 INTRODUCTION

While there is a wide variety of material responses beyond the classical displacement–strain–stress field theories, we will only focus here on three of the more common theoretical models. Our additional fields will include *thermal*, *pore-fluid*, and *electrical effects*, and we will limit the mechanical stress–strain response to small deformation elasticity. This greatly simplifies the presentation and allows us to easily develop the mathematical model and explore some solutions to actual problems. Of course many more theories have been established in the literature that would include a wide variety of additional material fields with more general stress–strain responses including finite deformation.

So, we will first investigate continuum theories for *thermoelastic solids* where our previous linear elastic model described in Section 6.2 will now be coupled with thermal fields of temperature, internal energy, heat flux, and entropy. This will increase the number of unknowns in the problem formulation and thus require development of additional thermodynamic constitutive relations in order to produce closure of the system equations. This theory is very useful in stress analysis of structures and machine parts that undergo significant temperature changes from ambient. If the temperature variation is sufficiently high, the stresses can reach levels that may lead to structural failure, especially for brittle materials.

Our second multiple field theory will be *poroelasticity*, which is an elastic continuum model of a fluid-saturated porous medium. This theory will then bring in additional concepts of porosity and fluid flow, thus requiring additional relations to determine all unknown problem variables. This continuum model has many applications for the response of soil, rock, ceramics, and biological materials.

Finally, we will explore *electro-elasticity or piezoelectric materials* whereby the mechanical linear elasticity fields are coupled with an electric charge field. Again new unknown variables will be created and thus additional electrical constitutive relations will be necessary. Many applications of this theory are found in sensors, actuators, and smart materials.

Continuum Mechanics Modeling of Material Behavior. http://dx.doi.org/10.1016/B978-0-12-811474-2.00007-1

7.2 THERMOELASTIC SOLIDS

Thermal effects within an elastic solid produce heat transfer via conduction, and this flow of thermal energy establishes a temperature field within the material. Most solids exhibit a deformation change with temperature variation, and thus the presence of a temperature distribution will generally induce stresses created from boundary or internal constraints. The continuum model will now require the energy equation, and with this coupling the unknowns of internal energy, heat flow, and temperature will now be present. Several new thermal constitutive laws will be needed to relate these new variables with various model unknowns. This will require some additional discussions on a few thermodynamic concepts that we have previously avoided. Starting with general elastic materials, we will develop the basic governing thermoelastic equations for isotropic materials and will investigate a few solutions to problems of engineering interest. As usual, our presentation will be brief, and more detailed information may be found in several studies devoted entirely to the subject such as Boley and Weiner (1960), and Nowinski (1978) and (1986), Kovalenko (1969).

Going back to our discussion on thermodynamics in Sections 5.6 and 5.7, the sum of the internal energy ε and irreversible heat energy $-\theta s$ is known as the *Helmholtz free energy density* Ψ:

$$\Psi = \varepsilon - s\theta \tag{7.2.1}$$

For elastic materials, we can assume that the free energy is only a function of the strain and temperature, $\Psi = \Psi(\varepsilon_{ij}, \theta)$. Note that we expect that the internal energy and entropy would also have the same functional form. Under this condition, the time rate of change of Ψ can be expressed using the chain rule

$$\dot{\Psi} = \left(\frac{\partial \Psi}{\partial \varepsilon_{ij}}\right)_{\theta} \dot{\varepsilon}_{ij} + \left(\frac{\partial \Psi}{\partial \theta}\right)_{\varepsilon_{ij}} \dot{\theta} \tag{7.2.2}$$

The energy equation (5.6.12) can be expressed by

$$\rho\dot{\varepsilon} = \rho(\dot{\Psi} + \dot{s}\theta + s\dot{\theta}) = T_{ij}\dot{\varepsilon}_{ij} - q_{i,i} + \rho h \tag{7.2.3}$$

Solving for $\dot{\Psi}$, we obtain

$$\rho\dot{\Psi} = T_{ij}\dot{\varepsilon}_{ij} - q_{i,i} + \rho h - \rho(\dot{s}\theta + s\dot{\theta}) \tag{7.2.4}$$

Substituting (7.2.2) into (7.2.4) gives

$$\left(\rho\frac{\partial \Psi}{\partial \varepsilon_{ij}} - T_{ij}\right)\dot{\varepsilon}_{ij} + \rho\left(\frac{\partial \Psi}{\partial \theta} + s\right)\dot{\theta} + q_{i,i} - \rho(h - \dot{s}\theta) = 0 \tag{7.2.5}$$

In order for relation (7.25) to be true for all $\dot{\varepsilon}_{ij}$ and $\dot{\theta}$, the following must be true:

$$\begin{aligned} T_{ij} &= \rho\frac{\partial \Psi}{\partial \varepsilon_{ij}} \\ s &= -\frac{\partial \Psi}{\partial \theta} \\ q_{i,i} &= \rho(h - \dot{s}\theta \end{aligned} \tag{7.2.6}$$

Eqs. $(7.2.6)_{1,2}$ are often referred to as the *thermodynamic potential relations*, and $(7.2.6)_1$ should be compared with the nonthermal hyperelastic equation (6.2.6) $T_{ij} = \partial U / \partial \varepsilon_{ij}$. Also, as previously discussed in Section 6.2, for linear elasticity, the mass density is a constant.

The free energy is a thermodynamic potential that is a function of the strain and temperature fields. We can define Ψ_0 as *the natural or reference state* with zero strain when the temperature is at θ_o. Using small strain theory and small temperature differences from reference $\Theta = \theta - \theta_o$, we can express the free energy potential per unit volume as a power series in the two variables ε_{ij} and Θ:

$$\rho\Psi(\varepsilon,\Theta) = \rho\Psi_0 + \frac{1}{2} C_{ijkl}\varepsilon_{ij}\varepsilon_{kl} - \beta_{ij}\varepsilon_{ij}\Theta - \frac{c_v}{2\theta_o}\Theta^2 \tag{7.2.7}$$

where C_{ijkl} is a general fourth-order tensor (later shown to be the elasticity tensor) characterizing the mechanical properties of the material, β_{ij} are material constants which account for the coupling of the strain and temperature fields, and c_v is the thermal constant called the *specific heat per unit volume at constant strain*. Note that terms linear in the strain and temperature have been dropped, since they would violate the minimum requirement of the free energy in the natural state.

Using the constitutive from (7.2.7) with properties $(7.2.6)_{1,2}$ yields

$$\begin{aligned} T_{ij} &= C_{ijkl}\varepsilon_{kl} - \beta_{ij}\Theta \\ s &= \beta_{ij}\varepsilon_{ij} + \frac{c_v}{\theta_o}\Theta \end{aligned} \tag{7.2.8}$$

Constitutive law $(7.2.8)_1$ is known as the *Duhamel–Neumann relation* for linear thermoelastic solids, while $(7.2.8)_2$ defines entropy. It can be seen that in the natural state both the stresses and entropy vanish.

Heat transfer in solids normally occurs by means of *conduction*, transferring heat from regions with higher temperature to regions of lower temperature. This process is spontaneous and irreversible and is connected with an increase in entropy. The temperature gradient is thus taken to be a thermodynamic force which causes heat to flow, and thus there must exist a conduction constitutive law which relates the heat flux vector to the temperature gradient. This relation is known as *Fourier's Law* and is given by

$$q_i = -k_{ij}\Theta,_j, \quad \boldsymbol{q} = -\boldsymbol{k}\nabla\Theta \tag{7.2.9}$$

where k_{ij} is the *conductivity tensor*. The minus sign is used to indicate heat flow from hot to cold. Using this conduction relation, we can write

$$q_{i,i} - q_{j,j} = 0 \Rightarrow -(k_{ij} - k_{ji})\Theta_{,ij} \Rightarrow k_{ij} = k_{ji} \tag{7.2.10}$$

and thus the conductivity tensor is symmetric. Furthermore, using the relation (5.7.12), we obtain

$$\boldsymbol{q}\cdot\nabla\Theta \le 0 \Rightarrow k_{ij}\Theta_{,i}\Theta_{,j} \ge 0 \tag{7.2.11}$$

and so the conductivity tensor is also positive definite (see Exercise 2.20). This property indicates that every principal minor of $\det(k_{ij})$ is greater than zero, which leads to the result

$$k_{\underline{ii}} \ge 0, \quad k_{\underline{ii}}k_{\underline{jj}} \ge k_{ij}^2 \tag{7.2.12}$$

Using the Fourier conduction law (7.2.9) and $(7.2.8)_2$, in the energy equation $(7.2.6)_3$ gives

$$-k_{ij}\Theta,_{ij}=\rho h-\beta_{ij}\dot{\varepsilon}_{ij}\theta-\frac{c_v}{\theta_o}\dot{\Theta}\theta$$

Linearizing this relation, by letting $\theta=\theta_o$ then produces the *linearized energy equation*

$$k_{ij}\Theta,_{ij}-c_v\dot{\Theta}-\beta_{ij}\theta_o\dot{\varepsilon}_{ij}+\rho h=0 \tag{7.2.13}$$

The third term $\beta_{ij}\theta_o\dot{\varepsilon}_{ij}$ represents a *coupling term* between the thermal and strain fields. It has been shown that this term is normally quite small and may be neglected for most applications where the strain rates are small (see Boley and Weiner, 1960). Under this assumption, the theory becomes *uncoupled* and the energy equation reduces to the *heat equation*

$$k_{ij}\Theta,_j-c_v\dot{\Theta}+\rho h=0 \tag{7.2.14}$$

which can be solve independently for the temperature distribution.

For the isotropic material case, the various material tensors reduce to

$$\begin{aligned} C_{ijkl}&=\lambda\delta_{ij}\delta_{kl}+\mu(\delta_{ik}\delta_{jl}+\delta_{il}\delta_{jk})\\ k_{ij}&=k_T\delta_{ij},\quad \beta_{ij}=(3\lambda+2\mu)\alpha_T\delta_{ij}\end{aligned} \tag{7.2.15}$$

where λ and μ are the Lamé and shear moduli, respectively, α_T is the coefficient of thermal expansion, and k_T is the thermal conductivity constant. So, the isotropic thermoelastic constitutive relations become

$$\begin{aligned} T_{ij}&=\lambda\varepsilon_{kk}\delta_{ij}+2\mu\varepsilon_{ij}-(3\lambda+2\mu)\alpha_T\delta_{ij}\Theta\\ s&=(3\lambda+2\mu)\alpha_T\varepsilon_{kk}+\frac{c_v}{\theta_o}\Theta\\ q_i&=-k_T\Theta_{,i}\end{aligned} \tag{7.2.16}$$

Hooke's law $(7.2.16)_1$ can easily be inverted to express the strain in terms of stress

$$\varepsilon_{ij}=\frac{1+\nu}{E}T_{ij}-\frac{\nu}{E}T_{kk}\delta_{ij}+\alpha_T\Theta\delta_{ij} \tag{7.2.17}$$

Notice that Hooke's law in this form implies an additive decomposition of mechanical and thermal strains $\varepsilon_{ij}=\varepsilon_{ij}^{mech}+\varepsilon_{ij}^{thermal}$, which is an expected result from a linear theory.

Our thermodynamic presentation here was only for the linear elastic material model. Other presentations have been given for more general materials including finite deformation and electro-magneto-mechanical materials; see Asaro and Lubarda (2006), Holzapfel (2006), and Bechtel and Lowe (2015) for further details.

7.2.1 GENERAL FORMULATION

Some parts of the ensuing presentation will be identical to the isothermal formulation, whereas other results create new terms or equations. We will pay special attention to these new contributions and recognize them in the field equations and boundary conditions. Following our common pattern, we present a general formulation of isotropic uncoupled linear thermoelasticity necessary for problem solution. Collecting the previous constitutive relations and adding in the other usual relations for small deformation theory then give the following list of governing field equations:

$$\text{Strain–displacement relations:}\quad \varepsilon_{ij} = \frac{1}{2}(u_{i,j} + u_{j,i}) \tag{7.2.18}$$

$$\text{Compatibility relations:}\quad \varepsilon_{ij,kl} + \varepsilon_{kl,ij} - \varepsilon_{ik,jl} - \varepsilon_{jl,ik} = 0 \tag{7.2.19}$$

$$\text{Equilibrium equations:}\quad T_{ij,j} + F_i = 0, \quad F_i = \rho b_i \tag{7.2.20}$$

$$\text{Energy equation:}\quad k_T \Theta,_{kk} - c_v \dot{\Theta} + \rho h = 0 \tag{7.2.21}$$

$$\text{Constitutive law:}\quad \begin{aligned} T_{ij} &= (\lambda + \mu)\varepsilon_{kk}\delta_{ij} + 2\mu\varepsilon_{ij} - (3\lambda + 2\mu)\alpha_{\mathrm{T}}\Theta\delta_{ij} \\ \varepsilon_{ij} &= \frac{1+\nu}{E}T_{ij} - \frac{\nu}{E}T_{kk}\delta_{ij} + \alpha_{\mathrm{T}}\Theta\delta_{ij} \end{aligned} \tag{7.2.22}$$

These 16 relations make up the fundamental set of field equations for the 16 unknowns u_i, ε_{ij}, T_{ij}, and Θ. All material constants, body forces, and any thermal source terms are assumed to be given *a priori*. Recall that the compatibility equations are used for the stress formulation in which displacements and strains are eliminated. Notice that these equations are very similar to those in the linear elasticity model given by relations (6.2.28)–(6.2.31). Here, we have temperature terms in Hooke's law and the energy equation. For the uncoupled case, the energy equation can be solved for the temperature distribution *independent of the stress field calculations*. Once determined, the temperatures can then be appropriately placed in the remaining field equations. As discussed in Section 6.2 for isothermal elasticity, the basic problem types and boundary condition forms can be categorized in the same way. However, we must also add thermal boundary conditions of temperature or heat flux specification to solve the heat equation for the temperature distribution.

As previously done for the isothermal case, it will prove to be helpful for problem solution to further reduce the general thermoelastic governing field equations to either a *stress* or *displacement formulation*. Details are similar to those outlined in Section 6.2.2, and we only give the final results

$$T_{ij,kk} + \frac{1}{1+\nu}T_{kk,ij} + E\alpha_{\mathrm{T}}\left(\frac{1}{1-\nu}\Theta_{,kk}\delta_{ij} + \frac{1}{1+\nu}\Theta_{,ij}\right) = -\frac{\nu}{1-\nu}\delta_{ij}F_{k,k} - F_{i,j} - F_{j,i} \tag{7.2.23}$$

$$\mu u_{i,kk} + (\lambda + \mu)u_{k,ki} - (3\lambda + 2\mu)\,\alpha_{\mathrm{T}}\Theta_{,i} + F_i = 0 \tag{7.2.24}$$

where (7.2.23) are the compatibility equations in terms of stress and temperature, whereas (7.2.24) are the equilibrium equations in terms of displacement and temperature. It should be noted that in addition to Hooke's law and the energy equation, the temperature field will also be present in any traction boundary conditions since tractions are related to the stresses. Although other general formulation concepts could be presented, we will end this discussion and move on to the solution of a few basic thermoelastic problems.

7.2.2 PROBLEM SOLUTIONS

We now will explore the solution to some thermoelastic boundary value problems for the two-dimensional uncoupled steady-state case. We shall further assume that body forces and any thermal source terms are zero. The basic two-dimensional formulation follows in a similar fashion as done previously for the isothermal problems in Section 6.2.3.

7.2.2.1 Cartesian coordinate formulation

Again we focus only on the two-dimensional (x,y) *plane stress* case with in-plane stresses being functions of in-plane coordinates:

$$T_{xx} = T_{xx}(x,y), \quad T_{yy} = T_{yy}(x,y), \quad T_{xy} = T_{xy}(x,y), \quad T_{zz} = T_{xz} = T_{yz} = 0 \tag{7.2.25}$$

This particular assumption is most applicable to bodies thin in the out-of-plane direction (see Fig. 6.10) and loaded only with in-plane forces. Likewise, the temperature field is also assumed to be two-dimensional $\Theta = \Theta(x,y)$.

Under these conditions, the strains come from Hooke's law

$$\begin{aligned} &\varepsilon_x = \frac{1}{E}(T_{xx} - \nu T_{yy}) + \alpha_T \Theta, \quad \varepsilon_y = \frac{1}{E}(T_{yy} - \nu T_{xx}) + \alpha_T \Theta \\ &\varepsilon_{xy} = \frac{1+\nu}{E} T_{xy}, \quad \varepsilon_z = -\frac{\nu}{E}(T_{xx} + T_{yy}) + \alpha_T \Theta, \quad \varepsilon_{xz} = \varepsilon_{yz} = 0 \end{aligned} \tag{7.2.26}$$

The equilibrium equations reduce to

$$\begin{aligned} &\frac{\partial T_{xx}}{\partial x} + \frac{\partial T_{xy}}{\partial y} = 0 \\ &\frac{\partial T_{xy}}{\partial x} + \frac{\partial T_{yy}}{\partial y} = 0 \end{aligned} \tag{7.2.27}$$

while the strain compatibility relations condense to the single relation

$$\frac{\partial^2 \varepsilon_x}{\partial y^2} + \frac{\partial^2 \varepsilon_y}{\partial x^2} = 2\frac{\partial^2 \varepsilon_{xy}}{\partial x \partial y} \tag{7.2.28}$$

The displacement equilibrium equations (7.2.24) become

$$\begin{aligned} &\mu \nabla^2 u + \frac{E}{2(1-\nu)}\frac{\partial}{\partial x}\left(\frac{\partial u}{\partial x} + \frac{\partial v}{\partial y}\right) - \frac{E\alpha_T}{1-\nu}\frac{\partial \Theta}{\partial x} = 0 \\ &\mu \nabla^2 v + \frac{E}{2(1-\nu)}\frac{\partial}{\partial y}\left(\frac{\partial u}{\partial x} + \frac{\partial v}{\partial y}\right) - \frac{E\alpha_T}{1-\nu}\frac{\partial \Theta}{\partial y} = 0 \end{aligned} \tag{7.2.29}$$

and the compatibility relations in terms of stress reduce to

$$\nabla^2(T_{xx} + T_{yy}) + E\alpha_T \nabla^2 \Theta = 0 \tag{7.2.30}$$

The tractions can be expressed by

$$
\begin{aligned}
t_x^n &= T_{xx}n_x + T_{xy}n_y = \left[\frac{E}{1-\nu^2}\left(\frac{\partial v}{\partial x}+\nu\frac{\partial u}{\partial y}\right)-\frac{E\alpha_T}{1-\nu}\Theta\right]n_x + \left[\frac{E}{2(1+\nu)}\left(\frac{\partial u}{\partial y}+\frac{\partial v}{\partial x}\right)\right]n_y \\
t_y^n &= T_{xy}n_x + T_{yy}n_y = \left[\frac{E}{2(1+\nu)}\left(\frac{\partial u}{\partial y}+\frac{\partial v}{\partial x}\right)\right]n_x + \left[\frac{E}{1-\nu^2}\left(\frac{\partial v}{\partial y}+\nu\frac{\partial u}{\partial x}\right)-\frac{E\alpha_T}{1-\nu}\Theta\right]n_y
\end{aligned}
\tag{7.2.31}
$$

For the steady-state case, the energy or heat equation reduces to the Laplace equation

$$
\frac{\partial^2\Theta}{\partial x^2}+\frac{\partial^2\Theta}{\partial y^2}=\nabla^2\Theta=0 \tag{7.2.32}
$$

It is observed that for thermal plane stress, the temperature effect is equivalent to adding an additional body force $-\frac{E\alpha_T}{1-\nu}\nabla\Theta$ to Navier's equations (7.2.29) and adding a traction term $-\frac{E\alpha_T}{1-\nu}\Theta\boldsymbol{n}$ to the applied boundary tractions (7.2.31). This concept can be generalized to three-dimensional theory.

The problem is then formulated in a two-dimensional region R in the x,y-plane. The energy equation (7.2.32) provides the governing equation for the steady temperature field with boundary conditions specifying either temperature or heat flux over the boundary of R. The displacement formulation is specified by relations (7.2.29) with displacement boundary conditions on u and v. The stress formulation includes the three governing equations (7.2.27) and (7.2.30) with appropriate stress or traction boundary conditions. The solution then involves the determination of the temperature and in-plane displacements, strains, and stresses $\{\Theta, u, v, \varepsilon_x, \varepsilon_y, \varepsilon_{xy}, T_{xx}, T_{yy}, T_{xy}\}$ in R. Out-of-plane strain ε_z can be determined from the in-plane stresses using Hooke's law (7.2.26).

Similar to isothermal elasticity, introducing the Airy stress function approach will greatly help in finding analytical solutions to problems of interest. Thus, we again use the representation

$$
T_{xx}=\frac{\partial^2\phi}{\partial y^2},\quad T_{yy}=\frac{\partial^2\phi}{\partial x^2},\quad T_{xy}=-\frac{\partial^2\phi}{\partial x\partial y} \tag{7.2.33}
$$

which satisfies the equilibrium equations identically, and the compatibility relation (7.2.30) becomes

$$
\frac{\partial^4\phi}{\partial x^4}+2\frac{\partial^4\phi}{\partial x^2\partial y^2}+\frac{\partial^4\phi}{\partial y^4}+E\alpha_T\nabla^2\Theta=\nabla^4\phi+E\alpha_T\nabla^2\Theta=0 \tag{7.2.34}
$$

Thus, the plane stress problem of linear thermoelasticity has been reduced to a single equation in terms of the Airy stress function, and this will allow solutions to be found for a large variety of problems. Note that this approach can also be used to formulate and solve the other two-dimensional deformation model *plane strain* (see Sadd, 2014).

7.2.2.2 Polar coordinate formulation

It is often desirable to formulate the two-dimensional thermoelastic problem in polar coordinates, thus allowing solutions with radial or angular symmetry to be easily solved. Example 2.18.1 and Appendix A provide help to determine many of the needed relations. For the plane stress case,

$$T_{rr} = T_{rr}(r,\theta),\quad T_{\theta\theta} = T_{\theta\theta}(r,\theta),\quad T_{r\theta} = T_{r\theta}(r,\theta),\quad T_{zz} = T_{rz} = T_{\theta z} = 0 \tag{7.2.35}$$

The in-plane displacements in the radial and tangential directions are given by u_r and u_θ, and all field quantities will depend only on the in-plane coordinates r and θ. The strain displacement relations become

$$\begin{gathered}\varepsilon_r = \frac{\partial u_r}{\partial r},\quad \varepsilon_\theta = \frac{u_r}{r} + \frac{1}{r}\frac{\partial u_\theta}{\partial \theta}\\ \varepsilon_{r\theta} = \frac{1}{2}\left(\frac{1}{r}\frac{\partial u_r}{\partial \theta} + \frac{\partial u_\theta}{\partial r} - \frac{u_\theta}{r}\right)\end{gathered} \tag{7.2.36}$$

Hooke's law now reads

$$\begin{gathered}T_{rr} = \frac{E}{1-v^2}[\varepsilon_r + v\varepsilon_\theta - (1+v)\,\alpha_T(\Theta-\Theta_o)]\\ T_{\theta\theta} = \frac{E}{1-v^2}[\varepsilon_\theta + v\varepsilon_r - (1+v)\,\alpha_T(\Theta-\Theta_o)]\\ T_{r\theta} = \frac{E}{1+v}\varepsilon_{r\theta}\end{gathered} \tag{7.2.37}$$

In the absence of body forces, the equilibrium equations reduce to

$$\begin{gathered}\frac{\partial T_{rr}}{\partial r} + \frac{1}{r}\frac{\partial T_{r\theta}}{\partial \theta} + \frac{T_{rr} - T_{\theta\theta}}{r} = 0\\ \frac{\partial T_{r\theta}}{\partial r} + \frac{1}{r}\frac{\partial T_{\theta\theta}}{\partial \theta} + \frac{2T_{r\theta}}{r} = 0\end{gathered} \tag{7.2.38}$$

The Airy stress function definition now becomes

$$\begin{gathered}T_{rr} = \frac{1}{r}\frac{\partial \phi}{\partial r} + \frac{1}{r^2}\frac{\partial^2 \phi}{\partial \theta^2}\\ T_{\theta\theta} = \frac{\partial^2 \phi}{\partial r^2}\\ T_{r\theta} = -\frac{\partial}{\partial r}\left(\frac{1}{r}\frac{\partial \phi}{\partial \theta}\right)\end{gathered} \tag{7.2.39}$$

which again satisfies (7.2.38) identically. The governing stress function equation is still given by

$$\nabla^4\phi + E\alpha_T\nabla^2\Theta = 0 \tag{7.2.40}$$

with the Laplacian and biharmonic operators now of the form

$$\begin{gathered}\nabla^2 = \frac{\partial^2}{\partial r^2} + \frac{1}{r}\frac{\partial}{\partial r} + \frac{1}{r^2}\frac{\partial^2}{\partial \theta^2}\\ \nabla^4 = \nabla^2\nabla^2 = \left(\frac{\partial^2}{\partial r^2} + \frac{1}{r}\frac{\partial}{\partial r} + \frac{1}{r^2}\frac{\partial^2}{\partial \theta^2}\right)\left(\frac{\partial^2}{\partial r^2} + \frac{1}{r}\frac{\partial}{\partial r} + \frac{1}{r^2}\frac{\partial^2}{\partial \theta^2}\right)\end{gathered} \tag{7.2.41}$$

This now completes our two-dimensional formulation and we now move on to explore several solutions.

EXAMPLE 7.2.1 THERMAL STRESSES IN A RECTANGULAR STRIP

Consider the thermoelastic problem in a rectangular domain as shown in Fig. 7.1. We will assume that the vertical dimension is unbounded and thus the region may be described as an infinite strip of material. For this problem, assume that the temperature is independent of x and given by $\Theta = T_o \sin\beta y$, where T_o and β are constants. Note that $\nabla^2\Theta = 0$ and thus the given temperature already satisfies the heat equation. If the stresses at $x = \pm a$ are to vanish, determine the thermal plane stress field.

Solution: From the governing equation (7.2.34) $\Rightarrow$

$$\nabla^4\phi = E\alpha_T T_o \beta^2 \sin\beta y \tag{7.2.42}$$

A particular solution to this equation is given by $\phi_p = \dfrac{E\alpha_T T_o}{\beta^2}\sin\beta y$. The homogeneous solution may be found using standard separation of variables using $\phi_h = f(x)\sin\beta y$ which gives the auxiliary ordinary differential equation $f'''' - 2\beta^2 f'' + \beta^4 f = 0$ with general solution

$$f = C_1 \sinh\beta x + C_2 \cosh\beta x + C_3 x \sinh\beta x + C_4 x \cosh\beta x$$

where C_i are constants to be determined. Now since the temperature field was symmetric in x, we expect the stresses to also exhibit the same symmetry. Thus, the stress function must also be symmetric in x and so $C_1 = C_4 = 0$. Combining the particular and homogeneous solutions, the resulting stresses

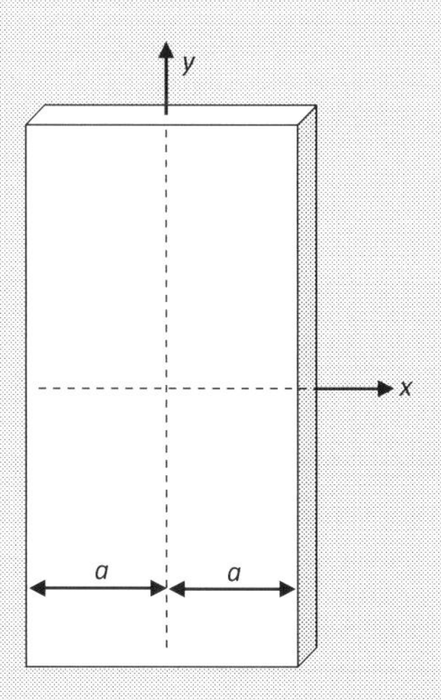

FIGURE 7.1

Thermoelastic rectangular strip problem.

then become

$$T_{xx} = -\beta^2[C_2 \cosh\beta x + C_3 x \sinh\beta x]\sin\beta y - E\alpha_T T_0 \sin\beta y$$
$$T_{yy} = \beta^2\left[C_2 \cosh\beta x + C_3\left(x\sinh\beta x + \frac{2}{\beta}\cosh\beta x\right)\right]\sin\beta y$$
$$T_{xy} = -\beta^2\left[C_2 \sinh\beta x + C_3\left(x\cosh\beta x + \frac{1}{\beta}\sinh\beta x\right)\right]\cos\beta y \qquad (7.2.43)$$

Applying the stress-free boundary conditions $T_{xx}(\pm a, y) = T_{xy}(\pm a, y) = 0 \Rightarrow$

$$C_2 \cosh\beta a + C_3 a \sinh\beta a = -\frac{E\alpha_T T_0}{\beta^2}$$
$$C_2 \sinh\beta a + C_3\left(a\cosh\beta a + \frac{1}{\beta}\sinh\beta a\right) = 0$$

Solving for the two remaining constants then gives

$$C_2 = -\frac{E\alpha_T T_0(a\beta\cosh\beta a + \sinh\beta a)}{\beta^2(a\beta + \sinh\beta a\cosh\beta a)}, \quad C_3 = \frac{E\alpha_T T_0 \sinh\beta a}{\beta(a\beta + \sinh\beta a \cosh\beta a} \qquad (7.2.44)$$

Note that by using superposition and Fourier methods, we could generate a more general temperature field and the corresponding stress solution.

EXAMPLE 7.2.2 AXISYMMETRIC PROBLEM SOLUTIONS

Determine the axisymmetric stress and displacement solutions for the general thermoelastic plane stress problem where all field quantities are functions only of the radial coordinate r.

Solution: For this solution class, we choose the following field variable forms:

$$T_{rr} = T_{rr}(r), \quad T_{\theta\theta} = T_{\theta\theta}(r), \quad T_{r\theta} = T_{r\theta}(r), \quad \Theta = \Theta(r), \quad \phi = \phi(r) \qquad (7.2.45)$$

Relations (7.2.39) then become

$$T_{rr} = \frac{1}{r}\frac{d\phi}{dr}, \quad T_{\theta\theta} = \frac{d^2\phi}{dr^2} = \frac{d}{dr}(rT_{rr}), \quad T_{r\theta} = 0 \qquad (7.2.46)$$

The governing equation in terms of the stress function (7.2.40) simplifies to

$$\frac{1}{r}\frac{d}{dr}\left\{r\frac{d}{dr}\left[\frac{1}{r}\frac{d}{dr}\left(r\frac{d\phi}{dr}\right)\right]\right\} + E\alpha_T\frac{1}{r}\frac{d}{dr}\left(r\frac{d\Theta}{dr}\right) = 0 \qquad (7.2.47)$$

This relation can be recast in terms of the radial stress by using $(7.2.46)_1$ giving the result

$$\frac{1}{r}\frac{d}{dr}\left\{r\frac{d}{dr}\left[\frac{1}{r}\frac{d}{dr}\left(r^2T_{rr}\right)\right]\right\}=-E\alpha_T\frac{1}{r}\frac{d}{dr}\left(r\frac{d\Theta}{dr}\right) \tag{7.2.48}$$

and the equation can be directly integrated to give

$$T_{rr}=\frac{C_3}{r^2}+C_2+\frac{C_1}{4}(2\log r-1)-\frac{E\alpha_T}{r^2}\int\Theta r\,dr \tag{7.2.49}$$

The constants of integration C_i are normally determined from the boundary conditions, and the temperature Θ appearing in the integral is again the temperature difference from the reference state. Using the boundedness condition, constants C_1 and C_3 must be set to zero for domains that include the origin. Combining this result with $(7.2.46)_2$ gives the general solution for the stress field.

Next, consider the displacement formulation and solution. For the axisymmetric case, $u_r=u(r)$ and $u_\theta=0$. Going back to the equilibrium equations (7.2.38), it is observed that the second equation vanishes identically. Using Hooke's law and the strain displacement relations in the first equilibrium equation gives

$$\frac{d}{dr}\left[\frac{1}{r}\frac{d}{dr}(ru)\right]=(1+\nu)\alpha_T\frac{d\Theta}{dr} \tag{7.2.50}$$

This equation can again be directly integrated giving the displacement solution

$$u=A_1r+\frac{A_2}{r}+\frac{(1+\nu)\alpha_T}{r}\int\Theta r\,dr \tag{7.2.51}$$

The constants of integration A_i are again determined from the boundary conditions. The general displacement solution (7.2.51) can then be used to determine the strains from relations (7.2.36) and stresses from Hooke's law (7.2.37). It is found that the stresses developed from this displacement solution do not contain the logarithmic term found in relation (7.2.49). Thus, the logarithmic term is considered inconsistent with a continuous single-valued displacement field and is commonly dropped for most but not all problem solutions.

EXAMPLE 7.2.3 CIRCULAR THERMOELASTIC PLATE PROBLEM

Determine the axisymmetric thermal stresses in an annular circular plate as shown in Fig. 7.2. Assume that the inner and outer boundaries are stress-free, while the temperature boundary conditions are $\Theta(r_i)=\theta_i, \Theta(r_o)=0$. Also examine the case $r_i \to 0$ and determine the solid plate solution with thermal boundary condition $\Theta(r_o)=\theta_o$.

Solution: The general stress solution was given in Eq. (7.2.49), and thus (dropping the log-term) we have

$$T_{rr}=\frac{C_3}{r^2}+C_2-\frac{E\alpha_T}{r^2}\int \Theta r\,dr \tag{7.2.52}$$

Using the stress-free boundary conditions $T_{rr}(r_i)=T_{rr}(r_o)=0$ determines the two constants C_2 and C_3. Incorporating these results, the stresses become

$$\begin{aligned} T_{rr}&=\frac{E\alpha_T}{r^2}\left\{\frac{r^2-r_i^2}{r_o^2-r_i^2}\int_{r_i}^{r_o}\Theta r\,dr-\int_{r_i}^{r}\Theta r\,dr\right\}\\ T_{\theta\theta}&=\frac{E\alpha_T}{r^2}\left\{\frac{r^2+r_i^2}{r_o^2-r_i^2}\int_{r_i}^{r_o}\Theta r\,dr+\int_{r_i}^{r}\Theta r\,dr-\Theta r^2\right\}\end{aligned} \tag{7.2.53}$$

and the corresponding displacement solution is given by

$$u=\frac{\alpha_T}{r}\left\{(1+\nu)\int_{r_i}^{r}\Theta r\,dr+\frac{(1-\nu)r^2+(1+\nu)r_i^2}{r_o^2-r_i^2}\int_{r_i}^{r_o}\Theta r\,dr\right\} \tag{7.2.54}$$

In order to explicitly determine the stress and displacement fields, the temperature distribution must be determined. As mentioned, this

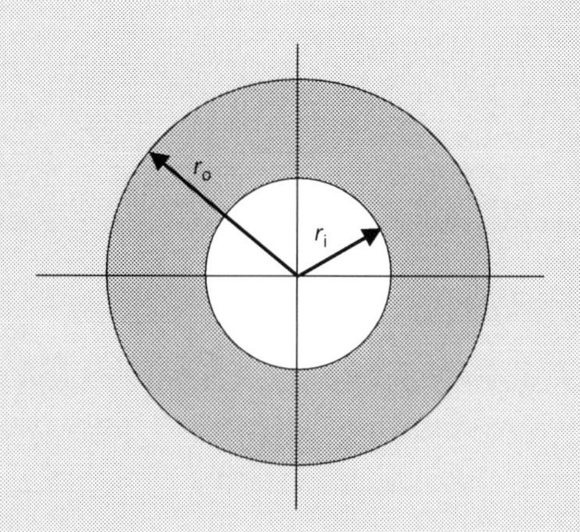

FIGURE 7.2

Annular plate geometry.

is calculated from the energy or heat conduction equation, and for the steady-state case was given by (7.2.32). For the axisymmetric problem, this reduces to

$$\frac{1}{r}\frac{d}{dr}\left(r\frac{d\Theta}{dr}\right)=0 \tag{7.2.55}$$

This equation can be integrated directly giving the solution

$$\Theta = A_1 \log r + A_2 \tag{7.2.56}$$

Using the given thermal boundary conditions $\Theta(r_i)=\theta_i, \Theta(r_o)=0$, the temperature solution is obtained as

$$\Theta = \frac{\theta_i}{\log\left(\frac{r_i}{r_o}\right)}\log\left(\frac{r}{r_o}\right) = \frac{\theta_i}{\log\left(\frac{r_o}{r_i}\right)}\log\left(\frac{r_o}{r}\right) \tag{7.2.57}$$

For the case $\theta_i > 0$, this distribution is shown in Fig. 7.3.

Substituting this temperature distribution into the stress solution (7.2.53) gives

$$\begin{aligned} T_{rr} &= \frac{E\alpha_T\theta_i}{2\log(r_o/r_i)}\left\{-\log\left(\frac{r_o}{r}\right)-\frac{r_i^2}{r_o^2-r_i^2}\left(1-\frac{r_o^2}{r^2}\right)\log\left(\frac{r_o}{r_i}\right)\right\} \\ T_{\theta\theta} &= \frac{E\alpha_T\theta_i}{2\log(r_o/r_i)}\left\{1-\log\left(\frac{r_o}{r}\right)-\frac{r_i^2}{r_o^2-r_i^2}\left(1+\frac{r_o^2}{r^2}\right)\log\left(\frac{r_o}{r_i}\right)\right\} \end{aligned} \tag{7.2.58}$$

When $\theta_i > 0$ and $T_{rr} < 0$, the hoop stress $T_{\theta\theta}$ takes on maximum values at the inner and outer boundaries of the plate. For the specific case $r_o/r_i = 3$, the stress distribution is illustrated in Fig. 7.4. MATLAB Code C-17 was used for the calculations and plotting.

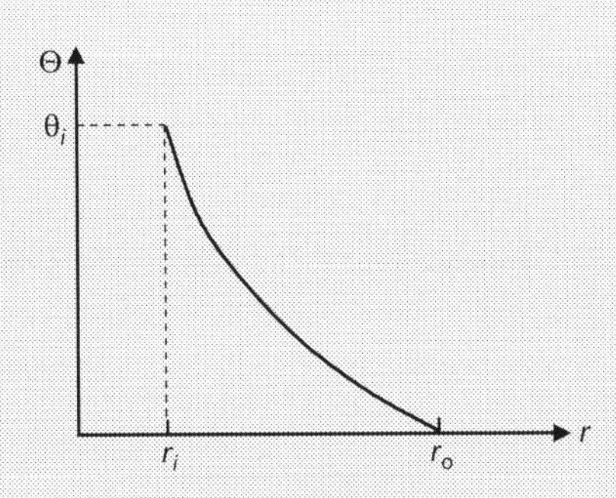

FIGURE 7.3

Temperature distribution in annual plate.

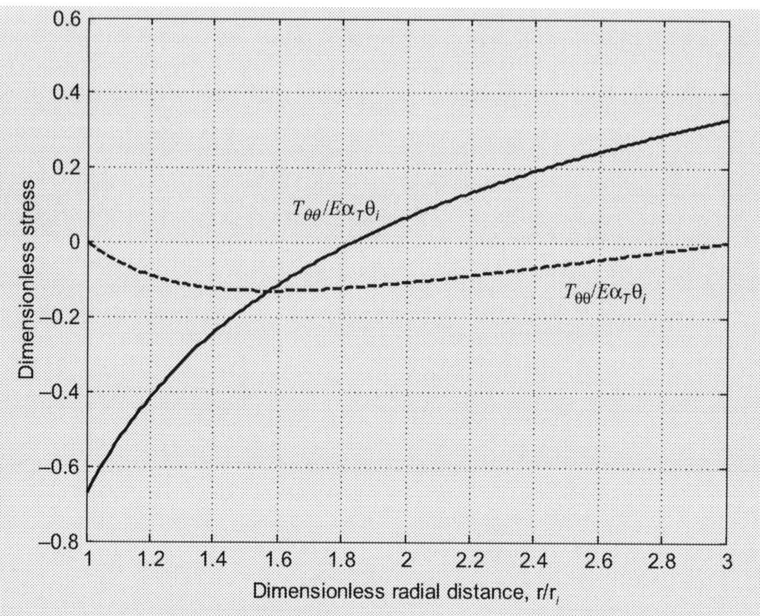

FIGURE 7.4

Thermoelastic stress distribution in annular plate.

Finally, by allowing the inner radius r_i to zero, we obtain the solution for a solid circular plate. For this case, the constant C_3 in solution (7.2.52) must be set to zero for finite stresses at the origin. The resulting stress field for zero boundary tractions becomes

$$
\begin{aligned}
T_{rr} &= E\alpha_{\mathrm{T}}\left\{\frac{1}{r_o^2}\int_0^{r_o}\Theta r\,dr-\frac{1}{r^2}\int_0^{r}\Theta r\,dr\right\}\\
T_{\theta\theta} &= E\alpha_{\mathrm{T}}\left\{\frac{1}{r_o^2}\int_0^{r_o}\Theta r\,dr+\frac{1}{r^2}\int_0^{r}\Theta r\,dr-\Theta\right\}
\end{aligned}
\tag{7.2.59}
$$

The behavior of the integral term $\frac{1}{r^2}\int_0^r \Theta r\,dr$ at the origin can be investigated using l'Hospital's rule, and it can be shown that $\lim_{r\to 0}\left(\frac{1}{r^2}\int_0^r \Theta r\,dr\right)=\frac{1}{2}\Theta(0)$, thus indicating bounded behavior at the plate's center. Using temperature boundary condition $\Theta(r_o) = \theta_o$, the general solution (7.2.55) predicts a uniform temperature $\Theta = \theta_o$ throughout the entire plate. For this case, relations (7.2.58) give $T_{rr} = T_{\theta\theta} = 0$, and thus the plate is stress-free. This particular result verifies the fact that a steady temperature distribution in a simply connected region with zero traction boundary conditions gives rise to zero stresses.

7.3 POROELASTICITY

Considerable applications of continuum mechanics have been made for porous materials containing an internal connected network of pore space that contains a viscous pore fluid. Examples of such materials include soils, rock, foams, powders, ceramics, and biological substances. Fig. 6.3 illustrated a few such material types and a more detailed photograph of a sponge sample is shown in Fig. 7.5. The usual density functions used in classical continuum mechanics must be modified to account for such microstructures. In order to apply continuum mechanics to such materials, it is normally assumed that the pore structures are uniformly distributed within the material and occur at length scales at least a few orders of magnitude less than that of the basic problem. This will then allow many of the usual classical methods of formulation to be extended to such materials. This concept is related to our discussion back in Section 1.1 on the definition of the representative volume element.

We consider the case of a porous isotropic linear elastic material in which pore fluid is freely allowed to move within the completely interconnected porous microstructure. The pore fluid motion is governed by diffusion mechanisms and these will be coupled with deformation within the elastic skeleton. For example, a consolidation of the solid skeleton will produce a rise in fluid pore pressure; likewise, a rise in pore pressure will produce material dilatational deformation. As done in the previous thermoelastic presentation, we will focus our attention on how this coupling integrates into the various model field equations. We will also explore a few basic solutions to problems of interest.

Pioneering work in formulating this theory was initiated by Terzaghi (1943), but it was Biot (collection of papers edited by Tolstoy, 1992) who over a period of 30 years

FIGURE 7.5

Sponge sample as an example of poroelastic material.

developed most of the fundamental details of poroelasticity. A few other good sources for further more detailed information can be found in Wang (2000) and Cheng (2016).

7.3.1 CONSTITUTIVE LAWS AND GENERAL FORMULATION

The theory assumes a coherent elastic solid skeleton and a freely moving pore fluid such that the solid and fluid phases are fully connected and uniformly distributed. This allows the construction of a continuum mechanics theory relating the various field quantities of interest within a governing set of coupled field equations. The pore fluid motion is governed by diffusion mechanisms and these will be coupled with linear elastic deformation behaviors of the solid skeleton. As previously mentioned, a consolidation of the solid skeleton will produce a rise in fluid pore pressure, which in turn will produce material dilation. Therefore, the total medium stress will be a combination of that within the solid and fluid portions as illustrated in Fig. 7.6. It is instructive to point out that there exists an analogy between our previous thermoelasticity model and that of poroelasticity, and Norris (1992) presents a detailed account of this fact. Temperature corresponds to pore fluid pressure, heat flux corresponds to fluid flow, and entropy corresponds to fluid mass.

New parameters for poroelasticity include a *specific discharge vector* q_i, which relates the relative fluid motion with respect to the solid, and is defined as the rate of fluid volume crossing a unit area specified by a normal in the x_i direction. Also the change in fluid content within the porous material is specified by the quantity ζ which represents the *increment in fluid volume per unit material volume.* For no source terms, the fluid mass balance relation can then be expressed as

$$\frac{\partial \zeta}{\partial t} = -q_{i,i} \tag{7.3.1}$$

Note that positive ζ corresponds to an increase in the fluid in the porous solid. Neglecting body forces, the fluid transport within the interstitial space is governed by the *Darcy law*

$$q_i = -\kappa p_{,i} \tag{7.3.2}$$

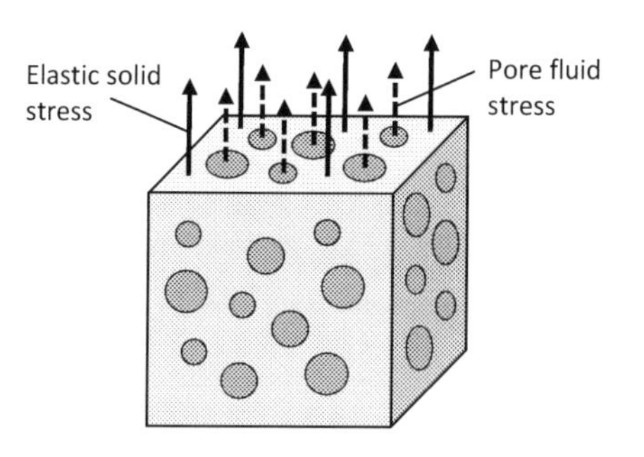

FIGURE 7.6

Dual stress distribution in poroelastic media.

where κ is the *permeability coefficient* (generally a function of the pore geometry and fluid viscosity) and p is the *fluid pore pressure*. Relation (7.3.2) is analogous to Fourier's law of heat conduction (7.2.9) and can be derived from the Navier–Stokes equations (6.4.13) by dropping the inertia terms. Combining relations (7.3.1) and (7.3.2) gives

$$\frac{\partial \zeta}{\partial t} = \kappa \nabla^2 p \tag{7.3.3}$$

Incorporating this fluid diffusion model within an isotropic elastic theory, Biot proposed a coupled linear constitutive law of the form

$$\begin{aligned} \varepsilon_{ij} &= \frac{1}{2\mu} T_{ij} - \left(\frac{1}{6\mu} - \frac{1}{9k} \right) T_{kk} \delta_{ij} + \frac{1}{3H} p \delta_{ij} \\ \zeta &= \frac{1}{3H} T_{kk} + \frac{1}{R} p \end{aligned} \tag{7.3.4}$$

where μ and k are, respectively, the usual shear and bulk elastic moduli of the *drained elastic skeleton*, and the additional material constants H and R characterize the coupling between the solid and fluid stress and strain fields. Stress–strain relation $(7.3.4)_1$ can be inverted and recast as

$$T_{ij} + \alpha p \delta_{ij} = (\lambda + \mu) \varepsilon_{kk} \delta_{ij} + 2\mu \varepsilon_{ij} \tag{7.3.5}$$

where λ is the usual Lamé's constant of the drained elastic skeleton and $\alpha = k/H$ and is commonly referred to as the *Biot coefficient of effective stress*. Note that all drained elastic constants satisfy relations presented back in Section 6.2.1. In this form, the constitutive relation is similar to that of a drained elastic solid with the left-hand side often referred to as an *effective stress*. It should be observed that by dropping the pore pressure term, relations $(7.3.4)_1$ and (7.3.5) reduce to the classical isotropic elasticity constitutive law (6.2.31). Constitutive relations (7.3.4) can be easily decomposed into volumetric and deviatoric equations as

$$\begin{aligned} \varepsilon_{kk} &= -\frac{P}{k} + \frac{p}{H} \\ \hat{\varepsilon}_{ij} &= \frac{1}{2\mu} \hat{T}_{ij} \\ \zeta &= -\frac{P}{H} + \frac{p}{R}, \quad P = -\frac{1}{3} T_{kk} \end{aligned} \tag{7.3.6}$$

In order to understand the behavior of fluid-filled porous materials, it is important to consider its special response during *undrained* and *drained conditions*. These two situations represent limiting behaviors with undrained conditions corresponding to the case where the fluid is trapped in the porous solid such that $\zeta = 0$, while the drained case corresponds to zero pore pressure, $p = 0$. It can be shown from Eqs. (7.3.6) that under undrained conditions the volumetric response can be written as

$$\varepsilon_{kk} = \varepsilon = -\frac{P}{K_u}, \quad \text{where } K_u = k\left[1 + \frac{kR}{H^2 - kR}\right], P = -\frac{1}{3} T_{kk} \tag{7.3.7}$$

Note that K_u may be thought of as the *undrained bulk modulus* of the material. For the drained case, the volumetric relationship from (7.3.6) becomes

$$\varepsilon_{kk} = \varepsilon = -\frac{P}{k} \tag{7.3.8}$$

Thus, under either drained or undrained conditions, the poroelastic model produces an elastic response with the undrained case having a higher volumetric stiffness ($K_u > k$). Substituting relation (7.3.8), with $p = 0$, yields

$$\zeta = \alpha\varepsilon \tag{7.3.9}$$

It can be observed that α may be interpreted as the ratio of fluid volume change in a material element to the volume change of the element itself under drained conditions. In regard to the time-dependent nature of the problem, see (7.3.3), the case of rapid loadings would correspond to the situation where pore fluid would have little time to flow, thus indicating a more undrained condition. On the other hand, for very slow loading situations, the pore pressure will have sufficient time to equilibrate with the boundary pressure, and if this boundary value is zero, then this case will correspond to a drained condition.

Linear isotropic poroelasticity is thus described by the coupled constitutive laws (7.3.4) for the solid skeleton and fluid, the Darcy relation (7.3.2), and the continuity equation (7.3.1). We also need the small deformation strain–displacement relations

$$\varepsilon_{ij} = \frac{1}{2}(u_{i,j} + u_{j,i}) \tag{7.3.10}$$

and limiting the discussion to quasi-static problems, we also include the equations of equilibrium

$$T_{ij,j} + F_i = 0 \tag{7.3.11}$$

A set of five material constants (μ, k, R, H, and κ) are needed to fully characterize the material, and it is common in the literature to use physically meaningful material constants (e.g. K_u and α) in place of R and H. Mechanical (tractions or displacements) and fluid drainage boundary conditions are required for solution to specific problems.

7.3.2 PROBLEM SOLUTIONS

The solution to problems in linear poroelasticity closely follows the procedures in linear elasticity presented in Sections 6.2.2 and 6.2.3. Stress and displacement formulations can be developed. For the stress formulation following the usual procedures, we can generate the Beltrami–Michell equations

$$T_{ij,kk} + \frac{1}{1+\nu}T_{kk,ij} + \frac{(1-2\nu)\alpha}{1-\nu}\left(\delta_{ij}p_{,kk} + \frac{1-\nu}{1+\nu}p_{,ij}\right) = -\frac{\nu}{1-\nu}\delta_{ij}F_{k,k} - F_{i,j} - F_{j,i} \tag{7.3.12}$$

and for the displacement formulation Navier's equations (equilibrium in terms of displacements) would become

$$\mu u_{i,kk} + (\lambda + \mu)u_{k,ki} - \alpha p_{,i} + F_i = 0 \tag{7.3.13}$$

As with the thermoelastic case, we again see that the coupling term $-\alpha p_{,i}$ acts as an added body force in the momentum balance relation. To help with problem solution, the common elasticity schemes of stress functions and displacement potentials along with two-dimensional models of plane strain and plane stress can be used.

EXAMPLE 7.3.1 ONE DIMENSIONAL CONSOLIDATION

Consider the one-dimensional poroelastic deformation of a material layer in the domain $0 \le x \le h; -\infty \le y, z \le \infty$ as shown in Fig. 7.7. The layer rests on a rigid impermeable base and its top surface has zero pore pressure and carries a suddenly applied uniform normal loading, p_o. Determine the time-dependent vertical displacement.

Solution: We assume for this problem deformations only in the x-direction where the only nonzero strain will be ε_x. Under such uniaxial strain, the stress field reduces to

$$\begin{aligned} T_{xx} &= (\lambda + 3\mu)\varepsilon_x - \alpha p \\ T_{yy} = T_{zz} &= (\lambda + \mu)\varepsilon_x - \alpha p \end{aligned} \tag{7.3.14}$$

while the equilibrium equations become

$$\frac{\partial T_{xx}}{\partial x} = 0 \tag{7.3.15}$$

Thus, the stress T_{xx} does not depend on x. Now since the problem loading is constant in time, this implies that the stress will also be independent of time, and thus will be constant. Boundary condition $T_{xx}(0,t) = -p_o H(t)$, thus gives $T_{xx} = -p_o$.

Navier's equations (7.3.13) reduce to the single relation

$$(\lambda + 2\mu)\frac{\partial^2 u}{\partial x^2} - \alpha\frac{\partial p}{\partial x} = 0 \tag{7.3.16}$$

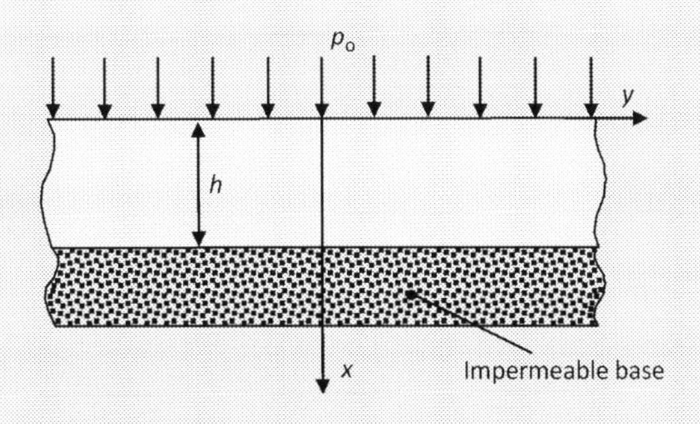

FIGURE 7.7

One-dimensional consolidation problem.

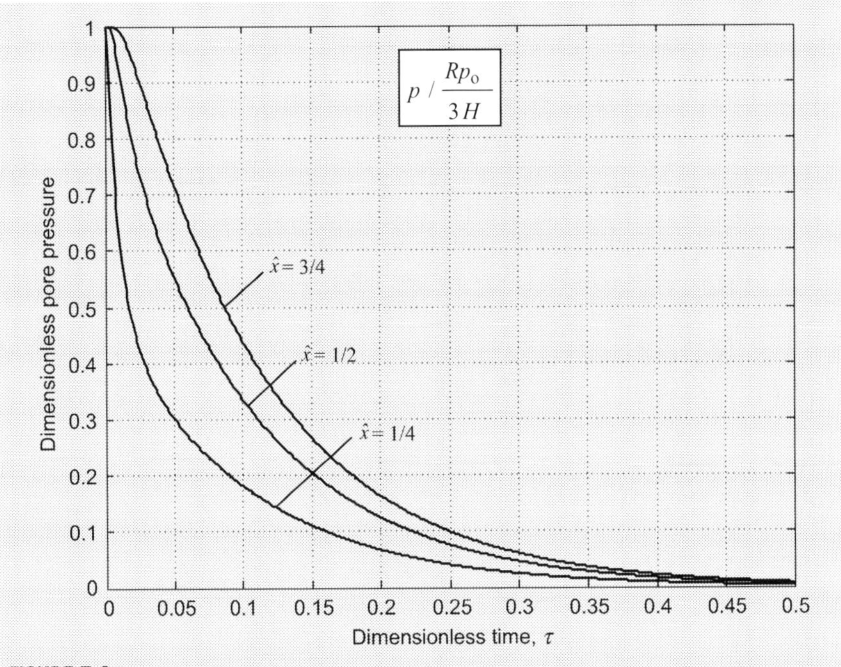

FIGURE 7.8

Pore pressure behavior for one-dimensional consolidation problem.

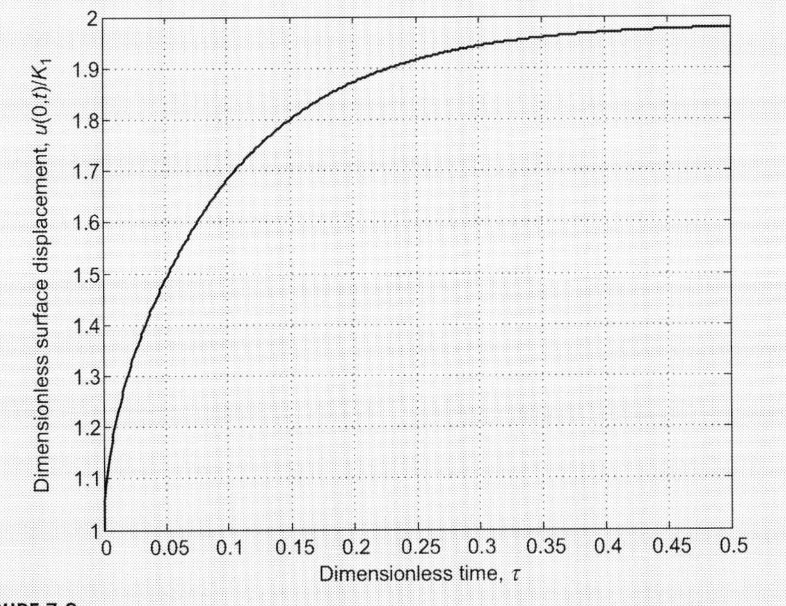

FIGURE 7.9

Surface displacement behavior for one-dimensional consolidation problem ($K_2 = 1$).

EXAMPLE 7.3.2 MANDEL'S CONSOLIDATION PROBLEM

Consider the problem originally presented by Mandel (1953) of a long rectangular cross-section poroelastic medium sandwiched between two impermeable plates as shown in Fig. 7.10. A compressive force $F=2p_o aH(t)$ is suddenly applied to the material as shown. The right and left boundaries are assumed to be stress-free and drained. The geometry is such that the deformation is only in the x,y-plane and the pore fluid moves only in the x-direction. Thus, the stress components and pore pressure are functions of x and t. Determine the pore pressure distribution and show that it results in a nonmonotonic response.

Solution: The problem may be modeled using a *two-dimensional plane strain* formulation where the displacements are of the form $u=u(x,y), v=v(x,y), w=0$. From the two-dimensional equilibrium equations, we find that $\frac{\partial T_{xx}}{\partial x}=0$ which implies that T_{xx} is a constant, but since it is to vanish at $x=\pm a$, $T_{xx}=0$.

For the plane strain case, the poroelastic Beltrami–Michell compatibility equations (see elasticity discussion in Section 6.2.3) reduce to

$$\nabla^2(T_{xx}+T_{yy}+2\eta p)=0 \tag{7.3.23}$$

where $\eta=\frac{1-2\nu}{2(1-\nu)}\alpha$. Since T_{xx} vanishes and T_{yy} and p depend only on x and t, (7.3.23) implies

$$T_{yy}+2\eta p=g(t) \tag{7.3.24}$$

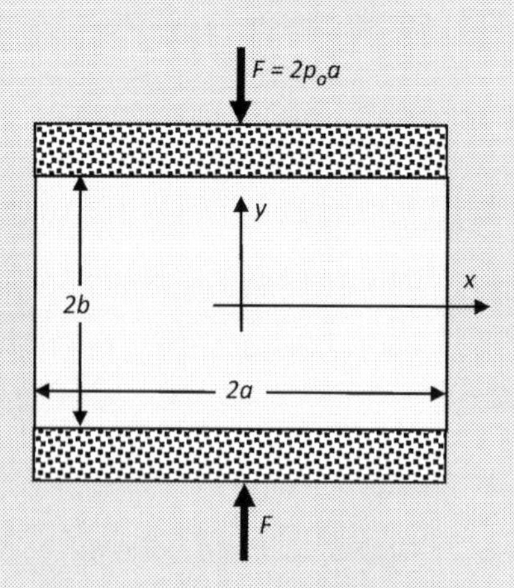

FIGURE 7.10

Mandel's consolidation problem geometry.

where the arbitrary function of time $g(t)=T_{yy}(a,t)$, since the pore pressure vanishes a $x=a$.

For plane strain, relation $(7.3.4)_2$ can be written as

$$\zeta=\frac{\alpha(1+\nu)}{3K}(T_{xx}+T_{yy}+A_1 p) \tag{7.3.25}$$

with $A_1=\dfrac{3}{B(1+\nu_u)}$.

In general, using the Beltrami–Michell compatibility equation (7.3.12) for the poroelastic case, one can show that

$$\frac{\partial\zeta}{\partial t}=c\nabla^2\zeta \tag{7.3.26}$$

where $c=\dfrac{2\kappa\mu(1-\nu)(\nu_u-\nu)}{\alpha^2(1-2\nu)^2(1-\nu_u)}$. Note that definitions of the Poisson's ratios were given previously in relations (7.3.22). Diffusion equation (7.3.26) for ζ becomes uncoupled which is not the case for the pore pressure equation.

Using results (7.3.23) and (7.3.24) with $T_{xx}=0$ gives the result

$$\frac{\partial}{\partial t}(T_{yy}+A_1 p)=c\frac{\partial^2}{\partial x^2}(T_{yy}+A_1 p) \tag{7.3.27}$$

Result (7.3.27) can be solved using Laplace transforms (see Wang, 2000 for details). The pore pressure terms can be eliminated using the Beltrami–Michell relation (7.3.23) and the inverse Laplace transform is handled by some basic contour integration. The final result for the pore pressure is given by

$$p(x,t)=\frac{2p_o}{A_1}\sum_{n=1}^{\infty}\frac{\sin\lambda_n}{\lambda_n-\sin\lambda_n\cos\lambda_n}\left(\cos\frac{\lambda_n x}{a}-\cos\lambda_n\right)e^{-\lambda_n^2 ct/a^2} \tag{7.3.28}$$

where λ_n are the roots of the equation $\dfrac{\tan\lambda}{\lambda}=\dfrac{3}{2\eta B(1+\nu_u)}=\dfrac{1-\nu}{\nu_u-\nu}$. Using MATLAB Code C-19 for the calculations and plotting, Fig. 7.11 illustrates the pore pressure distribution along the positive x-axis for several different values of dimensionless times $\tau=ct/a^2$. For small times, the pressure is not completely monotonic in time. This behavior is normally explained by the fact that there is a contraction at the drained edges of the sample and this will induce pore pressure buildup in the interior. See Cheng and Detournay (1988) for more complete solution details and interpretation.

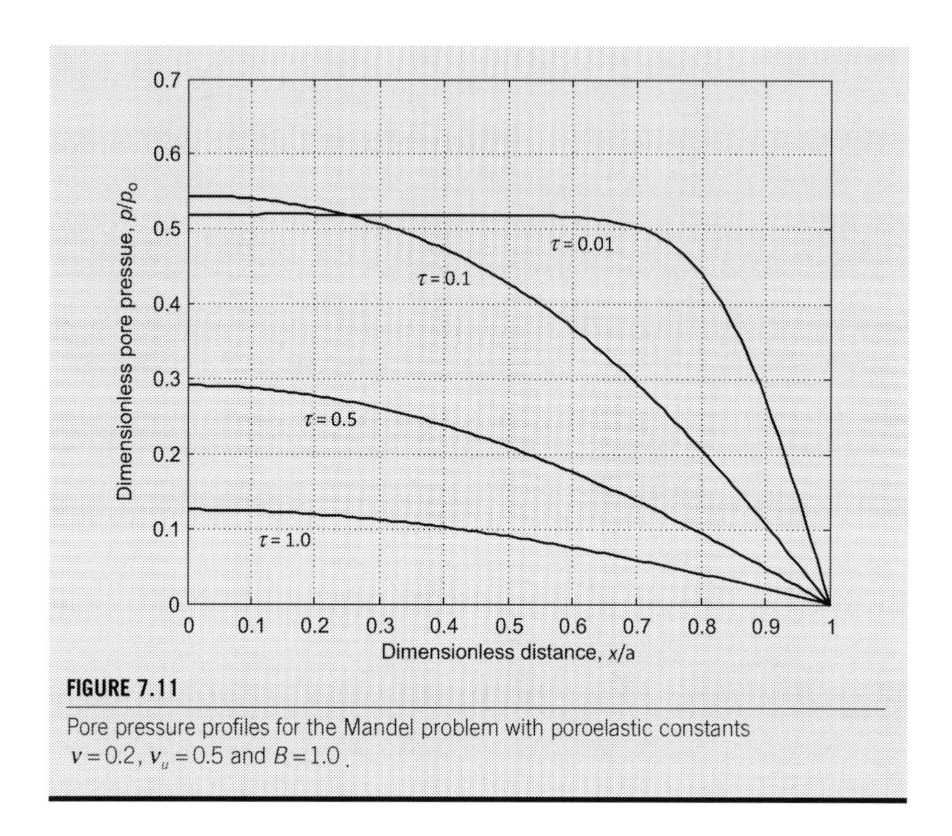

FIGURE 7.11

Pore pressure profiles for the Mandel problem with poroelastic constants $\nu = 0.2$, $\nu_u = 0.5$ and $B = 1.0$.

7.4 ELECTROELASTICITY

As our final multicontinuum theory we now wish to explore *electroelasticity* whereby the basic linear elasticity model will be coupled with an electrical field. The basic issue is that when an electric field is applied to some materials, they experience a deformation or strain. Such behavior is due to *piezoelectric or electrostrictive effects*. Electrostriction is a property of dielectric materials and is caused by a small displacement of ions in the crystal lattice exposed to an external electric field. Such displacements accumulate in the material and result in an overall strain in the direction of the field. The piezoelectric effect is the electric charge that accumulates in certain solid materials such as crystals, certain ceramics, and biological matter in response to applied mechanical stress. The piezoelectric effect is a *reversible process* in that materials exhibiting the internal generation of electrical charge resulting from an applied mechanical force also exhibit the reverse effect of the internal generation of a mechanical strain resulting from an applied electrical field. Both the piezoelectric or electrostrictive effects were discovered over a century ago and thus much research and application has been done on such materials. A large variety of both natural and synthetic piezoelectric crystalline and ceramic materials exist. Applications in this field include sensors, actuators, motors, smart materials, and many new microscale devices. We will only present a brief overview of the continuum mechanics modeling of such materials, and more detailed presentations can be found in Eringen and Maugin (1990), Kuang (2014), Peric (2004), and Bechtel and Lowe (2015).

7.4.1 CONSTITUTIVE LAWS AND GENERAL FORMULATION

Similar to our previous theories in this chapter, we now present a linear theory of electroelasticity for small deformations under quasi-static conditions with no body forces. We further assume that there are no magnetic fields, free currents, or electric charges. From Maxwell–Faraday theory of electrostatics with no mechanical deformation, we can define a linear electrical constitutive relation

$$D_i = \breve{\varepsilon}_{ik} E_k \tag{7.4.1}$$

where D_i is the *electric displacement*, $\breve{\varepsilon}_{ik}$ is the *dielectric permittivity material tensor*, and E_k is the *electric field strength*. Furthermore, these fields must satisfy

$$\begin{aligned} &D_{i,i} = 0, \quad \nabla \cdot \boldsymbol{D} = 0 \\ &\varepsilon_{ijk} E_{k,j} = 0, \quad \nabla \times \boldsymbol{E} = 0 \Rightarrow \boldsymbol{E} = -\nabla \phi^e \end{aligned} \tag{7.4.2}$$

where ϕ^e is the electric potential function.

Introducing the mechanical deformation, Hooke's law then couples the stress–strain with the electric field, and vice versa. The most common way to determine these coupled constitutive relations is to use energy schemes previously discussed. Following Pak (1992), the internal energy ε stored in a linear electroelastic material can be written as

$$\varepsilon = \frac{1}{2} T_{ij} \varepsilon_{ij} + \frac{1}{2} E_i D_i \tag{7.4.3}$$

If we wish to derive the governing equations in terms of E_i instead of D_i, we need to introduce another thermodynamic potential. This can be done by defining an *electric enthalpy density*

$$h = \varepsilon - E_i D_i \tag{7.4.4}$$

This form then becomes the correct potential when the mechanical displacement and the electrical potential are taken to be the independent variables.

For linear electroelastic materials, the electric enthalpy can be expressed by

$$h = \frac{1}{2} C_{ijkl} \varepsilon_{ij} \varepsilon_{kl} - \frac{1}{2} \breve{\varepsilon}_{ij} E_i E_j - e_{ikl} \varepsilon_{kl} E_i \tag{7.4.5}$$

where e_{ikl} is the *piezoelectric material tensor*. Using the usual thermodynamic procedures, the constitutive relations come from the potential relations

$$\begin{aligned} T_{ij} &= \frac{\partial h}{\partial \varepsilon_{ij}} = C_{ijkl} \varepsilon_{kl} - e_{kij} E_k \\ D_i &= -\frac{\partial h}{\partial E_i} = \breve{\varepsilon}_{ik} E_k + e_{ijk} \varepsilon_{jk} \end{aligned} \tag{7.4.6}$$

Kuang, (2014) shows that actually several different forms of these couple equations can be derived. Typical boundary conditions for this problem are

$$\begin{aligned} T_{ij}n_j &= t_i \quad \text{(specified traction)} \\ D_i n_i &= -q_s \quad \text{(specified surface charge)} \end{aligned} \tag{7.4.7}$$

In addition to the symmetries for C_{ijkl} from Section 6.2.1, the electromaterial tensors generally satisfy the conditions $\breve{\varepsilon}_{ij} = \breve{\varepsilon}_{ji}, e_{ijk} = e_{ikj}$. Thus, for the general problem, there exists 21 elastic constants, 18 piezoelectric constants, and 6 permittivity constants. However, for most applications, there is considerable symmetry and this number of independent material constants reduce to more reasonable numbers.

Most applications of electroelasticity involve anisotropic crystal, ceramic, and composite materials and thus require the anisotropic constitutive forms in (7.4.6). As used previously for linear elasticity, we can use the Voigt notational scheme to express these electroconstitutive relations. Kuang (2014) provides a specific example for a transversely isotropic ceramic (such as lead zirconate titanate, PZT). For the case where x_3 is the electrical poling direction and axis of symmetry, x_1,x_2 is the isotropic plane and relations (7.4.6) would take the form

$$\begin{bmatrix} T_{11} \\ T_{22} \\ T_{33} \\ T_{23} \\ T_{31} \\ T_{12} \end{bmatrix} = \begin{bmatrix} C_{11} & C_{12} & C_{13} & 0 & 0 & 0 \\ \cdot & C_{11} & C_{13} & 0 & 0 & 0 \\ \cdot & \cdot & C_{33} & 0 & 0 & 0 \\ \cdot & \cdot & \cdot & C_{44} & 0 & 0 \\ \cdot & \cdot & \cdot & \cdot & C_{44} & 0 \\ \cdot & \cdot & \cdot & \cdot & \cdot & C_{66} \end{bmatrix} \begin{bmatrix} \varepsilon_{11} \\ \varepsilon_{22} \\ \varepsilon_{33} \\ 2\varepsilon_{23} \\ 2\varepsilon_{31} \\ 2\varepsilon_{12} \end{bmatrix} - \begin{bmatrix} 0 & 0 & e_{31} \\ 0 & 0 & e_{32} \\ 0 & 0 & e_{33} \\ 0 & e_{24} & 0 \\ e_{15} & 0 & 0 \\ 0 & 0 & 0 \end{bmatrix} \begin{bmatrix} E_1 \\ E_2 \\ E_3 \end{bmatrix} \tag{7.4.8}$$

with $C_{66} = (C_{11} - C_{22})/2$.

$$\begin{bmatrix} D_1 \\ D_2 \\ D_3 \end{bmatrix} = \begin{bmatrix} \breve{\varepsilon}_{11} & 0 & 0 \\ 0 & \breve{\varepsilon}_{22} & 0 \\ 0 & 0 & \breve{\varepsilon}_{33} \end{bmatrix} \begin{bmatrix} E_1 \\ E_2 \\ E_3 \end{bmatrix} + \begin{bmatrix} 0 & 0 & e_{31} \\ 0 & 0 & e_{32} \\ 0 & 0 & e_{33} \\ 0 & e_{24} & 0 \\ e_{15} & 0 & 0 \\ 0 & 0 & 0 \end{bmatrix}^T \begin{bmatrix} \varepsilon_{11} \\ \varepsilon_{22} \\ \varepsilon_{33} \\ 2\varepsilon_{23} \\ 2\varepsilon_{31} \\ 2\varepsilon_{12} \end{bmatrix} \tag{7.4.9}$$

For two-dimensional plane strain or plane stress problems, considerable reduction in this large system will occur and a variety of analytical solutions for the stress, strain, displacement, and electric fields have been developed. However, since the solution methods are lengthy and detailed, we will not go into this material. We thus end our presentation on this interesting multifield coupled continuum mechanics topic.

REFERENCES

Asaro, R.J., Lubarda, V.A., 2006. Mechanics of Solids and Materials. Cambridge University Press, Cambridge.

Boley, B.A., Weiner, J.H., 1960. Theory of Thermal Stresses. John Wiley, New York.

Bechtel, S.E., Lowe, R.L., 2015. Fundamentals of Continuum Mechanics with Applications to Mechanical, Thermomechanical and Smart Materials. Elsevier, Amsterdam, the Netherlands.
Biot, M.A., 1991. In: Tolstoy, I. (Ed.), Acoustics, Elasticity and Thermodynamics of Porous Media: Twenty-One Papers by M.A. Biot. Acoustical Society of America.
Cheng, A.H.D., 2016. Poroelasticity: Theory and Applications of Transport in Porous Media. Springer International Publishing, Switzerland.
Cheng, A.H.D., Detournay, E., 1988. A Direct Boundary Element Method for Plane Strain Poroelasticity. Int. J. Num. Anal. Methods Geomech. 12, 551–572.
Eringen, A.C., Maugin, G.A., 1990. Electrodynamics of Continua, vol. 1. Springer, New York.
Holzapfel, G.A., 2006. Nonlinear Solid Mechanics: A Continuum Approach for Engineering. John Wiley, West Sussex.
Kovalenko, A.D., 1969. Thermoelasticity. Noordhoff, Groningen, the Netherlands.
Kuang, Z.B., 2014. Theory of Electroelasticity. Springer, New York.
Mandel, J., 1953. Consolidation des sols. Geotechnique 3, 287–299.
Norris, A., 1992. On the Correspondence between Poroelasticity and Thermoelasticity. J. Appl. Phys. 71, 1138–1141.
Nowacki, J.L., 1978. Theory of Thermoelasticity with Applications. Sijthoff-Noordhoff, Groningen, the Netherlands.
Nowacki, W., 1986. Thermoelasticity. Pergamon Press, Oxford.
Pak, Y.E., 1992. Linear electro-elastic fracture mechanics of piezoelectric materials. Int. J. Fracture 52, 79–100.
Peric, L., 2004. Couple tensors of piezoelectric materials state, e-book. MPI Ultrasonics. http://www.mpi-ultrasonics.com
Sadd, M.H., 2014. Elasticity: Theory, Applications and Numerics. Third ed Elsevier, Waltham, MA.
Terzaghi, K., 1943. Theoretical Soil Mechanics. John Wiley, New York.
Wang, H.F., 2000. Theory of Linear Poroelasticity with Applications to Geomechanics and Hydrogeology. Princeton University Press, Princeton, NJ.

EXERCISES

7.1 Starting from the energy relation $(7.2.6)_3$, explicitly develop the linearize energy equation (7.2.13).

7.2 Invert Hooke's law $(7.2.16)_1$ to express the strain in terms of stress

$$\varepsilon_{ij} = \frac{1+\nu}{E}T_{ij} - \frac{\nu}{E}T_{kk}\delta_{ij} + \alpha_T \Theta \delta_{ij}$$

7.3 Develop Navier's thermoelastic equilibrium equations in terms of the displacements (7.2.24)

$$\mu u_{i,kk} + (\lambda+\mu)u_{k,ki} - (3\lambda+2\mu)\alpha_T \Theta_{,i} + F_i = 0$$

7.4 For an anisotropic solid, Fourier's conduction law was given by $q_i = -k_{ij}\Theta_{,j}$ and it was shown that the conductivity tensor was symmetric, $k_{ij} = k_{ji}$. Show that for a two-dimensional problem in the x,y-plane, the heat conduction equation for uncoupled steady-state conditions becomes

$$k_{xx}\frac{\partial^2\Theta}{\partial x^2}+2k_{xy}\frac{\partial^2\Theta}{\partial x\partial y}+k_{yy}\frac{\partial^2\Theta}{\partial y^2}=0$$

7.5 Justify the plane stress thermoelastic relations (7.2.29) and (7.2.30).

7.6 For the plane stress case, invert relations (7.2.26) and determine the nonzero stress forms in terms of the strains and thus verify the traction relations given by (7.2.31).

7.7 For Example 7.2.1, show that the stresses are given by relation (7.2.43) and also verify that under the zero traction boundary conditions, constants C_2 and C_3 are expressed by (7.2.44).

7.8 Using the general displacement solution (7.2.51), solve the thermoelastic problem of a solid circular elastic plate $0 \le r \le a$ with a totally restrained boundary edge at $r = a$. Next show that for the case of a uniform temperature distribution, displacements and hence stresses will be zero.

7.9 Consider the axisymmetric plane stress problem of a solid circular plate of radius a with a *constant internal heat generation* specified by h_o. The steady-state conduction equation thus becomes

$$\frac{\partial^2\Theta}{\partial r^2}+\frac{1}{r}\frac{\partial\Theta}{\partial r}+h_o=0$$

Determine the temperature distribution for the case with boundary condition $\Theta(\boldsymbol{a}) = \theta_o$ and then show that the resulting thermal stresses for the case with zero boundary stress are given by

$$T_{rr}=\frac{E\alpha_T h_o}{16}(r^2-a^2),\quad T_{\theta\theta}=\frac{E\alpha_T h_o}{16}(3r^2-a^2)$$

Such solutions are useful to determine the thermal stresses in structures made of radioactive materials.

7.10 Invert the poroelastic constitutive law $(7.3.4)_1$ to develop the form (7.3.5).

7.11 Verify the undrained volumetric response (7.3.7).

7.12 Develop the Beltrami–Michell equation (7.3.12) for the general poroelastic case.

7.13 Using the Beltrami–Michell equation (7.3.12), verify the result

$$\nabla^2(T_{kk}+4\eta^* p)=-\frac{1+\nu}{1-\nu}F_{k,k}$$

where η^* is defined in terms of α and ν.

7.14 Develop Navier's equation (7.3.13) for the general poroelastic case.

7.15 Justify for the two-dimensional plane strain case (defined in Example 7.3.2) that the Beltrami–Michell compatibility equation for poroelastic materials is given by (7.3.23).

7.16 Combine constitutive relations (7.4.8) and (7.4.9) into a single matrix equation of the form7.17

$$\begin{bmatrix} T_{11} \\ T_{22} \\ T_{33} \\ T_{23} \\ T_{31} \\ T_{12} \\ D_1 \\ D_2 \\ D_3 \end{bmatrix} = [9\times 9] \begin{bmatrix} \varepsilon_{11} \\ \varepsilon_{22} \\ \varepsilon_{33} \\ 2\varepsilon_{23} \\ 2\varepsilon_{31} \\ 2\varepsilon_{12} \\ E_1 \\ E_2 \\ E_3 \end{bmatrix}$$

7.17 Consider the electroelastic material given by Eqs. (7.4.8) and (7.4.9). For case of x_2, x_3 plane strain with $\varepsilon_{11} = \varepsilon_{13} = \varepsilon_{12} = 0$ and $E_1 = 0$, determine the reduced forms of these relations. The result should reduce to a single 5×5 system as per the form from Exercise 7.16.

CHAPTER

8 General constitutive relations and formulation of nonlinear theories of solids and fluids

We now wish to expand our material modeling and explore several types of nonlinear continuum mechanics theories for both solids and fluids. In order to investigate such theoretical models, we now use finite deformation kinematics and will, in general, incorporate nonlinear constitutive relations. Recall from Chapter 3 that we established many different deformation/strain tensors, and likewise in Chapter 4 we presented several different stress tensors. Thus, we now face a choice in what particular strain measure and stress tensor to use within a given constitutive law. This choice will be guided by the fact that we want our relations to satisfy the principle of objectivity or frame-indifference. This principle was discussed in Sections 2.9, 3.6, 3.14, and 4.9, and thus we already have some guidance from this previous work. We start our presentation with some general principles of constitutive equation development and then move into general simple materials. This will be followed by exploring several particular material theories including nonlinear elasticity, nonlinear and non-Newtonian viscous fluids, and nonlinear viscoelastic materials.

8.1 INTRODUCTION AND GENERAL CONSTITUTIVE AXIOMS

Before going into specific details, we first explore constitutive equations in a much more general way. A constitutive equation can be described as a relation between independent variables and dependent response variables. Independent variables would commonly include the density, material motion, temperature, and their spatial and temporal derivatives and history. Dependent variables would normally be the stress, strain energy, Helmholtz free energy, heat flux, internal energy, and entropy. Thus, we could write a general system of constitutive equations of the form

$$\begin{aligned}
\boldsymbol{T} &= \boldsymbol{T}(\rho, \chi, \theta, \text{and derivatives and history}) \\
U &= U(\rho, \chi, \theta, \text{and derivatives and history}) \\
\Psi &= \Psi(\rho, \chi, \theta, \text{and derivatives and history}) \\
\boldsymbol{q} &= \boldsymbol{q}(\rho, \chi, \theta, \text{and derivatives and history}) \\
\varepsilon &= \varepsilon(\rho, \chi, \theta, \text{and derivatives and history}) \\
s &= s(\rho, \chi, \theta, \text{and derivatives and history})
\end{aligned} \tag{8.1.1}$$

Continuum Mechanics Modeling of Material Behavior. http://dx.doi.org/10.1016/B978-0-12-811474-2.00008-3

Note that spatial derivatives of the motion would lead to the deformation gradient and other strain tensors. This general form provides considerable leeway as to which specific independent variables, derivatives, and history to include in the constitutive model.

Starting with the pioneering work by Noll (1958) and later enhanced and summarized by the classic publications of Truesdell and Toupin (1960) and Truesdell and Noll (1965), several *principles* normally taken as *axioms* have been proposed which provide guidance and restrictions on the development of constitutive relations. The list of axioms that have been developed commonly includes the following:

Consistency Constitutive equations must be consistent or admissible with the principles of balance of mass, momentum, energy, and second law of thermodynamics.
Coordinate Invariance Constitutive equations must be set in tensor language to ensure that they are *invariant* with respect to coordinate transformations.
Just Setting Combined with basic conservation equations, constitutive relations should provide a *unique solution* to a problem with meaningful boundary and initial conditions.
Material Frame-Indifference The material response must be the *same for all observers*. This principle is sometimes known as *material objectivity*.
Equipresence For a given material type, a variable present as an independent variable in one constitutive equation should be present in all, unless there is a contradiction with another constitutive axiom.
Determinism Only the *present and past history* of the independent variables in a constitutive equation can affect the present response, and normally we assume that the material response should exhibit *fading memory* such that recent history has more effect than past history.
Symmetry Constitutive relations must be consistent with known symmetries of the material.
Local Action The material response at point $\boldsymbol{X}$ is only a function of the independent variables (and derivatives and history) in the *neighborhood* of the point.

These concepts act as a *guide* in formulating proper constitutive equations for material behavior, leading to a wide variety of nonlinear, history-dependent relations. The last axiom of local action can be dropped, thus allowing the development *nonlocal continuum models*. Most of our previous work with the classical material theories presented in Chapters 6 and 7 followed these axioms. Due to the limited scope of the text, we will only explore in detail a small variety of nonlinear constitutive models. It should also be mentioned that not all useful constitutive theories satisfy all of these conditions, and thus some of the previous concepts may be relaxed for *special* situations.

EXAMPLE 8.1.1 USE OF SECOND LAW TO MODIFY GENERAL CONSTITUTIVE RELATIONS FOR THERMOVISCOUS FLUIDS

Consider a proposed set of constitutive relations for a general thermoviscous fluid that includes the rate of deformation tensor, mass density, and entropy:

$$\begin{aligned} \boldsymbol{T} &= \boldsymbol{T}\left(\boldsymbol{D}, \frac{1}{\rho}, s\right) \\ \varepsilon &= \varepsilon\left(\boldsymbol{D}, \frac{1}{\rho}, s\right) \\ \boldsymbol{q} &= \boldsymbol{q}\left(\boldsymbol{D}, \frac{1}{\rho}, s\right) \end{aligned} \tag{8.1.2}$$

Note that this set satisfies the axiom of Equipresence. Investigate any limitations coming from application of the second law of thermodynamics.

Solution: Using the second law in the form (5.7.10),

$$\begin{aligned} &\rho(\theta\dot{s} - \dot{\varepsilon}) + T_{ij}D_{ij} - \frac{1}{\theta}(\boldsymbol{q} \cdot \nabla\theta) \geq 0 \Rightarrow \\ &\rho\left(\theta - \frac{\partial \varepsilon}{\partial s}\right)\dot{s} - \rho\frac{\partial \varepsilon}{\partial D_{ij}}\dot{D}_{ij} + \frac{\partial \varepsilon}{\partial(1/\rho)}\frac{\dot{\rho}}{\rho} + T_{ij}D_{ij} - \frac{1}{\theta}(q_k\theta_{,k}) \geq 0 \end{aligned} \tag{8.1.3}$$

The above equation must hold for arbitrary values of $\dot{s}$ and $\dot{D}_{ij}$, and thus we must have

$$\theta = \frac{\partial \varepsilon}{\partial s} \quad \text{and} \quad \frac{\partial \varepsilon}{\partial D_{ij}} = 0$$

This implies that the internal energy ε is independent of $\boldsymbol{D}$ and so θ is also independent of $\boldsymbol{D}$.

Next, defining the *thermodynamic pressure* p by the relation $p = -\frac{\partial \varepsilon}{\partial(1/\rho)}$, and using the continuity equation (5.2.4), Eq. (8.1.3) can be expressed as

$$(T_{ij} - p\delta_{ij})D_{ij} - \frac{1}{\theta}(q_k\theta_{,k}) \geq 0 \tag{8.1.4}$$

Now since $\boldsymbol{T}$, p, and $\boldsymbol{q}$ are independent of $\theta_{,k}$, this requires that the heat flux $\boldsymbol{q}$ must vanish identically. Therefore, this class of thermoviscous fluids can only respond *adiabatically*, that is, with no heat flow. This finding is of course troubling since there is no physical reason for this adiabatic response. Furthermore, the fact that the internal energy is independent of the rate of deformation is also unrealistic. We conclude that the problem must lay with our original assumption of the forms of the constitutive laws (8.1.2). Clearly a more general form is needed for a realistic constitutive description of the fluid. Such a more general approach would likely be to include the temperature gradient $\theta_{,k}$ in the list of independent variables in (8.1.2).

More detailed studies on these types of problems are provided in Allen (2016), Chapter 6.

8.2 GENERAL SIMPLE MATERIALS

Several decades ago, Noll (1958) developed a general constitutive theory of *simple materials*. His purpose was to unify and clarify the large variety of constitutive laws that had been previously developed up to that point in time. His approach was part of the rebirth of continuum mechanics, often called *rational mechanics*, and this work served as the foundation for a large part of subsequent research in the field. Noll employed the general axioms of determinism, local action, and material frame-indifference. Starting with determinism, he first established a mechanical theory where the stress at location $\boldsymbol{X}$ and time t was given by the history of material motion $\boldsymbol{x} = \chi(\boldsymbol{X}, t)$. Then invoking the concept of local action, he limited the material motion to only the local deformation gradient $\boldsymbol{F}(\boldsymbol{X}, t)$, and this leads to the constitutive relation for simple materials:

$$\boldsymbol{T}(\boldsymbol{X},t) = \mathop{\mathfrak{F}}_{\tau=-\infty}^{\tau=t} (\boldsymbol{F}(\boldsymbol{X},\tau)) \tag{8.2.1}$$

where $\mathfrak{F}$ is a tensor-valued *history response functional* which is a function of the deformation gradient function over the history time variable $-\infty \le \tau \le t$. Note that if the material is nonhomogeneous, the response functional could depend explicitly on $\boldsymbol{X}$. Using our previous discussions in Chapters 3 and 4, the requirement that form (8.2.1) satisfies material frame-indifference can be written as

$$\boldsymbol{Q}(t) \mathop{\mathfrak{F}}_{\tau=-\infty}^{\tau=t} (\boldsymbol{F}(\boldsymbol{X},\tau)) \boldsymbol{Q}^T(t) = \mathop{\mathfrak{F}}_{\tau=-\infty}^{\tau=t} (\boldsymbol{Q}(\tau)\boldsymbol{F}(\boldsymbol{X},\tau)) \tag{8.2.2}$$

Considering the local deformation relative to location $\boldsymbol{X}$, it can be shown that constitutive form (8.2.1) can be expressed in terms of the relative deformation gradient (see Section 3.15) as

$$\boldsymbol{T}(t) = \mathop{\tilde{\mathfrak{F}}}_{s=0}^{s=\infty} (\boldsymbol{F}_t(t-s)) \tag{8.2.3}$$

Notice that simple materials are not characterized by just material constants nor just material functions, but rather the material functional $\tilde{\mathfrak{F}}$ must be specified. Such a general specification is very challenging and a large body of research has addressed this issue.

Noll also proposed a general *material classification* scheme based on *symmetry groups* which are related to changes in reference configuration that result in equivalent descriptions of the material's constitutive response. Without going into any details of the theory, we can list a few of his conclusions based on the stress constitutive response functional:

- a material is *isotropic* if there is a reference configuration such that proper frame rotations have no effect on the constitutive response functional;
- a material is a *fluid* if it is isotropic and the constitutive response functional is unaffected by any/all density-preserving changes in reference configuration;
- a material is a *solid* if changes in the reference configuration, other than rigid rotations, affect the constitutive response functional.

Allen (2016) provides a more detailed description of these concepts.

In order to apply this type of theoretical constitutive formulation to problem solution, more reduction in the general form (8.2.1) must be made. Such reductions in the response functional can be determined for special material types such as fluids and solids. We will pursue this in the remaining sections of this chapter.

8.3 NONLINEAR FINITE ELASTICITY

We choose the entry point into nonlinear material behavior with the elastic response. For this case, nonlinearity enters the mathematical model though finite deformation relations and the constitutive laws. Unlike linear elasticity from Section 6.2, we will now have to account for differences between current and reference configurations. Because the material is elastic, it will return to its reference configuration (original zero loading configuration) when the loadings are removed. This type of behavior will thus eliminate any history-dependent response in the constitutive law. We will limit our study to include only static mechanical effects and focus on incompressible materials. Our presentation will be brief and more detailed information can be found in Green and Adkins (1970), Ogden (1984), and Holzapfel (2006).

8.3.1 CONSTITUTIVE LAWS AND GENERAL FORMULATION

We now revisit elastic materials but this time will consider nonlinear models under finite deformations. For an elastic solid, the principles of determinism and local action imply that the stress is determined by the *present value of the local deformation gradient tensor*. Thus, the general history functional in relation (8.2.1) would reduce to a simple function

$$\boldsymbol{T}(\boldsymbol{X},t)=\boldsymbol{f}(\boldsymbol{F}(\boldsymbol{X},t)) \tag{8.3.1}$$

We could use this simple material starting form and pursue some detailed mathematical steps to get into our final useful constitutive relation. However, a shorter more direct scheme will get us where we want to go and so we pursue that method. The available finite strain deformation tensors previously discussed in Sections 3.5 and 3.6 included $\boldsymbol{F}$, $\boldsymbol{E}$, $\boldsymbol{C}$, and $\boldsymbol{B}$. However, it was only the left Cauchy–Green strain tensor that satisfied objectivity, $\boldsymbol{B}^*=\boldsymbol{QBQ}^T$. Thus, in order to satisfy the axiom of material frame-indifference, we must choose an objective strain tensor for use in the constitutive form. This implies

$$\boldsymbol{T}=\boldsymbol{g}(\boldsymbol{B}) \tag{8.3.2}$$

If we invoke the objectivity test for relation (8.3.2), we find that

$$\boldsymbol{Qg}(\boldsymbol{B})\boldsymbol{Q}^T=\boldsymbol{g}(\boldsymbol{QBQ}^T) \tag{8.3.3}$$

and thus conclude that the response function $\boldsymbol{g}$ must be an *isotropic function* of the strain measure $\boldsymbol{B}$. Now from our previous work in Section 2.14, the tensor-valued representation theorem (2.14.3) gives the specific form for the response function, and thus the constitutive law becomes

$$\boldsymbol{T} = \boldsymbol{g}(\boldsymbol{B}) = \phi_0 \boldsymbol{I} + \phi_1 \boldsymbol{B} + \phi_2 \boldsymbol{B}^2 \tag{8.3.4}$$

where the coefficients ϕ_i are functions of the invariants of $\boldsymbol{B}$, $\phi_i = \phi_i(I_B, II_B, III_B)$. Using the Cayley–Hamilton Theorem, see result (2.13.7), we can eliminate the $\boldsymbol{B}^2$ term and write (8.3.4) as

$$\boldsymbol{T} = \alpha_0 \boldsymbol{I} + \alpha_1 \boldsymbol{B} + \alpha_2 \boldsymbol{B}^{-1} \tag{8.3.5}$$

where again $\alpha_i = \alpha_i(I_B, II_B, III_B)$.

Next, if we consider *incompressible materials*, $III_B = \det \boldsymbol{B} = \det(\boldsymbol{F}\boldsymbol{F}^{\mathrm{T}}) = (\det \boldsymbol{F})^2 = 1$, and thus $\alpha_i = \alpha_i(I_B, II_B)$. Also since it can be shown (Exercise 2.19) that $I_{B^{-1}} = II_B / III_B$, and so for this case $II_B = I_{B^{-1}}$ and therefore $\alpha_i = \alpha_i(I_B, I_{B^{-1}})$. In general, for incompressible materials, the stress can only be determined up to an arbitrary hydrostatic pressure. This is because any pressure applied to an incompressible material will not produce volumetric change and thus the stress cannot be uniquely determined from the strains. Combining all of this reduces constitutive relation (8.3.5) to

$$\boldsymbol{T} = -p\boldsymbol{I} + \alpha_1 \boldsymbol{B} + \alpha_2 \boldsymbol{B}^{-1} \tag{8.3.6}$$

We could also proceed to develop the finite elastic constitutive law from strain energy concepts as previously given for the linear elastic case by relation (6.2.6). For the finite strain case, the Cauchy stress can be expressed as (Asaro and Lubarda, 2006)

$$T_{ij} = B_{ik} \frac{\partial U}{\partial B_{kj}} + \frac{\partial U}{\partial B_{ik}} B_{kj} \tag{8.3.7}$$

Choosing the isotropic incompressible case, the strain energy function will only depend on the first two invariants, $U = U(I_B, II_B)$, and thus

$$\frac{\partial U}{\partial B_{ij}} = \frac{\partial U}{\partial I_B}\frac{\partial I_B}{\partial B_{ij}} + \frac{\partial U}{\partial II_B}\frac{\partial II_B}{\partial B_{ij}} \tag{8.3.8}$$

Completing the calculations, we can develop the constitutive law using either the $\boldsymbol{B}^2$ or $\boldsymbol{B}^{-1}$ form as

$$\begin{aligned} T_{ij} + p\delta_{ij} &= 2\left(\frac{\partial U}{\partial I_B} + I_B \frac{\partial U}{\partial II_B}\right) B_{ij} - 2\frac{\partial U}{\partial II_B} B_{ik} B_{kj} \\ T_{ij} + p\delta_{ij} &= 2\frac{\partial U}{\partial I_B} B_{ij} - 2\frac{\partial U}{\partial II_B} B_{ij}^{-1} \end{aligned} \tag{8.3.9}$$

where we have absorbed all hydrostatic terms into p. Forms (8.3.9) represent the *hyperelastic* development and correspond completely with the previous relations using the Cauchy elastic scheme.

Over the years, specific simplified forms of the strain energy function have been developed. For the incompressible case, these have been created starting with the particular general form

$$U = U(I_B, II_B) = \sum_{m,n=0}^{\infty} C_{mn}(I_B - 3)^m (II_B - 3)^n \tag{8.3.10}$$

where C_{mn} are constants, and $C_{00} = 0$. Notice that in the reference configuration, $B_{ij} = \delta_{ij}$ and thus $I_B = II_B = 3$. Thus, this form automatically satisfies the condition that the strain energy vanishes in the reference configuration. Based on this, several specific models have been developed and two of the more common examples are

- *neo-Hookean*: $U = C_{10}(I_B - 3) \Rightarrow \alpha_1 = 2C_{10}, \alpha_2 = 0$;
- *Mooney–Rivlin model*: $U = C_{10}(I_B - 3) + C_{01}(II_B - 3) \Rightarrow \alpha_1 = 2C_{10}, \alpha_2 = -2C_{01}$.

Other forms have also been proposed in the literature (see, e.g. Ogden, 1972; Arruda and Boyce, 1993).

Our study will be limited to only static mechanical behaviors, and thus the stress field will be required to satisfy the equilibrium equations either in terms of the Cauchy stress in the current configuration (5.3.8) or in terms of the Piola–Kirchhoff stress in the reference configuration (5.3.13). Boundary conditions will be basically the same as those used in linear elasticity: boundary specification of either the displacement (motion) or tractions.

8.3.2 PROBLEM SOLUTIONS

In order to illustrate some of the features of nonlinear elasticity, we now explore the solution to several example problems. Many applications of finite elasticity are used to model rubber and polymeric materials which are nearly incompressible. Thus, we will restrict the following examples to this type of material and employ constitutive relation (8.3.6).

EXAMPLE 8.3.1 UNIAXIAL EXTENSION OF AN INCOMPRESSIBLE NONLINEAR ELASTIC BAR

Consider a nonlinear elastic bar under uniform uniaxial loading as shown in Fig. 8.1. Assume that the material is incompressible and that the deformation is uniform and purely extensional. Determine the Cauchy stress field form. Next, eliminate the arbitrary pressure term and find the Cauchy and first Piola–Kirchhoff axial stress in terms of the stretch ratio. Plot and compare these behaviors for the case of neo-Hookean and Mooney–Rivlin constitutive models.

Solution: From Example 3.6.2, extensional motion and its deformation gradient tensor are given by

$$\begin{aligned} x_1 &= \lambda_1 X_1 \\ x_2 &= \lambda_2 X_2 \\ x_3 &= \lambda_3 X_3 \end{aligned} \Rightarrow \boldsymbol{F} = \frac{\partial \boldsymbol{x}}{\partial \boldsymbol{X}} = \begin{bmatrix} \lambda_1 & 0 & 0 \\ 0 & \lambda_2 & 0 \\ 0 & 0 & \lambda_3 \end{bmatrix}$$

where λ_i are the stretch ratios. Since the deformation is uniform in the X_2 and X_3 directions, $\lambda_2 = \lambda_3$. Assuming sample incompressibility, we must have $\det \boldsymbol{F} = \lambda_1\lambda_2\lambda_3 = 1$ and this implies that

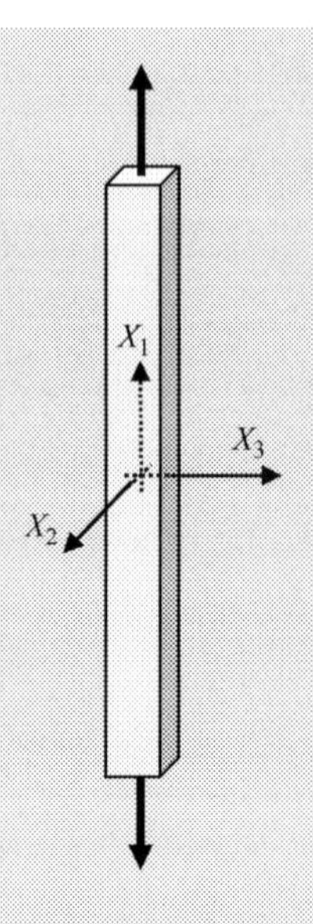

FIGURE 8.1

Uniaxial extension of elastic sample.

$$\lambda_1 = \lambda, \quad \text{and} \quad \lambda_2 = \lambda_3 = \frac{1}{\sqrt{\lambda}}$$

and thus the entire deformation field is determined by the single stretch ratio parameter λ. Under these conditions, the deformation tensors then become

$$\boldsymbol{F} = \begin{bmatrix} \lambda & 0 & 0 \\ 0 & \lambda^{-1/2} & 0 \\ 0 & 0 & \lambda^{-1/2} \end{bmatrix} \Rightarrow$$

$$\boldsymbol{B} = \boldsymbol{F}\boldsymbol{F}^{\mathrm{T}} = \begin{bmatrix} \lambda & 0 & 0 \\ 0 & \lambda^{-1/2} & 0 \\ 0 & 0 & \lambda^{-1/2} \end{bmatrix} \begin{bmatrix} \lambda & 0 & 0 \\ 0 & \lambda^{-1/2} & 0 \\ 0 & 0 & \lambda^{-1/2} \end{bmatrix} = \begin{bmatrix} \lambda^2 & 0 & 0 \\ 0 & 1/\lambda & 0 \\ 0 & 0 & 1/\lambda \end{bmatrix}$$

$$\boldsymbol{B}^{-1} = \begin{bmatrix} \lambda^{-2} & 0 & 0 \\ 0 & \lambda & 0 \\ 0 & 0 & \lambda \end{bmatrix}$$

The invariants are given by $I_B = \lambda^2 + 2/\lambda$ and $I_{B^{-1}} = \lambda^{-2} + 2\lambda$. The stresses follow from (8.3.6): $\boldsymbol{T} = -p\boldsymbol{I} + \alpha_1\boldsymbol{B} + \alpha_2\boldsymbol{B}^{-1} \Rightarrow$

$$\begin{aligned} T_{11} &= -p + \alpha_1\lambda^2 + \alpha_2\lambda^{-2} \\ T_{22} &= T_{33} = -p + \alpha_1\frac{1}{\lambda} + \alpha_2\lambda \\ T_{12} &= T_{23} = T_{31} = 0 \end{aligned} \tag{8.3.11}$$

It is observed that the stresses are all constant (homogeneous) and thus they will automatically satisfy the equations of equilibrium. Furthermore, since the sides of the sample are stress-free, $T_{22} = T_{33} = 0$ everywhere in the sample. Using this fact in relations (8.3.11) allows us to solve for the hydrostatic pressure

$$p = \alpha_1 \frac{1}{\lambda} + \alpha_2 \lambda$$

Substituting this result back into $(8.3.11)_1$, we can solve for the axial stress

$$T_{11} = \left(\lambda^2 - \frac{1}{\lambda}\right)\left(\alpha_1 - \frac{\alpha_2}{\lambda}\right) \tag{8.3.12}$$

The first Piola–Kichhoff stresses are found using relation (4.7.5) with $J = 1$:

$$\boldsymbol{T}^o = \boldsymbol{T}(\boldsymbol{F}^{-1})^T = \begin{bmatrix} T_{11} & 0 & 0 \\ 0 & 0 & 0 \\ 0 & 0 & 0 \end{bmatrix} \begin{bmatrix} 1/\lambda & 0 & 0 \\ 0 & \lambda^{1/2} & 0 \\ 0 & 0 & \lambda^{1/2} \end{bmatrix} = \begin{bmatrix} T_{11}/\lambda & 0 & 0 \\ 0 & 0 & 0 \\ 0 & 0 & 0 \end{bmatrix} \tag{8.3.13}$$

Thus, the only nonzero PK1 stress is $T_{11}^o = T_{11}/\lambda$. Note that referring back to Example 3.7.1, Nanson's formula (3.7.3) was used for the extension problem. Applying those results for this case, it is found that the reference cross-section area is related to the current area by the relation $A_o = \lambda A$. It should be pointed out that these results are consistent with the requirement that the total force on the bar cross-section be equal in both the reference and current configurations, that is, $T_{11}^o A_o = T_{11} A$.

Next, we wish to compare the Cauchy and PK1 axial stress predictions for the neo-Hookean and Mooney–Rivlin models. Generated using MATLAB Code C-20, Fig. 8.2 shows plots of each stress for particular choices of material constants. Note that since λ is dimensionless, the stress units in the figure correspond to the units associated with the choice of material constants α_i. Both stresses generally show similar nonlinear behavior versus the stretch λ. However, because of the scaling factor λ, there is considerable difference between the magnitudes of the two stress measures, with $T_{11} > T_{11}^o$.

Finally, we wish to explore some comparisons with experimental data for this extensional deformation problem. Early work by Rivlin and Saunders (1951) provided some basic experimental data on vulcanized rubber material, which is shown in Fig. 8.3A. This data was replotted and curve fit as illustrated in Fig. 8.3B. Using these numerical results (based on data), a Cauchy stress versus stretch plot was constructed and is shown in Fig. 8.3C. For comparison, a Mooney–Rivlin model (with parameters selected to best match with the data prediction) is also shown in this figure. All results were done using MATLAB Code C-22 with the data set shown in Appendix C. Although the Mooney–Rivlin predictions match closely with the data, in general, for other types of deformations, this will not be so, and other more sophisticated models generally result in better comparisons.

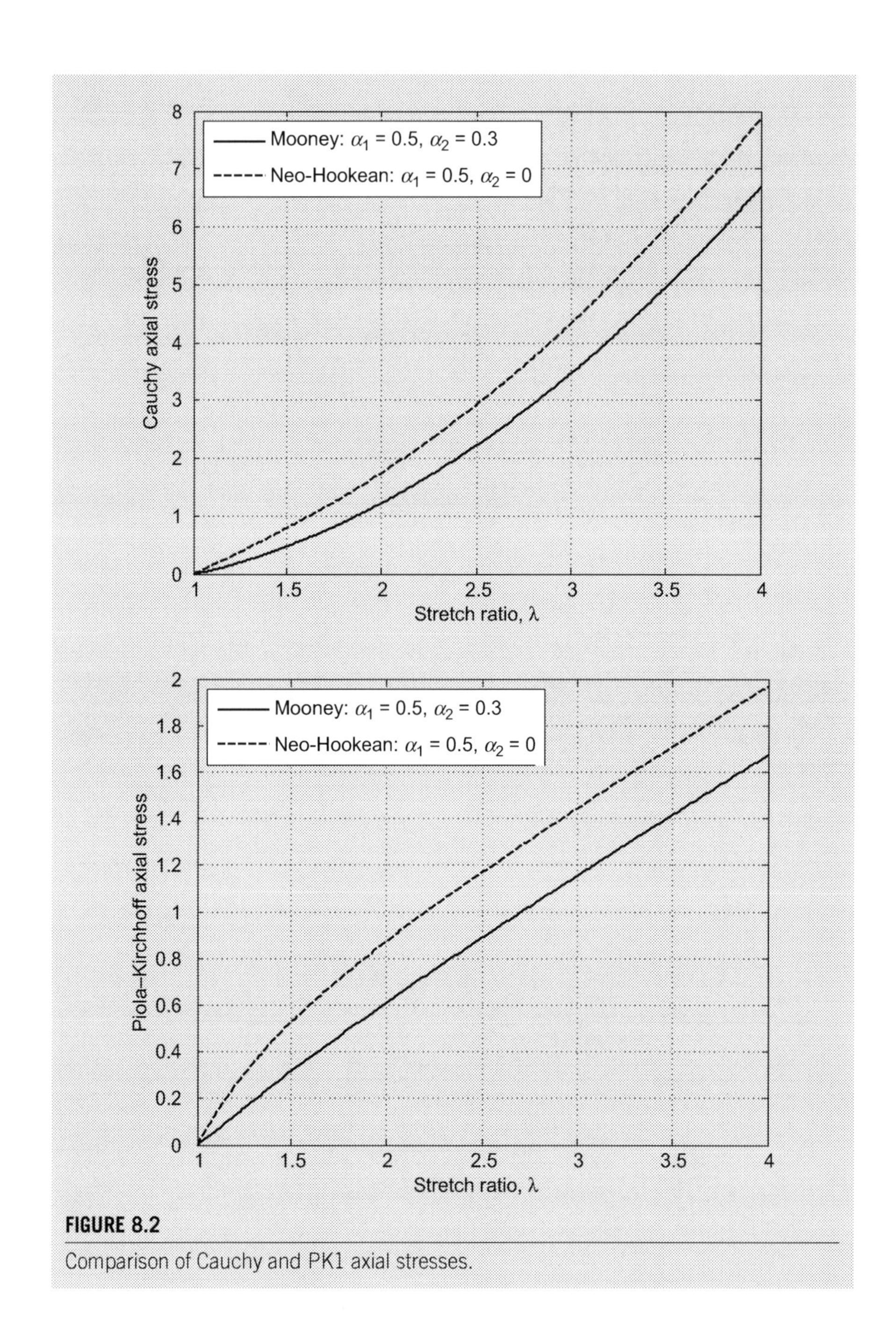

FIGURE 8.2

Comparison of Cauchy and PK1 axial stresses.

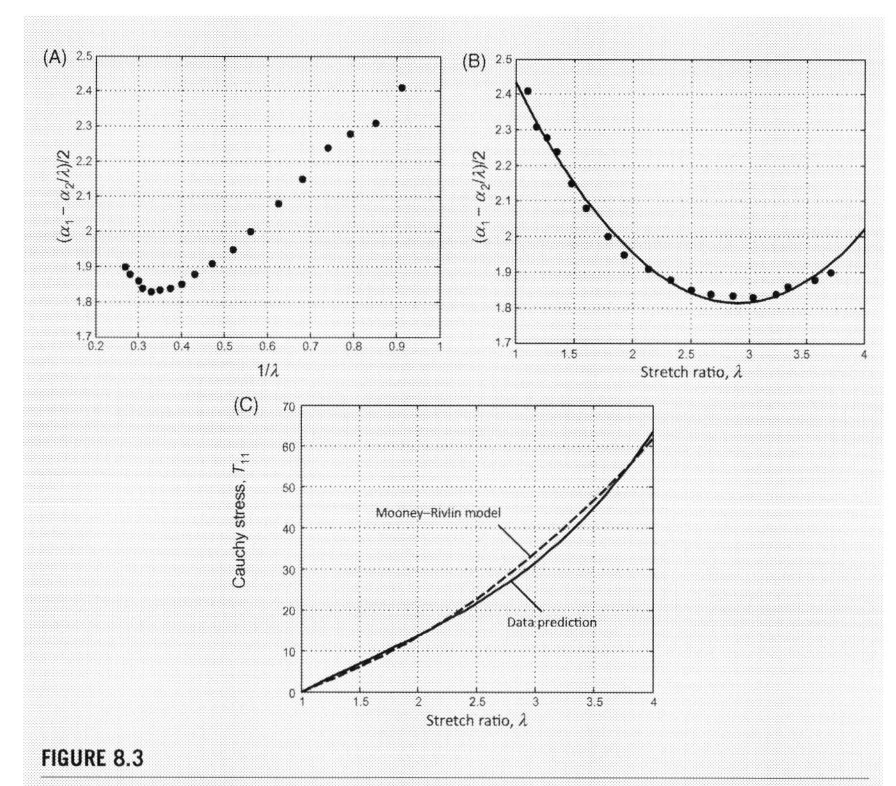

FIGURE 8.3

Comparison of Mooney–Rivlin model with experimental prediction from Rivlin and Saunders (1951): (A) Rivlin and Saunders (1951) data (reprinted with permission of Royal Society); (B) polynomial curve fit to data; (C) comparison of stress predictions.

EXAMPLE 8.3.2 SIMPLE SHEAR DEFORMATION OF A NONLINEAR ELASTIC SAMPLE

A block of a nonlinear incompressible elastic material is to undergo simple shearing deformation as shown in Fig. 8.4. Determine the resulting stress field required to support this deformation and explore other features of the problem which differ from the corresponding linear elastic problem.

Solution: The deformation and several strain tensors were previously given in Example 3.6.2. Collecting this information

$$\begin{aligned} x_1 &= X_1 + \gamma X_2 \\ x_2 &= X_2 \qquad \Rightarrow \\ x_3 &= X_3 \end{aligned}$$

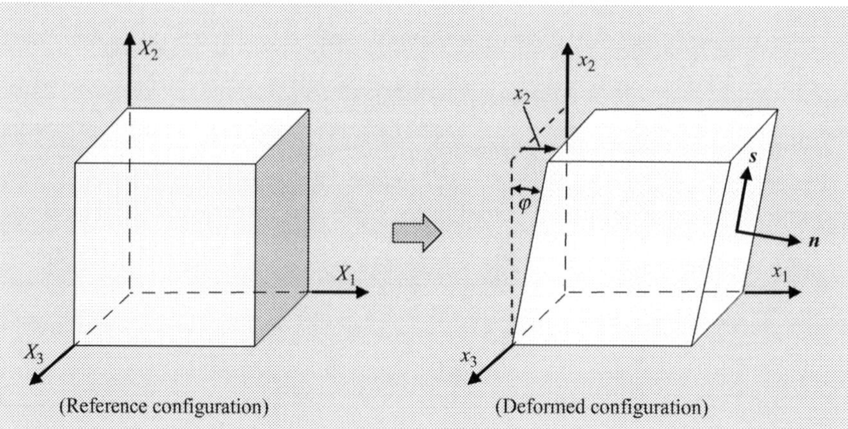

FIGURE 8.4

Simple shear deformation.

$$F = \begin{bmatrix} 1 & \gamma & 0 \\ 0 & 1 & 0 \\ 0 & 0 & 1 \end{bmatrix}, \quad B = \begin{bmatrix} 1+\gamma^2 & \gamma & 0 \\ \gamma & 1 & 0 \\ 0 & 0 & 1 \end{bmatrix}, \quad B^{-1} = \begin{bmatrix} 1 & -\gamma & 0 \\ -\gamma & 1+\gamma^2 & 0 \\ 0 & 0 & 1 \end{bmatrix}$$

The invariants are given by $I_B = I_{B^{-1}} = 3+\gamma^2$ and this implies that $\alpha_i = \alpha_i(\gamma^2)$. Note that $\det F = 1$, and hence the deformation is isochoric. The stresses follow from (8.3.6):

$$T = -pI + \alpha_1 B + \alpha_2 B^{-1} \Rightarrow$$

$$\begin{aligned} T_{11} &= -p + \alpha_1(1+\gamma^2) + \alpha_2 \\ T_{22} &= -p + \alpha_1 + \alpha_2(1+\gamma^2) \\ T_{33} &= -p + \alpha_1 + \alpha_2 \\ T_{12} &= (\alpha_1 - \alpha_2)\gamma \\ T_{23} &= T_{31} = 0 \end{aligned} \tag{8.3.14}$$

We can assume a plane stress condition and thus choose $T_{33} = 0$. This allows the determination of the hydrostatic pressure $p = \alpha_1 + \alpha_2$, and thus the nonzero stresses become

$$\begin{aligned} T_{11} &= \alpha_1\gamma^2 \\ T_{22} &= \alpha_2\gamma^2 \\ T_{12} &= (\alpha_1 - \alpha_2)\gamma \end{aligned} \tag{8.3.15}$$

Notice that this system, either (8.3.14) or (8.3.15) produces the *universal relation* of Rivlin

$$\gamma = \frac{T_{11} - T_{22}}{T_{12}} \tag{8.3.16}$$

which is independent of the material functions α_i. As in the previous example, the stresses are all constant (homogeneous) and thus they will automatically satisfy the equations of equilibrium.

The quantity $\alpha_1 - \alpha_2$ represents the generalized shear modulus for the nonlinear elastic problem, and since $\alpha_i = \alpha_i(\gamma^2)$, we can write a power series expansion for small γ as

$$\alpha_1 - \alpha_2 = \mu + \bar{\mu}\gamma^2 + O(\gamma^4), \quad \gamma \ll 1$$

where μ and $\bar{\mu}$ are constants. This allows the representation of the shear stress as

$$T_{12} = \mu\gamma + \bar{\mu}\gamma^3 + O(\gamma^5) \tag{8.3.17}$$

Thus, the shear stress is an odd function of the shear magnitude γ and the first term corresponds to the linear elastic case. Using a similar analysis, we can show that the normal stresses T_{11} and T_{22} will be even functions of γ with the first-order term $O(\gamma^2)$. Notice that from the linear elastic case, $T_{22} = 0$, but for nonlinear elasticity $T_{22} \neq 0$. Also the case here with $T_{11} \neq T_{22}$ is sometimes called the *Poynting effect.*

Consider next the tractions on the inclined face with normal vector $\boldsymbol{n}$ and tangent vector $\boldsymbol{s}$ as shown in Fig. 8.4. Using the fact that $\sin\varphi = \dfrac{\gamma}{\sqrt{1+\gamma^2}}, \cos\varphi = \dfrac{1}{\sqrt{1+\gamma^2}}$, these unit vectors are easily determined as

$$\boldsymbol{n} = \frac{1}{\sqrt{1+\gamma^2}}(\boldsymbol{e}_1 - \gamma\boldsymbol{e}_2), \quad \boldsymbol{s} = \frac{1}{\sqrt{1+\gamma^2}}(\gamma\boldsymbol{e}_1 + \boldsymbol{e}_2) \tag{8.3.18}$$

The traction vector on this plane is given by

$$\boldsymbol{t} = \boldsymbol{T}\boldsymbol{n} = \begin{bmatrix} T_{11} & T_{12} & 0 \\ T_{12} & T_{22} & 0 \\ 0 & 0 & 0 \end{bmatrix} \begin{bmatrix} 1/\sqrt{1+\gamma^2} \\ -\gamma/\sqrt{1+\gamma^2} \\ 0 \end{bmatrix} = \frac{(T_{11} - \gamma T_{12})}{\sqrt{1+\gamma^2}}\boldsymbol{e}_1 + \frac{(T_{12} - \gamma T_{22})}{\sqrt{1+\gamma^2}}\boldsymbol{e}_2 \tag{8.3.19}$$

and thus the normal and shear traction components are

$$N = \boldsymbol{t}\cdot\boldsymbol{n} = T_{22} - \frac{\gamma T_{12}}{1+\gamma^2}, \quad S = \boldsymbol{t}\cdot\boldsymbol{s} = \frac{T_{12}}{1+\gamma^2} \tag{8.3.20}$$

where we have used relation (8.3.16). Note as $\gamma \to 0$, relations (8.3.20) give $N \to T_{22}$ and $S \to T_{12}$.

EXAMPLE 8.3.3 TORSION OF A NONLINEAR ELASTIC CYLINDER

Consider a nonlinear elastic cylinder of circular section with a radius a in the reference configuration as shown in Fig. 8.5. The cylinder is to undergo torsional deformation and thus using cylindrical coordinates, the deformation can be stated as

$$r = R, \quad \theta = \Theta + \kappa Z, \quad z = Z \tag{8.3.21}$$

where (r, θ, z) are the spatial coordinates, (R, Θ, Z) are the reference coordinates, and κ represents the angle of twist per unit length. The cylinder is assumed to be traction-free along the lateral sides $r = a$. Determine the resulting stress field that corresponds to this deformation.

Solution: For this case, the deformation gradient can be calculated using relation (3.17.10):

$$\boldsymbol{F} = \begin{bmatrix} \dfrac{\partial r}{\partial R} & \dfrac{1}{R}\dfrac{\partial r}{\partial \Theta} & \dfrac{\partial r}{\partial Z} \\ r\dfrac{\partial \theta}{\partial R} & \dfrac{r}{R}\dfrac{\partial \theta}{\partial \Theta} & r\dfrac{\partial \theta}{\partial Z} \\ \dfrac{\partial z}{\partial R} & \dfrac{1}{R}\dfrac{\partial z}{\partial \Theta} & \dfrac{\partial z}{\partial Z} \end{bmatrix} = \begin{bmatrix} 1 & 0 & 0 \\ 0 & 1 & \kappa r \\ 0 & 0 & 1 \end{bmatrix}$$

Note that $\det F = 1$. The left Cauchy–Green strain tensor becomes

$$\boldsymbol{B} = \boldsymbol{F}\boldsymbol{F}^{\mathrm{T}} = \begin{bmatrix} 1 & 0 & 0 \\ 0 & 1 & \kappa r \\ 0 & 0 & 1 \end{bmatrix} \begin{bmatrix} 1 & 0 & 0 \\ 0 & 1 & 0 \\ 0 & \kappa r & 1 \end{bmatrix} = \begin{bmatrix} 1 & 0 & 0 \\ 0 & 1+\kappa^2 r^2 & \kappa r \\ 0 & \kappa r & 1 \end{bmatrix}$$

and its inverse is found to be

$$\boldsymbol{B}^{-1} = \begin{bmatrix} 1 & 0 & 0 \\ 0 & 1 & -\kappa r \\ 0 & -\kappa r & 1+\kappa^2 r^2 \end{bmatrix}$$

The invariants are given by $I_B = I_{B^{-1}} = 3 + \kappa^2 r^2$ and this implies that $\alpha_i = \alpha_i(\kappa^2 r^2)$, and thus the stresses will only be functions of the radial coordinate. The stresses follow from (8.3.6):

$$\boldsymbol{T} = -p\boldsymbol{I} + \alpha_1 \boldsymbol{B} + \alpha_2 \boldsymbol{B}^{-1} \Rightarrow$$

$$\begin{aligned} T_{rr} &= -p + \alpha_1 + \alpha_2 \\ T_{\theta\theta} &= -p + \alpha_1(1+\kappa^2 r^2) + \alpha_2 \\ T_{zz} &= -p + \alpha_1 + \alpha_2(1+\kappa^2 r^2) \\ T_{\theta z} &= (\alpha_1 - \alpha_2)\kappa r \\ T_{r\theta} &= T_{rz} = 0 \end{aligned} \tag{8.3.22}$$

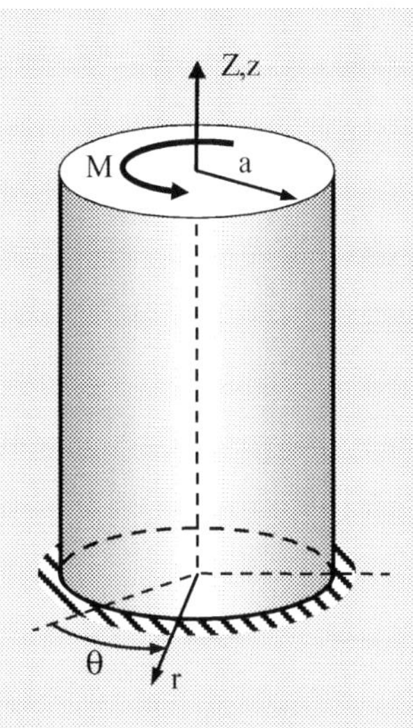

FIGURE 8.5

Torsion of an elastic cylinder.

The equations of equilibrium reduce to

$$\begin{aligned}&\frac{\partial T_{rr}}{\partial r}+\frac{T_{rr}-T_{\theta\theta}}{r}=0\\&\frac{\partial p}{\partial \theta}=\frac{\partial p}{\partial z}=0\end{aligned}\tag{8.3.23}$$

From the last two equations, $p = p(r)$, and thus the first equation becomes

$$-\frac{dp}{dr}+\frac{d}{dr}(\alpha_1+\alpha_2)-\alpha_1\kappa^2 r=0$$

and this can be integrated to give a form for the pressure

$$p(r)=(\alpha_1+\alpha_2)-\int^{r}\alpha_1\kappa^2 r'\,\mathrm{d}r'+K\tag{8.3.24}$$

where K is the arbitrary constant of integration. Now the traction-free boundary condition on the lateral sides $r = a$ becomes

$$T_{rr}(a)=0\Rightarrow p(a)=(\alpha_1+\alpha_2)_{r=a}\tag{8.3.25}$$

Using relation (8.3.24) in (8.3.25) gives

$$K=\int^{a}\alpha_1\kappa^2 r'\,\mathrm{d}r'\tag{8.3.26}$$

and thus the pressure relation becomes

$$p(r)=(\alpha_1+\alpha_2)+\int_{r}^{a}\alpha_1\kappa^2 r'\,\mathrm{d}r'\tag{8.3.27}$$

The nonzero stresses then become

$$
\begin{aligned}
T_{rr} &= -\int_r^a \alpha_1 \kappa^2 r' \, dr' \\
T_{\theta\theta} &= \alpha_1 \kappa^2 r^2 - \int_r^a \alpha_1 \kappa^2 r' \, dr' \\
T_{zz} &= \alpha_2 \kappa^2 r^2 - \int_r^a \alpha_1 \kappa^2 r' \, dr' \\
T_{\theta z} &= (\alpha_1 - \alpha_2)\kappa r
\end{aligned}
\tag{8.3.28}
$$

The resultant torsional moment M on the free end can be calculated by

$$
M = \int_0^a T_{\theta z} 2\pi r^2 \, dr = 2\pi\kappa \int_0^a (\alpha_1 - \alpha_2) r^3 \, dr \tag{8.3.29}
$$

From (8.3.28), the T_{zz} stress component at any section can be expressed by $T_{zz} = \alpha_2 \kappa^2 r^2 + T_{rr}$, and thus the total normal force on any section is given by

$$
N = \int_0^a T_{zz} 2\pi r \, dr = 2\pi \int_0^a (\alpha_2 \kappa^2 r^3 + r T_{rr}) dr \tag{8.3.30}
$$

Now using integration by parts and the equilibrium equation, we can evaluate the integral

$$
\begin{aligned}
\int_0^a r T_{rr} \, dr &= \left[\frac{1}{2} r^2 T_{rr} \right]_0^a - \int_0^a \left(\frac{1}{2} r^2 \frac{dT_{rr}}{dr} \right) dr \\
&= \int_0^a \left(\frac{1}{2} r^2 \frac{T_{rr} - T_{\theta\theta}}{r} \right) dr \\
&= -\frac{1}{2} \int_0^a \alpha_1 \kappa^2 r^3 \, dr
\end{aligned}
$$

and using this result, the normal force becomes

$$
N = 2\pi \int_0^a \alpha_2 \kappa^2 r^3 \, dr - \pi \int_0^a \alpha_1 \kappa^2 r^3 \, dr \tag{8.3.31}
$$

Note that unlike the linear elastic model, the normal force here will not be zero, and in fact most specific nonlinear models would predict $N < 0$.

Given the specific forms of the constitutive function $\alpha_i = \alpha_i(\kappa^2 r^2)$, all stresses, and the twisting moment and normal force can be calculated. Note that for the neo-Hookean model ($\alpha_1 = \text{constant}, \alpha_2 = 0$), and thus the shear stress $T_{\theta z} = \alpha_1 \kappa r$ which is the same form as linear elasticity; however, $N = -\pi \alpha_1 \kappa^2 \dfrac{a^4}{4} \neq 0$.

Some additional nonlinear elastic problems are given in the exercises.

8.4 NONLINEAR VISCOUS FLUIDS

Our previous discussion on incompressible fluid dynamics in Section 6.4 presented the linear viscous or Newtonian model governed by the constitutive relation

$$\boldsymbol{T} + p\boldsymbol{I} = 2\mu_o \boldsymbol{D} \tag{8.4.1}$$

where μ_o is the constant material viscosity parameter and for incompressible fluids the rate of deformation tensor satisfies $\operatorname{tr} \boldsymbol{D} = 0$. The quantity $\boldsymbol{T} - p\boldsymbol{I}$ is often called the *extra stress*. This constitutive law leads to the Navier–Stokes equations and results in accurate predictions for many fluid mechanics problems. Typical fundamental predictions of this model include a linear shear stress–shear strain behavior (6.4.8) and a Hagen–Poiseuille flow with the volumetric flow rate proportional to the pressure gradient (6.4.33). However, there exist many fluids (polymeric melts, paints, gels, colloidal suspensions, biological, and food liquids) with behaviors that do not match these predictions. Because of this, more sophisticated constitutive fluid laws have been developed to provide appropriate theoretical modeling of such flows. These nonclassical fluid models are generally referred to as *non-Newtonian* and they commonly incorporate nonlinear and memory effects in the constitutive relations. Because of the memory effects, these models often include a viscoelastic response. Originally, much of this non-Newtonian study was done under the name of *rheology* defined as *the study of flow and deformation of matter* (see discussion in Section 1.4). We will now present a few basic models focusing on such incompressible fluids. Further details on this topic can be found in Truesdell and Noll (1965), Coleman et al. (1966), Truesdell (1974), and Bird et al. (1987). Although some past simple models only modify the viscosity in relation (8.4.1), we explore more fundamental theories that alter the constitutive relation in much more general and meaningful ways.

8.4.1 REINER–RIVLIN FLUID

Following along a similar approach as that was used for the elastic solids, it would be logical to propose a general nonlinear incompressible fluid model where the stress (less the hydrostatic pressure) would be expressed in terms of a nonlinear function of the rate of deformation tensor:

$$\boldsymbol{T} + p\boldsymbol{I} = \boldsymbol{g}(\boldsymbol{D}) \tag{8.4.2}$$

Note that although the velocity and the velocity gradient are not objective, the rate of deformation does satisfy the objectivity requirement. Since fluids are generally assumed to be isotropic, the material function $\boldsymbol{g}$ must be an isotropic function of the rate of deformation, and so using arguments analogous to those in the previous section

$$\boldsymbol{T} + p\boldsymbol{I} = \beta_1 \boldsymbol{D} + \beta_2 \boldsymbol{D}^2 \tag{8.4.3}$$

where $\beta_i = \beta_i(II_D, III_D)$. Note that for an incompressible fluid, $I_D = 0$. Constitutive form (8.4.3) is commonly called a *Reiner–Rivlin fluid* (R–R fluid), named after their independent development back in the 1950s.

If we consider again the shear flow shown in Fig. 6.22 which is specified by $v = \{v_1(x_2), 0, 0\}$, this yields a rate of deformation tensor and its square as

$$D = \frac{1}{2}\begin{bmatrix} 0 & \dot{\gamma} & 0 \\ \dot{\gamma} & 0 & 0 \\ 0 & 0 & 0 \end{bmatrix}, \quad D^2 = \frac{1}{4}\begin{bmatrix} \dot{\gamma}^2 & 0 & 0 \\ 0 & \dot{\gamma}^2 & 0 \\ 0 & 0 & 0 \end{bmatrix} \tag{8.4.4}$$

where $\dot{\gamma} = \dfrac{\partial v_1}{\partial x_2}$. Using the R–R fluid constitutive law (8.4.3) yields the following stresses:

$$\begin{aligned} T_{11} &= T_{22} = -p + \frac{\beta_2}{4}\dot{\gamma}^2, \quad T_{33} = -p \\ T_{12} &= \frac{\beta_1}{2}\dot{\gamma}, \quad T_{23} = T_{31} = 0 \end{aligned} \tag{8.4.5}$$

where $\beta_i = \beta_i(\dot{\gamma}^2)$. For this case, we note that the traditional form for the viscosity is now given by $\mu = T_{12}/\dot{\gamma} = \beta_1(\dot{\gamma}^2)/2$, which is no longer a constant. Furthermore, while the Newtonian model predicted zero normal stresses, the R–R fluid indicates nonzero T_{11} and T_{22}. Thus, we find *normal stress effects* present in a solely shearing type of flow geometry. These effects will also be present in more complicated flow fields such as in a rotating cylindrical viscometer or in pipe flow. Many of these results match qualitatively with experimental observations of non-Newtonian flows.

However, a major negative issue of our findings in this shear flow example is that $T_{11} = T_{22}$. This result occurs regardless of the material parameters and is thus universal for this flow. The result is unexpected since the two normal stresses act on totally different planes. Sadly this result does not match with experimental results, thereby indicating that our R–R constitutive model is not expected to consistently give quantitative predictions that correspond with real material behavior. We conclude that even though our constitutive creation concepts seemed reasonable, evidently some aspects of the nonlinear fluid behavior were not properly account for. This situation leads us to consider other more general constitutive schemes.

8.4.2 SIMPLE INCOMPRESSIBLE FLUID

Recall Noll's development of the simple material given by relation (8.2.3). For an incompressible fluid, this form would become

$$T + pI = \underset{s=0}{\overset{s=\infty}{\mathfrak{F}}}\left(F_t(t-s)\right) \tag{8.4.6}$$

It can be shown that this relation is reducible to

$$T + pI = \underset{s=0}{\overset{s=\infty}{\hat{\mathfrak{F}}}}\left(G_t(t-s)\right) \tag{8.4.7}$$

where $G_t(t-s) = C_t(t-s) - I$ and $C_t(t-s)$ is the relative right Cauchy–Green strain (3.15.6). We assume that when the deformation goes to zero, the stress reduces to just hydrostatic pressure and thus $\underset{s=0}{\overset{s=\infty}{\tilde{\Im}}}(\mathbf{0}) = \underset{s=0}{\overset{s=\infty}{\hat{\Im}}}(\mathbf{0}) = 0$.

Green and Rivlin (1957) proved that this general constitutive functional can be expressed as a series of integrals, thus expressing (8.4.7) in the form

$$T + pI = \int_0^\infty M_1(s) G_t(t-s)\,ds + \int_0^\infty \int_0^\infty M_2(s_1, s_2) G_t(t-s_2)\,ds_1\,ds_2 + \cdots \tag{8.4.8}$$

This integral representation can be truncated to model finite linear and infinitesimal theories, and if we retain only the first term

$$T + pI = \int_0^\infty M_1(s) G_t(t-s)\,ds \tag{8.4.9}$$

which would represent a finite linear viscoelastic fluid. Compare this with the linear integral viscoelastic relation (6.5.38). Although the general formulation (8.4.8) is elegant, it is difficult to apply to specific flow problems, and thus we will not pursue further details of this integral representation. However, we will explore the general simple fluid model as represented by (8.4.7) in more detail later.

8.4.3 RIVLIN–ERICKSEN FLUID

Rivlin and Ericksen (1955) and later Truesdell and Noll (1965) presented a differential rate type representation scheme for nonlinear isotropic materials. We could assume for simple materials that the deformation gradient history $F_t(t-s)$ has constitutive influence on the stress only over a short period of time for small s. Further assuming F has continuous time derivatives, we could expand it in terms of a Taylor series near $s = 0$. For such a case, the response functional $\tilde{\Im}$ reduces to an ordinary nonlinear function f, that is, $T = f(F, \dot{F}, \ddot{F}, \dddot{F}, \ldots)$. Such a constitutive case is often referred to as a *material of the differential type.* For the incompressible fluid case, the constitutive form simplifies to

$$T + p\mathbf{I} = f(A_1, A_2, A_3, \ldots, A_n) \tag{8.4.10}$$

where $A_1, A_2, A_3, \ldots, A_n$ are the Rivlin–Ericksen tensors previously defined in Section 3.16:

$$A^{(n)} = \frac{D^n}{D\tau^n}[C_t(\tau)]_{\tau=t} = \frac{D^n}{D\tau^n}\left[F_t^{\mathrm{T}}(\tau) F_t(\tau)\right]_{\tau=t} \tag{8.4.11}$$

or

$$\begin{aligned} A_1 &= L + L^T = 2D \\ A_2 &= \frac{D}{Dt} A^{(1)} + A^{(1)} L + L^T A^{(1)} \\ &\vdots \\ A_n &= \frac{D}{Dt} A^{(n-1)} + A^{(n-1)} L + L^T A^{(n-1)} \end{aligned} \tag{8.4.12}$$

Since we are dealing with isotropic materials, the response function must satisfy the usual mathematical isotropic property

$$\boldsymbol{Q}\boldsymbol{f}(\boldsymbol{A}_1,\boldsymbol{A}_2,\boldsymbol{A}_3,\ldots)\boldsymbol{Q}^T = \boldsymbol{f}(\boldsymbol{Q}\boldsymbol{A}_1\boldsymbol{Q}^T,\boldsymbol{Q}\boldsymbol{A}_2\boldsymbol{Q}^T,\boldsymbol{Q}\boldsymbol{A}_3\boldsymbol{Q}^T,\ldots,\boldsymbol{Q}\boldsymbol{A}_n\boldsymbol{Q}^T) \tag{8.4.13}$$

We can use this constitutive scheme including various orders n and can then apply appropriate representation theorems from Section 2.14 to express the specific constitutive forms. Truncation at zero order gives $\boldsymbol{T}=-p\boldsymbol{I}$ (inviscid model), while truncation at first order yields the Newtonian case $\boldsymbol{T}=-p\boldsymbol{I}+2\mu_o\boldsymbol{D}$. The results for order $n = 2$ (sometimes called *complexity* 2) in general give

$$\begin{aligned}\boldsymbol{T}&=-p\boldsymbol{I}+\boldsymbol{f}(\boldsymbol{A}_1,\boldsymbol{A}_2)\\&=-p\boldsymbol{I}+\mu_1\boldsymbol{A}_1+\mu_2\boldsymbol{A}_1^2+\mu_3\boldsymbol{A}_2+\mu_4\boldsymbol{A}_2^2\\&\quad+\mu_5(\boldsymbol{A}_1\boldsymbol{A}_2+\boldsymbol{A}_2\boldsymbol{A}_1)+\mu_6(\boldsymbol{A}_1\boldsymbol{A}_2^2+\boldsymbol{A}_2^2\boldsymbol{A}_1)\\&\quad+\mu_7(\boldsymbol{A}_1^2\boldsymbol{A}_2+\boldsymbol{A}_2\boldsymbol{A}_1^2)+\mu_8(\boldsymbol{A}_1^2\boldsymbol{A}_2^2+\boldsymbol{A}_2^2\boldsymbol{A}_1^2)\end{aligned} \tag{8.4.14}$$

where $\mu_i=\mu_i(tr\,\boldsymbol{A}_1^2,tr\,\boldsymbol{A}_1^3,tr\,\boldsymbol{A}_2,tr\,\boldsymbol{A}_2^2,tr\,\boldsymbol{A}_2^3,tr(\boldsymbol{A}_1\boldsymbol{A}_2),tr(\boldsymbol{A}_1^2\boldsymbol{A}_2),tr(\boldsymbol{A}_1\boldsymbol{A}_2^2),tr(\boldsymbol{A}_1^2\boldsymbol{A}_2^2))$

Note that (8.4.14) includes the strain rate terms of third and fourth order, for example, $\boldsymbol{A}_1\boldsymbol{A}_2^2$ and $\boldsymbol{A}_1^2\boldsymbol{A}_2^2$. This complexity of the general $n = 2$ case is often simplified by considering only the second-order terms

$$\boldsymbol{T}=-p\boldsymbol{I}+\mu_1\boldsymbol{A}_1+\mu_2\boldsymbol{A}_1^2+\mu_3\boldsymbol{A}_2 \tag{8.4.15}$$

where $\mu_i=\mu_i(II_D,III_D)$. This constitutive model is normally called a *second-order fluid*.

EXAMPLE 8.4.1 STEADY SIMPLE SHEARING FLOW FOR FLUID MODELS

Consider steady simple shearing flow with a velocity field $\boldsymbol{v}=\{\kappa x_2,0,0\}$, where κ is the rate of shear. Determine the stress fields for the Newtonian, Reiner–Rivlin, and second-order fluid models, and compare the results.

Solution: For this flow field, the various required kinematical tensors are given by

$$\begin{aligned}&\boldsymbol{D}=\frac{1}{2}\boldsymbol{A}_1=\frac{\kappa}{2}\begin{bmatrix}0&1&0\\1&0&0\\0&0&0\end{bmatrix},\quad \boldsymbol{D}^2=\frac{\kappa^2}{4}\begin{bmatrix}1&0&0\\0&1&0\\0&0&0\end{bmatrix}\\&\boldsymbol{A}_2=2\kappa^2\begin{bmatrix}0&0&0\\0&1&0\\0&0&0\end{bmatrix},\quad \boldsymbol{A}_1^2=\kappa^2\begin{bmatrix}1&0&0\\0&1&0\\0&0&0\end{bmatrix}\end{aligned} \tag{8.4.16}$$

For the Newtonian case, $\boldsymbol{T}=-p\boldsymbol{I}+2\mu_o\boldsymbol{D}$, and the stresses become

$$\begin{aligned}&T_{11}=T_{22}=T_{33}=-p\\&T_{12}=\mu_o\kappa,\quad T_{23}=T_{31}=0\end{aligned} \tag{8.4.17}$$

For the R–R fluid, $\boldsymbol{T}=-p\boldsymbol{I}+\beta_1\boldsymbol{D}+\beta_2\boldsymbol{D}^2$, and the stresses are given by

$$T_{11} = T_{22} = -p + \frac{\beta_2}{4}\kappa^2, \quad T_{33} = -p$$
$$T_{12} = \frac{\beta_1}{2}\kappa, \quad T_{23} = T_{31} = 0 \tag{8.4.18}$$

Finally, the second-order fluid model $\boldsymbol{T} = -p\boldsymbol{I} + \mu_1\boldsymbol{A}_1 + \mu_2\boldsymbol{A}_1^2 + \mu_3\boldsymbol{A}_2$ gives

$$\begin{aligned} T_{11} &= -p + \mu_2\kappa^2 \\ T_{22} &= -p + \mu_2\kappa^2 + 2\mu_3\kappa^2 \\ T_{33} &= -p \\ T_{12} &= \mu_1\kappa, \quad T_{23} = T_{31} = 0 \end{aligned} \tag{8.4.19}$$

We observe that both nonlinear theories predict normal stress effects under the shear flow. As previously mentioned, the R–R model predicts $T_{11} = T_{22}$, whereas second-order fluid results give the more realist case of $T_{11} \neq T_{22}$. While the Newtonian model indicates constant viscosity, both nonlinear models predict a variable viscosity coefficient, T_{12}/κ.

8.4.4 VISCOMETRIC FLOWS OF INCOMPRESSIBLE SIMPLE FLUIDS

Coleman et al. (1966) and Truesdell (1974) summarize considerable research on non-Newtonian fluid flow. They present a detailed look at viscometric flows of simple fluids, and this work has led to very useful schemes to characterize such fluids. A *viscometric flow* (sometimes called a *laminar shear flow*) is generally defined by a velocity field of the form

$$v_1 = 0, \quad v_2 = \kappa x_1, \quad v_3 = 0 \tag{8.4.20}$$

where κ is the rate of shear. We have of course already used this type of deformation in our previous examples and have discussed that this flow is commonly used as an approximation of the velocity in various types of viscometers (devices used to measure fluid properties). We now wish to explore more details about this general type of deformation and look at its application for general non-Newtonian memory fluids.

We will generalize the situation by considering a local flow field form (8.4.20) not only in the Cartesian system, but also in other orthogonal systems that might involve curvilinear flow. Using some of our past kinematics from Chapter 3, a viscometric flow (8.4.20) will yield the following deformation tensors:

$$\boldsymbol{F} = \begin{bmatrix} 1 & 0 & 0 \\ \kappa t & 1 & 0 \\ 0 & 0 & 1 \end{bmatrix} \Rightarrow \boldsymbol{F}_t(t-s) = \begin{bmatrix} 1 & 0 & 0 \\ -\kappa s & 1 & 0 \\ 0 & 0 & 1 \end{bmatrix} \Rightarrow \boldsymbol{C}_t(t-s) = \begin{bmatrix} 1+\kappa^2 s^2 & -\kappa s & 0 \\ -\kappa s & 1 & 0 \\ 0 & 0 & 1 \end{bmatrix} \tag{8.4.21}$$

and the Rivlin–Ericksen tensors

$$\boldsymbol{A}_1 = \kappa\begin{bmatrix} 0 & 1 & 0 \\ 1 & 0 & 0 \\ 0 & 0 & 0 \end{bmatrix}, \quad \boldsymbol{A}_2 = 2\kappa^2\begin{bmatrix} 1 & 0 & 0 \\ 0 & 0 & 0 \\ 0 & 0 & 0 \end{bmatrix}, \quad \boldsymbol{A}_n = 0 \quad (n \geq 3) \tag{8.4.22}$$

Note that for this case

$$C_t(t-s) = I - sA_1 + \frac{s^2}{2}A_2 \tag{8.4.23}$$

In terms of a matrix N,

$$N = \begin{bmatrix} 0 & 0 & 0 \\ 1 & 0 & 0 \\ 0 & 0 & 0 \end{bmatrix} \tag{8.4.24}$$

we can express the Rivlin–Ericksen tensors as

$$\begin{aligned} A_1 &= \kappa(N + N^T) \\ A_2 &= 2\kappa^2 N^T N \end{aligned} \tag{8.4.25}$$

Relation (8.4.23) indicates that for viscometric flow $C_t(t-s)$ depends only on the two Rivlin–Ericksen tensors A_1 and A_2. For such a case, it has been shown that the general simple fluid constitutive relation (8.4.7) reduces to

$$T = -pI + f(A_1, A_2) \tag{8.4.26}$$

where the general functional has reduced to a simple function f. This of course is a major simplification. Now using the representation form (8.4.14) along with relations (8.4.25) yields

$$T = -pI + \tau(\kappa)(N + N^T) + \sigma_1(\kappa)N^T N + \sigma_2(\kappa)NN^T \tag{8.4.27}$$

where the coefficients $\tau(\kappa), \sigma_1(\kappa), \sigma_2(\kappa)$ are referred to as the *viscometric functions* defined by

$$\begin{aligned} &\tau(\kappa) = T_{12} && \ldots \text{ shear stress function} \\ &\sigma_1(\kappa) = T_{11} - T_{33} && \ldots \text{ first normal stress function} \\ &\sigma_2(\kappa) = T_{22} - T_{33} && \ldots \text{ second normal stress function} \end{aligned} \tag{8.4.28}$$

From constitutive law (8.4.27), we conclude that these three viscometric functions fully characterized the fluid under viscometric flow. The behavior of these viscometric functions with the rate of shear κ can be determined by experimentation, and typical qualitative results are shown in Fig. 8.6.

Note that the normal stress functions are sometimes defined in a slightly different way. Our scheme (8.4.28) is the form originally defined by Truesdell and Noll (1965). Written in terms of the original parameters μ_i in relation (8.4.14), these functions may be expressed as

$$\begin{aligned} \tau(\kappa) &= \mu_1\kappa + 2\mu_5\kappa^3 + 4\mu_6\kappa^5 \\ \sigma_1(\kappa) &= \mu_2\kappa^2 + 2\mu_3\kappa^2 + 4\mu_4\kappa^4 + 4\mu_7\kappa^4 + 8\mu_8\kappa^6 \\ \sigma_2(\kappa) &= \mu_2\kappa^2 \end{aligned} \tag{8.4.29}$$

Since it can be shown that the material parameters μ_i will be even functions of κ, the shear stress function will be an odd function of κ, whereas the two normal stress functions will be even functions of this variable. For the case where rate of shear $\kappa \to 0$, all

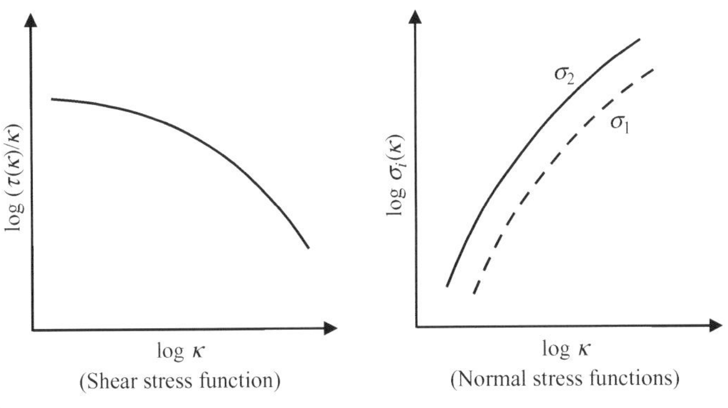

FIGURE 8.6

Typical behaviors of viscometric functions.

viscometric functions should vanish $\tau(0) = \sigma_1(0) = \sigma_2(0) = 0$. Based on these arguments, the viscometric functions could be expanded in a Taylor series for small κ:

$$\begin{aligned}\tau(\kappa) &= \mu_0\kappa + a_1\kappa^3 + a_2\kappa^5 + \cdots \\ \sigma_1(\kappa) &= b_1\kappa^2 + b_2\kappa^4 + \cdots \\ \sigma_2(\kappa) &= c_1\kappa^2 + c_2\kappa^4 + \cdots\end{aligned} \tag{8.4.30}$$

for constants a_i, b_i, and c_i. Note that first-order results in κ give Newtonian behavior with $\tau = T_{12} \approx \mu_0\kappa$ and $\sigma_1 = \sigma_2 = 0$. The next order of behavior would give the beginnings of non-Newtonian response, and for small κ these effects would be $O(\kappa^2)$ in the normal stresses and $O(\kappa^3)$ in the shear stress.

The individual stresses coming from relation (8.4.27) are given by

$$\begin{aligned}T_{11} &= -p + \sigma_1(\kappa) \\ T_{22} &= -p + \sigma_2(\kappa) \\ T_{33} &= -p \\ T_{12} &= \tau(\kappa), \quad T_{23} = T_{31} = 0\end{aligned} \tag{8.4.31}$$

Stresses from these rectilinear viscometric flows are thus constants and will automatically satisfy the equations of motion.

As previously mentioned, we are also interested in applying this viscometric flow geometry to other orthogonal flow fields. Using standard coordinate frame transformation, we can easily justify that several other flow geometries are also viscometric. For example, consider the two flow cases (Hagen–Poiseuille and circular Couette) shown in Fig. 8.7. For the Hagen–Poiseulle flow, the velocity field $\boldsymbol{v} = v_z(r)\boldsymbol{e}_z$ in the cylindrical coordinate system can be transformed into the standard Cartesian form by the particular orthogonal transformation

$$\boldsymbol{v}^* = \boldsymbol{Q}\boldsymbol{v} = \begin{bmatrix} 1 & 0 & 0 \\ 0 & 0 & 1 \\ 0 & -1 & 0 \end{bmatrix} \begin{bmatrix} 0 \\ 0 \\ v_z(r) \end{bmatrix} = \begin{bmatrix} 0 \\ v_z(r) \\ 0 \end{bmatrix} \tag{8.4.32}$$

and the new right-handed basis vectors become $\boldsymbol{n}_i = \{\boldsymbol{e}_r, \boldsymbol{e}_z, -\boldsymbol{e}_\theta\}$.

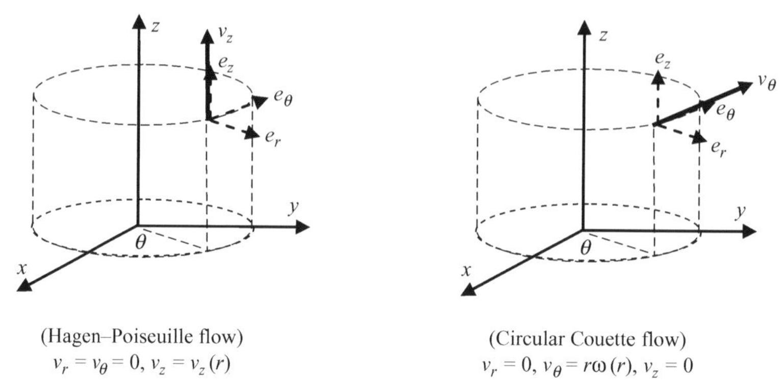

FIGURE 8.7

Viscometric curvilinear flows.

Likewise, the circular Couette flow $\boldsymbol{v} = r\omega(r)\boldsymbol{e}_\theta$ can also be transformed into the standard Cartesian form using a different $\boldsymbol{Q}$ matrix as

$$\boldsymbol{v}^* = \boldsymbol{Q}\boldsymbol{v} = \begin{bmatrix} 1 & 0 & 0 \\ 0 & 1 & 0 \\ 0 & 0 & 1 \end{bmatrix} \begin{bmatrix} 0 \\ r\omega(r) \\ 0 \end{bmatrix} = \begin{bmatrix} 0 \\ r\omega(r) \\ 0 \end{bmatrix} \tag{8.4.33}$$

and the new basis vectors become $\boldsymbol{n}_i = \{\boldsymbol{e}_r, \boldsymbol{e}_\theta, \boldsymbol{e}_z\}$. Note that for this case, the transformation matrix is simply the unit tensor. So, we have shown that the two curvilinear flows are actually viscometric flows with respect to different local orthogonal coordinates. This result allows various flows in standard viscometers to be used to characterize the various viscometric functions. Truesdell and Noll (1965) and Coleman et al. (1966) discuss several such experimental schemes.

EXAMPLE 8.4.2 CIRCULAR COUETTE FLOW

Circular Couette flow $v_r = 0, v_\theta = r\omega(r), v_z = 0$ often exists in the region between two infinitely long concentric cylinders turning with different angular velocities as shown in Fig. 8.8. For this flow geometry using cylindrical coordinates, determine the two Rivlin–Ericksen tensors and the corresponding stress field and the velocity distribution v_θ.

Solution: In order to use our existing viscometric flow theory, we must use the transformation (8.4.33) and adjust our basis vectors $\boldsymbol{n}_i$ such that $\boldsymbol{n}_i = \{\boldsymbol{e}_r, \boldsymbol{e}_\theta, \boldsymbol{e}_z\}$. Note that for this case, it is just a unity transformation. Using these cylindrical coordinates,

$$\boldsymbol{D} = \frac{1}{2}\boldsymbol{A}_1 = \frac{1}{2}\begin{bmatrix} 0 & \kappa(r) & 0 \\ \kappa(r) & 0 & 0 \\ 0 & 0 & 0 \end{bmatrix},\quad \boldsymbol{A}_2 = \begin{bmatrix} 2\kappa^2(r) & 0 & 0 \\ 0 & 0 & 0 \\ 0 & 0 & 0 \end{bmatrix} \tag{8.4.34}$$

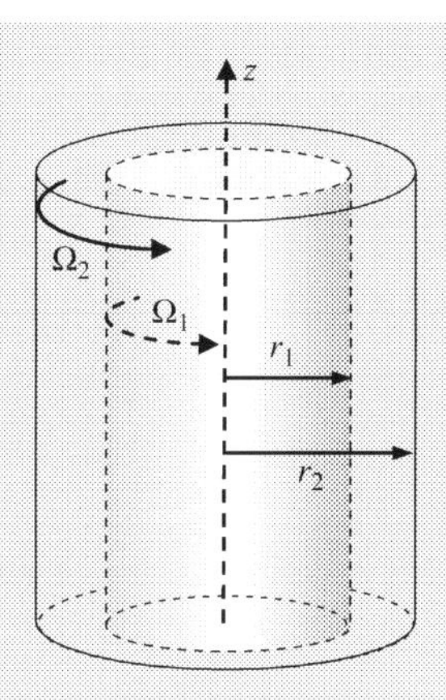

FIGURE 8.8

Circular couette flow geometry.

where $\kappa(r) = r\dfrac{d\omega}{dr}$. Note the identical form as (8.4.22). Using (8.4.27) in the cylindrical system

$$\begin{aligned} T_{r\theta} &= \tau(\kappa) \\ T_{rr} &= -p + \sigma_1(\kappa) \\ T_{\theta\theta} &= -p + \sigma_2(\kappa) \\ T_{zz} &= -p \\ T_{\theta z} &= T_{zr} = 0 \end{aligned} \tag{8.4.35}$$

We assume that the three viscometric functions are known and then wish to determine the velocity distribution $v_\theta = r\omega(r)$ and the stress components. In cylindrical coordinates, the equations of motion reduce to

$$\begin{aligned} -\frac{dp}{dr} + \frac{dT_{rr}}{dr} + \frac{T_{rr} - T_{\theta\theta}}{r} &= -\rho r\omega^2 \\ \frac{1}{r}\frac{d}{dr}(r^2 T_{r\theta}) &= 0 \end{aligned} \tag{8.4.36}$$

Relation $(8.4.36)_2$ is easily integrated giving $T_{r\theta} = C/r^2$, where C is the constant of integration. Then applying the torque M per unit height condition $M = (2\pi r T_{r\theta})r$ gives that $C = M/2\pi$ and thus the shear stress is determined:

$$T_{r\theta} = \tau(\kappa(r)) = \frac{M}{2\pi r^2} \tag{8.4.37}$$

Following Coleman et al. (1966), we now define the inverse function λ such that $\kappa(r) = \lambda(S)$ with $S(r) = \tau(\kappa) = M/2\pi r^2$. Since we have assumed that $\tau(\kappa)$ is known, we likewise assume that λ is also known. Next, consider the expression

$$\frac{d\omega}{dr} = \frac{d\omega}{dS}\frac{dS}{dr} = \frac{d\omega}{dS}\left(-\frac{2M}{2\pi r^3}\right) = -\frac{2}{r}S\frac{d\omega}{dS} \tag{8.4.38}$$

and thus

$$\kappa(r)=\lambda(S)=r\frac{d\omega}{dr}=-2S\frac{d\omega}{dS}\Rightarrow d\omega=-\frac{\lambda(S)}{2S}dS \tag{8.4.39}$$

Then integrating this previous result from inner radius r_1 to variable radius r gives

$$\omega(r)=\Omega_1-\frac{1}{2}\int_{M/2\pi r_1^2}^{M/2\pi r^2}\frac{\lambda(S)}{S}dS \tag{8.4.40}$$

where Ω_1 is the angular velocity of the inner cylinder. With $\lambda(S)$ known, (8.4.40) then gives $\omega(r)$ and thus the velocity $v_\theta = r\omega(r)$ can be determined.

The first equation of motion $(8.4.36)_1$ can be integrated (see Coleman et al., 1966 for details). This leads to an expression for T_{rr} which can be used to determine the difference between normal tractions on the inner and outer cylinders:

$$T_{rr}(r_2)-T_{rr}(r_1)=\int_{r_1}^{r_2}\left\{\frac{1}{r}\left[\hat{\sigma}_2\left(\frac{M}{2\pi r^2}\right)-\hat{\sigma}_1\left(\frac{M}{2\pi r^2}\right)\right]-\rho r\omega^2\right\}dr \tag{8.4.41}$$

where $\hat{\sigma}_1(S)=\sigma_1(\lambda(S))$ and $\hat{\sigma}_2(S)=\sigma_2(\lambda(S))$ are the *modified normal stress functions*. Since experimental techniques can collect data on normal tractions on the inner and outer cylinders, relation (8.4.41) provides a scheme to determine the difference between these modified normal stress functions $\hat{\sigma}_2-\hat{\sigma}_1$.

For Newtonian flows, $\sigma_1=\sigma_2=0$, while the Reiner–Rivlin fluid predicted $\sigma_1=\sigma_2$. For either of these cases, (8.4.41) implies $T_{rr}(r_2)-T_{rr}(r_1)<0$ or expressed in terms of a wall pressure $(p=-T_{rr})$ we find $p(r_2)>p(r_1)$. Thus, the wall pressure is higher on the outer cylinder and this would correspond with a free surface which would slope upwards as r increases as shown in Fig. 8.9A. Now for the non-Newtonian case, it has been found for most fluids that $\sigma_2>\sigma_1$ as shown in Fig. 8.6. Thus, for this case applying similar arguments related to relation (8.4.41), we conclude that it may be possible for $T_{rr}(r_2)-T_{rr}(r_1)>0$ implying that $p(r_2)<p(r_1)$. With higher pressure on the inner cylinder, the free surface profile would now slope opposite to the Newtonian flow as shown in Fig. 8.9B. This non-Newtonian behavior has often been observed and is commonly called the *rod climbing effect*. A simple web search will show several photographs and videos of such behavior for real fluids.

Additional features and analysis of this flow example along with several other curvilinear flows are presented in Coleman et al. (1966).

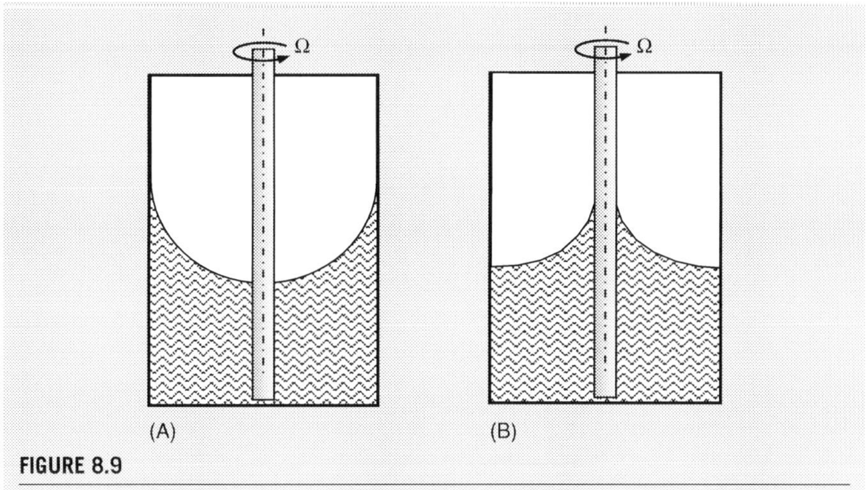

FIGURE 8.9

Typical free surface variations in circular Couette flow ($\Omega_1=\Omega$, $\Omega_2=0$): (A) Newtonian fluid—no normal stresses and free surface shape from centrifugal force alone; (B) non-Newtonian fluid—normal stress effects modify free surface shape.

Many more specific constitutive relations for non-Newtonian flow have been presented in the literature. Oldroyd (1984) has developed an extensive body of work using a *convective coordinate* approach whereby the coordinates are embedded in the material as it undergoes deformation and flow. Studies by Lodge (1964) and Bird et al. (1987) provide more details on Oldroyd's approach and several other flow models. We will not pursue these models here, and instead move on to investigate nonlinear viscoelastic behavior in the final section of this chapter.

8.5 NONLINEAR INTEGRAL VISCOELASTIC CONSTITUTIVE MODELS

Although we have already seen some nonlinear viscoelastic constitutive relations in the previous section dealing with non-Newtonian viscous fluids, we now wish to further explore this type of response for general materials including both solids and fluids. Over the years, numerous single integral constitutive relations have been developed for nonlinear viscoelastic behavior. These approaches likely originated from the linear viscoelastic integral forms that were developed decades earlier (Section 6.5). Also, it should be noted that a multiple integral form came out of the general simple fluid model discussed in Section 8.4.2. Here, we will focus on just a couple of the more common integral models, and Bird et al. (1987) provide a nice summary of many other constitutive models of this type.

8.5.1 INTEGRAL MODELS USING A SINGLE DEFORMATION TENSOR

Recall that a single integral memory constitutive equation was previously given by relation (8.4.9). This form was the result of a truncation of the multiple integral series representation of the general simple fluid model of Noll. This relation can be rewritten in a somewhat more general expression as

$$\boldsymbol{T} + p\boldsymbol{I} = \int_0^\infty \mu(I_i, t, s)\tilde{\boldsymbol{E}}_t(t-s)\,ds \tag{8.5.1}$$

where $\tilde{\boldsymbol{E}}_t(t-s)$ is some relative strain tensor, often incorporating $\boldsymbol{C}_t(t-s)$, $\boldsymbol{C}_t^{-1}(t-s)$, or $\boldsymbol{G}_t(t-s)$, and $\mu(I_i, t, s)$ is a material memory function that could depend on particular invariants and time constants. These single-integral models have been presented by Lodge (1964), Bird et al. (1987), and others. Normally $\mu(I_i, t, s)$ is a decreasing function of the elapsed time and often its form is expressed by

$$\mu(I_i, t, s) = \sum_i \frac{\mu_o(I_i)}{\tau_i^2} e^{-s/\tau_i} \tag{8.5.2}$$

Models like these can actually be connected to *convected Maxwell models* somewhat similar to Example 6.5.6.

Consider the case with $\tilde{\boldsymbol{E}}_t(t-s) = \boldsymbol{C}_t^{-1}(t-s)$ under shearing flow deformation $v_1 = 0, v_2 = \kappa x_1, v_3 = 0$. This gives

$$\boldsymbol{C}_t^{-1}(t-s) = \begin{bmatrix} 1 & \kappa s & 0 \\ \kappa s & 1+\kappa^2 s^2 & 0 \\ 0 & 0 & 1 \end{bmatrix} \Rightarrow \begin{aligned} &T_{11} = T_{33} = -p + \int_0^\infty \mu(s)\,ds \\ &T_{22} = -p + \int_0^\infty (1+\kappa^2 s^2)\mu(s)\,ds \\ &T_{12} = \kappa \int_0^\infty s\mu(s)\,ds \\ &T_{23} = T_{31} = 0 \end{aligned} \tag{8.5.3}$$

and hence for this case the viscometric functions become

$$\begin{aligned} \tau(\kappa) &= T_{12} = \kappa \int_0^\infty s\mu(s)\,ds \\ \sigma_1(\kappa) &= T_{11} - T_{33} = 0 \\ \sigma_2(\kappa) &= T_{22} - T_{33} = \kappa^2 \int_0^\infty s^2 \mu(s)\,ds \end{aligned} \tag{8.5.4}$$

Because the first normal stress function vanishes for all choices of the material function μ, this model will, in general, not match with experimental observations.

Furthermore, if the material function is taken as only the first term in relation (8.5.2) and no invariant dependency is included, the simplified expression becomes $\mu(I_i, t, s) = \frac{\mu_o}{\tau_1^2} e^{-s/\tau_1}$, and the stresses and viscometric functions reduce to

$$\begin{aligned} &T_{11} = T_{33} = -p + \frac{\mu_o}{\tau_1} \\ &T_{22} = -p + \frac{\mu_o}{\tau_1}(1 + 2\kappa^2 \tau_1^2) \\ &T_{12} = \mu_o \kappa \\ &T_{23} = T_{31} = 0 \end{aligned} \Rightarrow \begin{aligned} \tau(\kappa) &= T_{12} = \mu_o \kappa \\ \sigma_1(\kappa) &= T_{11} - T_{33} = 0 \\ \sigma_2(\kappa) &= T_{22} - T_{33} = 2\kappa^2 \mu_o \tau_1 \end{aligned} \tag{8.5.5}$$

Note for this simplified case, the shear stress and shear stress function become the Newtonian values. Many other specific forms of constitutive relation (8.5.1) are given in Bird et al. (1987).

8.5.2 K-BKZ INTEGRAL MODELS

A remarkable integral constitutive law was co-developed and first presented by Kaye (1962) and Bernstein et al. (1963). The constitutive law has been known as the *K-BKZ theory for incompressible nonlinear viscoelastic materials*. Over the last 50 plus years, this constitutive relation has proved to be very useful in predicting experimental observations and has resulted in several hundred publications (see recent review by Mitsoulis, 2013). The basic constitutive assumptions coming from the BKZ group are that the material is incompressible, with no preferred reference configuration. The existence of a stored energy function is retained under the idea that the material wants to return to its past shapes; however, it does this by having a stronger desire to return to shapes in its immediate history than those in its more distant history. Thus, we have a fading memory built into the constitutive concept. The basic theory was initially developed under isothermal conditions, and we will limit our discussion to this case. However, the theory has now been extended to problems containing temperature variation.

Although construction of the K-BKZ theory could be made using some of the fundamentals previously described, we take a somewhat different path. This constitutive concept for incompressible viscoelastic materials could be thought of as coming from a further generalization our previous relation (8.5.1):

$$\boldsymbol{T} + p\boldsymbol{I} = \int_0^\infty m(s)\boldsymbol{H}(\boldsymbol{C}_t(t-s))\,ds \tag{8.5.6}$$

where $m(s)$ is a material function of the elapsed time and $\boldsymbol{H}(\boldsymbol{C}_t(t-s))$ is an isotropic tensor-valued function of the objective relative right Cauchy–Green strain tensor. Similar to the finite elastic case in Section 8.3.1, we can employ the standard representation theory and express $\boldsymbol{H}$ as a polynomial in terms of $\boldsymbol{C}$ and $\boldsymbol{C}^{-1}$. Thus, rewrite (8.5.6) as

$$\boldsymbol{T} + p\boldsymbol{I} = \int_0^\infty m(s)[\phi_1(I_1,I_2)\boldsymbol{C}_t(t-s) + \phi_2(I_1,I_2)\boldsymbol{C}_t^{-1}(t-s)]ds \tag{8.5.7}$$

and the invariants are simply the traces of the two strain tensors. Thus, we have developed an *integral constitutive form that uses two deformation tensors*.

The K-BKZ theory is more commonly expressed by

$$\boldsymbol{T} + p\boldsymbol{I} = 2\int_0^\infty \left[\frac{\partial U}{\partial I_1}\boldsymbol{C}_t^{-1}(t-s) - \frac{\partial U}{\partial I_2}\boldsymbol{C}_t(t-s)\right]ds \tag{8.5.8}$$

where $U = U(s, I_1, I_2)$ with

$$I_1 = \operatorname{tr}\boldsymbol{C}_t^{-1}(t-s), \quad I_2 = \operatorname{tr}\boldsymbol{C}_t(t-s) \tag{8.5.9}$$

This form should be compared with relation $(8.3.9)_2$ used for the nonlinear elastic constitutive law. It is noted that U plays the role of an elastic stored energy function similar to our constitutive work for elastic materials. However, U is also a relaxing time-dependent potential function with nonequilibrium properties related through the invariants I_1 and I_2 and the elapsed time s. Thus, we can also interpret U as a *memory function*. These issues represent fundamental aspects of this elastic fluid theory. Several variants of the K-BKZ relation have been given in the literature and many of these have been presented in the review article by Mitsoulis (2013).

Specific forms for the material potential function U have been proposed, and one of the more original forms was given by Zapas (1966):

$$\begin{aligned} U = -\frac{\alpha'(t)}{2}(I_1-3)^2 - \frac{9}{2}\beta'(t)\log\left(\frac{I_1+I_2+3}{9}\right) \\ -24(\beta'(t)-c'(t))\log\left(\frac{I_1+15}{I_2+15}\right) - c'(t)(I_1-3) \end{aligned} \tag{8.5.10}$$

where $\alpha(t), \beta(t), c(t)$ are time functions that are positive and monotonically decreasing. Other proposed forms commonly include *separable formulations* where $U(s, I_1, I_2) = M(s)W(I_1, I_2)$.

We now explore a couple of examples that apply this theory to some standard flow/deformation problems.

EXAMPLE 8.5.1 STEADY SIMPLE SHEARING OF A K-BKZ VISCOELASTIC FLUID

Consider the steady shearing flow deformation $v_1 = 0, v_2 = \kappa x_1, v_3 = 0$ of a K-BKZ viscoelastic fluid. Determine the resulting stress components and the viscometric functions.

Solution: For this volume-preserving flow, the strain tensors have been computed in previous examples:

$$\boldsymbol{C}_t^{-1}(t-s) = \begin{bmatrix} 1 & \kappa s & 0 \\ \kappa s & 1+\kappa^2 s^2 & 0 \\ 0 & 0 & 1 \end{bmatrix}, \quad \boldsymbol{C}_t(t-s) = \begin{bmatrix} 1+\kappa^2 s^2 & -\kappa s & 0 \\ -\kappa s & 1 & 0 \\ 0 & 0 & 1 \end{bmatrix} \tag{8.5.11}$$

and thus $I_1 = I_2 = 3 + \kappa^2 s^2$. The K-BKZ constitutive relation (8.5.8) gives

$$\begin{aligned} T_{11} &= -p + 2\int_0^\infty \left[\frac{\partial U}{\partial I_1} - \frac{\partial U}{\partial I_2}(1+\kappa^2 s^2)\right] ds \\ T_{22} &= -p + 2\int_0^\infty \left[\frac{\partial U}{\partial I_1}(1+\kappa^2 s^2) - \frac{\partial U}{\partial I_2}\right] ds \\ T_{33} &= -p + 2\int_0^\infty \left[\frac{\partial U}{\partial I_1} - \frac{\partial U}{\partial I_2}\right] ds \\ T_{12} &= 2\kappa \int_0^\infty s\left[\frac{\partial U}{\partial I_1} + \frac{\partial U}{\partial I_2}\right] ds, \quad T_{23} = T_{31} = 0 \end{aligned} \tag{8.5.12}$$

The viscometric functions follow from (8.4.28):

$$\tau(\kappa) = T_{12} = 2\kappa \int_0^\infty s \left[\frac{\partial U}{\partial I_1} + \frac{\partial U}{\partial I_2} \right] ds = \int_0^\infty \frac{\partial U}{\partial (\kappa s)} ds$$

$$\sigma_1(\kappa) = T_{11} - T_{33} = -2\kappa^2 \int_0^\infty s^2 \frac{\partial U}{\partial I_2} ds \tag{8.5.13}$$

$$\sigma_2(\kappa) = T_{22} - T_{33} = 2\kappa^2 \int_0^\infty s^2 \frac{\partial U}{\partial I_1} ds$$

where we have used the chain rule to get $\frac{\partial U}{\partial I_1} + \frac{\partial U}{\partial I_2} = \frac{\partial U}{\partial(\kappa s)} \frac{\partial(\kappa s)}{\partial I_1} + \frac{\partial U}{\partial(\kappa s)} \frac{\partial(\kappa s)}{\partial I_2}$ $= \frac{1}{\kappa s} \frac{\partial U}{\partial(\kappa s)}$. Since these stresses are all homogeneous and time independent (steady flow), they will identically satisfy the equations of motion.

EXAMPLE 8.5.2 UNIAXIAL EXTENSION OF AN INCOMPRESSIBLE NONLINEAR K-BKZ VISCOELASTIC MATERIAL

Determine the stress components in an incompressible K-BKZ viscoelastic material that is subjected to a uniaxial deformation in the x_1-direction.

Solution: Using the previous results in Example 3.15.1, the relative isochoric motion is given by

$$\xi_1(\tau) = \frac{\lambda(\tau)}{\lambda(t)} x_1(t), \quad \xi_2(\tau) = \sqrt{\frac{\lambda(t)}{\lambda(\tau)}} x_2(t), \quad \xi_3(\tau) = \sqrt{\frac{\lambda(t)}{\lambda(\tau)}} x_3(t) \tag{8.5.14}$$

where λ is the primary stretch ratio. Thus, the deformation gradient and right Cauchy–Green strain tensors become

$$\boldsymbol{F}_t(\tau) = \frac{\partial \xi}{\partial \boldsymbol{x}} = \begin{bmatrix} \frac{\lambda(\tau)}{\lambda(t)} & 0 & 0 \\ 0 & \sqrt{\frac{\lambda(t)}{\lambda(\tau)}} & 0 \\ 0 & 0 & \sqrt{\frac{\lambda(t)}{\lambda(\tau)}} \end{bmatrix} \tag{8.5.15}$$

$$\boldsymbol{C}_t(\tau) = \begin{bmatrix} \left(\frac{\lambda(\tau)}{\lambda(t)}\right)^2 & 0 & 0 \\ 0 & \frac{\lambda(t)}{\lambda(\tau)} & 0 \\ 0 & 0 & \frac{\lambda(t)}{\lambda(\tau)} \end{bmatrix}, \quad \boldsymbol{C}_t^{-1}(\tau) = \begin{bmatrix} \left(\frac{\lambda(t)}{\lambda(\tau)}\right)^2 & 0 & 0 \\ 0 & \frac{\lambda(\tau)}{\lambda(t)} & 0 \\ 0 & 0 & \frac{\lambda(\tau)}{\lambda(t)} \end{bmatrix} \tag{8.5.16}$$

Thus, the invariants become $I_1 = \frac{\lambda^2(t)}{\lambda^2(\tau)} + 2\frac{\lambda(\tau)}{\lambda(t)}, I_2 = \frac{\lambda^2(\tau)}{\lambda^2(t)} + 2\frac{\lambda(t)}{\lambda(\tau)}$.

The stresses follow from (8.5.8):

$$\begin{aligned} T_{11} &= -p + 2\int_{-\infty}^{t}\left[\frac{\partial U}{\partial I_1}\frac{\lambda^2(t)}{\lambda^2(\tau)} - \frac{\partial U}{\partial I_2}\frac{\lambda^2(\tau)}{\lambda^2(t)}\right]d\tau \\ T_{22} = T_{33} &= -p + 2\int_{-\infty}^{t}\left[\frac{\partial U}{\partial I_1}\frac{\lambda(\tau)}{\lambda(t)} - \frac{\partial U}{\partial I_2}\frac{\lambda(t)}{\lambda(\tau)}\right]d\tau \\ T_{12} &= T_{23} = T_{31} = 0 \end{aligned} \tag{8.5.17}$$

We can define a stress difference as

$$\sigma \equiv T_{11} - T_{22} = T_{11} - T_{33} = \int_{-\infty}^{t}\left[\frac{\lambda^2(t)}{\lambda^2(\tau)} - \frac{\lambda(\tau)}{\lambda(t)}\right]h\left(\frac{\lambda(t)}{\lambda(\tau)}, t-\tau\right)d\tau \tag{8.5.18}$$

where

$$h\left(\frac{\lambda(t)}{\lambda(\tau)}, t-\tau\right) = 2\left[\frac{\partial U}{\partial I_1} + \frac{\lambda(\tau)}{\lambda(t)}\frac{\partial U}{\partial I_2}\right] \tag{8.5.19}$$

For a single-step stress relaxation, with $\lambda(t)=1, t<0$ and $\lambda(t)=\lambda=\text{constant}$, $t \geq 0$, relation (8.5.18) gives

$$\sigma = \left(\lambda^2 - \frac{1}{\lambda}\right)H(\lambda, t) \tag{8.5.20}$$

where

$$H(\lambda, t) = \int_{t}^{\infty} h(\lambda, \xi)d\xi \Rightarrow h(\lambda, t) = -\frac{\partial H(\lambda, t)}{\partial t} \tag{8.5.21}$$

Relations (8.5.20) and (8.5.21) then imply that data from a stress relaxation experiment allow the determination of the material functions H and h and thus allow the calculation of the stress response to any other uniaxial deformation history.

Although many other additional nonlinear continuum mechanics theories including plastic and viscoplastic response could be presented, in order to keep the text of reasonable length, we end our discussion of this general topic.

REFERENCES

Allen, M.B., 2016. Continuum Mechanics: The Birthplace of Mathematical Models. John Wiley, Hoboken.

Arruda, E.M., Boyce, M.C., 1993. A three dimensional constitutive model for large stretch behavior of rubber elastic materials. J. Mech. Phys. Sol. 41, 389–412.

Asaro, R.J., Lubarda, V.A., 2006. Mechanics of Solids and Materials. Cambridge University Press, Cambridge.

Bernstein, B., Kearsley, E.A., Zapas, L.J., 1963. A study of stress relaxation with finite strain. Trans. Soc. Rheol. 7, 391–410.

Bird, R.B., Armstrong, R.C., Hassager, O., 1987. Dynamics of Polymeric Liquids. John Wiley, New York.

Coleman, B.D., Markovitz, H., Noll, W., 1966. Viscometric Flows of Non-Newtonian Fluids. Springer, New York.

Green, A.E., Adkins, J.E., 1970. Large Elastic Deformations, Second ed. Oxford University Press, London.

Green, A.E., Rivlin, R.S., 1957. The mechanics of non-linear materials with memory. Arch. Rat. Mech. Anal. 1, 1–21.

Holzapfel, G.A., 2006. Nonlinear Solid Mechanics: A Continuum Approach for Engineering. John Wiley, West Sussex.

Kaye, A., 1962. Non-Newtonian Flow in Incompressible Fluids. Note No. 134. College of Aeronautics, Cranford, UK.

Lodge, A.S., 1964. Elastic Liquids. Academic Press, New York.

Mitsoulis, E., 2013. 50 Years of the K-BKZ Constitutive Relation for Polymers. ISRN Polymer Science, 2013, .

Noll, W., 1958. A mathematical theory of the mechanical behavior of continuous media. Arch. Rat Mech. Anal. 2, 197–226.

Ogden, R.W., 1972. Large deformation isotropic elasticity—on the correlation of theory and experiment for incompressible rubberlike solids. Proc. Roy. Soc. Lond. A 326, 565–584.

Ogden, R.W., 1984. Non-Linear Elastic Deformations. Ellis-Horwood, Chichester, UK.

Oldroyd, J.G., 1984. An approach to non-Newtonian fluid mechanics. J. Non-Newtonian Fluid Mech. 14, 9–46.

Rivlin, R.S., Ericksen, J.L., 1955. Stress–deformation relations for isotropic materials. Arch. Rat. Mech. Anal. 4, 323–425.

Rivlin, R.S., Saunders, D.W., 1951. Large elastic deformations of isotropic materials. VII. Experiments on the deformation of rubber. Phil. Trans. Roy. Soc. Lond. Ser. A 243, 251–288.

Truesdell, C.A., Toupin, R.A., 1960. Classical field theories. Flugge, S. (Ed.), Encyclopedia of Physics, III, First ed. Springer, Berlin.

Truesdell, C.A., Noll, W., 1965. Nonlinear field theories. Flugge, S. (Ed.), Encyclopedia of Physics, III, Third ed Springer, Berlin.

Truesdell, C.A., 1974. The meaning of viscometry in fluid dynamics. Annu. Rev. Fluid Mech. 6, 111–146.

Zapas, L.J., 1966. Viscoelastic behavior under large deformations. J. Res. Natl. Bureau Standards 70A, 525–532.

EXERCISES

8.1 Starting with the general nonlinear elastic constitutive form (8.3.1) $\boldsymbol{T} = \boldsymbol{f}(\boldsymbol{F})$, first show that frame-indifference would imply $\boldsymbol{QTQ}^T = \boldsymbol{f}(\boldsymbol{QF})$. Next consider the Polar Decomposition Theorem for the deformation gradient (3.6.10), $\boldsymbol{F} = \boldsymbol{RU}$ where $\boldsymbol{R}$ is the rotation tensor and $\boldsymbol{U}$ is the right stretch tensor. Choosing $\boldsymbol{Q} = \boldsymbol{R}^T$ show that $\boldsymbol{T} = \boldsymbol{Rf}(\boldsymbol{U})\boldsymbol{R}^T$. Finally, show that the constitutive form can be expressed in terms of the second Piola–Kirchhoff stress and the right Cauchy–Green strain tensor $\boldsymbol{S} = \boldsymbol{h}(\boldsymbol{C})$ for an appropriately defined function $\boldsymbol{h}$. Note this additional form is often referred to as the *reference constitutive relation*.

8.2 Verify the constitutive form $(8.3.9)_1$.

8.3 Explicitly determine the Cauchy and PK1 axial stress predictions for the neo-Hookean and Mooney–Rivlin models in Example 8.3.1.

8.4 Determine the second Piola–Kirchhoff stress tensor PK2 for the extensional deformation problem in Example 8.3.1.

8.5 The finite elastic uniaxial extension data from Rivlin and Saunders (1951) is again shown in the figure. Similar to Example 8.3.1, we wish to curve-fit the data this time using a linear line as shown. Clearly this linear approximation is only valid for extension ratios $1 \le \lambda \le 2.5$. First determine the linear relation to properly fit the data. Next using the MATLAB Code C-21 in Appendix C, insert the fitting relation to plot the stress versus extension ratio for the data relation. Finally, using a trial-and-error scheme, select the α_1 and α_2 parameters to make the same stress plots for the Mooney and neo-Hookean models that will closely match with the data predictions. It is suggested that you start with $\alpha_1 = 3, \alpha_2 = 0.5$.

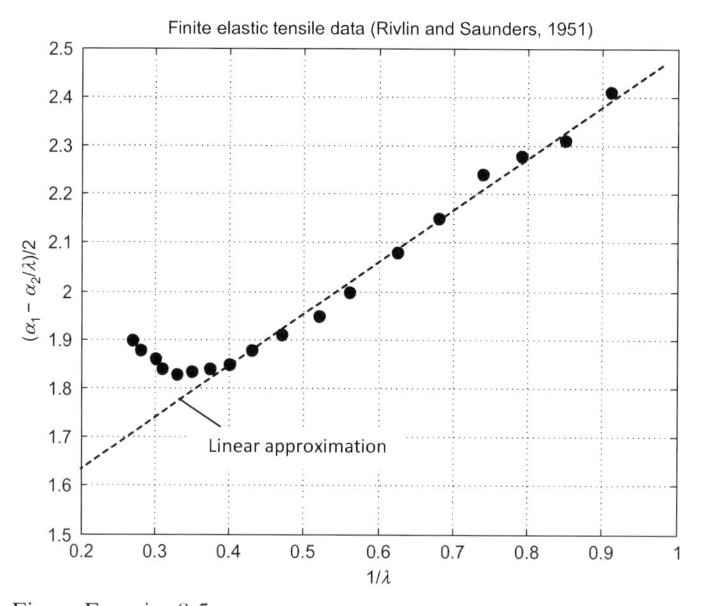

Figure Exercise 8.5

8.6 Verify result (8.3.20) in Example 8.3.2.

8.7 For the simple shear deformation problem of Example 8.3.2, determine the Piola–Kirchhoff stress tensors PK1 and PK2.

8.8 For the simple shear deformation problem of Example 8.3.2, if the strain energy form (8.3.10) is given by $U = C_{10}(I_B - 3) + C_{20}(I_B - 3)^2$, determine the stress field. Using unit values for C_{10} and C_{20}, plot the response T_{12} versus γ and compare with corresponding neo-Hookean response.

8.9 For the torsional Example 8.3.3, evaluate all stresses, the twisting moment, and normal axial force for the neo-Hookean solid. Compare these values with corresponding predictions from linear elasticity.

8.10 Consider the nonlinear elastic problem that includes the torsion, extension, and inflation of an incompressible solid cylinder:

$$r = \lambda_r R, \quad \theta = \Theta + \kappa Z, \quad z = \lambda_z Z$$

where $\lambda_r, \kappa, \lambda_z$ are constants satisfying the incompressibility constraint $\lambda_r^2 \lambda_z = 1$.
Following the solution steps used in Example 8.3.3, determine the stresses, twisting moment, and normal axial stress.

8.11 An incompressible elastic sheet is to be stretched uniformly in the x_1,x_2-plane by forces F_1 and F_2 as shown. Assume that the deformation is given by

$$x_1 = \lambda_1 X_1, \quad x_2 = \lambda_2 X_2, \quad x_3 = \mu X_3$$

where the stretches $\lambda_1, \lambda_2, \mu$ are all constants. First determine the incompressibility constraint, and then calculate all Cartesian stresses. Finally, using the traction boundary conditions $F_1 = T_{11} 2hb, F_2 = T_{22} 2ha$, determine the specific forms of the two material functions α_1 and α_2.

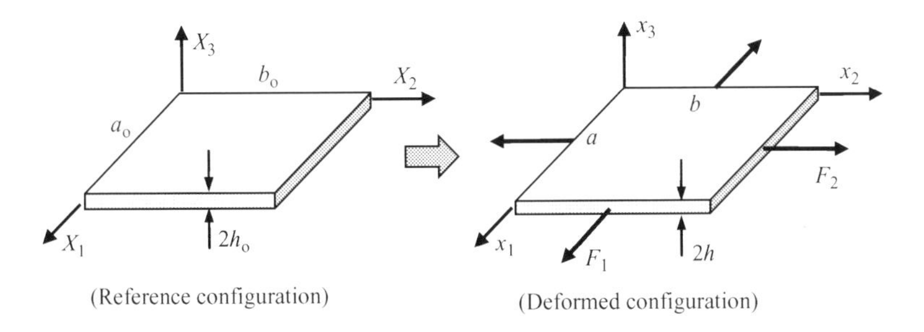

Figure Exercise 8.11

8.12 A fluid moves with the velocity field $v_1 = kx_1, v_2 = -kx_2, v_3 = 0$. Verify that this motion is isochoric. Next determine the rate of deformation tensor and the stress field using the Reiner–Rivlin fluid model.

8.13 Under steady simple shear flow, the Reiner–Rivlin fluid stresses were given by

$$T_{11} + p = T_{22} + p = \frac{\beta_2}{4}\kappa^2, \quad T_{12} = \frac{\beta_1}{2}\kappa$$

where $\beta_i = \beta_i(\kappa^2)$. Choose the specific forms $\beta_1/2 = \mu_o + \mu_1\kappa^2$ and $\beta_2/4 = \eta_o + \eta_1\kappa^2$ where μ_i and η_i are constants. Calculate the stress components and make log–log plots of $(T_{11} + p)/\eta_o$ and T_{12}/μ_o versus κ for the case of $\mu_1/\mu_o = -0.2$ and $\eta_1/\eta_o = -0.1$.

8.14 Determine the viscometric functions $\tau(\kappa), \sigma_1(\kappa), \sigma_2(\kappa)$ for the second-order fluid model under simple shearing flow in Example 8.4.1.

8.15 For the velocity field specified in Exercise 8.12, determine the rate of deformation tensor and the two nonzero Rivlin–Ericksen tensors. Next, calculate the stresses predicted by the second-order fluid model.

8.16 Starting with representation (8.4.14) and using the forms (8.4.25) for the Rivlin–Ericksen tensors, formally derive the viscometric flow representation (8.4.27) with the definitions of the viscometric functions given by (8.4.29).

8.17 A fluid of *grade three* that satisfies the axiom of consistency has been found to be

$$\boldsymbol{T} = -p\boldsymbol{I} + \mu_1\boldsymbol{A}_1 + \mu_2\boldsymbol{A}_1^2 + \mu_3\boldsymbol{A}_2 + \beta_3(\text{tr}\,\boldsymbol{A}_1^2)\boldsymbol{A}_1$$

Determine the stress field for a flow $v_1 = 2kx_1, \quad v_2 = -kx_2, \quad v_3 = -kx_3$.

8.18 For the general constitutive form (8.5.1), consider the case with a general material function $\mu = \mu(s)$ and $\tilde{\boldsymbol{E}}_t(t-s) = \boldsymbol{G}_t(t-s) = \boldsymbol{C}_t(t-s) - \boldsymbol{I}$. Assuming the material to be under shearing flow deformation $v_1 = 0, v_2 = \kappa x_1, v_3 = 0$, determine the stresses and the viscometric functions $\tau(\kappa), \sigma_1(\kappa), \sigma_2(\kappa)$.

8.19 Using the material memory function form $\mu(I_i, t, s) = \frac{\mu_o}{\tau_1^2}e^{-s/\tau_1}$, explicitly develop the stresses given by equations (8.5.5).

8.20 Show that the shear viscometric function for the K-BKZ theory under shear flow in Example 8.5.1 is given by

$$\tau(\kappa) = T_{12} = 2\kappa\int_0^\infty s\left[\frac{\partial U}{\partial I_1} + \frac{\partial U}{\partial I_2}\right]\mathrm{d}s = \int_0^\infty \frac{\partial U}{\partial(\kappa s)}\mathrm{d}s$$

8.21 For the case of shear flow $v_1 = 0, v_2 = \kappa x_1, v_3 = 0$, show that the proposed form for the K-BKZ material potential function U given by relation (8.5.10) will reduce to

$$U = -\frac{\alpha'(t)}{2}\kappa^4 s^4 - \frac{9}{2}\beta'(t)\log\left(1 + \frac{2}{9}\kappa^2 s^2\right) - c'(t)\kappa^2 s^2$$

8.22 For the shear flow case, use the results from Exercises 8.20 and 8.21 to show that the shear viscometric function is given by

$$\tau(\kappa) = T_{12} = -2\kappa \int_0^\infty \left[\alpha'(s)\kappa^2 s^2 + \frac{9\beta'(s)}{9+2\kappa^2 s^2} + c'(s) \right] s\,ds$$

8.23 Using the results of Exercise 8.22, the apparent viscosity is given by

$$\eta(\kappa) = \left| \frac{T_{12}}{\kappa} \right| = \left| 2\int_0^\infty \left[\alpha'(s)\kappa^2 s^2 + \frac{9\beta'(s)}{9+2\kappa^2 s^2} + c'(s) \right] s\,ds \right|$$

Choosing exponential forms for the material functions

$$\alpha(t) = A_o e^{-a_o t}, \quad \beta(t) = B_o e^{-b_o t}, \quad c(t) = C_o e^{-c_o t}$$

show that the viscosity is given by

$$\eta(\kappa) = 2\left[\frac{6A_o}{a_o^3}\kappa^2 + \frac{9B_o b_o}{2\kappa^2} g\left(\frac{3b_o}{\kappa\sqrt{2}} \right) + \frac{C_o}{c_o} \right]$$

where g is an auxiliary function associated with the sine and cosine integrals

$$g(z) = \int_0^\infty \frac{t\,e^{-zt}}{1+t^2}\,dt$$

8.24 Verify the computations for the kinematical tensors in relations (8.5.14)–(8.5.16) in uniaxial extension Example 8.5.2.

8.25 Verify the analysis steps in relations (8.5.18)–(8.5.21) for the uniaxial extension Example 8.5.2.

8.26 Consider a K-BKZ material under *biaxial deformation* in the x_1,x_2-plane

$$\xi_1(\tau) = \frac{\lambda_1(\tau)}{\lambda_1(t)} x_1(t), \quad \xi_2(\tau) = \frac{\lambda_2(\tau)}{\lambda_2(t)} x_2(t), \quad \xi_3(\tau) = \frac{\lambda_3(\tau)}{\lambda_3(t)} x_3(t)$$

(a) Show that if the deformation is to be isochoric $\dfrac{\lambda_3(\tau)}{\lambda_3(t)} = \dfrac{\lambda_1(t)}{\lambda_1(\tau)} \dfrac{\lambda_2(t)}{\lambda_2(\tau)}$, and thus we can eliminate the stretch in the x_3-direction.

(b) Determine the kinematical deformation tensors $\boldsymbol{F}_t(\tau), \boldsymbol{C}_t(\tau), \boldsymbol{C}_t^{-1}(\tau)$, and the two invariants $I_1 = \operatorname{tr} \boldsymbol{C}_t^{-1}(\tau), I_2 = \operatorname{tr} \boldsymbol{C}_t(\tau)$.

(c) Finally, calculate the stress field associated with this deformation.

CHAPTER

Constitutive relations and formulation of theories incorporating material microstructure

As mentioned in in Section 6.1, all real materials have internal microstructure at various length scales, and some examples were shown in Figs. 6.1–6.3. Recently, there has been considerable demand on structural performance and this has led to the development of many new heterogeneous synthetic materials with complex microstructures that help provide the desired material properties. Our previous continuum mechanics theories have generally been developed for problems with length scales several orders of magnitude larger than these microstructural features. In this fashion, microstructure is then averaged over these heterogeneous material phases to allow standard continuum mechanics to be employed. We now wish to explore theories that attempt to model some of these microstructural effects by embedding additional features into the classical field equations. Over the last few decades, numerous theories have been developed. Since this is a textbook, we will limit our study and primarily explore several micromechanical continuum theories coupled with linear elasticity. In this fashion, we can easily develop the basic formulations and find a few closed-form micromechanical solutions to compare with corresponding classical predictions. This will provide the reader with some basic micromechanical background on formulation concepts and solution results. Within the elasticity context, we will discuss micropolar, distributed voids, doublet mechanics, higher-gradient theories, fabric tensor models, and some fundamental aspects of damage mechanics.

9.1 INTRODUCTION TO MICROMECHANICS MATERIAL MODELING

In the last several decades, there has been considerable interest in micromechanical modeling of materials. This interest has been motivated by the fact that all real materials and more importantly numerous recent synthetic materials have heterogeneous microstructures that affect the overall load–deformation response. Material heterogeneity occurs at many different length scales, and for continuum mechanics modeling these scales should normally be at least an order or two smaller than the behaviors sought. Based on this, these heterogeneous features are normally referred to as microstructure. Examples of this commonly include atoms, molecules, particulate and

Continuum Mechanics Modeling of Material Behavior. http://dx.doi.org/10.1016/B978-0-12-811474-2.00009-5

fiber composites, soil, rock, concretes, granular materials, porous and cellular solids, and many others.

Micromechanical modeling is often categorized into two general camps that are labeled as *discrete* or *continuum* approaches. Discrete models generally included detailed simulation of the various microstructural phases and geometries, and examples would include atomistic modeling, molecular dynamics, particulate discrete element methods, and many other numerical schemes based on detailed finite-element analysis of specific heterogeneity. Generally, this approach requires extensive numerical and computational effort and is often limited to small length and time scales. On the other hand, continuum micromechanical modeling incorporates additional, continuously distributed fields of various types that bring new features into the theory. These additional relations are then combined with some of the classical equations (kinematics, stress, balance laws) to establish a modified continuum theory. This latter approach will be our focus in this chapter.

One of the fundamental goals of micromechanical material modeling is to develop theories to predict the response of the heterogeneous material on the basis of the geometries and properties of the individual phases. A simple example of this concept is illustrated in Fig. 9.1 which shows a particulate composite material containing particles embedded in a matrix material. The deformation response of this two-phase composite will depend on several features of the material makeup including: particle and matrix material moduli, particular geometries related to particle size and distribution, and bonding between the two phases. The question is: can we develop an *equivalent continuum* that would properly model the composite?

In regard to material behavior, *homogenization* is concerned with developing equivalent continuum theories for materials that contain various heterogeneous microstructural phases and geometries. This is most commonly done by assuming that we can define a statistically equivalent homogeneous medium characterized by a *representative volume element* (RVE) or for ordered microstructure by a *repeating unit cell*

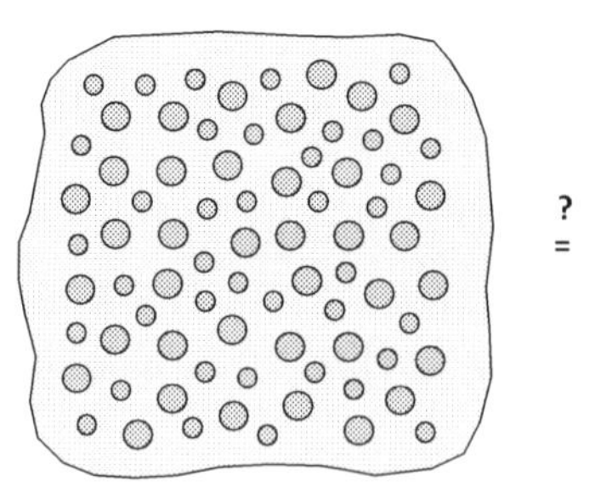

(Heterogeneous particulate composite)
- Moduli of particles and matrix
- Particle size and distribution
- Phase interface conditions

(Equivalent homogeneous material)

FIGURE 9.1

Heterogeneous particulate composite material.

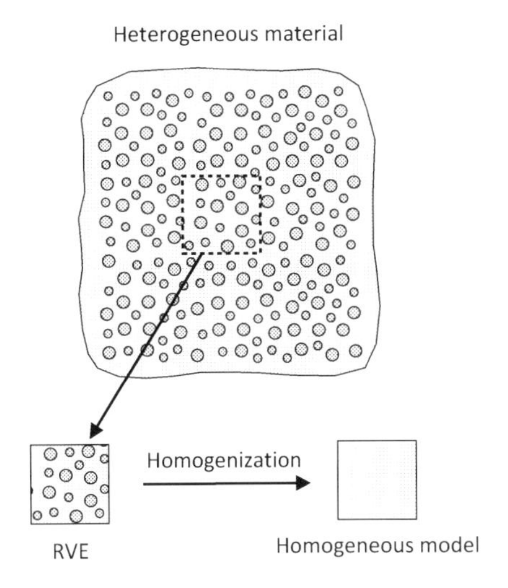

FIGURE 9.2

Concepts of modeling heterogeneous materials.

(RUC). Fig. 9.2 graphically illustrates the basic steps. Hill (1963) was one of the originators of the RVE concept for composite materials. A more general definition of RVE would be that it represents a subvolume of a heterogeneous material by statistically including all microstructural heterogeneities (grains, inclusions, voids, fibers, etc.), thereby creating a homogenized sample. We wish for the RVE to remain small enough to be considered as a volume element in the continuum mechanics sense in order to describe properties at a point. Also, the RVE is expected to properly allow typical prescribed boundary conditions to create mean strain or stress on the material element. Based on microscopic RVE and macroscopic length scales L, we normally need the relation $L_{micro} << L_{RVE} << L_{macro}$ to be satisfied. Once the RVE is determined, a homogenization process is applied to develop the homogeneous model material element. Homogenization is commonly accomplished using either analytical/mathematical methods or numerical/computational schemes. Some of this activity has been described as *meso-mechanical modeling*, that is, being in the middle between micro and macro. Nemat-Nasser and Hori (1993), Charalambakis (2010), Fish and Kuznetsov (2012), and Nguyen et al. (2011) provide reviews and further background on these topics.

Recent analytical studies have constructed various generalizations of classical continuum theories by introducing different kinds of microbehaviors into the basic theory. These models generally go deeply into the basics of continuum mechanics in order to capture these new phenomena. Microbehaviors commonly come from enhanced kinematics, conservation principles, and constitutive relations. Some researchers (Fish and Kuznetsov, 2012) classify these schemes into two categories. One type, sometimes referred to as *higher-order continua*, is characterized by having additional microdegrees of freedom independent of the usual continuum mechanics macro displacements and rotations. The second type of theory may be called *higher grade continua* and would employ higher-order spatial derivatives of the displacement, thereby generating new types of strain tensors. In the Noll sense (see Section 8.4.2), this type of model becomes a *nonsimple* material. We will pursue examples of each of these cases for linear elastic materials. Geometric material microstructure will commonly introduce *theoretical length scales* into the new models, a feature not found in classical continuum mechanics. As we briefly explore the micromechanical theories, we will look for these new parameters and investigate and compare their effect on problem solution. Ultimately, these theoretical length scales should be somehow connected to length scales associated with the actual material microstructure.

9.2 MICROPOLAR ELASTICITY

The response of many heterogeneous materials has indicated dependency on microscale length parameters such as embedded particle displacement and/or rotation. This would indicate that a continuum model of this material might need additional distributed microstructural degrees of freedom and hence a higher-order continua. Solids exhibiting such behavior include a large variety of cemented particulate materials such as particulate composites, ceramics, and various concretes. This concept can be qualitatively illustrated by considering a simple *lattice model* of a particulate material as shown in Fig. 9.3. Using such a scheme, the macro load transfer within the heterogeneous particulate solid is modeled using the microforces and moments between adjacent particles (see Chang and Ma, 1991; Sadd et al., 1992, 2004). Depending on the microstructural packing geometry (sometimes referred to as *fabric*), this method establishes a lattice network that can be thought of as an interconnect series of elastic bar or beam elements interconnected at particle centers. Thus, the network represents, in some way, the material microstructure and brings microstructural dimensions such as the grid size into the model. Furthermore, the elastic network will establish internal bending moments and forces, which will depend on internal degrees of freedom (e.g. rotations) at each connecting point in the microstructure

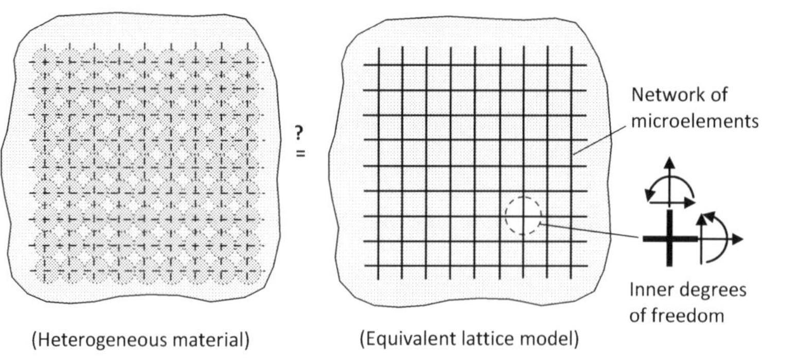

FIGURE 9.3

Heterogeneous material with internal microstructure.

as shown. These internal rotations would be, in a sense, independent of the overall macrodeformations.

This concept then suggests that an elastic continuum theory including an independent rotation field with concentrated pointwise moments might be suitable for modeling such heterogeneous materials. These approaches have been formulated under the names *Cosserat continuum, oriented media, asymmetric elasticity, micropolar*, or *couple-stress theories*. The Cosserat continuum developed in 1909 was historically one of the first models of this category. However, in the next 50 years, very little activity occurred in this field. Renewed interest began during the 1960s, and this produced numerous articles on theoretical refinements and particular analytical and computational applications. Toupin (1964) and Mindlin (1964) developed a general *linear elasticity theory with microstructure* that allowed the stress to depend on both the strain and an additional *kinematic microdeformation* tensor. This and related research has led to the development *couple-stress* and *micropolar theories.* These approaches allow material deformation to include additional independent *microrotational* degrees of freedom. The studies by Eringen (1968) and Kunin (1983) provide detailed background on much of this work, whereas Nowacki (1986) presents a comprehensive account on dynamic and thermoelastic applications of such theories. Later, Eringen (1999) provided additional extensions of the basic micropolar approach. He used the concept of *deformation of material directors* to categorize the microcontinuum into three types: micromorphic with deformable directors, microstretch with directors that only stretch, and micropolar with rigid directors that only rotate. Our study will only include the micropolar case and only for linear elastic materials.

Micropolar theory for linear, elastic small deformation theory incorporates an additional internal degree of freedom called the *microrotation* and allows for the existence *body and surface couples*. For this approach, the new kinematic strain–deformation relation is expressed as

$$\varepsilon_{ij} = u_{j,i} - \varepsilon_{ijl}\phi_l \tag{9.2.1}$$

where ε_{ij} is usual infinitesimal strain tensor, ε_{ijl} is the alternating symbol, u_i is the displacement vector, and ϕ_i is the *microrotation vector*. Note that this new kinematic variable ϕ_i is independent of the displacement u_i and thus is not, in general, the same as the usual macrorotation vector, that is,

$$\omega_i = \frac{1}{2}\varepsilon_{ijk}u_{k,j} \neq \phi_i \tag{9.2.2}$$

Later in our discussion, we will relax this restriction and develop a more specialized theory that normally allows for simpler analytical problem solution.

The existence of body and surface couples (moments) included in the new theory will introduce additional terms in the equilibrium equations. Skipping the derivation details, the linear and angular equilibrium equations thus become

$$\begin{gathered} T_{ji,j} + F_i = 0 \\ m_{ji,j} + \varepsilon_{ijk}T_{jk} + C_i = 0 \end{gathered} \tag{9.2.3}$$

where T_{ij} is the usual Cauchy stress tensor, F_i is the body force per unit volume, m_{ij} is the *surface moment tensor* normally referred to as the *couple-stress tensor*, and C_i is the *body couple* per unit volume. Notice that as a consequence of including couple-stresses and body couples, relation $(9.2.3)_2$ implies that *the stress tensor T_{ij} will no longer be symmetric*.

For linear elastic isotropic materials, the constitutive relations for a micropolar material are given by

$$\begin{aligned} T_{ij} &= \lambda \varepsilon_{kk}\delta_{ij} + (\mu+\kappa)\varepsilon_{ij} + \mu\varepsilon_{ji} \\ m_{ij} &= \alpha\phi_{k,k}\delta_{ij} + \beta\phi_{i,j} + \gamma\phi_{j,i} \end{aligned} \tag{9.2.4}$$

where $\lambda, \mu, \kappa, \alpha, \beta, \gamma$ are the micropolar elastic moduli. Note that classical elasticity relations correspond to the case where $\kappa = \alpha = \beta = \gamma = 0$. The requirement of a positive definite strain energy function puts the following restrictions on these moduli:

$$\begin{aligned} &0 \le 3\lambda + 2\mu + \kappa, \quad 0 \le 2\mu + \kappa, \quad 0 \le \kappa \\ &0 \le 3\alpha + \beta + \gamma, \quad -\gamma \le \beta \le \gamma, \quad 0 \le \gamma \end{aligned} \tag{9.2.5}$$

Relations (9.2.1) and (9.2.4) can be substituted into the equilibrium equations (9.2.3) to establish two sets of governing field equations in terms of the displacements and microrotations. Appropriate boundary conditions to accompany these field equations are somewhat more problematic. For example, it is not completely clear as to how to specify the microrotation ϕ_i and/or couple-stress m_{ij} on domain boundaries. Some developments on this subject have determined particular field combinations whose boundary specification guarantees a unique solution to the problem.

Two-dimensional couple-stress theory

Rather than continuing on with the general three-dimensional equations, we will now move directly into two-dimensional problems under plane strain conditions. In addition to the usual assumption $u = u(x,y), v = v(x,y), w = 0$, we include the restrictions on the microrotation, $\phi_x = \phi_y = 0, \phi_z = \phi_z(x,y)$. Furthermore, relation (9.2.2) will be relaxed and *the microrotation will be allowed to coincide with the macrorotation*:

$$\phi_i = \omega_i = \frac{1}{2}\varepsilon_{ijk}u_{k,j} \tag{9.2.6}$$

This particular theory is then a special case of micropolar elasticity and is commonly referred to *couple-stress theory*. Eringen (1968) refers to this theory as *indeterminate* since the antisymmetric part of the stress tensor is not determined solely by the constitutive relations.

Stresses on a typical in-plane element are shown in Fig. 9.4. Notice the similarity of this force system with the microstructural system illustrated previously in Fig. 9.3. For the two-dimensional case with no body forces or body couples, the equilibrium equations (9.2.3) reduce to

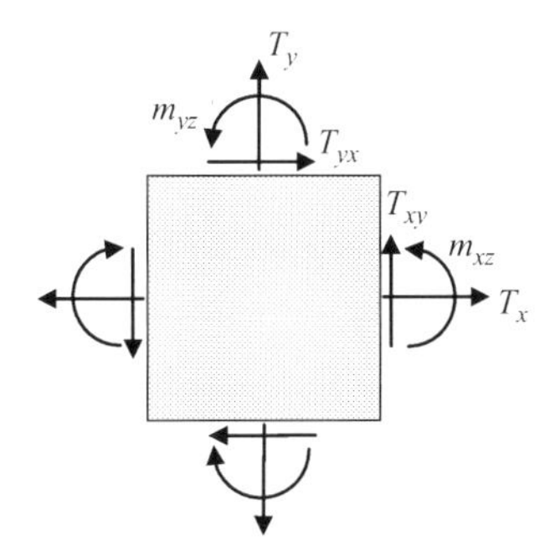

FIGURE 9.4

Couple-stresses on two-dimensional element.

$$\begin{aligned}\frac{\partial T_{xx}}{\partial x}+\frac{\partial T_{yx}}{\partial y}=0\\ \frac{\partial T_{xy}}{\partial x}+\frac{\partial T_{yy}}{\partial y}=0\\ \frac{\partial m_{xz}}{\partial x}+\frac{\partial m_{yz}}{\partial y}+T_{xy}-T_{yx}=0\end{aligned}\tag{9.2.7}$$

The in-plane strains can be expressed as

$$\begin{aligned}\varepsilon_x=\frac{\partial u}{\partial x},\quad \varepsilon_y=\frac{\partial v}{\partial y}\\ \varepsilon_{xy}=\frac{\partial v}{\partial x}-\phi_z,\quad \varepsilon_{yx}=\frac{\partial u}{\partial y}+\phi_z\end{aligned}\tag{9.2.8}$$

while using (9.2.6) gives

$$\phi_z=\frac{1}{2}\left(\frac{\partial v}{\partial x}-\frac{\partial u}{\partial y}\right)\tag{9.2.9}$$

Notice that by substituting (9.2.9) into $(9.2.8)_2$ gives the result $\varepsilon_{xy}=\varepsilon_{yx}$.

The constitutive equations (9.2.4) yield the following forms for the stress components:

$$\begin{aligned}T_{xx}=\lambda(\varepsilon_x+\varepsilon_y)+(2\mu+\kappa)\varepsilon_x\\ T_{yy}=\lambda(\varepsilon_x+\varepsilon_y)+(2\mu+\kappa)\varepsilon_y\\ T_{xy}=(2\mu+\kappa)\varepsilon_{xy}=T_{yx}\\ m_{xz}=\gamma\frac{\partial\phi_z}{\partial x},\quad m_{yz}=\gamma\frac{\partial\phi_z}{\partial y}\end{aligned}\tag{9.2.10}$$

In regard to the last pair of equations in (9.2.10), some authors (Mindlin, 1963; Boresi and Chong, 2000) define the gradients of the rotation ϕ_z as the *curvatures*. Thus, they establish a linear constitutive law between the couple-stresses and curvatures. This approach is completely equivalent to the current method. It is to be noted from (9.2.10) that under the assumptions of couple-stress theory, we find the unpleasant situation that the antisymmetric part of the stress tensor disappears from the constitutive relations. In order to remedy this, we can solve for the antisymmetric stress term from the moment equilibrium equation $(9.2.7)_3$ to get

$$T_{[xy]}=\frac{1}{2}(T_{xy}-T_{yx})=-\frac{1}{2}\left(\frac{\partial m_{xz}}{\partial x}+\frac{\partial m_{yz}}{\partial y}\right)=-\frac{\gamma}{2}\nabla^2\phi_z \tag{9.2.11}$$

By cross-differentiation, we can eliminate the displacements from (9.2.8) and (9.2.9) and develop the particular compatibility equations for this case

$$\begin{gathered}\frac{\partial^2\varepsilon_x}{\partial y^2}+\frac{\partial^2\varepsilon_y}{\partial x^2}=2\frac{\partial^2\varepsilon_{xy}}{\partial x\partial y}\\ \frac{\partial^2\phi_z}{\partial x\partial y}=\frac{\partial^2\phi_z}{\partial y\partial x}\\ \frac{\partial\phi_z}{\partial x}=\frac{\partial\varepsilon_{xy}}{\partial x}-\frac{\partial\varepsilon_x}{\partial y}\\ \frac{\partial\phi_z}{\partial y}=\frac{\partial\varepsilon_y}{\partial x}-\frac{\partial\varepsilon_{xy}}{\partial y}\end{gathered} \tag{9.2.12}$$

Using the constitutive forms (9.2.10), these relations may be expressed in terms of the stresses as

$$\begin{gathered}\frac{\partial^2 T_{xx}}{\partial y^2}+\frac{\partial^2 T_{yy}}{\partial x^2}-\nu\nabla^2(T_{xx}+T_{yy})=\frac{\partial^2}{\partial x\partial y}(T_{xy}+T_{yx})\\ \frac{\partial m_{xz}}{\partial y}=\frac{\partial m_{yz}}{\partial x}\\ m_{xz}=l^2\frac{\partial}{\partial x}(T_{xy}+T_{yx})-2l^2\frac{\partial}{\partial y}[T_{xx}-\nu(T_{xx}+T_{yy})]\\ m_{yz}=2l^2\frac{\partial}{\partial x}[T_{yy}-\nu(T_{xx}+T_{yy})]-l^2\frac{\partial}{\partial y}(T_{xy}+T_{yx})\end{gathered} \tag{9.2.13}$$

where $\nu=\lambda/(2\lambda+2\mu+\kappa)$, and $l=\sqrt{\gamma/(4\mu+2\kappa)}$ is a material constant with units of *length*. Notice that this result then introduces a *length scale* into the problem. If $l=0$, the couple-stress effects are eliminated and the problem reduces to classical elasticity. It should also be pointed out that only three of the four equations in set (9.2.13) are independent as the second relation can be established from the other equations.

Proceeding along similar lines as classical elasticity, we introduce a stress function approach Carlson (1966) to solve (9.2.13). A self-equilibrated form satisfying (9.2.7) identically can be written as

$$\begin{gathered}T_{xx}=\frac{\partial^2\Phi}{\partial y^2}-\frac{\partial^2\Psi}{\partial x\partial y},\quad T_{yy}=\frac{\partial^2\Phi}{\partial x^2}+\frac{\partial^2\Psi}{\partial x\partial y}\\ T_{xy}=-\frac{\partial^2\Phi}{\partial x\partial y}-\frac{\partial^2\Psi}{\partial y^2},\quad T_{yx}=-\frac{\partial^2\Phi}{\partial x\partial y}+\frac{\partial^2\Psi}{\partial x^2}\\ m_{xz}=\frac{\partial\Psi}{\partial x},\quad m_{yz}=\frac{\partial\Psi}{\partial y}\end{gathered} \tag{9.2.14}$$

where $\Phi = \Phi(x, y)$ and $\Psi = \Psi(x, y)$ are the stress functions for this case. If Ψ is taken to be zero, the representation reduces to the usual Airy form (6.2.47) with no couple-stresses. Using form (9.2.14) in the compatibility equations (9.2.13) produces

$$\begin{gathered} \nabla^4\Phi = 0 \\ \frac{\partial}{\partial x}(\Psi - l^2\nabla^2\Psi) = -2(1-\nu)l^2\frac{\partial}{\partial y}(\nabla^2\Phi) \\ \frac{\partial}{\partial y}(\Psi - l^2\nabla^2\Psi) = 2(1-\nu)l^2\frac{\partial}{\partial x}(\nabla^2\Phi) \end{gathered} \tag{9.2.15}$$

Differentiating the second equation with respect to x and the third with respect to y and adding will eliminate the Φ function and give the following result:

$$\nabla^2\Psi - l^2\nabla^4\Psi = 0 \tag{9.2.16}$$

Thus, the two stress functions satisfy governing equations $(9.2.15)_1$ and (9.2.16). We will now consider a specific application of this theory for the following stress concentration problem.

EXAMPLE 9.2.1 MICROPOLAR ELASTICITY STRESS CONCENTRATION AROUND A CIRCULAR HOLE UNDER UNIFORM UNIAXIAL LOADING

We now wish to investigate the effects of couple-stress theory on the two-dimensional stress distribution around a circular hole in an infinite medium under uniform tension T at infinity. Recall that this problem was previously solved for the nonpolar case in Example 6.2.4 and the problem geometry is shown in Fig. 6.12. The hole is to have radius a, and the far-field stress is directed along the x-direction as shown. The solution for this case will be first developed for the micropolar model and then the additional simplification for couple-stress theory will be incorporated. This solution was first presented by Kaloni and Ariman (1967) and later by Eringen (1968).

Solution: As expected, for this problem, the plane strain formulation and solution is best done in polar coordinates. For this system, the equilibrium equations become

$$\begin{gathered} \frac{\partial T_{rr}}{\partial r} + \frac{1}{r}\frac{\partial T_{\theta r}}{\partial \theta} + \frac{T_{rr} - T_{\theta\theta}}{r} = 0 \\ \frac{\partial T_{r\theta}}{\partial r} + \frac{1}{r}\frac{\partial T_{\theta\theta}}{\partial \theta} + \frac{T_{r\theta} - T_{\theta r}}{r} = 0 \\ \frac{\partial m_{rz}}{\partial r} + \frac{1}{r}\frac{\partial m_{\theta z}}{\partial \theta} + \frac{m_{rz}}{r} + T_{r\theta} - T_{\theta r} = 0 \end{gathered} \tag{9.2.17}$$

while the strain–deformation relations are

$$\varepsilon_r = \frac{\partial u_r}{\partial r}, \quad \varepsilon_\theta = \frac{1}{r}\left(\frac{\partial u_\theta}{\partial \theta} + u_r\right)$$
$$\varepsilon_{r\theta} = \frac{\partial u_\theta}{\partial r} - \phi_z, \quad \varepsilon_{\theta r} = \frac{1}{r}\left(\frac{\partial u_r}{\partial \theta} - u_\theta\right) + \phi_z \tag{9.2.18}$$

The constitutive equations in polar coordinates read as

$$T_{rr} = \lambda(\varepsilon_r + \varepsilon_\theta) + (2\mu + \kappa)\varepsilon_r$$
$$T_{\theta\theta} = \lambda(\varepsilon_r + \varepsilon_\theta) + (2\mu + \kappa)\varepsilon_\theta$$
$$T_{r\theta} = (\mu + \kappa)\varepsilon_{r\theta} + \mu\varepsilon_{\theta r}, \quad T_{\theta r} = (\mu + \kappa)\varepsilon_{\theta r} + \mu\varepsilon_{r\theta} \tag{9.2.19}$$
$$m_{rz} = \gamma\frac{\partial \phi_z}{\partial r}, \quad m_{\theta z} = \gamma\frac{1}{r}\frac{\partial \phi_z}{\partial \theta}$$

and the strain–compatibility relations take the form

$$\frac{\partial \varepsilon_{\theta r}}{\partial r} - \frac{1}{r}\frac{\partial \varepsilon_r}{\partial \theta} + \frac{\varepsilon_{\theta r} - \varepsilon_{r\theta}}{r} - \frac{\partial \phi_z}{\partial r} = 0$$
$$\frac{\partial \varepsilon_\theta}{\partial r} - \frac{1}{r}\frac{\partial \varepsilon_{r\theta}}{\partial \theta} + \frac{\varepsilon_\theta - \varepsilon_r}{r} - \frac{1}{r}\frac{\partial \phi_z}{\partial \theta} = 0 \tag{9.2.20}$$
$$\frac{\partial m_{\theta z}}{\partial r} - \frac{1}{r}\frac{\partial m_{rz}}{\partial \theta} + \frac{m_{\theta z}}{r} = 0$$

For the polar coordinate case, the stress–stress function relations become

$$T_{rr} = \frac{1}{r}\frac{\partial \Phi}{\partial r} + \frac{1}{r^2}\frac{\partial^2 \Phi}{\partial \theta^2} - \frac{1}{r}\frac{\partial^2 \Psi}{\partial r \partial \theta} + \frac{1}{r^2}\frac{\partial \Psi}{\partial \theta}$$
$$T_{\theta\theta} = \frac{1}{r^2}\frac{\partial^2 \Phi}{\partial r^2} + \frac{1}{r}\frac{\partial^2 \Psi}{\partial r \partial \theta} - \frac{1}{r^2}\frac{\partial \Psi}{\partial \theta}$$
$$T_{r\theta} = -\frac{1}{r}\frac{\partial^2 \Phi}{\partial r \partial \theta} + \frac{1}{r^2}\frac{\partial \Phi}{\partial \theta} - \frac{1}{r}\frac{\partial \Psi}{\partial r} - \frac{1}{r^2}\frac{\partial^2 \Psi}{\partial \theta^2} \tag{9.2.21}$$
$$T_{\theta r} = -\frac{1}{r}\frac{\partial^2 \Phi}{\partial r \partial \theta} + \frac{1}{r^2}\frac{\partial \Phi}{\partial \theta} + \frac{\partial^2 \Psi}{\partial r^2}$$
$$m_{rz} = \frac{\partial \Psi}{\partial r}, \quad m_{\theta z} = \frac{1}{r}\frac{\partial \Psi}{\partial \theta}$$

Using constitutive relations (9.2.19), the compatibility equations (9.2.20) can be expressed in terms of stresses, and combining this result with (9.2.21) will yield the governing equations for the stress functions in polar coordinates:

$$\frac{\partial}{\partial r}(\Psi - l_1^2\nabla^2\Psi) = -2(1-\nu)l_2^2\frac{1}{r}\frac{\partial}{\partial \theta}(\nabla^2\Phi)$$
$$\frac{1}{r}\frac{\partial}{\partial \theta}(\Psi - l_1^2\nabla^2\Psi) = 2(1-\nu)l_2^2\frac{\partial}{\partial r}(\nabla^2\Phi) \tag{9.2.22}$$

where

$$l_1^2 = \frac{\gamma(\mu+\kappa)}{\kappa(2\mu+\kappa)}, \quad l_2^2 = \frac{\gamma}{2(2\mu+\kappa)}$$
$$\nabla^2 = \frac{\partial^r}{\partial r^2} + \frac{1}{r}\frac{\partial}{\partial r} + \frac{1}{r^2}\frac{\partial^2}{\partial \theta^2} \tag{9.2.23}$$

Note that for the micropolar case, *two length parameters* l_1 and l_2 appear in the theory.

The appropriate solutions to equations (9.2.22) for the problem under study are given by

$$\Phi = \frac{T}{4}r^2(1-\cos 2\theta) + A_1 \log r + \left(\frac{A_2}{r^2} + A_3\right)\cos 2\theta$$
$$\Psi = \left(\frac{A_4}{r^2} + A_5 K_2\left(\frac{r}{l_1}\right)\right)\sin 2\theta \tag{9.2.24}$$

where K_n is the *modified Bessel function of the second kind or order n*, and A_i are constants to be determined with $A_4 = 8(1-\nu)l_1^2 A_3$. Using this stress function solution, the components of the stress and couple-stress then follow from (9.2.21) to be

$$T_{rr} = \frac{T}{2}(1+\cos 2\theta) + \frac{A_1}{r^2} - \left(\frac{6A_2}{r^4} + \frac{4A_3}{r^2} - \frac{6A_4}{r^4}\right)\cos 2\theta$$
$$+\frac{2A_5}{l_1 r}\left[\frac{3l_1}{r}K_o(r/l_1) + \left(1 + \frac{6l_1^2}{r^2}\right)K_1(r/l_1)\right]\cos 2\theta$$
$$T_{\theta\theta} = \frac{T}{2}(1-\cos 2\theta) - \frac{A_1}{r^2} + \left(\frac{6A_2}{r^4} - \frac{6A_4}{r^4}\right)\cos 2\theta$$
$$-\frac{2A_5}{l_1 r}\left[\frac{3l_1}{r}K_o(r/l_1) + \left(1 + \frac{6l_1^2}{r^2}\right)K_1(r/l_1)\right]\cos 2\theta$$
$$T_{r\theta} = -\left(\frac{T}{2} + \frac{6A_2}{r^4} + \frac{2A_3}{r^2} - \frac{6A_4}{r^4}\right)\sin 2\theta$$
$$+\frac{A_5}{l_1 r}\left[\frac{6l_1}{r}K_o(r/l_1) + \left(1 + \frac{12l_1^2}{r^2}\right)K_1(r/l_1)\right]\sin 2\theta \tag{9.2.25}$$
$$T_{\theta r} = -\left(\frac{T}{2} + \frac{6A_2}{r^4} + \frac{2A_3}{r^2} - \frac{6A_4}{r^4}\right)\sin 2\theta$$
$$+\frac{A_5}{l_1^2}\left[\left(1 + \frac{6l_1^2}{r^2}\right)K_o(r/l_1) + \left(\frac{3l_1}{r} + \frac{12l_1^3}{r^3}\right)K_1(r/l_1)\right]\sin 2\theta$$
$$m_{rz} = -\left\{\frac{2A_4}{r^3} + \frac{A_5}{l_1}\left[\frac{2l_1}{r}K_o(r/l_1) + \left(1 + \frac{4l_1^2}{r^2}\right)K_1(r/l_1)\right]\right\}\sin 2\theta$$
$$m_{\theta z} = \left\{\frac{2A_4}{r^3} + \frac{2A_5}{r}\left[K_o(r/l_1) + \frac{2l_1}{r}K_1(r/l_1)\right]\right\}\cos 2\theta$$

For boundary conditions, we use the usual forms for the nonpolar variables, while the couple stress m_{rz} is taken to vanish on the hole boundary and at infinity

$$\begin{aligned} T_{rr}(a,\theta) &= T_{r\theta}(a,\theta) = m_{rz}(a,\theta) = 0 \\ T_{rr}(\infty,\theta) &= \frac{T}{2}(1+\cos 2\theta) \\ T_{r\theta}(\infty,\theta) &= -\frac{T}{2}\sin 2\theta \\ m_{rz}(\infty,\theta) &= 0 \end{aligned} \tag{9.2.26}$$

Using these conditions, sufficient relations can be developed to determine the arbitrary constants A_i giving the results

$$\begin{aligned} A_1 &= -\frac{T}{2}a^2, \quad A_2 = -\frac{Ta^4(1-F)}{4(1+F)} \\ A_3 &= \frac{Ta^2}{2(1+F)}, \quad A_4 = \frac{4T(1-\nu)a^2 l_2^2}{1+F} \\ A_5 &= -\frac{Tal_1 F}{(1+F)K_1(a/l_1)} \end{aligned} \tag{9.2.27}$$

where

$$F = 8(1-\nu)\frac{l_2^2}{l_1^2}\left[4+\frac{a^2}{l_1^2}+\frac{2a}{l_1}\frac{K_o(a/l_1)}{K_1(a/l_1)}\right]^{-1} \tag{9.2.28}$$

and the problem is now solved.

Let us now investigate the maximum stress and discuss the nature of the concentration behavior in the vicinity of the hole and compare with classical elasticity predictions. As in the previous nonpolar case, the circumferential stress $T_{\theta\theta}$ on the hole boundary will be the maximum stress. From the (9.2.25),

$$T_{\theta\theta}(a,\theta) = T\left(1-\frac{2\cos 2\theta}{1+F}\right) \tag{9.2.29}$$

As expected, the maximum value of this quantity occurs at $\theta = \pm\pi/2$, and thus the stress concentration factor for the micropolar stress problem is given by

$$K = \frac{(T_{\theta\theta})_{\max}}{T} = \frac{3+F}{1+F} \tag{9.2.30}$$

Notice that for micropolar theory, the stress concentration depends on the material parameters and on the *size of the hole*. This problem has also been solved by Mindlin (1963) for couple-stress theory, and this result may be found from the current solution by letting $l_1 = l_2 = l$. Fig. 9.5 illustrates the behavior of the stress concentration factor as a function of a/l_1 for several cases of length ratio l_2/l_1 with $\nu = 0$. It is observed that the micropolar/couple-stress

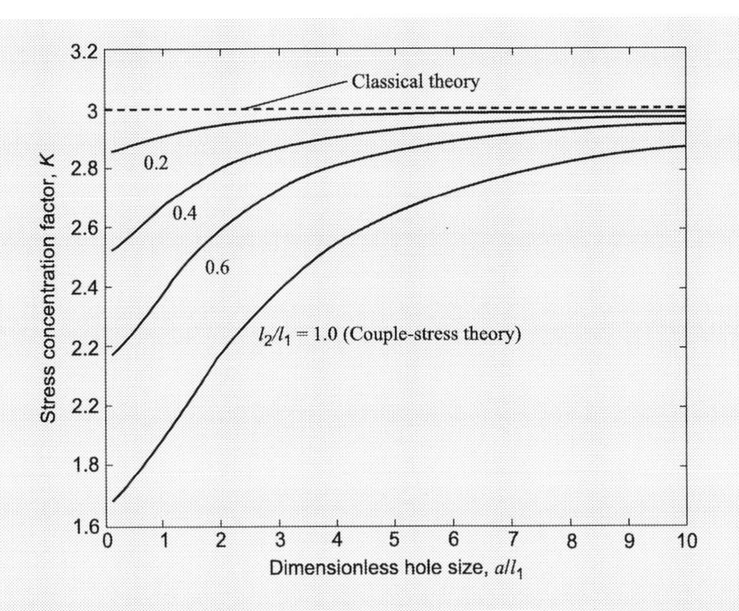

FIGURE 9.5

Stress concentration behavior for micropolar theory ($\nu = 0$).

concentration factors are *less than* that predicted by classical theory ($K = 3$), and differences between the theories depend on the ratio of the hole size to the microstructural length parameter l_l (or l). If the length parameter is small in comparison to the hole size, very little differences in the stress concentration predictions will occur. For the case $l_1 = l_2 = l = 0$, it can be shown that $F \to 0$, thus giving $K = 3$ which matches with the classical result. Mindlin (1963) also investigated other far-field loading conditions for this problem. He showed that for the case of equal biaxial loading, the couple-stress effects disappear completely, while for pure shear loading couple-stress effects produce a significant reduction in the stress concentration when compared to classical theory.

Originally, it was hoped that this solution could be used to explain the observed reduction in stress concentration factors for small holes in regions of high stress gradients. Unfortunately, it has been pointed out by several authors (Schijve, 1966; Ellis and Smith, 1967; Kaloni and Ariman, 1967) that for typical metals the reduction in the stress concentration for small holes cannot be accurately accounted for using couple-stress or micropolar theory. Additional similar solutions for stress concentrations around circular inclusions have been developed by Weitsman (1965) and Hartranft and Sih (1965).

More recent studies have had success in applying micropolar/couple-stress theory to fiber-reinforced composites (Sun and Yang, 1975) and granular materials (Chang and Ma, 1991). With respect to computational methods, micropolar finite-element techniques have been developed by Kennedy and Kim (1987) and Kennedy (1999). Many other published research papers have appeared dealing with various applications and developments of this microelasticity theory.

9.3 ELASTICITY THEORY WITH VOIDS

Another interesting micromechanics model has been proposed for porous and granular materials with distributed voids. Originally developed by Goodman and Cowin (1972) for more general materials, we focus here only on the linear elastic theory which was given by Cowin and Nunziato (1983). A series of application papers followed including Cowin and Puri (1983), Cowin (1984a,b), and Cowin (1985a). The theory is intended for elastic materials containing a uniform distribution of small voids as shown in Fig. 9.6. When the void volume vanishes, the material behavior reduces to classical elasticity theory. The primary feature of the new theory is the introduction of a volume fraction (related to void volume), which is taken as an *independent kinematic variable*. The other variables of displacement and strain are retained in their usual form for small deformations. The inclusion of the new variable requires additional microforces to provide proper equilibrium of the micropore volume.

The theory begins by expressing the material mass density as the following product:

$$\rho = \gamma v \tag{9.3.1}$$

where ρ is the bulk (overall) mass density, γ is the mass density of the matrix material, and v is the *matrix volume fraction* or *volume distribution function*. This function describes the way the medium is distributed in space and is taken to be an independent variable, thus introducing an additional kinematic degree of freedom in the theory. The linear theory with voids deals with small changes from a *stress- and strain-free reference configuration*. In this configuration, relation (9.3.1) can be written as $\rho_R = \gamma_R v_R$. The independent kinematical variables of this theory are the usual displacements u_i, and the *change in volume fraction* from the reference configuration expressed by

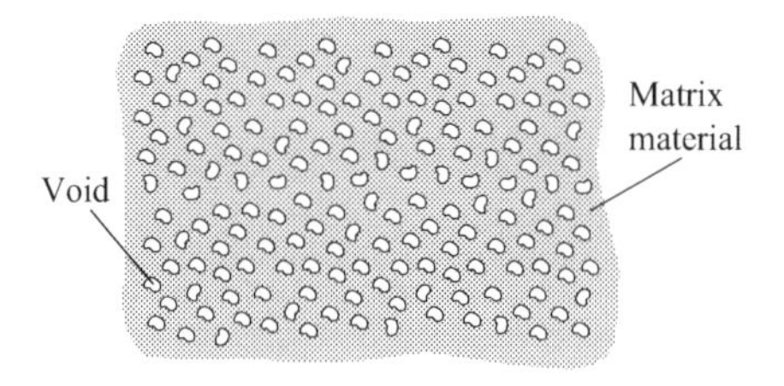

FIGURE 9.6

Solid continuum with distributed voids.

$$\phi = v - v_R \tag{9.3.2}$$

The strain displacement relations are those of classical small deformation elasticity

$$\varepsilon_{ij} = \frac{1}{2}(u_{i,j} + u_{j,i}) \tag{9.3.3}$$

and likewise for the equilibrium equations (with no body forces) reads

$$T_{ij,j} = 0 \tag{9.3.4}$$

The general development of this theory included external body forces and dynamic inertial terms, but these will not be retained in our brief presentation.

In order to develop the microequilibrium of the void volume, new micromechanics theory involving the balance of *equilibrated force* is introduced. Details of this development are beyond the scope of our presentation and we will only give the final results for the quasi-static case:

$$h_{i,i} + g = 0 \tag{9.3.5}$$

where h_i is the *equilibrated stress vector* and g is the *intrinsic equilibrated body force*. Simple physical meanings of these variables have proved difficult to provide. Cowin and Nunziato (1983) have indicated that these variables can be related to particular self-equilibrated singular force systems. In particular, h_i and g can be associated with double force systems, and the expression $h_{i,i}$ can be related to centers of dilatation (see Sadd, 2014, Chapter 15). This balance of equilibrated force has been re-examined by Fang et al. (2006) and Chen and Lan (2008), and these studies have provided some additional corrections and clarity.

The constitutive equations for linear isotropic elastic materials with voids provide relations for the stress tensor, equilibrated stress vector, and intrinsic body force of the form

$$\begin{gathered} T_{ij} = \lambda\varepsilon_{kk}\delta_{ij} + 2\mu\varepsilon_{ij} + \beta\phi\delta_{ij} \\ h_i = \alpha\phi_{,i} \\ g = -\omega\dot{\phi} - \xi\phi - \beta\varepsilon_{kk} \end{gathered} \tag{9.3.6}$$

where the material constants $\lambda, \mu, \alpha, \beta, \xi, \omega$ all depend on the reference fraction v_R and satisfy the inequalities

$$\mu \geq 0, \quad \alpha \geq 0, \quad \xi \geq 0, \quad \omega \geq 0, \quad 3\lambda + 2\mu \geq 0, \quad M = \frac{3\lambda + 2\mu}{\beta^2}\xi \geq 3 \tag{9.3.7}$$

Note that even though we have dropped dynamic inertial terms, constitutive relation $(9.3.6)_3$ includes a *time-dependent response* in the volume fraction. This fact indicates that the theory will predict a *viscoelastic* type of behavior (Cowin, 1985a) even for problems with time independent boundary conditions and homogeneous deformations.

For this theory, the boundary conditions on stress and displacement are the same as those of classical elasticity. The boundary conditions on the self-equilibrated stress vector are taken to have a *vanishing normal component*, that is, $h_i n_i = 0$ where n_i is the surface unit normal vector. Using this with the constitutive statement $(9.3.6)_2$ develops the boundary specification on the volume fraction

$$\phi_{,i} n_i = 0 \tag{9.3.8}$$

This completes our brief general presentation of the theory. We will now discuss the solution to the classical two-dimensional stress concentration problem around a circular hole discussed previously.

EXAMPLE 9.3.1 ELASTICITY WITH VOIDS STRESS CONCENTRATION AROUND A CIRCULAR HOLE UNDER UNIFORM UNIAXIAL LOADING

Consider again the stress concentration problem of a stress-free circular hole of radius a in an infinite plane under uniform tension as shown in Fig. 6.12. We have previously discussed this problem for the classical elastic case in Example 6.2.4 and for the micropolar/couple-stress case in Example 9.2.1. We will now outline the solution given by Cowin (1984b) and compare the results with the micropolar/couple-stress and classical solutions.

Solution: The problem is formulated under plane stress conditions:

$$T_{xx} = T_{xx}(x,y), \quad T_{yy} = T_{yy}(x,y), \quad T_{xy} = T_{xy}(x,y), \quad T_{zz} = T_{xz} = T_{yz} = 0$$

For this two-dimensional case, the constitutive relations reduce to

$$\begin{aligned} T_{ij} &= \frac{2\mu}{\lambda + 2\mu}(\lambda \varepsilon_{kk} + \beta\phi)\delta_{ij} + 2\mu\varepsilon_{ij} \\ g &= -\omega\dot{\phi} - \left(\xi - \frac{\beta^2}{\lambda + 2\mu}\right)\phi - \frac{2\mu\beta}{\lambda + 2\mu}\varepsilon_{kk} \end{aligned} \tag{9.3.9}$$

where all indices are taken over the limited range 1, 2. Using a stress formulation, the single nonzero compatibility relation becomes

$$T_{kk,mm} - \frac{\mu\beta}{\lambda + \mu}\phi_{,mm} = 0 \tag{9.3.10}$$

Introducing the usual Airy stress function (6.2.47) denoted here by ψ allows this relation to be written as

$$\nabla^4\psi - \frac{\mu\beta}{\lambda + \mu}\nabla^2\phi = 0 \tag{9.3.11}$$

For this case, relation (9.3.5) for balance of equilibrated forces reduces to

$$\alpha\nabla^2\phi - \frac{\alpha}{h^2}\phi - \omega\dot{\phi} = \frac{\beta}{3\lambda + 2\mu}\left(\nabla^2\psi - \frac{\mu\beta}{\lambda + \mu}\phi\right) \tag{9.3.12}$$

The parameter h is defined by

$$\frac{\alpha}{h^2}=\xi-\frac{\beta^2}{\lambda+\mu} \tag{9.3.13}$$

and has units of length and thus can be taken as a *microstructural length measure* for this particular theory.

Relations (9.3.11) and (9.3.12) now form the governing equations for the plane stress problem. The presence of the time-dependent derivative term in (9.3.12) requires some additional analysis. Using Laplace transform theory, Cowin (1984b) has shown that under steady boundary conditions, solutions for ϕ and ψ can be determined from the limiting case where $t\rightarrow\infty$ which is related to taking $\omega=0$. Thus, by taking the Laplace transform of (9.3.11) and (9.3.12) gives

$$\begin{gathered}\nabla^4\bar{\psi}-\frac{\mu\beta}{\lambda+\mu}\nabla^2\bar{\phi}=0\\ \alpha\nabla^2\bar{\phi}-\frac{\alpha}{\bar{h}^2}\bar{\phi}=\frac{\beta}{3\lambda+2\mu}\left(\nabla^2\bar{\psi}-\frac{\mu\beta}{\lambda+\mu}\bar{\phi}\right)\end{gathered} \tag{9.3.14}$$

where $\bar{\phi}=\bar{\phi}(s),\bar{\psi}=\bar{\psi}(s)$ are the standard Laplace transforms of ϕ,ψ, and s is the Laplace transform variable (see Section 6.5), and the parameter $\bar{h}$ is defined by $\frac{\alpha}{\bar{h}^2}=\frac{\alpha}{h^2}+\omega s$.

Boundary conditions on the problem follow from our previous discussions to be

$$T_{rr}=T_{r\theta}=\frac{\partial\bar{\phi}}{\partial r}=0\,\text{on}\,r=a$$

For the circular hole problem, the solution to system (9.3.14) is developed in polar coordinates. Guided by the results from classical elasticity, we look for solutions of the form $f_1(r)+f_2(r)\cos 2\theta$, where f_1 and f_2 are arbitrary functions of the radial coordinate. Employing this scheme, the properly bounded solution satisfying the boundary condition $\frac{\partial\bar{\phi}}{\partial r}=0$ at $r=a$ is found to be

$$\begin{gathered}\bar{\phi}=\frac{-\xi\bar{p}}{M\beta\omega s+\beta\xi(M-3)}+\frac{A_3(\lambda+\mu)}{\mu\bar{h}^2\beta}[\bar{F}(r)-1]\cos 2\theta\\ \bar{\psi}=\frac{\mu\beta}{\lambda+\mu}\bar{h}^2\bar{\phi}+\left[\frac{\bar{p}r^2}{4}+A_1\log r+\left(\frac{A_2}{r^2}+A_3-\frac{\bar{p}r^2}{4}\right)\cos 2\theta\right]\end{gathered} \tag{9.3.15}$$

where $\bar{F}$ is given by

$$\bar{F}(r)=1+\frac{4\mu\xi\bar{h}^2}{(\lambda+\mu)M\omega s+4\mu\xi N}\left[\frac{1}{r^2}+\frac{2\bar{h}K_2(r/\bar{h})}{a^3K_2'(a/\bar{h})}\right] \tag{9.316}$$

and $\bar{p}$ is the Laplace transform of the uniaxial stress at infinity, K_2 is the modified Bessel function of the second kind of order 2, $N = \frac{\lambda+\mu}{4\mu}(M-3) \geq 0$, and the constants A_1, A_2, A_3 are determined from the stress-free boundary conditions as

$$A_1 = -\frac{1}{2}\bar{p}a^2, \quad A_2 = -\frac{1}{4}\bar{p}a^4, \quad A_3 = \frac{\bar{p}a^2}{2\bar{F}(a)}$$

Note in relation (9.3.16), the bar on F indicates the dependency on the Laplace transform parameter s, and the bar is to be removed for the case where $s \to 0$ and $\bar{h}(s)$ is replaced by h.

This completes the solution for the Laplace-transformed volume fraction and Airy stress function. The transformed stress components can now be obtained from the Airy function using the usual relations

$$\bar{T}_{rr} = \frac{1}{r}\frac{\partial\bar{\psi}}{\partial r} + \frac{1}{r^2}\frac{\partial^2\bar{\psi}}{\partial\theta^2}, \quad \bar{T}_{\theta\theta} = \frac{\partial^2\bar{\psi}}{\partial r^2}, \quad \bar{T}_{r\theta} = \frac{1}{r^2}\frac{\partial\bar{\psi}}{\partial\theta} - \frac{1}{r}\frac{\partial^2\bar{\psi}}{\partial r\partial\theta} \tag{9.3.17}$$

We will now consider the case where the far-field tension T is a constant in time, and thus $\bar{p} = T/s$. Rather than formally inverting (inverse Laplace transformation) the resulting stress components generated from relations (9.3.17), Cowin develops the results for the cases of $t = 0$ and $t \to \infty$. It turns out that for the initial condition at $t = 0$, the stresses match those found from classical elasticity (see Example 6.2.4). However, for the final-value case ($t \to \infty$), which implies ($s \to 0$), the stresses are different than predictions from classical theory.

Focusing our attention only to the hoop stress, the elasticity with voids solution for the final-value case is determined as

$$T_{\theta\theta} = \frac{T}{2}\left\{\left(1+\frac{a^2}{r^2}\right) + \cos 2\theta\left[a^2\frac{F''(r)}{F(a)} - \left(1+3\frac{a^4}{r^4}\right)\right]\right\} \tag{9.3.18}$$

The maximum value of this stress is again found at $r = a$ and $\theta = \pm\pi/2$ and is given by

$$\begin{aligned}(T_{\theta\theta})_{\max} &= T_{\theta\theta}(a, \pm\pi/2) = T\left(3 - \frac{a^2}{2}\frac{F''(a)}{F(a)}\right)\\ &= T\left[3 + \left(2N + [1+(4+L^2)N]\frac{K_1(L)}{LK_o(L)}\right)^{-1}\right]\end{aligned} \tag{9.3.19}$$

where $L = a/h$. It is observed from this relation that the stress concentration factor $K = (T_{\theta\theta})_{\max}/T$ will always be greater than or equal to 3. Thus, the elasticity theory with voids predicts an *elevation* of the stress concentration when

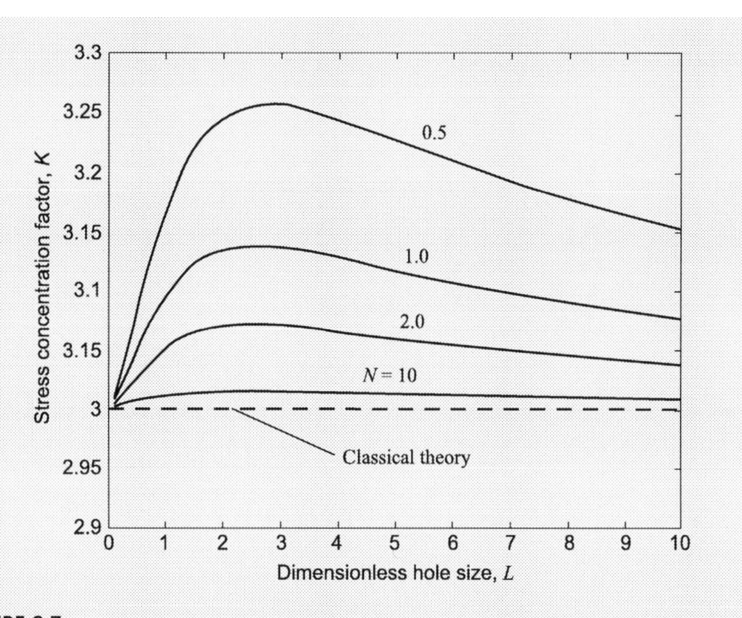

FIGURE 9.7

Stress concentration behavior for elastic material with voids.

compared to the classical result. The behavior of the stress concentration factor as a function of the dimensionless hole size L is shown in Fig. 9.7. It can be seen that the stress concentration factor reduces to the classical result as L approaches zero or infinity. For a particular value of the material parameter N, the stress concentration takes on a maximum value at a finite intermediate value of L.

It is interesting to compare these results with our previous study of the same stress concentration problem using the micropolar theory previously discussed. Recall in Example 9.2.1 we solved the identical problem for micropolar/couple-stress theories, and the results were shown in Fig. 9.5. Thus, Figs. 9.5 and 9.7 illustrate the stress concentration behavior as a function of a nondimensional ratio of hole radius divided by a microstructural length parameter. While the current model with voids indicates an elevation of stress concentration, the micropolar/couple-stress results show a decrease in this factor. Micropolar/couple-stress theory also predicts that the largest difference from the classical result occurs at a dimensionless hole size ratio of zero. However, for elasticity with voids, this difference occurs at a finite hole size ratio approximately between 2 and 3. It is apparent that micropolar theory (allowing independent microrotational deformation) will give fundamentally different results than the current void theory which allows for an independent microvolumetric deformation.

9.4 DOUBLET MECHANICS

We now wish to explore a micromechanical theory that has demonstrated applications for particulate materials in predicting observed behaviors that cannot be shown using classical continuum mechanics. The theory known as *doublet mechanics* (DM) was originally developed by Granik (1978). It has been applied to granular materials by Granik and Ferrari (1993) and Ferrari et al. (1997) and to asphalt concrete materials by Sadd and Dai (2004). DM is a micromechanical theory based on a discrete material model whereby solids are represented as arrays of points or nodes at finite distances. A pair of such nodes is referred to as a *doublet*, and the nodal spacing distances introduce *length scales* into the theory. Current applications of this scheme have normally used regular arrays of nodal spacing, thus generating a regular lattice microstructure with similarities to the micropolar model shown in Fig. 9.3. Each node in the array is allowed to have a translation and rotation, and increments of these variables are expanded in a Taylor series about the nodal point. The order at which the series is truncated defines the degree of approximation employed. The lowest order case using only a single term in the series will not contain any length scales, while using additional terms results in a multilength scale theory. The allowable kinematics develops microstrains of elongation, shear, and torsion (about the doublet axis). Through appropriate constitutive assumptions, these microstrains can be related to corresponding elongational, shear, and torsional microstresses.

Although not necessary, a *granular interpretation* of DM is commonly employed, in which the material is viewed as an assembly of circular or spherical particles. A pair of such particles represents a doublet as shown in Fig. 9.8. Corresponding to the doublet (A, B), there exists a *doublet* or *branch vector* ζ_α connecting the adjacent particle centers and defining the doublet axis α. The magnitude of this vector $\eta_\alpha = |\zeta_\alpha|$ is simply the sum of the two radii for particles in contact. However, in general, the particles need not be in contact, and the length scale η_α could be used to represent a more general microstructural feature. As mentioned, the kinematics allow relative elongational, shearing, and torsional motions between the particles, and this is used

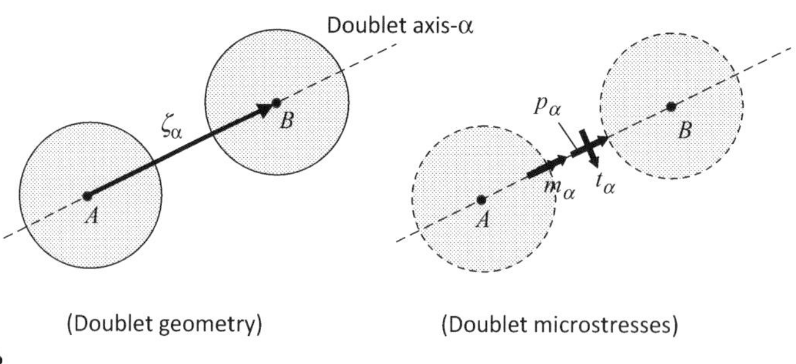

FIGURE 9.8

Doublet Mechanics geometry.

to develop *elongational microstress* $\boldsymbol{p}_\alpha$, *shear microstress* $\boldsymbol{t}_\alpha$, and *torsional microstress* $\boldsymbol{m}_\alpha$ as shown in Fig. 9.8. It should be pointed out that these microstresses are not second-order tensors in the usual continuum mechanics sense. Rather, they are vector quantities that represent the elastic microforces and microcouples of interaction between doublet particles. Their directions are dependent on the doublet axes that are determined by the material microstructure. Also, these microstresses are not continuously distributed but rather exist only at particular points in the medium being simulated by DM theory.

If $\boldsymbol{u}(\boldsymbol{x},t)$ is the displacement field coinciding with a particle displacement, then the increment function can be written as

$$\Delta \boldsymbol{u}_\alpha = \boldsymbol{u}(\boldsymbol{x}+\zeta_\alpha, t) - \boldsymbol{u}(\boldsymbol{x}, t) \tag{9.4.1}$$

where $\alpha = 1, \ldots, n$, and n is referred to as the *valence* of the lattice. Considering only the case where the doublet interactions are symmetric, it can be shown that the shear and torsional microdeformations and stresses vanish, and thus only extensional strains and stresses will exist. For this case, the extensional microstrain ε_α (representing the elongational deformation of the doublet vector) is defined by

$$\varepsilon_\alpha = \frac{\boldsymbol{q}_\alpha \cdot \Delta \boldsymbol{u}_\alpha}{\eta_\alpha} \tag{9.4.2}$$

where $\boldsymbol{q}_\alpha = \zeta_\alpha / \eta_\alpha$ is the unit vector in the α-direction. The increment function (9.4.1) can be expanded in a Taylor series as

$$\Delta \boldsymbol{u}_\alpha = \sum_{m=1}^{M} \frac{(\eta_\alpha)^m}{m!} (\boldsymbol{q}_\alpha \cdot \nabla)^m \, \boldsymbol{u}(\boldsymbol{x}, t) \tag{9.4.3}$$

Using this result in relation (9.4.2) develops the series expansion for the extensional microstrain

$$\varepsilon_\alpha = q_{\alpha i} \sum_{m=1}^{M} \frac{(\eta_\alpha)^{m-1}}{m!} q_{\alpha k_1} \cdots q_{\alpha k_m} \frac{\partial^m u_i}{\partial x_{k_1} \ldots \partial x_{k_m}} \tag{9.4.4}$$

where $q_{\alpha k}$ are the direction cosines of the doublet directions with respect to the coordinate system. As mentioned, the number of terms used in the series expansion of the local deformation field determines the order of approximation in DM theory. For the first-order case ($m = 1$), the scaling parameter η_α will drop from the formulation, and the elongational microstrain is reduced to

$$\varepsilon_\alpha = q_{\alpha i} q_{\alpha j} \varepsilon_{ij} \tag{9.4.5}$$

where ε_{ij} is the usual small strain tensor. For this case, it has been shown that the DM solution can be calculated directly from the corresponding continuum elasticity solution through the relation

$$T_{ij} = \sum_{\alpha=1}^{n} q_{\alpha i} q_{\alpha j} p_\alpha \tag{9.4.6}$$

For the two-dimensional case, this result can be expressed in matrix form

$$\boldsymbol{T} = \boldsymbol{Q}\boldsymbol{p} \Rightarrow \boldsymbol{p} = \boldsymbol{Q}^{-1}\boldsymbol{T} \tag{9.4.7}$$

where $\boldsymbol{T} = \{T_{xx}, T_{yy}, T_{xy}\}^T$ is the continuum elastic stress vector in Cartesian coordinates, $\boldsymbol{p}$ is the microstress vector, and $\boldsymbol{Q}$ is a transformation matrix. For plane problems, this transformation matrix can be written as

$$\boldsymbol{Q} = \begin{bmatrix} (q_{11})^2 & (q_{21})^2 & (q_{31})^2 \\ (q_{12})^2 & (q_{22})^2 & (q_{32})^2 \\ q_{11}q_{12} & q_{21}q_{22} & q_{31}q_{32} \end{bmatrix} \tag{9.4.8}$$

This result allows a straightforward development of first-order DM solutions for many problems of engineering interest.

EXAMPLE 9.4.1 DM SOLUTION OF THE ELASTICITY FLAMANT PROBLEM

We now wish to investigate a specific application of the DM model for a two-dimensional problem with regular particle packing microstructure. The case of interest is the Flamant problem of a concentrated force acting on the free surface of a semi-infinite solid as shown in Fig. 9.9. The classical elasticity solution to this problem was originally developed in Example 6.2.5, and the Cartesian stress distribution was given by relations (6.2.67):

$$\begin{aligned} T_{xx} &= -\frac{2Px^2y}{\pi(x^2+y^2)^2} \\ T_{yy} &= -\frac{2Py^3}{\pi(x^2+y^2)^2} \\ T_{xy} &= -\frac{2Pxy^2}{\pi(x^2+y^2)^2} \end{aligned} \tag{9.4.9}$$

This continuum mechanics solution specifies that the normal stresses are *everywhere compressive* in the half space ($y > 0$), and a plot of the distribution of normal and shear stresses on a surface y = constant was shown in Fig. 6.15.

The DM model of this problem is established by choosing a regular two-dimensional hexagonal packing as shown in Fig. 9.9. This geometrical microstructure establishes three doublet axes at angles $\gamma = 60°$ as shown. Using only first-order approximation, DM shear and torsional microstresses vanish, leaving only elongational microstress components (p_1, p_2, p_3) as shown. Positive elongational components correspond to tensile forces between particles.

For this fabric geometry, the transformation matrix (9.4.8) becomes

$$\boldsymbol{Q} = \begin{bmatrix} \cos^2\gamma & \cos^2\gamma & 1 \\ \sin^2\gamma & \sin^2\gamma & 0 \\ -\cos\gamma\sin\gamma & \cos\gamma\sin\gamma & 0 \end{bmatrix} \tag{9.4.10}$$

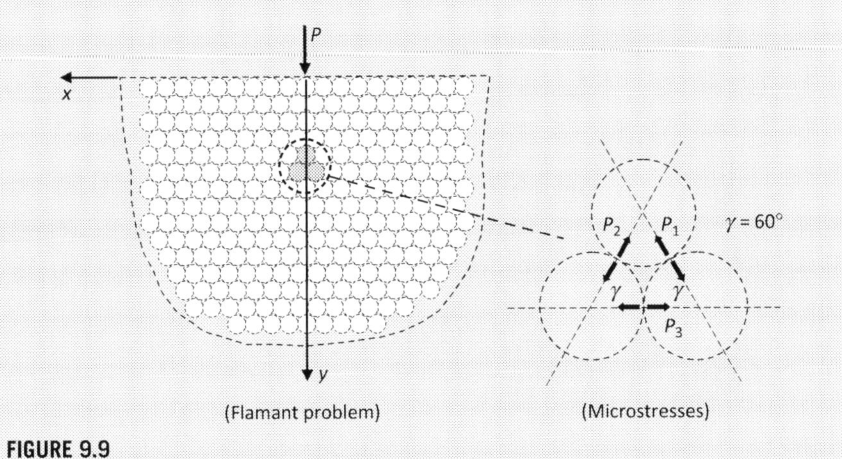

FIGURE 9.9

Flamant problem for the Doublet Mechanics model.

Using this transformation in relation (9.4.7) produces the following microstresses:

$$\begin{aligned} p_1 &= -\frac{4Py^2(\sqrt{3}x+y)}{3\pi(x^2+y^2)^2} \\ p_2 &= -\frac{4Py^2(\sqrt{3}x-y)}{3\pi(x^2+y^2)^2} \\ p_3 &= -\frac{2Py(3x^2-y^2)}{3\pi(x^2+y^2)^2} \end{aligned} \tag{9.4.11}$$

Although these DM microstresses actually exist only at discrete points and in specific directions as shown in Fig. 9.9, we will use these results to make continuous contour plots over the half-space domain under study. In this fashion, we can compare DM predictions with the corresponding classical elasticity results. Reviewing the stress fields given by (9.4.9) and (9.4.11), we can only directly compare the horizontal elasticity component with the DM microstress p_3. The other stress components act in different directions, and thus do not allow a simple direct comparison.

Fig. 9.10 illustrates contour plots of the elasticity T_{xx} and DM p_3 stress components. As mentioned previously, the classical elasticity results predict a totally compressive stress field as shown. Note, however, the difference in predictions from DM theory. There exists a symmetric region of tensile microstress below the loading point in region $y \geq \sqrt{3}\,|x|$. It has been pointed out in the literature that there exists experimental evidence of such tensile behavior in granular and particulate composite materials under similar surface loading, and Ferrari et al. (1997) refer to this issue as *Flamant's paradox*. It would appear that micromechanical effects are the mechanisms for the observed tensile behaviors, and DM theory offers a possible approach to predict this phenomenon. Note that our comparisons are of qualitative nature. Additional anomalous elastic behaviors have been reported for other plane elasticity problems (Ferrari et al., 1997; Sadd and Dai, 2004).

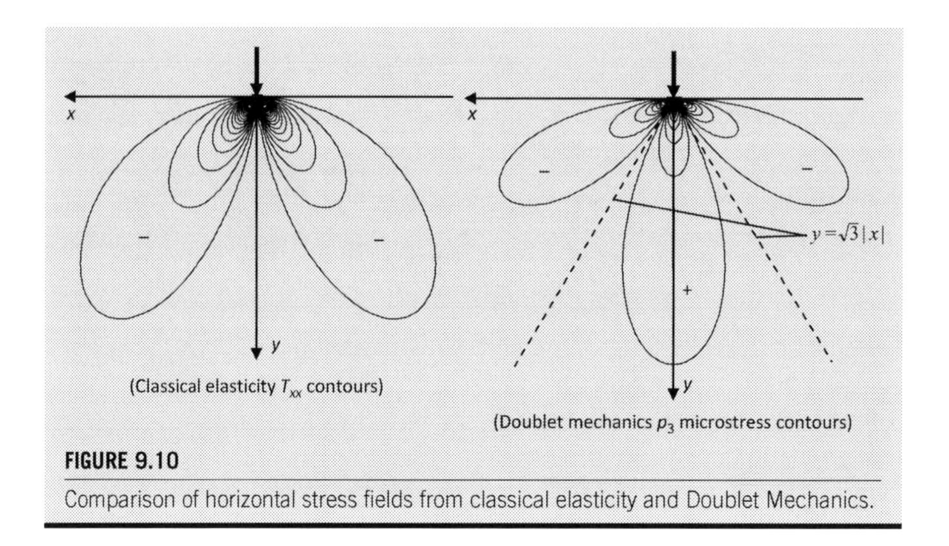

FIGURE 9.10

Comparison of horizontal stress fields from classical elasticity and Doublet Mechanics.

9.5 HIGHER GRADIENT ELASTICITY THEORIES

In classical elasticity, the strain energy can be expressed as a quadratic function of the infinitesimal strain tensor $U = U(\varepsilon_{ij})$, and thus the stress $T_{ij} = \partial U / \partial \varepsilon_{ij}$ will be a linear function of the strain. With interest in microstructural behavior and especially focusing on finding better ways to model singularities, localization, and size effects, researchers began to explore new expanded elasticity theories that would also include higher gradients of strain in the constitutive relation. These approaches have become known as *higher grade continua theories*. In addition to his work on micropolar elasticity, Mindlin (1965) was one of the originators of higher gradient theories, and he developed a linear elasticity theory that included the strain and its first and second derivatives in the constitutive relation for stress. It was anticipated that such theories would possibly remove various singularities in the solution field variables, thereby producing a more acceptable model of real materials. Starting in the 1980s, Afantis and coworkers began significant and very extensive research in this field now called *gradient elasticity theory*. Extensions to the inelastic response and other behaviors have also been included within the theory. Countless papers have been published in this field, and Aifantis (2003) and Askes and Aifantis (2011) have provided review articles on many of the more recent developments and applications of this higher-grade continua research.

For the isotropic case, Mindlin (1965) expressed the strain energy in terms of the usual strain and also included a sizeable list of first and second derivatives of the strain:

$$U = \frac{1}{2}\lambda \varepsilon_{ii}\varepsilon_{jj} + \mu \varepsilon_{ij}\varepsilon_{ij} + a_1 \varepsilon_{ij,j}\varepsilon_{ik,k} + \cdots + b_1 \varepsilon_{ii,jj}\varepsilon_{kk,ll} + \cdots \tag{9.5.1}$$

Using the usual hyperelastic relation, the stress can be computed for a particular case as

$$T_{ij} = \frac{\partial U}{\partial \varepsilon_{ij}} = \lambda \varepsilon_{kk}\delta_{ij} + 2\mu\varepsilon_{ij} + c_1\varepsilon_{kk,ll}\delta_{ij} + c_2\varepsilon_{ij,kk} + \frac{1}{2}c_3(\varepsilon_{kk,ij} + \varepsilon_{kk,ji}) \tag{9.5.2}$$

where λ and μ are the usual elastic constants and c_i being new material constants. Aifantis and others have generally simplified this constitutive form into

$$T_{ij} = \lambda\varepsilon_{kk}\delta_{ij} + 2\mu\varepsilon_{ij} - c[\lambda\varepsilon_{kk}\delta_{ij} + 2\mu\varepsilon_{ij}]_{,ll} \tag{9.5.3}$$

where c is the remaining single gradient material constant which has units of length squared, and hence $\sqrt{c}$ would represent a *length measure*. We thus now have a linear elastic theory that contains the usual strain terms *and second-order gradients in the strain tensor*. Over the years, other different and often more complicated constitutive forms have been developed. However, these are beyond the level and purpose of the text and thus we will limit our further study using only constitutive form (9.5.3). The remaining elasticity field equations of strain–displacement, compatibility, and equilibrium stay the same.

Following Ru and Aifantis (1993), we can establish a simple relationship between the gradient theory (9.5.3) and classical elasticity. First, noting that constitutive law (9.5.3) can be expressed in direct notational and operator form

$$\boldsymbol{T} = (1 - c\nabla^2)\boldsymbol{T}^0, \quad \boldsymbol{T}^0 = (\lambda \boldsymbol{I}tr + 2\mu)\varepsilon \tag{9.5.4}$$

where fields marked by $(\ )^0$ correspond to classical elasticity. Thus, in the absence of body forces, the equilibrium equations can be expressed by

$$\nabla \cdot \boldsymbol{T} = (1 - c\nabla^2)\nabla \cdot \boldsymbol{T}^0 = 0 \tag{9.5.5}$$

Considering the displacement form of the equilibrium equations (6.2.38), we can then write

$$(1 - c\nabla^2)\boldsymbol{L}\boldsymbol{u} = 0, \quad \boldsymbol{L} = \mu\nabla^2 + (\lambda + \mu)\nabla\nabla. \tag{9.5.6}$$

and it is noted that $\boldsymbol{L}\boldsymbol{u}^0 = 0$. Combining these results together generates the relation

$$(1 - c\nabla^2)\boldsymbol{u} = \boldsymbol{u}^0 \tag{9.5.7}$$

Thus, the gradient and classical solutions are related, and (9.5.7) provides a convenient solution scheme for gradient problems if one already knows the corresponding classical solution.

Special boundary conditions for this gradient theory are somewhat involved, and we will avoid a detailed development of this issue. Commonly an extra boundary condition for this theory is to also specify that the second normal derivative of the displacement vanishes, that is, $\partial^2 \boldsymbol{u}/\partial n^2 = 0$.

Before heading into the example problems, it should be pointed out that by incorporating higher strain gradients into constitutive relations, we inherently create a *non-simple material* in the Noll sense. This is so because by using higher-order derivatives, we imply a *nonlocal model*. For example, consider the one-dimensional case where the displacement difference can be expressed to first and second orders by the two relations:

$$u(x+dx)-u(x)=\frac{du}{dx}dx=\varepsilon\,dx$$
$$u(x+dx)-u(x)=\frac{du}{dx}dx+\frac{d^2u}{dx^2}(dx)^2=\varepsilon\,dx+\frac{d\varepsilon}{dx}(dx)^2 \tag{9.5.8}$$

The first-order case $(9.5.8)_1$ corresponds to usual modeling valid in a small neighborhood where $dx << 1$. However, the second case $(9.5.8)_2$ expresses the same quantity but includes the higher-order term $\frac{d\varepsilon}{dx}(dx)^2$ that incorporates a higher derivative of strain with $(dx)^2$, thereby implying the use of a larger value of dx. This fact implies that some nonlocal effects will now come into the analysis.

Nonlocal continuum elasticity theories have been developed by Eringen (1972, 1977, 1983). They postulate that the stress at location $\boldsymbol{x}$ depends not only on the strain at $\boldsymbol{x}$, but also on the strain in a neighborhood of $\boldsymbol{x}$. In this sense, the nonlocal theory incorporates *longer range actions* between material points, and such interactions have been found to exist in atomic lattices and in granular materials. For isotropic materials, the constitutive relation for the stress can thus be written as

$$T_{ij}(\boldsymbol{x})=\int_V(\lambda(|\boldsymbol{x}'-\boldsymbol{x}|)\varepsilon_{kk}(\boldsymbol{x}')\delta_{ij}+2\mu(|\boldsymbol{x}'-\boldsymbol{x}|)\varepsilon_{ij}(\boldsymbol{x}'))dV(\boldsymbol{x}') \tag{9.5.9}$$

where now the usual elastic constants λ and μ become spatial functions of the distance variable $|\boldsymbol{x}'-\boldsymbol{x}|$. The integral in relation (9.5.9) can be expanded in a series of spatial derivatives of the strain which gives constitutive forms similar to our gradient theories. Other methods have specified simplified forms for the kernel integrand functions that allow solutions to be made for particular problems.

Numerous solutions using the gradient elastic constitutive laws have been developed. We now will explore a couple of these and compare solutions with the corresponding classical elasticity case. This will allow us to examine the differences and to determine some of the effects specific of gradient theory.

EXAMPLE 9.5.1 GRADIENT ELASTICITY SOLUTION TO SCREW DISLOCATION IN THE *Z*-DIRECTION

Consider the elasticity problem of a screw dislocation in the z-direction. This is a type of internal defect in a solid that corresponds to a jump discontinuity of displacement. For a screw dislocation, the displacement field has the special form of antiplane strain where

$$u=v=0,\quad w=w(x,y) \tag{9.5.10}$$

The problem geometry and jump in displacement are shown in Fig. 9.11.

Solution: We follow the gradient elasticity solution that has been given by Gutkin and Aifantis (1996). For the antiplane strain displacement field, the only two nonzero strains are given by Sadd (2014):

$$\varepsilon_{xz} = \frac{1}{2}\frac{\partial w}{\partial x}, \quad \varepsilon_{yz} = \frac{1}{2}\frac{\partial w}{\partial y} \tag{9.5.11}$$

Using the gradient constitutive law (9.5.3), the only two nonzero stresses are

$$T_{xz} = \mu \frac{\partial}{\partial x}\left[w - c\left(\frac{\partial^2 w}{\partial x^2} + \frac{\partial^2 w}{\partial y^2}\right)\right], \quad T_{yz} = \mu \frac{\partial}{\partial y}\left[w - c\left(\frac{\partial^2 w}{\partial x^2} + \frac{\partial^2 w}{\partial y^2}\right)\right] \tag{9.5.12}$$

The equilibrium equations reduce to the single relation

$$\frac{\partial T_{xz}}{\partial x} + \frac{\partial T_{yz}}{\partial y} = 0 \tag{9.5.13}$$

and using (9.5.12) this gives the governing equation for the w-displacement

$$\frac{\partial^2 w}{\partial x^2} + \frac{\partial^2 w}{\partial y^2} - c\left(\frac{\partial^4 w}{\partial x^4} + 2\frac{\partial^4 w}{\partial x^2 \partial y^2} + \frac{\partial^4 w}{\partial y^4}\right) = 0 \tag{9.5.14}$$

This equation can be solved using Fourier transforms. We omit the details of such an analysis and only quote the final result

$$w = \frac{b}{2\pi}\left[\tan^{-1}\left(\frac{y}{x}\right) + \text{sgn}(y)\int_0^\infty \frac{\xi \sin \xi x}{\xi^2 + 1/c} e^{-|y|\sqrt{\xi^2 + 1/c}}\, d\xi\right] \tag{9.5.15}$$

where b denotes the *Burgers vector* specifying the magnitude of the discontinuous displacement (see Fig. 9.11). This displacement gives the following strain components:

$$\begin{aligned} \varepsilon_{xz} &= \frac{b}{4\pi}\left[-\frac{y}{r^2} + \frac{y}{r\sqrt{c}}K_1\left(\frac{r}{\sqrt{c}}\right)\right] \\ \varepsilon_{yz} &= \frac{b}{4\pi}\left[\frac{x}{r^2} + \frac{x}{r\sqrt{c}}K_1\left(\frac{r}{\sqrt{c}}\right)\right] \end{aligned} \tag{9.5.16}$$

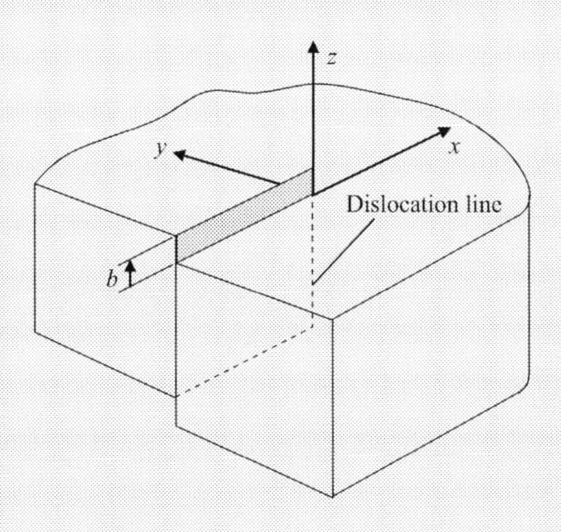

FIGURE 9.11

Screw dislocation geometry.

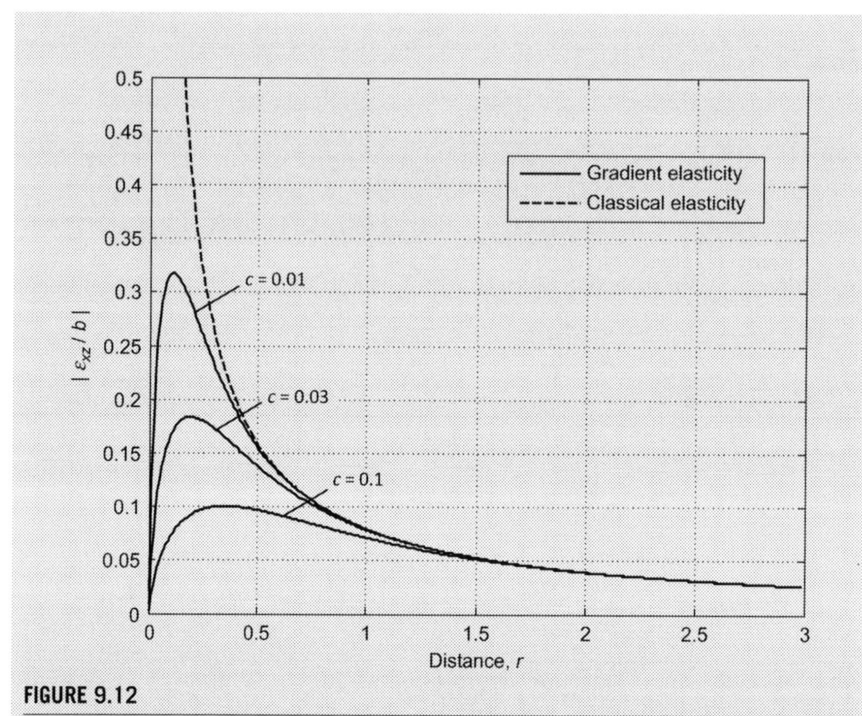

FIGURE 9.12

Local shear strain ε_{xz} behavior for a screw dislocation.

with $r=\sqrt{x^2+y^2}$ and K_1 being the modified Bessel function of the second kind of order 1. In regard to the classical elasticity solution, the corresponding results are given by Sadd (2014):

$$w=\frac{b}{2\pi}\tan^{-1}\left(\frac{y}{x}\right),\quad \varepsilon_{xz}=-\frac{b}{4\pi}\frac{y}{r^2},\quad \varepsilon_{yz}=\frac{b}{4\pi}\frac{x}{r^2} \tag{9.5.17}$$

From the structure of the displacement, strain, and stress solution fields, it is apparent that the gradient solution contains the classical solution plus an additive term due to the new model. This result is consistent with the findings of Ru and Aifantis (1993) for general boundary value problems of this theory. Using MATLAB Code C-23 for calculations and plotting, Fig. 9.12 illustrates the spatial behaviors of the strain component ε_{xz} for each theory. It is clearly seen that the gradient theory removes the singular behavior for the strain components (and also for the stresses). Note that the gradient strain and stress predications go to zero at the dislocation interface ($r = 0$). The removal of singular behaviors from classical elasticity is a common occurrence when using gradient theories.

EXAMPLE 9.5.2 GRADIENT ELASTICITY SOLUTION TO FLAMANT PROBLEM

Consider again the Flamant problem as shown in Fig. 6.14 which represents a point force or line load P acting normally to the free surface of an elastic half-space. Under a plane strain assumption, we wish to determine the gradient elasticity solution and compare particular features with the classical solution previously given in Example 6.2.5.

Solution: For this problem, we follow the solution developed by Li et al. (2004). They formulated the problem in two-dimensional *plane strain* with

$$u = u(x,y), \quad v = v(x,y), \quad w = 0 \tag{9.5.18}$$

which gives the following strain field:

$$\varepsilon_x = \frac{\partial u}{\partial x}, \quad \varepsilon_y = \frac{\partial v}{\partial y}, \quad \varepsilon_{xy} = \frac{1}{2}\left(\frac{\partial u}{\partial y} + \frac{\partial v}{\partial x}\right), \quad \varepsilon_z = \varepsilon_{xz} = \varepsilon_{yz} = 0 \tag{9.5.19}$$

For this deformation, the stresses follow from constitutive law (9.5.3):

$$\begin{aligned} T_{xx} &= \lambda(\varepsilon_x + \varepsilon_y) + 2\mu\varepsilon_x - c\left[\lambda(\varepsilon_x + \varepsilon_y) + 2\mu\varepsilon_x\right]_{,mm} \\ T_{yy} &= \lambda(\varepsilon_x + \varepsilon_y) + 2\mu\varepsilon_y - c\left[\lambda(\varepsilon_x + \varepsilon_y) + 2\mu\varepsilon_y\right]_{,mm} \\ T_{zz} &= \lambda(\varepsilon_x + \varepsilon_y) - c\left[\lambda(\varepsilon_x + \varepsilon_y)\right]_{,mm} \\ T_{xy} &= 2\mu\varepsilon_{xy} - 2\mu c\left[\lambda(\varepsilon_x + \varepsilon_y)\right]_{,mm} \\ T_{yz} &= T_{xz} = 0 \end{aligned} \tag{9.5.20}$$

Combining (9.5.19) into (9.5.20), and then inserting this result into the equilibrium equations gives

$$\begin{aligned} &\mu\nabla^2 u + (\lambda+\mu)\frac{\partial}{\partial x}\left(\frac{\partial u}{\partial x} + \frac{\partial v}{\partial y}\right) + c\nabla^2\left[\mu\nabla^2 u + (\lambda+\mu)\frac{\partial}{\partial x}\left(\frac{\partial u}{\partial x} + \frac{\partial v}{\partial y}\right)\right] = 0 \\ &\mu\nabla^2 v + (\lambda+\mu)\frac{\partial}{\partial y}\left(\frac{\partial u}{\partial x} + \frac{\partial v}{\partial y}\right) + c\nabla^2\left[\mu\nabla^2 v + (\lambda+\mu)\frac{\partial}{\partial y}\left(\frac{\partial u}{\partial x} + \frac{\partial v}{\partial y}\right)\right] = 0 \end{aligned} \tag{9.5.21}$$

This system of equations can again be solved by Fourier transforms, and details are provided in Li et al. (2004). Extracting the solution for the in-plane stresses

$$\begin{aligned} T_{xx} &= -\frac{P}{\pi}\int_0^\infty (1-\xi y)e^{-\xi y}\cos(\xi x)\,d\xi = -\frac{2Px^2 y}{\pi(x^2+y^2)^2} \\ T_{yy} &= -\frac{P}{\pi}\int_0^\infty (1+\xi y)e^{-\xi y}\cos(\xi x)\,d\xi = -\frac{2Py^3}{\pi(x^2+y^2)^2} \\ T_{xy} &= -\frac{P}{\pi}\int_0^\infty \xi y e^{-\xi y}\sin(\xi x)\,d\xi = -\frac{2Pxy^2}{\pi(x^2+y^2)^2} \end{aligned} \tag{9.5.22}$$

which surprisingly turn out to be the same as the classical elasticity results (see Eqs. (6.2.67)). Ru and Aifantis (1993) have shown that gradient problems of this type and boundary condition have this general feature of stress field solutions matching the classical prediction.

It should be pointed out that a more recent study of this problem by Lazar and Maugin (2006) using a different set of boundary conditions claims that *both* the gradient elastic stresses and displacements are free of singularities. Since Li et al. (2004) did not provide a finalized detailed displacement solution, we will choose the equivalent results from Lazar and Maugin (2006). They give the following relation for the vertical displacement:

$$v(x,y) = \frac{P}{2\pi\mu}\left[2(1-\nu)(\log r + K_0(r/\sqrt{c}) - \frac{x^2}{r^2} + \frac{x^2-y^2}{r^2}\left(\frac{2c}{r^2} - K_2(r/\sqrt{c})\right)\right] \quad (9.5.23)$$

where again $K_{0,2}$ are the modified Bessel functions of the second kind of orders 0 and 2.

We now wish to further explore this result on free surface $y = 0$. The corresponding prediction from classical elasticity is given by

$$v(x,0) = \frac{P(1-\nu)}{\pi\mu}\log x + v_o \quad (9.5.24)$$

where v_o is an arbitrary rigid-body vertical displacement, and in fact solution (9.5.23) should also include such a term. Note the logarithmic singularity in the classical prediction.

Using MATLAB Code C-24 for calculations and plotting, these vertical surface displacement relations are shown in Fig. 9.13 for the case $\nu = 0.3$ and $c = 0.1$. The displacements are normalized with respect to the factor $\dfrac{P(1-\nu)}{\pi\mu}$. Unlike the stresses, displacement results indicate that the gradient solution does not match with the classical prediction. More importantly, the logarithmic singularity at the loading point in classical elasticity (Sadd, 2014) is not found, and the gradient theory predicts a finite value directly under the loading point. Again we see that gradient elasticity eliminates a singular behavior found in classical theory.

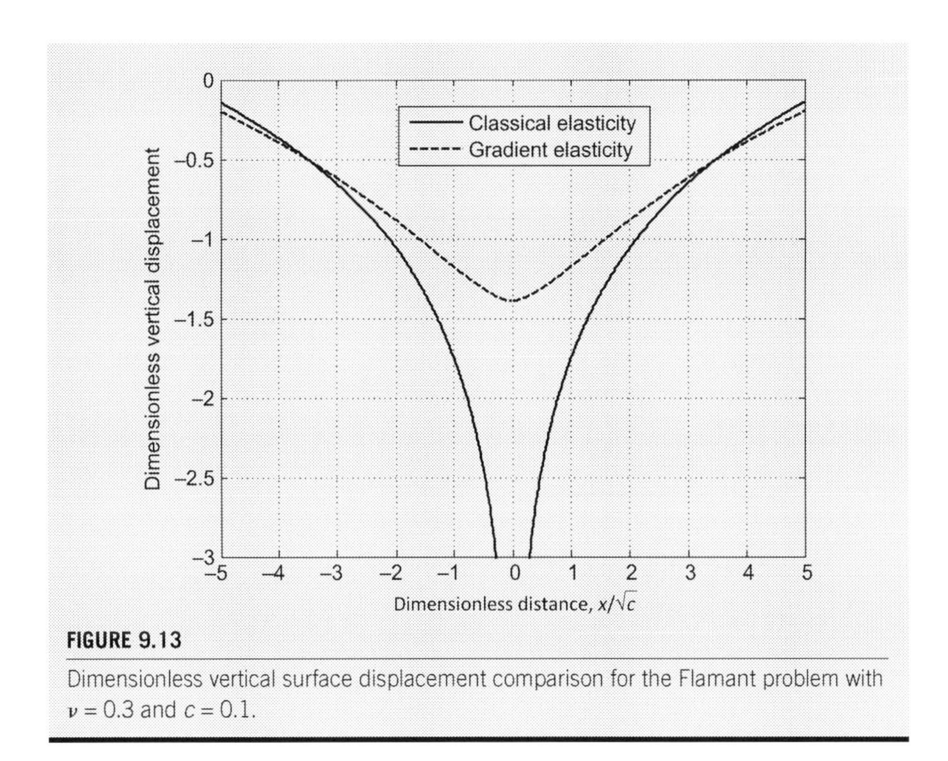

FIGURE 9.13

Dimensionless vertical surface displacement comparison for the Flamant problem with $\nu = 0.3$ and $c = 0.1$.

Examination of the published literature on gradient elasticity indicates a variety of constitutive choices which lead to different results for a given problem. In some cases, singular behaviors are removed from the classical predictions, whereas in other cases they are not. Many additional problems in gradient elasticity have appeared in the literature and no doubt new developments of this theory will continue. With respect to other continuum mechanics areas, gradient theories have also had many applications in plasticity, viscoplasticity, and localization problems.

9.6 FABRIC THEORIES FOR GRANULAR MATERIALS

Granular materials are an important class of discrete media that are composed of a large collection of independent particles or grains as shown in Fig. 9.14. Such materials include powders, sands, rocks, food grains, and other types of particulate systems. These materials will have void space that could be filled with air or a fluid. Each particle has immediate neighbors that make contact and thus affect the forces and motions (displacements and rotations) of the particle. This internal force/moment system within the granular assembly is primarily dependent on the packing geometry. Because of these and other features, many claim that granular materials are one of the most complicated media known, exhibiting localization, surface instabilities, solid–fluid transition, dilatancy, initial and induced anisotropy, microstructural response, and other complex behaviors.

FIGURE 9.14

Examples of granular material.

The behavior of a granular material is inherently related to the internal microstructural packing geometry of the grains. The term *fabric* is often used to describe this geometry, and some typical fabric measures are shown in Fig. 9.15 that include the *particle size and shape*, *branch vectors* $\boldsymbol{l}$ that go between adjacent particle mass centers, *contact normal vectors* $\boldsymbol{n}$ that act perpendicular to surface contact points, and *void geometry* that would measure particular aspects of each void. Experimental schemes using photoelasticity on transparent model granular assemblies have shown that material fabric plays an important role in load transfer. Observations indicate that granular microstructure produces well-defined loading chains that transfer most of the forces through the material. A simple example of this effect is shown in Fig. 9.16 where two granular packing geometries are loaded by an identical central vertical load. Photoelastic data on these two assemblies would indicate that although some load transfer would be distributed in other

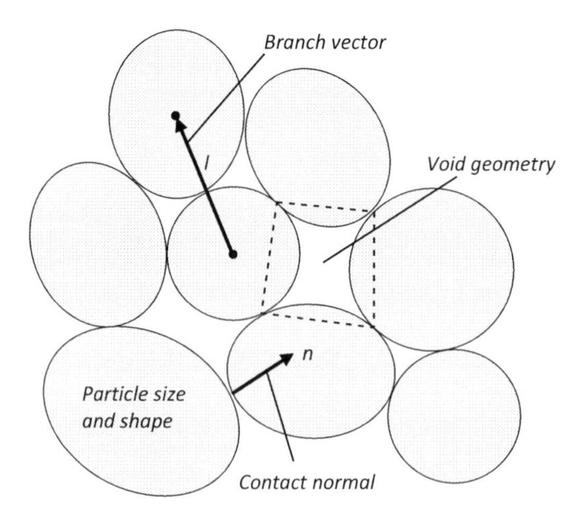

FIGURE 9.15

Typical local fabric in granular media.

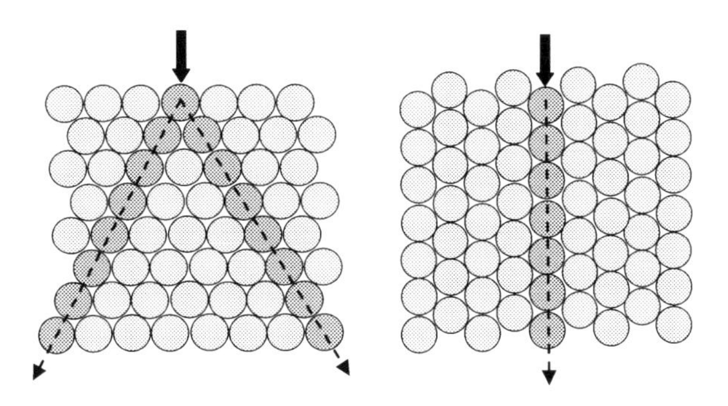

FIGURE 9.16

Fabric effects on load transfer in granular materials.

grains, the primary load transfer would occur along the shaded particles as shown. Thus, load transfer and stress distribution in granular materials is highly dependent on the material fabric. This situation is in strong contrast to the way a continuous medium would transfer such applied loading (see the linear elastic Flamant problem in Example 6.2.5).

Continuum descriptions of granular materials have employed a very large collection of schemes including various micropolar, higher gradient, and void and fabric tensor theories. Computational methods have used *discrete element modeling* with great success (Cundall and Strack, 1979). Here, we will briefly present some basic concepts of using fabric tensor theories to simulate dry granular material behavior under static equilibrium conditions.

We start with an interesting application developed by Cowin (1985b) that established a relationship between the continuum elasticity tensor and a fabric tensor. The beginning step is the assumption that there exists a *symmetric second-order fabric tensor* $\boldsymbol{M}$ that characterizes the geometric components of a multistructural and/or multiphase material. While the existence of $\boldsymbol{M}$ is assumed, the precise nature of how this fabric tensor is related to the material microstructure is not specified. We will explore this issue in some detail later in the presentation. For the elastic case, we thus choose a constitutive form

$$\boldsymbol{T} = \boldsymbol{f}(\boldsymbol{\varepsilon}, \boldsymbol{M}) \tag{9.6.1}$$

where $\boldsymbol{\varepsilon}$ is the usual small deformation strain tensor and $\boldsymbol{M}$ is the fabric tensor. Since all of these tensors are symmetric and second order, using the principle of frame indifference

$$\boldsymbol{Q}\boldsymbol{f}(\boldsymbol{\varepsilon}, \boldsymbol{M})\boldsymbol{Q}^{\mathrm{T}} = \boldsymbol{f}(\boldsymbol{Q}\boldsymbol{\varepsilon}\boldsymbol{Q}^{\mathrm{T}}, \boldsymbol{Q}\boldsymbol{M}\boldsymbol{Q}^{\mathrm{T}}) \tag{9.6.2}$$

We have seen relation (9.6.2) before in Section 8.4.3 dealing with Rivlin–Ericksen fluids of grade 2. As before, we can apply the representation theorem (2.14.5) to write

$$\begin{aligned} \boldsymbol{T} = {} & f_1\boldsymbol{I} + f_2\boldsymbol{M} + f_3\boldsymbol{M}^2 + f_4\boldsymbol{\varepsilon} + f_5\boldsymbol{\varepsilon}^2 + f_6(\boldsymbol{M}\boldsymbol{\varepsilon} + \boldsymbol{\varepsilon}\boldsymbol{M}) \\ & + f_7(\boldsymbol{M}^2\boldsymbol{\varepsilon} + \boldsymbol{\varepsilon}\boldsymbol{M}^2) + f_8(\boldsymbol{M}\boldsymbol{\varepsilon}^2 + \boldsymbol{\varepsilon}^2\boldsymbol{M}) + f_9(\boldsymbol{M}^2\boldsymbol{\varepsilon}^2 + \boldsymbol{\varepsilon}^2\boldsymbol{M}^2) \end{aligned} \tag{9.6.3}$$

where the terms f_i are functions of the invariants $tr\,\boldsymbol{M}$, $tr\,\boldsymbol{M}^2$, $tr\,\boldsymbol{M}^3$, $tr\,\boldsymbol{\varepsilon}$, $tr\,\boldsymbol{\varepsilon}^2$, $tr\,\boldsymbol{\varepsilon}^3$, $tr(\boldsymbol{M\varepsilon})$, $tr(\boldsymbol{M}^2\boldsymbol{\varepsilon})$, $tr(\boldsymbol{M\varepsilon}^2)$, and $tr(\boldsymbol{M}^2\boldsymbol{\varepsilon}^2)$. This general form must meet the requirement that $\boldsymbol{T}$ be linear in $\boldsymbol{\varepsilon}$ and vanish when $\boldsymbol{\varepsilon} = 0$, thus giving the reduction

$$\boldsymbol{T} = f_1\boldsymbol{I} + f_2\boldsymbol{M} + f_3\boldsymbol{M}^2 + f_4\boldsymbol{\varepsilon} + f_6(\boldsymbol{M\varepsilon} + \boldsymbol{\varepsilon M}) + f_7(\boldsymbol{M}^2\boldsymbol{\varepsilon} + \boldsymbol{\varepsilon M}^2) \tag{9.6.4}$$

with the material functions f_i reducing to the forms

$$\begin{aligned} f_1 &= a_1\,tr\,\boldsymbol{\varepsilon} + a_2\,\mathrm{tr}\,\boldsymbol{M\varepsilon} + a_3\,tr\,\boldsymbol{M}^2\boldsymbol{\varepsilon} \\ f_2 &= d_1\,tr\,\boldsymbol{\varepsilon} + b_1\,\mathrm{tr}\,\boldsymbol{M\varepsilon} + b_2\,tr\,\boldsymbol{M}^2\boldsymbol{\varepsilon} \\ f_3 &= d_2\,tr\,\boldsymbol{\varepsilon} + d_3\,tr\,\boldsymbol{M\varepsilon} + b_3\,tr\,\boldsymbol{M}^2\boldsymbol{\varepsilon} \\ f_4 &= 2c_1, \quad f_6 = 2c_2, \quad f_7 = 2c_3 \end{aligned} \tag{9.6.5}$$

where a_i, b_i, c_i, and d_i are functions of $tr\,\boldsymbol{M}$, $tr\,\boldsymbol{M}^2$, and $tr\,\boldsymbol{M}^3$. Using these results, we can now express the stress in index notation as

$$\begin{aligned} T_{ij} = {} & \delta_{ij}(a_1\varepsilon_{kk} + a_2M_{rp}\varepsilon_{pr} + a_3M_{rq}M_{qp}\varepsilon_{pr}) \\ & + M_{ij}(d_1\varepsilon_{kk} + b_1M_{rp}\varepsilon_{pr} + b_2M_{rq}M_{qp}\varepsilon_{pr}) \\ & + M_{is}M_{sj}(d_2\varepsilon_{kk} + d_3M_{rp}\varepsilon_{pr} + b_3M_{rq}M_{qp}\varepsilon_{pr}) \\ & + 2c_1\varepsilon_{ij} + 2c_2(M_{ir}\varepsilon_{rj} + \varepsilon_{ir}M_{rj}) \\ & + 2c_3(M_{ip}M_{pr}\varepsilon_{rj} + \varepsilon_{ir}M_{rp}M_{pj}) \end{aligned} \tag{9.6.6}$$

Now for linear elastic materials, the constitutive form in terms of the elasticity tensor $\boldsymbol{C}$ is given by

$$T_{ij} = C_{ijkl}\varepsilon_{kj}, \quad \text{with symmetries } C_{ijkl} = C_{jikl} = C_{ijlk} = C_{klij} \tag{9.6.7}$$

Comparing relations (9.6.6) with (9.6.7) and invoking the appropriate symmetries gives the following relationship between the elasticity and fabric tensors:

$$\begin{aligned} C_{ijkl} = {} & a_1\delta_{ij}\delta_{kl} + a_2(M_{ij}\delta_{kl} + \delta_{ij}M_{kl}) + a_3(\delta_{ij}M_{kq}M_{ql} + \delta_{kl}M_{iq}M_{qj}) \\ & + b_1M_{ij}M_{kl} + b_2(M_{ij}M_{kq}M_{ql} + M_{is}M_{sj}M_{kl}) + b_3M_{is}M_{sj}M_{kq}M_{ql} \\ & + c_1(\delta_{ki}\delta_{lj} + \delta_{li}\delta_{kj}) + c_2(M_{ik}\delta_{lj} + M_{kj}\delta_{li} + M_{il}\delta_{kj} + M_{lj}\delta_{ki}) \\ & + c_3(M_{ir}M_{rk}\delta_{lj} + M_{kr}M_{rj}\delta_{li} + M_{ir}M_{rl}\delta_{kj} + M_{lr}M_{rj}\delta_{ik}) \end{aligned} \tag{9.6.8}$$

and thus the final constitutive form for the stress in terms of the strain and fabric tensor is given by

$$\begin{aligned} T_{ij} = {} & a_1\delta_{ij}\varepsilon_{kk} + a_2(M_{ij}\varepsilon_{kk} + \delta_{ij}M_{kl}\varepsilon_{kl}) + a_3(\delta_{ij}M_{kq}M_{ql}\varepsilon_{kl} + M_{iq}M_{qj}\varepsilon_{kk}) \\ & + b_1M_{ij}M_{kl}\varepsilon_{kl} + b_2(M_{ij}M_{kq}M_{ql}\varepsilon_{kl} + M_{is}M_{sj}M_{kl}\varepsilon_{kl}) + b_3M_{is}M_{sj}M_{kq}M_{ql}\varepsilon_{kl} \\ & + 2c_1\varepsilon_{kk} + 2c_2(M_{ik}\varepsilon_{kj} + M_{kj}\varepsilon_{ki}) + 2c_3(M_{ir}M_{rk}\varepsilon_{kj} + M_{kr}M_{rj}\varepsilon_{ki}) \end{aligned} \tag{9.6.9}$$

As a special case, consider the situation where the fabric tensor is isotropic so that $M_{ij} = M_o\delta_{ij}$. For this case, relation (9.6.8) gives

$$\begin{aligned} & C_{1111} = C_{2222} = C_{3333} = a_1 + 2c_1 + 2(a_2 + 2c_2)M_o + (2a_3 + b_1 + 4c_3)M_o^2 + 2b_2M_o^3 + b_3M_o^4 \\ & C_{1122} = C_{1133} = C_{2233} = a_1 + 2a_2M_o + (2a_3 + b_1)M_o^2 + 2b_2M_o^3 + b_3M_o^4 \\ & C_{1212} = C_{1313} = C_{2323} = c_1 + 2c_2M_o + 2c_3M_o^2 \end{aligned} \tag{9.6.10}$$

Now the two Lamé elastic constants in classical elasticity can be expressed as

$$\begin{aligned} C_{1111} &= C_{2222} = C_{3333} = \lambda + 2\mu \\ C_{1122} &= C_{1133} = C_{2233} = \lambda \\ C_{1212} &= C_{1313} = C_{2323} = \mu \end{aligned} \tag{9.6.11}$$

and so these elastic constants can be written in terms of the fabric tensor component M_o:

$$\begin{aligned} \lambda &= a_1 + 2a_2 M_o + (2a_3 + b_1) M_o^2 + 2b_2 M_o^3 + b_3 M_o^4 \\ \mu &= c_1 + 2c_2 M_o + 2c_3 M_o^2 \end{aligned} \tag{9.6.12}$$

Other material symmetries such as orthotropic and transversely isotropic have been discussed by Cowin (1985b). However, he points out that not all possible elastic material symmetries can be represented by relation (9.6.8). Of course to apply these elegant relations, the specific nature of the fabric tensor must be determined.

Next, we move on to explore the development of particular fabric tensors and definitions of stress associated with granular materials. Major work in this area has been done by many researchers including Mehrabadi et al. (1982), Nicot and Darve (2005), Li et al. (2009), and Chang and Liu (2013). Recall that for such materials the stress is transmitted in the form of force chains through the contacting particles and this fact couples the material fabric to the load transfer. Different fabric tensors and stress definitions have been proposed, incorporating different microstructural features as shown previously in Fig. 9.15, and normally this is done on a statistical basis. What can complicate things is the fact that loading can cause new contacts to appear and old contacts to disappear, thereby changing fabric during loading. For the simplified approach, here we will disregard such situations. We will also assume that all particles are convex and thus will only transmit contact forces and not moments.

Fabric is a rather general term, and thus over the years many different geometrical parameters have been proposed to create a second-order fabric tensor. Many attempts have used some type of outer or dyadic product of a particular vector field. A few such examples are shown in Fig. 9.17 and include *particle orientation vectors* $\mathbf{v}^p$, *contact normal vectors* $\boldsymbol{n}$, and *branch vectors* $\boldsymbol{l}$. All of these vectors carry the directional ambiguity of being only defined uniquely to a line of action, and thus each would carry a ± prefix sign.

The particle orientation vector would characterize the particles nonspherical or noncircular shape through a geometric calculation that would result in an oriented vector along the principle size direction. Numerical methods exist to carry out such calculations. The magnitude of this vector can be scaled to the amount of size difference between the maximum and minimum particle lengths. However, this vector is commonly normalized as are the contact normal and branch vectors. Based on these unit vectors, fabric tensors can be expressed as an average over an appropriate RVE as

$$\begin{aligned} M_{ij}^{\mathrm{p}} &= \frac{1}{N^{\mathrm{p}}} \sum_V n_i^{\mathrm{p}} n_j^{\mathrm{p}} \\ M_{ij}^{\mathrm{c}} &= \frac{1}{N^{\mathrm{c}}} \sum_V n_i^{\mathrm{c}} n_j^{\mathrm{c}} \\ M_{ij}^{l} &= \frac{1}{N^{l}} \sum_V n_i^{l} n_j^{l} \end{aligned} \tag{9.6.13}$$

where $n_i^{\mathrm{p}}, n_i^{\mathrm{c}}, n_i^{l}$ are the unit particle orientation, contact, and branch vectors, $N^{\mathrm{p}}, N^{\mathrm{c}}, N^{l}$ are the total number of particles, contacts, and branch vectors in V, respectively. Note that each of these fabric tensors is symmetric with a unit trace.

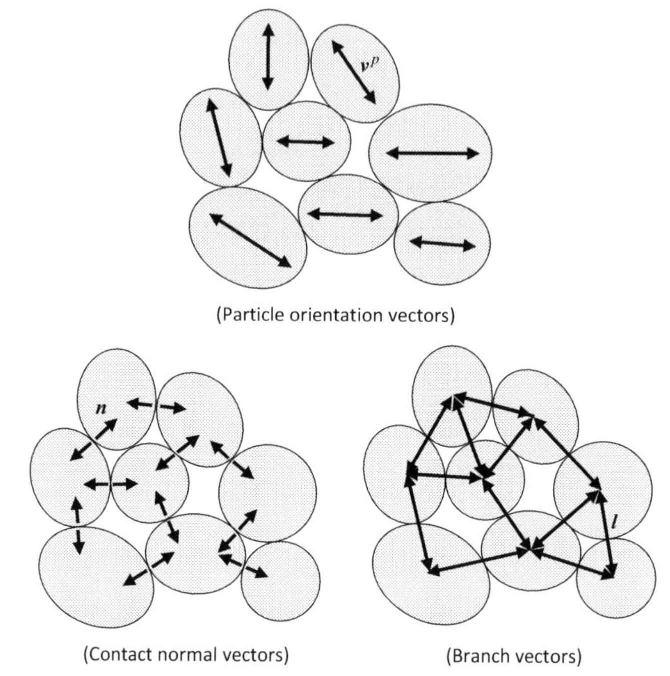

FIGURE 9.17

Example fabric vector distributions in granular media.

Now for regular granular packings like those (HCP) shown in Fig. 9.16, the fabric is completely determined if we assume no further re-arrangement with loading. Thus, the stress–strain response of such idealized cases can easily be determined based on the individual particle contact force deformation response. However, we wish to explore the more general case of a granular system with random packing of spherical particles, and this generally requires the use of a *statistical representation* of material fabric. Following the procedures outlined by Chang and Liu (2013), the *distribution of contact normals* can be expressed by a spherical harmonic Fourier expansion in the local spherical coordinate system defined in Fig. 9.18. Including just the first two terms yields

$$\xi(\phi,\theta)=\frac{1}{4\pi}\left[1+\frac{a}{4}(3\cos 2\phi+1)+3b\sin^2\phi\cos 2\theta\right] \tag{9.6.14}$$

where a and b are two constants related to the material fabric. The distribution function $\xi(\phi,\theta)$ is a probability density function that satisfies the usual condition

$$\int_0^{2\pi}\int_0^{\pi}\xi(\phi,\theta)\sin\phi\, d\phi\, d\theta=1 \tag{9.6.15}$$

At a typical point on the particle's surface, the three orthogonal unit vectors that coincide with the spherical basis vectors are given by

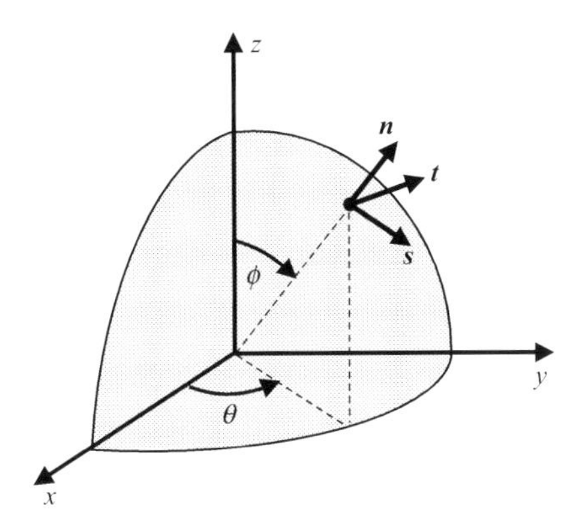

FIGURE 9.18

Local particle coordinate system.

$$\begin{aligned} \boldsymbol{n} &= \cos\phi \boldsymbol{e}_3 + \sin\phi\cos\theta \boldsymbol{e}_1 + \sin\phi\sin\theta \boldsymbol{e}_2 \\ \boldsymbol{s} &= -\sin\phi \boldsymbol{e}_3 + \cos\phi\cos\theta \boldsymbol{e}_1 + \cos\phi\sin\theta \boldsymbol{e}_2 \\ \boldsymbol{t} &= -\sin\theta \boldsymbol{e}_1 + \cos\theta \boldsymbol{e}_2 \end{aligned} \tag{9.6.16}$$

Using relation $(9.6.16)_1$, Eq. (9.6.14) can be alternatively expressed as the Cartesian tensor expression

$$\xi(\boldsymbol{n}) = \frac{1}{4\pi} D_{ij} n_i n_j \tag{9.6.17}$$

where D_{ij} is called the *contact density tensor* and is given by

$$\boldsymbol{D} = \begin{bmatrix} 1+a & 0 & 0 \\ 0 & 1-a/2+3b & 0 \\ 0 & 0 & 1-a/2-3b \end{bmatrix} \tag{9.6.18}$$

Now we choose for our material fabric tensor to be the contact normal distribution defined by $(9.6.13)_2$, and thus $M_{ij} = \frac{1}{N^c}\sum_V n_i n_j$ can be expressed in terms of an integrated statistical form as

$$M_{ij} = \int_0^{2\pi}\int_0^{\pi} \xi(\phi,\theta) n_i n_j \sin\phi \, d\phi \, d\theta \tag{9.6.19}$$

Using the probability density function form from relation (9.6.14) allows (9.6.19) to be written as

$$\boldsymbol{M} = \frac{1}{15}\begin{bmatrix} 5+2a & 0 & 0 \\ 0 & 5-a+6b & 0 \\ 0 & 0 & 5-a-6b \end{bmatrix} \tag{9.6.20}$$

The force distribution in an idealized granular material is such that the *contact force or force fabric* is characterized by the orientation distribution of *interparticle contact forces*. These forces can be described by their components along the local ***n, s, t*** directions

$$\begin{aligned} f_n &= \bar{f} A_{ij} n_i n_j \\ f_s &= a_s \bar{f} A_{ij} n_i s_j \\ f_t &= a_t \bar{f} A_{ij} n_i t_j \end{aligned} \tag{9.6.21}$$

where $\overline{f}$ is the *mean force* and A_{ij} is the *force–fabric tensor* given by the expressions

$$\overline{f} = \frac{1}{4\pi}\int_0^{2\pi}\int_0^{\pi} f_n \sin\phi\, d\phi\, d\theta \tag{9.6.22}$$

$$\boldsymbol{A} = \begin{bmatrix} 1+a_n & 0 & 0 \\ 0 & 1-a_n/2+3b_n & 0 \\ 0 & 0 & 1-a_n/2-3b_n \end{bmatrix} \tag{9.6.23}$$

and the constants a_n, b_n, a_s, and a_t defined the force–fabric distribution.

For this type of granular material modeling, the Cauchy stress is commonly expressed as an average outer or dyadic product of the branch vectors with the particle contact force

$$T_{ij} == \frac{1}{N^c}\sum_V l_i^c f_j^c \tag{9.6.24}$$

For spherical particles, we assume an average branch vector length $\overline{l}$, and incorporating the previous relations, we can express (9.6.24) as

$$T_{ij} = A_{pq} D_{mn} T_{ijpqmn} \tag{9.6.25}$$

where

$$T_{ijpqmn} = m_v \overline{l}\, \overline{f} \int_0^{2\pi}\int_0^{\pi} (n_p n_q n_i n_j + a_s n_p s_q s_i n_j + a_t n_p t_q t_i n_j) n_m n_n \sin\phi\, d\phi\, d\theta \tag{9.6.26}$$

where m_v is the total number of contacts per unit volume (RVE). Under the stated modeling assumptions, relations (9.6.25) and (9.6.26) then give a constitutive relationship between the stress and material and force fabric. Chang and Liu (2013) conducted numerical discrete element modeling for some biaxial and triaxial test configurations and found good agreement with this fabric modeling scheme.

We have presented only a couple of studies that employ continuum mechanics to model the response of granular materials. Many other micromechanical models have been applied to such materials that incorporate fabric in different ways. Some studies have explored plastic and flow response of granular media. This has been an active research field, and because of the complexity of granular materials, this research is expected to continue.

9.7 CONTINUUM DAMAGE MECHANICS

We now wish to explore yet another aspect of micromechanical material behavior which is related to the internal damage within solid media. We assume that this damage initially occurs at length scales several orders of magnitude smaller than that used to analyze the problem. Thus, the damage evolution would be referred to as a micromechanical process. Internal damage identification and characterization is of course very important in the development of safe and reliable materials and structures. Continuum damage mechanics (CDM) is concerned with the macrolevel

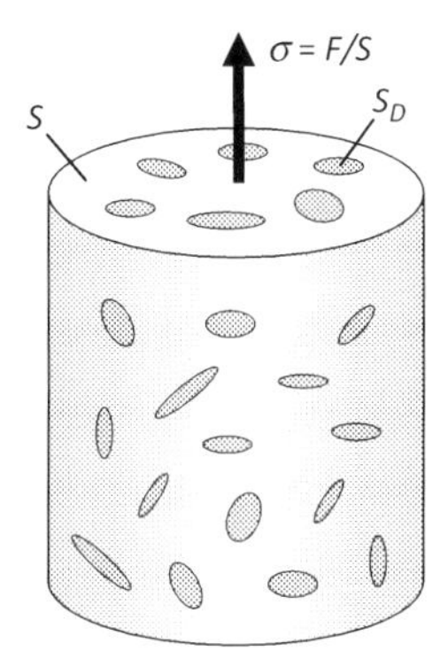

FIGURE 9.19

One-dimensional deformation of damaged material.

characterization and modeling of the effects of distributed defects on the overall material behavior. These defects commonly take the form of microcracks and microcavities, breakage of bonds, collapse of cells and fibers, interfacial decohesions, etc., and these occur in most every material. The goal of CDM is to properly model such discontinuous defects by some type of continuum theory. Due to the statistical nature of material defects, use of an RVE and homogenization methods are often used in developing CDM models.

As usual, our presentation here will only be very brief with just enough information to provide an introduction to the field. This requires that we focus discussion again only on the elastic response. This area of study was initiated in the late 1950s and has now developed into a mature applied science. Major texts including Kachanov (1986), Lemaitre (1996), Krajcinovic (1996), and Voyiadjis and Kattan (2005) as well as review articles such as Chaboche (1988a,b) and Kondo et al. (2007) that provide more details on this topic.

In order to begin our presentation, it will be useful if we first explore the simple one-dimensional uniaxial loading case of a damaged material as illustrated in Fig. 9.19. The material sample (RVE) is assumed to have a distributed collection of damaged zones, and in particular these affect the surface area S by reducing the effective area to be S–S_D as shown, where S_D is the total defect area on the plane under study. We then can define a *surface damage variable D* as

$$D = \frac{S_D}{S} \tag{9.7.1}$$

Note that the damage parameter is bounded by $0 \leq D \leq 1$, where $D = 0$ corresponds to the undamaged case, while $D = 1$ represents the fully damaged material. Using this simple concept, the standard uniaxial stress expression $\sigma = F/S$ is therefore modified to

$$\bar{\sigma} = \frac{F}{S - S_D} = \frac{F}{S(1 - S_D/S)} = \frac{\sigma}{1 - D} \tag{9.7.2}$$

where $\bar{\sigma}$ is normally called the *effective stress*. Because we have subtracted out the defect area, this stress is sometimes referred to as acting on a fictitious undamaged material. Each of these stresses are related to the strain by a uniaxial Hooke's law $\sigma = E\varepsilon$ and $\bar{\sigma} = \bar{E}\bar{\varepsilon}$, where E is the modulus of the damaged material and $\bar{E}$ is the modulus of the effective (undamaged) material. Linking these relations is commonly done by using a *strain equivalence principle* whereby it is assumed that the strains in the damaged and undamaged configurations are the same. Thus, we can write

$$\varepsilon = \bar{\varepsilon} \Rightarrow \frac{\sigma}{E} = \frac{\bar{\sigma}}{\bar{E}} \tag{9.7.3}$$

Then combining relations (9.7.2) and (9.7.3) gives the modulus of the damaged material

$$E = (1 - D)\bar{E} \tag{9.7.4}$$

and the modified constitutive relation can thus be expressed as

$$\sigma = (1 - D)\bar{E}\varepsilon \tag{9.7.5}$$

Note that as expected, $E \leq \bar{E}$.

Now in order to use this uniaxial relation (9.7.4), we would need a *damage evolution law* which would predict the functional relationship between the damage parameter D and the strain

$$D = \hat{f}(\varepsilon) \tag{9.7.6}$$

This function could be determined from a standard uniaxial tension test. However, such a form would only be valid for monotonic loading, since during an unloading/reloading phase, we expect the damage parameter would keep its maximum value that had been reached before. A common scheme to handle this unloading/reloading problem is to introduce a variable say κ which corresponds to the maximum level of strain reached in the material before current time t:

$$\kappa(t) = \max \varepsilon(\tau), \quad \tau \leq t \tag{9.7.7}$$

Thus, we can then rewrite (9.7.6) as

$$D = \hat{f}(\kappa) \tag{9.7.8}$$

which is now valid for all load histories.

It is often convenient to introduce a *limit state function* as the difference between the strain and its previous maximum value

$$f(\varepsilon, \kappa) = \varepsilon - \kappa \tag{9.7.9}$$

Then the following conditions are often connected with relation (9.7.9):

$$f \leq 0, \quad \dot{\kappa} \geq 0, \quad \dot{\kappa} f = 0 \tag{9.7.10}$$

The first condition implies that the strain ε can never be greater than κ and the second condition indicates that κ cannot decrease.

Using this uniaxial theory, we can examine a typical elastic stress–strain curve for a damaging material as shown in Fig. 9.20. We note that the stress–strain law (9.7.5) is of classical form with a secant modulus $E_s = E = (1-D)\bar{E}$ associated with the damaged material. The material response becomes nonlinear because of the damage, but unloading is done in a linear fashion with modulus E_s.

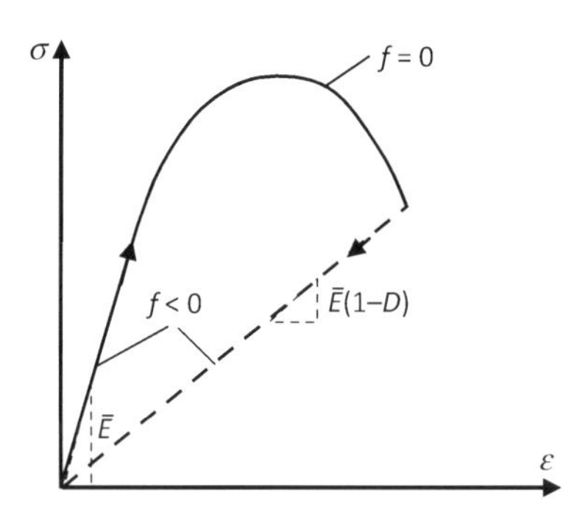

FIGURE 9.20

Uniaxial elastic stress–strain response of a damaged material during loading–unloading.

EXAMPLE 9.7.1 NUMERICAL EVALUATION AND PLOT OF UNIAXIAL DAMAGE MODEL

Using the previous uniaxial continuum damage model with a damage evolution relation $D = \hat{f}(\varepsilon) = 1 - e^{-m\varepsilon}$ (m = constant), calculate and plot the damage evolution and stress–strain response for a monotonic loading situation.

Solution: The MATLAB code for this example is easily constructed; see code C-25 in Appendix C. The code makes two plots: damage parameter evolution with strain, and the stress–strain results for the cases m = 0.5, 1.0, 1.5. These results are shown in Fig. 9.21. It is clearly demonstrated that as the damage parameter increases, the stress response is significantly reduced. This result is of course the expected behavior.

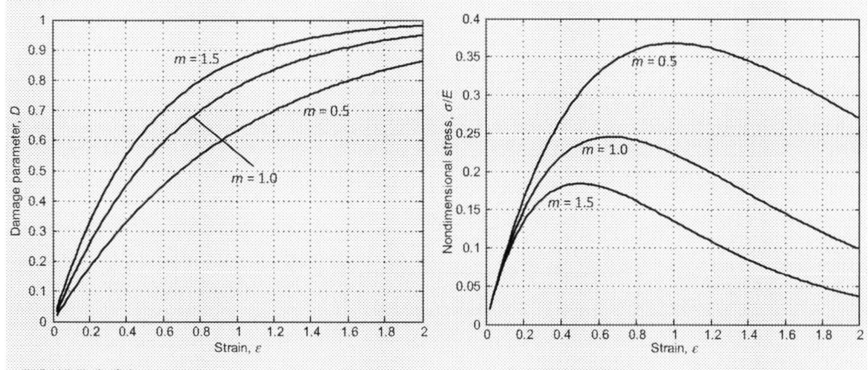

FIGURE 9.21

Numerical results for uniaxial damage model with $D = \hat{f}(\varepsilon) = 1 - e^{-m\varepsilon}$.

We can extend this simple uniaxial model to a three-dimensional elastic case where the elasticity tensor C_{ijkl} for the damaged material can be expressed in terms of the undamaged material tensor C^{o}_{ijkl} as

$$C_{ijkl} = (1-D)C^{o}_{ijkl} \tag{9.7.11}$$

and thus the constitutive relation becomes

$$T_{ij} = (1-D)C^{o}_{ijkl}\varepsilon_{kl} \tag{9.7.12}$$

The corresponding effective stress is then given through the usual equation

$$T_{ij} = (1-D)\overline{T}_{ij} \tag{9.7.13}$$

Now for the three-dimensional case, the damage evolution relation (9.7.9) can be generalized to

$$f(\varepsilon,\kappa) = \varepsilon_{eq}(\varepsilon) - \kappa \tag{9.7.14}$$

where the equivalent strain could be chosen as

$$\varepsilon_{eq}(\varepsilon) = \sqrt{\varepsilon_{ij}\varepsilon_{ij}} \tag{9.7.15}$$

or another choice sometimes proposed is based on the elastic strain energy

$$\varepsilon_{eq}(\varepsilon) = \sqrt{C^{o}_{ijkl}\varepsilon_{ij}\varepsilon_{kl}} \tag{9.7.16}$$

Note that using the simple scalar product approach used in form (9.7.11) places the damage parameter equally on all elastic moduli C^{o}_{ijkl}. This approach is generally referred to as *isotropic damage theory* and would likely have difficulty predicting behaviors for anisotropic materials.

Often CDM theory is generated from thermodynamic principles of irreversible processes. Following similar formulations as used in Sections 7.2 and 7.4, we can start with an appropriate definition of the free energy in terms of the strain and the damage parameter D:

$$\Psi = \Psi(\varepsilon, D) = \frac{1}{2\rho}(1-D)C^{o}_{ijkl}\varepsilon_{ij}\varepsilon_{kl} \tag{9.7.17}$$

where ρ is the mass density, and we have used the usual strain equivalency principle. The stress is then given by

$$T_{ij} = \rho\frac{\partial \Psi}{\partial \varepsilon_{ij}} = (1-D)C^{o}_{ijkl}\varepsilon_{kl} \tag{9.7.18}$$

which is the same as Eq. (9.7.12).

For the isotropic case, we can express the strain in terms of the stress as

$$\varepsilon_{ij} = \frac{1+\nu}{E}\frac{T_{ij}}{(1-D)} - \frac{\nu}{E}\frac{T_{kk}}{(1-D)}\delta_{ij} \tag{9.7.19}$$

It is noted from this constitutive form that only the modulus of elasticity is affected ($\bar{E} = (1 - D)E$) by the damage theory and Poisson's ratio remains the same.

Next, we consider another thermodynamic potential

$$Y = -\rho \frac{\partial \Psi}{\partial D} = \frac{1}{2} C^{o}_{ijkl} \varepsilon_{ij} \varepsilon_{kl} \tag{9.7.20}$$

which is referred to as the *damage energy release rate*. Using relation (9.7.17), we can express Y in terms of the elastic strain energy $U = \frac{1}{2} C_{ijkl} \varepsilon_{ij} \varepsilon_{kl}$ as

$$Y = \frac{U}{1 - D} \tag{9.7.21}$$

Next, consider the case at *constant stress* which implies

$$dT_{ij} = C^{o}_{ijkl} [(1 - D) d\varepsilon_{kl} - \varepsilon_{kl}\, dD] = 0 \Rightarrow d\varepsilon_{kl} = \frac{dD}{1 - D} \varepsilon_{kl} \tag{9.7.22}$$

For this case, the elastic strain energy is given by

$$\begin{aligned} dU &= T_{ij}\, d\varepsilon_{ij} = T_{ij} \varepsilon_{ij} \frac{dD}{1 - D} \\ &= (1 - D) C^{o}_{ijkl} \varepsilon_{kl} \varepsilon_{ij} \frac{dD}{1 - D} \\ &= C^{o}_{ijkl} \varepsilon_{ij} \varepsilon_{kl}\, dD \end{aligned} \tag{9.7.23}$$

and thus

$$\begin{aligned} \left. \frac{dU}{dD} \right|_{T=\text{const.}} &= C^{o}_{ijkl} \varepsilon_{ij} \varepsilon_{kl} = 2Y \Rightarrow \\ Y &= \frac{1}{2} \left. \frac{dU}{dD} \right|_{T=\text{const.}} \end{aligned} \tag{9.7.24}$$

For the elastic case, this implies that Y is the strain energy release rate through loss of stiffness which occurs due to damage.

It is sometimes useful to express the general form (9.7.21) for the damage energy release rate in terms of the hydrostatic and effective (von Mises) stresses. Using the results from Section 4.5, we decompose the stress tensor into spherical and deviatoric parts and also use the definition of the effective stress σ_e:

$$\begin{aligned} T_{ij} &= \sigma_H \delta_{ij} + \hat{T}_{ij} \\ \sigma_e &= \sqrt{\frac{3}{2} \hat{T}_{ij} \hat{T}_{ij}} \end{aligned} \tag{9.7.25}$$

where the hydrostatic stress is given by $\sigma_H = \frac{1}{3} T_{kk}$. Using these results, relation (9.7.21) can be expressed as

$$Y = \frac{U}{1 - D} = \frac{\sigma_e^2}{2\bar{E}(1 - D)^2} \left[\frac{2}{3}(1 + \nu) + 3(1 - 2\nu) \left(\frac{\sigma_H}{\sigma_e} \right)^2 \right] \tag{9.7.26}$$

The ratio $\left(\dfrac{\sigma_H}{\sigma_e}\right)$ is referred to as the *triaxiality ratio* and it plays a role in the rupture of materials.

Because of the limitations of using only a scalar damage parameter, more general theories have been developed which incorporate a *damage tensor model*. Following the concept first shown in relation (9.7.13), we can express the effective stress relation in a more general scheme

$$\bar{T}_{ij} = M_{ijkl} T_{kl} \tag{9.7.27}$$

where M_{ijkl} is a *fourth-order damage effect tensor*. If $\boldsymbol{M}$ is left to be completely general, the effective stress could become nonsymmetric. This situation is normally avoided. Several different forms of $\boldsymbol{M}$ have been proposed in the literature, and often these have expressed the fourth-order tensor in terms of a second-order damage tensor. This scheme makes it easier to characterize with material parameters. One such form provided by Voyiadjis and Kattan (2005) is

$$M_{ijkl} = (\delta_{ik} - \phi_{ik})^{-1/2} (\delta_{jl} - \phi_{jl})^{-1/2} \tag{9.7.28}$$

where ϕ_{ij} is a *second-order damage tensor*. Further details on this and other types of tensor formulations can be found in the quoted references.

We have only presented a brief outline of CDM that illustrates how continuum mechanics plays an important role in the formulation. Many additional details on damage evolution through physical behaviors including fracture mechanics, dislocation motions, fatigue, plasticity, creep, etc. are commonly incorporated into the theory. This material is well beyond the scope of the text and can be found in the quoted literature.

REFERENCES

Aifantis, E.C., 2003. Update on a class of gradient theories. Mech. Mater. 35, 259–280.

Askes, H., Aifantis, E.C., 2011. Gradient elasticity in statics and dynamics: an overview of formulations, length scale identification procedures, finite element implementations and new results. Int. J. Solids Struct. 48, 1962–1990.

Boresi, A.P., Chong, K.P., 2000. Elasticity in Engineering Mechanics. John Wiley, New York.

Carlson, D.E., 1966. Stress functions for plane problems with couple stresses. J. Appl. Math. Phys. 17, 789–792.

Chaboche, J.L., 1988a. Continuum damage mechanics: Part I. General concepts. J. Appl. Mech. 55, 59–64.

Chaboche, J.L., 1988b. Continuum damage mechanics: Part II. Damage growth, crack initiation, and crack growth. J. Appl. Mech. 55, 65–72.

Chang, C.S., Liu, Y., 2013. Stress Fabric Granular Mater. 2.

Chang, C.S., Ma, L.A., 1991. Micromechanical-base micro-polar theory for deformation of granular solids. Int. J. Solids Struct. 28, 67–86.

Charalambakis, N., 2010. Homogenization techniques and micromechanics. A survey and perspectives. Appl. Mech. Rev. 63.

Chen, K.C., Lan, J.Y., 2008. Microcontinuum derivation of Goodman-Cowin theory for granular materials. Con. Mech. Thermodyn. 20, 331–345.
Cowin, S.C., Nunziato, J.W., 1983. Linear elastic materials with voids. J. Elasticity 13, 125–147.
Cowin, S.C., Puri, P., 1983. The classical pressure vessel problems for linear elastic materials with voids. J. Elasticity 13, 157–163.
Cowin, S.C., 1984a. A note on the problem of pure bending for a linear elastic material with voids. J. Elasticity 14, 227.
Cowin, S.C., 1984b. The stresses around a hole in a linear elastic material with voids. Q. J. Mech. Appl. Math. 37, 441–465.
Cowin, S.C., 1985a. The viscoelastic behavior of linear elastic materials with voids. J. Elasticity 15, 185–191.
Cowin, S.C., 1985b. The relationship between the elasticity tensor and the fabric tensor. Mech. Matl. 4, 137–147.
Cundall, P.A., Strack, O.D., 1979. A discrete numerical model for granular assemblies. Geotchnique 29, 47–65.
Ellis, R.W., Smith, C.W., 1967. A thin-plate analysis and experimental evaluation of couple-stress effects. Exp. Mech. 7, 372–380.
Eringen, A.C., 1968. Theory of micropolar elasticity. Liebowitz, H. (Ed.), Fracture, 2, Academic Press, New York, pp. 662–729.
Eringen, A.C., 1972. Linear theory of nonlocal elasticity and dispersion of plane waves. Int. J. Eng. Sci. 10, 425–435.
Eringen, A.C., 1977. Edge dislocation in nonlocal elasticity. Int. J. Eng. Sci. 15, 177–183.
Eringen, A.C., 1983. On differential equations of nonlocal elasticity and solution of screw dislocation and surface waves. Int. J. Appl. Phys. 54, 4703–4710.
Eringen, A.C., 1999. Microcontinuum Field Theories. I. Foundations and Solids. Springer, New York.
Fang, C., Wang, Y., Hutter, K., 2006. A thermo-mechanical continuum theory with internal length for cohesionless granular materials. Con. Mech. Thermodyn. 17, 545–576.
Ferrari, M., Granik, V.T., Imam, A., Nadeau, J., 1997. Advances in Doublet Mechanics. Springer, Berlin.
Fish, J., Kuznetsov, S., 2012. From homogenization to generalized continua. Intl. J. Comp. Meth. Eng. Mech. 13, 77–87.
Goodman, M.A., Cowin, S.C., 1972. A continuum theory for granular materials. Arch. Rat. Mech. Anal. 44, 249–266.
Granik, V.T., 1978. Microstructural mechanics of granular media. Technique Report IM/MGU 78-.[241]. Inst. Mech. of Moscow State University [in Russian].
Granik, V.T., Ferrari, M., 1993. Microstructural mechanics of granular media. Mech. Mater. 15, 301–322.
Gutkin, M.Y., Aifantis, E.C., 1996. Screw dislocation in gradient elasticity. Scripta Matl. 35, 1353–1358.
Hartranft, R.J., Sih, G.C., 1965. The effect of couple-stress on the stress concentration of a circular inclusion. J. Appl. Mech. 32, 429–431.
Hill, R., 1963. Elastic properties of reinforced solids: some theoretical principles. J. Mech. Phys. Solids 11, 357–372.
Kachanov, L.M., 1986. Introduction to Continuum Damage Mechanics. Springer, Dordrecht, the Netherlands.
Kaloni, P.N., Ariman, T., 1967. Stress concentration effects in micropolar elasticity. J. Appl. Math. Phys. 18, 136–141.

Kennedy, T.C., Kim, J.B., 1987. Finite element analysis of a crack in a micropolar elastic material. Raghavan, R., CoKonis, T.J. (Eds.), Computers in Engineering, 3, ASME, pp. 439–444.

Kennedy, T.C., 1999. Modeling failure in notched plates with micropolar strain softening. Composite Struct. 44, 71–79.

Kondo, D., Welemane, H., Cormery, F., 2007. Basic concepts and models in continuum damage mechanics. Rev. Eur. Genie 11, 927–943.

Krajcinovic, D., 1996. Damage Mechanics. North-Holland, Amsterdam.

Kunin, I.A., 1983. Elastic Media with Microstructure. II. Three-Dimensional Models. Springer, Berlin.

Lazar, M., Maugin, G.A., 2006. A note of line forces in gradient elasticity. Mech. Res. Commun. 33, 674–680.

Lemaitre, J., 1996. A Course on Damage Mechanics. Springer, Berlin.

Li, S., Miskioglu, I., Altan, B.S., 2004. Solution to line loading of a semi-infinite solid in gradient elasticity. Int. J. Solid Struct. 41, 3395–3410.

Li, X., Yu, H.S., Li, X.S., 2009. Macro-micro relations in granular mechanics. Int. J. Solids Struct. 46, 4331–4341.

Mehrabadi, M.M., Nemat-Nasser, S., Oda, M., 1982. On statistical description of stress and fabric in granular materials. Int. J. Num. Anal. Meth. Geomech. 6, 95–108.

Mindlin, R.D., 1963. Influence of couple-stress on stress concentrations. Exp. Mech. 3, 1–7.

Mindlin, R.D., 1964. Microstructure in linear elasticity. Arch. Rat. Mech. Anal. 16, 51–78.

Mindlin, R.D., 1965. Second gradient of strain and surface tension in linear elasticity. Int. J. Solids Struct. 1, 417–438.

Nemat-Nasser, S., Hori, H., 1993. Micromechanics: overall properties of heterogeneous materials. Applied Mathematics and Mechanics Series, Vol 37. North-Holland, Amsterdam.

Nguyen, V.P., Stroeven, M., Sluys, L.J., 2011. Multiscale continuous and discontinuous modeling of heterogeneous materials: a review on recent developments. J. Multiscale Model. 3, 1–42.

Nicot, F., Darve, F., 2005. A multi-scale approach to granular materials. Mech. Mater. 37, 980–986.

Nowacki, W., 1986. Theory of Asymmetric Elasticity. Pergamon Press, Oxford.

Ru, C.Q., Aifantis, E.C., 1993. A simple approach to solve boundary-value problems in gradient elasticity. Acta Mech. 101, 59–68.

Sadd, M.H., 2014. Elasticity: Theory, Applications and Numerics, Third ed Elsevier, Waltham, MA.

Sadd, M.H., Dai, Q., 2004. A comparison of micromechanical modeling of asphalt materials using finite elements and doublet mechanics. Mech. Mater. 37, 641–662.

Sadd, M.H., Dai, Q., Parmameswaran, Shukla, A., 2004. Microstructural simulation of asphalt materials: modeling and experimental studies. J. Mater. Civil Eng. 16, 107–115.

Sadd, M.H., Qiu, L., Boardman, W.G., Shukla, A., 1992. Modelling wave propagation in granular media using elastic networks. Int. J. Rock Mech. Min. Sci. Geomech. 29, 161–170.

Schijve, J., 1966. Note of couple stresses. J. Mech. Phys. Solids 14, 113–120.

Sun, C.T., Yang, T.Y., 1975. A couple-stress theory for gridwork-reinforced media. J. Elasticity 5, 45–58.

Toupin, R.A., 1964. Theories of elasticity with couple-stress. Arch. Rat. Mech. Anal. 17, 85–112.

Weitsman, Y., 1965. Couple-stress effects on stress concentration around a cylindrical inclusion in a field of uniaxial tension. J. Appl. Mech. 32, 424–428.

Voyiadjis, G.Z., Kattan, P.I., 2005. Damage Mechanics. Taylor & Francis, Boca Raton, FL.

Since T_{xx} is constant, the diffusion equations (7.3.3) and $(7.3.4)_2$ give

$$\frac{1}{R}\frac{\partial p}{\partial t}-\kappa\frac{\partial^2 p}{\partial x^2}=0 \tag{7.3.17}$$

and this simplification removes the coupling between pore pressure and stress and thus provides a single equation to solve for the pore pressure p. The boundary conditions for this problem are

$$p(0,t)=0,\quad \frac{\partial p(h,t)}{\partial x}=0 \tag{7.3.18}$$

while the initial condition is taken as an undrained situation and hence using $(7.3.6)_3$, we get

$$p(x,0)=\frac{R}{3H}p_o \tag{7.3.19}$$

The solution to Eq. (7.3.17) can be found by standard separation of variables (see Wang, 2000 for details) and is given by

$$p(x,t)=\frac{Rp_o}{3H}\sum_{n=1,3,5,\ldots}^{\infty}\frac{4}{n\pi}\sin\left(\frac{n\pi\hat{x}}{2}\right)e^{-n^2\pi^2\tau} \tag{7.3.20}$$

where we have introduced dimensionless distance $\hat{x}=x/h$ and dimensionless time $\tau=\kappa t/4Rh^2$.

The displacement is found from integrating Eq. (7.3.16), and Wang gives the result of the displacement of the top of the layer as

$$u(0,t)=K_1\left[1+K_2\sum_{n=1,3,5,\ldots}^{\infty}\frac{8}{n^2\pi^2}\left(1-e^{-n^2\pi^2\tau}\right)\right] \tag{7.3.21}$$

with $K_1=\dfrac{p_o h(1-2\nu_u)}{2\mu(1-\nu_u)}, K_2=\dfrac{\nu_u-\nu}{(1-\nu_u)(1-2\nu_u)}$, and the drained and undrained Poisson's ratios given by

$$\nu=\frac{3k-2\mu}{2(3k+\mu)},\quad \nu_u=\frac{3K_u-2\mu}{2(3K_u+\mu)} \tag{7.3.22}$$

The time behavior of the pore pressure is shown in Fig. 7.8 for three different locations within the layer. At all locations, the pore pressure starts at the same value and exhibits a monotonic decrease to zero. Fig. 7.9 illustrates the surface displacement $u(0,t)/K_1$ versus time response for the case with $K_2=1$. Although it is difficult to see from the graph, there exists an initial undrained elastic displacement followed by a time-dependent component. The layer deforms initially with an elastic response determined by the undrained Poisson's ratio ν_u, and as time progresses it consolidates into another elastic condition characterized by the drained value ν. All calculations and plotting were done using MATLAB Code C-18.

EXERCISES

9.1 Verify the two-dimensional couple-stress constitutive relations (9.2.10).

9.2 Justify the compatibility equations (9.2.12).

9.3 Verify that the two-dimensional stress–stress function relations (9.2.14) are a self-equilibrated form.

9.4 For the couple-stress theory, show that the two stress functions satisfy

$$\nabla^4\Phi = 0, \quad \nabla^2\Psi - l^2\nabla^4\Psi = 0$$

9.5 Using the general stress relations (9.2.25) for the stress concentration problem of Example 9.2.1, show that the circumferential stress on the boundary of the hole is given by

$$T_{\theta\theta}(a,\theta) = T\left(1 - \frac{2\cos 2\theta}{1+F}\right)$$

Verify that this expression gives a maximum at $\theta = \pm\pi/2$, and explicitly show that this value will reduce to the classical case of $3\boldsymbol{T}$ by choosing $l_1 = l_2 = l = 0$.

9.6 For isotropic couple-stress theory, it has been proposed to use a hyperelastic formulation such that the strain energy function may be specified by

$$U = \frac{1}{2}\lambda\varepsilon_{kk}\varepsilon_{mm} + \mu(\varepsilon_{km}\varepsilon_{km} + l^2\chi_{km}\chi_{km})$$

where λ and μ are the usual elastic constants, ε_{ij} is the strain tensor, l^2 is a length parameter, and χ_{ij} is the *symmetric curvature tensor* defined by $\chi_{ij} = \frac{1}{2}(\omega_{i,j} + \omega_{j,i})$ with the rotation vector given by $\omega_i = \frac{1}{2}\varepsilon_{ijk}u_{k,j}$. Using the relations $T_{ij} = \frac{\partial U}{\partial \varepsilon_{ij}}$ and $m_{ij} = \frac{\partial U}{\partial \chi_{ij}}$, show that the constitutive relations for the stress and couple-stress are given by

$$T_{ij} = \lambda\varepsilon_{kk}\delta_{ij} + 2\mu\varepsilon_{ij}, \quad m_{ij} = 2\mu l^2\chi_{ij}$$

9.7 For the formulation given in Exercise 9.6, consider a torsional displacement field $u_1 = 0, u_2 = -\kappa x_1 x_3, u_3 = \kappa x_1 x_2$, where κ is a constant. Determine the strains, rotations, symmetric curvature tensor, and then using the constitutive relations calculate the stresses and couple-stresses.

9.8 Starting with the general relations (9.3.6), verify that the two-dimensional plane stress constitutive equations for elastic materials with voids are given by (9.3.9).

9.9 For elastic materials with voids, using the usual single strain compatibility equation, develop the stress and stress-function compatibility forms (9.3.10) and (9.3.11).

9.10 Using the inequalities (9.3.7), verify that the length parameter h^2 defined by relation (9.3.13) is positive.

9.11 Compare the hoop stress $T_{\theta\theta}(r, \pi/2)$ predictions from elasticity with voids given by relation (9.3.18) with the corresponding results from classical theory. Choosing $N = 1/2$ and $L = 2$, for the elastic material with voids, make a comparative plot of $T_{\theta\theta}(r, \pi/2)/T$ versus r/a for these two theories.

9.12 Verify the doublet mechanics $\boldsymbol{Q}$ transformational matrix given by relation (9.4.10) in Example 9.4.1.

9.13 For the doublet mechanics Flamant solution in Example 9.4.1, develop contour plots (similar to Fig. 9.10) for the microstresses p_1 and p_2. Are there zones where these microstresses are tensile?

9.14 Consider the gradient elasticity problem under a one-dimensional deformation field of the form $u = u(x), v = w = 0$. Using constitutive form (9.5.3), determine the stress components. Next show that equilibrium equations reduce to $\frac{dT_x}{dx} = 0$, and this will lead to the equation

$$\frac{d^4u}{dx^4} - \frac{1}{c}\frac{d^2u}{dx^2} = 0$$

Finally show that the solution for the displacement is given by

$$u = c_1 + c_2 x + c_3 \sinh\left(\frac{x}{\sqrt{c}}\right) + c_4 \cosh\left(\frac{x}{\sqrt{c}}\right)$$

9.15 Starting with the given displacement form (9.5.10), explicitly justify relations (9.5.11)–(9.5.14) in the gradient dislocation Example 9.5.1.

9.16 Using integral tables verify that the gradient Flamant stress integral solution reduces to the classical elasticity form as per Eq. (9.5.22).

9.17 Justify the reduced fabric constitutive form (9.6.6).

9.18 For the isotropic fabric tensor case $M_{ij} = M_o \delta_{ij}$, explicitly justify relations (9.6.10)–(9.6.12).

9.19 Consider the case of a fabric tensor given by

$$M_{ij} = \begin{bmatrix} M_1 & 0 & 0 \\ 0 & M_2 & 0 \\ 0 & 0 & M_3 \end{bmatrix}$$

Using relation (9.6.8), determine the components of the elasticity tensor and show that there will be only nine nonzero terms, thus leading to an orthotropic constitutive form.

9.20 Consider the two-dimensional fabric tensor formulation in Section 9.6. Using the distribution function $\xi(\phi) = \frac{1}{2\pi}(1 + a\cos 2\phi)$, first show that it satisfies the probability relation $\int_0^{2\pi} \xi(\phi)d\phi = 1$. Next noting that the normal vector is now given by $\boldsymbol{n} = (\cos\phi, \sin\phi)$, and $\xi(\boldsymbol{n}) = \frac{1}{2\pi} D_{ij}n_i n_j$, $M_{ij} = \int_0^{2\pi} \xi(\phi)n_i n_j \, d\phi$, verify that the contact density and fabric tensors are given by $\boldsymbol{D} = \begin{bmatrix} 1+a & 0 \\ 0 & 1-a \end{bmatrix}$, $\boldsymbol{M} = \frac{1}{2}\begin{bmatrix} 1+a/2 & 0 \\ 0 & 1-a/2 \end{bmatrix}$.

9.21 An elastic damage model is sometimes generated from the response of a system of N parallel bars as shown. Each bar is assume to have identical length, material stiffness $k = k_i = A_i E_i / l$, and deformation δ. The load deflection response for the ith bar is then given by $P_i = k_i \delta$, and from equilibrium the total load of the entire assembly is $P = \sum_{i=1}^{N} P_i$. The bars are totally elastic and fail by complete rupture with a given rupture strength distribution among the bars. For the case where if n bars fail, show that $P = k(1-D)\delta$, where the damage parameter is given by $D = n/N$ and $K = Nk$. Note how this result has the same form as relation (9.7.5).

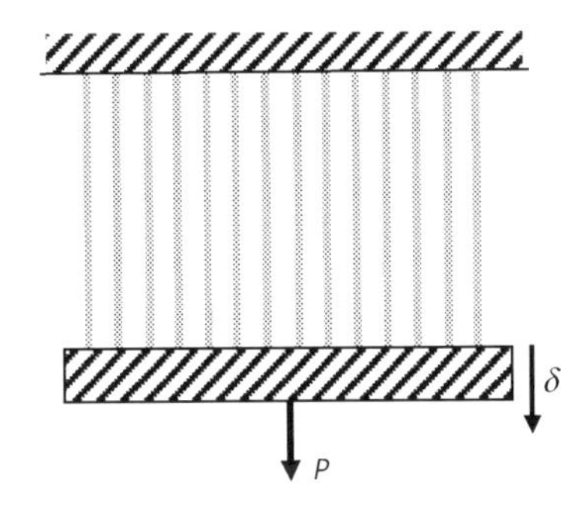

9.22 For uniaxial damage theory, instead of using equivalency of the strain, consider the case of *elastic strain energy equivalency*. For this case, the form of the strain energy is taken to be the same in both the damaged and undamaged (effective) configurations and thus $\frac{1}{2}\sigma\varepsilon = \frac{\sigma^2}{2E} = \frac{\bar{\sigma}^2}{2\bar{E}}$. Show that the new relation for the damaged modulus is given by $E = (1-D)^2 \bar{E}$.

9.23 Similar to Example 9.7.1, consider the uniaxial damage response for the case with a damage evolution relation $D = \hat{f}(\varepsilon) = \frac{\varepsilon^2}{m}$ with $0 \le \varepsilon \le 2$ and $m > 4$. Make plots of the damage parameter evolution with strain, and the stress–strain results for the cases $m = 5$, 10, and 20. Note that this case has a smaller damage evolution when compared to the text example.

9.24 Explicitly verify relation (9.7.26) for the damage energy release rate.

APPENDIX

Basic Field Equations in Cartesian, Cylindrical, and Spherical Coordinates

For convenience, some of the basic three-dimensional field equations of continuum mechanics are listed here for Cartesian, cylindrical, and spherical coordinate systems. This collection will help save time from searching these results in the various chapters of the text. Cylindrical and spherical coordinates previously shown in Figs. 2.6 and 2.7 are related to the basic Cartesian system as shown in Fig. A.1. For convenience, the Cartesian notation x_1, x_2, x_3 is replaced by x, y, z.

A.1 SMALL STRAIN–DISPLACEMENT RELATIONS

CARTESIAN COORDINATES $\boldsymbol{u} = (u, v, w)$

$$
\begin{aligned}
&\varepsilon_x = \frac{\partial u}{\partial x}, \quad \varepsilon_y = \frac{\partial v}{\partial y}, \quad \varepsilon_z = \frac{\partial w}{\partial z} \\
&\varepsilon_{xy} = \frac{1}{2}\left(\frac{\partial u}{\partial y} + \frac{\partial v}{\partial x}\right) \\
&\varepsilon_{yz} = \frac{1}{2}\left(\frac{\partial v}{\partial z} + \frac{\partial w}{\partial y}\right) \\
&\varepsilon_{zx} = \frac{1}{2}\left(\frac{\partial w}{\partial x} + \frac{\partial u}{\partial z}\right)
\end{aligned}
\tag{A.1}
$$

CYLINDRICAL COORDINATES

$$
\begin{aligned}
&\varepsilon_r = \frac{\partial u_r}{\partial r}, \quad \varepsilon_\theta = \frac{1}{r}\left(u_r + \frac{\partial u_\theta}{\partial \theta}\right), \quad \varepsilon_z = \frac{\partial u_z}{\partial z} \\
&\varepsilon_{r\theta} = \frac{1}{2}\left(\frac{1}{r}\frac{\partial u_r}{\partial \theta} + \frac{\partial u_\theta}{\partial r} - \frac{u_\theta}{r}\right) \\
&\varepsilon_{\theta z} = \frac{1}{2}\left(\frac{\partial u_\theta}{\partial z} + \frac{1}{r}\frac{\partial u_z}{\partial \theta}\right) \\
&\varepsilon_{zr} = \frac{1}{2}\left(\frac{\partial u_r}{\partial z} + \frac{\partial u_z}{\partial r}\right)
\end{aligned}
\tag{A.2}
$$

Continuum Mechanics Modeling of Material Behavior. http://dx.doi.org/10.1016/B978-0-12-811474-2.00010-1

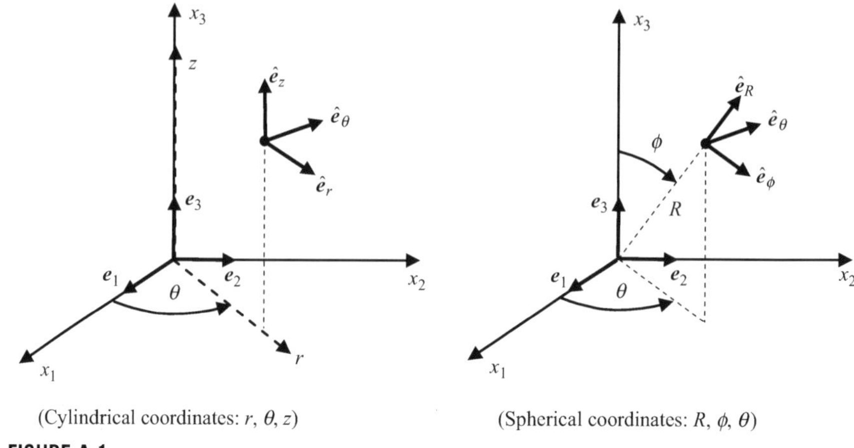

FIGURE A.1

Cylindrical and spherical coordinate systems.

SPHERICAL COORDINATES

$$
\begin{aligned}
\varepsilon_R &= \frac{\partial u_R}{\partial R}, \quad \varepsilon_\phi = \frac{1}{R}\left(u_R + \frac{\partial u_\phi}{\partial \phi}\right) \\
\varepsilon_\theta &= \frac{1}{R\sin\phi}\left(\frac{\partial u_\theta}{\partial \theta} + \sin\phi u_R + \cos\phi u_\phi\right) \\
\varepsilon_{R\phi} &= \frac{1}{2}\left(\frac{1}{R}\frac{\partial u_R}{\partial \phi} + \frac{\partial u_\phi}{\partial R} - \frac{u_\phi}{R}\right) \\
\varepsilon_{\phi\theta} &= \frac{1}{2R}\left(\frac{1}{\sin\phi}\frac{\partial u_\phi}{\partial \theta} + \frac{\partial u_\theta}{\partial \phi} - \cot\phi u_\theta\right) \\
\varepsilon_{\theta R} &= \frac{1}{2}\left(\frac{1}{R\sin\phi}\frac{\partial u_R}{\partial \theta} + \frac{\partial u_\theta}{\partial R} - \frac{u_\theta}{R}\right)
\end{aligned}
\tag{A.3}
$$

A.2 DEFORMATION GRADIENT TENSOR FOR FINITE DEFORMATION

CARTESIAN COORDINATES $x = \chi(X, t)$

$$
\boldsymbol{F} = \frac{\partial \boldsymbol{x}}{\partial \boldsymbol{X}} = \begin{bmatrix}
\frac{\partial x_1}{\partial X_1} & \frac{\partial x_1}{\partial X_2} & \frac{\partial x_1}{\partial X_3} \\
\frac{\partial x_2}{\partial X_1} & \frac{\partial x_2}{\partial X_2} & \frac{\partial x_2}{\partial X_3} \\
\frac{\partial x_3}{\partial X_1} & \frac{\partial x_3}{\partial X_2} & \frac{\partial x_3}{\partial X_3}
\end{bmatrix}
\tag{A.4}
$$

CYLINDRICAL COORDINATES

$$r = r(R,\Theta,Z),\quad \theta = \theta(R,\Theta,Z),\quad z = z(R,\Theta,Z)$$

$$\boldsymbol{F} = \begin{bmatrix} \dfrac{\partial r}{\partial R} & \dfrac{1}{R}\dfrac{\partial r}{\partial \Theta} & \dfrac{\partial r}{\partial Z} \\ r\dfrac{\partial \theta}{\partial R} & \dfrac{r}{R}\dfrac{\partial \theta}{\partial \Theta} & r\dfrac{\partial \theta}{\partial Z} \\ \dfrac{\partial z}{\partial R} & \dfrac{1}{R}\dfrac{\partial z}{\partial \Theta} & \dfrac{\partial z}{\partial Z} \end{bmatrix} \tag{A.5}$$

$$r = r(X,Y,Z),\quad \theta = \theta(X,Y,Z),\quad z = z(X,Y,Z)$$

$$\boldsymbol{F} = \begin{bmatrix} \dfrac{\partial r}{\partial X} & \dfrac{\partial r}{\partial Y} & \dfrac{\partial r}{\partial Z} \\ r\dfrac{\partial \theta}{\partial X} & r\dfrac{\partial \theta}{\partial Y} & r\dfrac{\partial \theta}{\partial Z} \\ \dfrac{\partial z}{\partial X} & \dfrac{\partial z}{\partial Y} & \dfrac{\partial z}{\partial Z} \end{bmatrix} \tag{A.6}$$

SPHERICAL COORDINATES

$$R = R(R_o,\Phi,\Theta),\quad \phi = \phi(R_o,\Phi,\Theta),\quad \theta = \theta(R_o,\Phi,\Theta)$$

$$\boldsymbol{F} = \begin{bmatrix} \dfrac{\partial R}{\partial R_o} & \dfrac{1}{R_o}\dfrac{\partial R}{\partial \Phi} & \dfrac{1}{R_o \sin\Phi}\dfrac{\partial R}{\partial \Theta} \\ R\dfrac{\partial \phi}{\partial R_o} & \dfrac{R}{R_o}\dfrac{\partial \phi}{\partial \Phi} & \dfrac{R}{R_o \sin\Phi}\dfrac{\partial \phi}{\partial \Theta} \\ R\sin\phi\dfrac{\partial \theta}{\partial R_o} & \dfrac{R\sin\phi}{R_o}\dfrac{\partial \theta}{\partial \Phi} & \dfrac{R\sin\phi}{R_o \sin\Phi}\dfrac{\partial \theta}{\partial \Theta} \end{bmatrix} \tag{A.7}$$

A.3 EQUATIONS OF MOTION (CONSERVATION OF LINEAR MOMENTUM)

CARTESIAN COORDINATES

$$\begin{aligned} \frac{\partial T_{xx}}{\partial x} + \frac{\partial T_{yx}}{\partial y} + \frac{\partial T_{zx}}{\partial z} + F_x &= \rho\left(\frac{\partial v_x}{\partial t} + v_x\frac{\partial v_x}{\partial x} + v_y\frac{\partial v_x}{\partial y} + v_z\frac{\partial v_x}{\partial z}\right) \\ \frac{\partial T_{xy}}{\partial x} + \frac{\partial T_{yy}}{\partial y} + \frac{\partial T_{zy}}{\partial z} + F_y &= \rho\left(\frac{\partial v_y}{\partial t} + v_x\frac{\partial v_y}{\partial x} + v_y\frac{\partial v_y}{\partial y} + v_z\frac{\partial v_y}{\partial z}\right) \\ \frac{\partial T_{xz}}{\partial x} + \frac{\partial T_{yz}}{\partial y} + \frac{\partial T_{zz}}{\partial z} + F_z &= \rho\left(\frac{\partial v_z}{\partial t} + v_x\frac{\partial v_z}{\partial x} + v_y\frac{\partial v_z}{\partial y} + v_z\frac{\partial v_z}{\partial z}\right) \end{aligned} \tag{A.8}$$

CYLINDRICAL COORDINATES

$$\begin{aligned}
&\frac{\partial T_{rr}}{\partial r}+\frac{1}{r}\frac{\partial T_{r\theta}}{\partial \theta}+\frac{\partial T_{rz}}{\partial z}+\frac{1}{r}(T_{rr}-T_{\theta\theta})+F_r=\rho\left(\frac{\partial v_r}{\partial t}+v_r\frac{\partial v_r}{\partial r}+\frac{v_\theta}{r}\frac{\partial v_r}{\partial \theta}+v_z\frac{\partial v_r}{\partial z}-\frac{v_\theta^2}{r}\right)\\
&\frac{\partial T_{r\theta}}{\partial r}+\frac{1}{r}\frac{\partial T_{\theta\theta}}{\partial \theta}+\frac{\partial T_{\theta z}}{\partial z}+\frac{2}{r}T_{r\theta}+F_\theta=\rho\left(\frac{\partial v_\theta}{\partial t}+v_r\frac{\partial v_\theta}{\partial r}+\frac{v_\theta}{r}\frac{\partial v_\theta}{\partial \theta}+v_z\frac{\partial v_\theta}{\partial z}+\frac{v_r v_\theta}{r}\right)\\
&\frac{\partial T_{rz}}{\partial r}+\frac{1}{r}\frac{\partial T_{\theta z}}{\partial \theta}+\frac{\partial T_{zz}}{\partial z}+\frac{1}{r}T_{rz}+F_z=\rho\left(\frac{\partial v_z}{\partial t}+v_r\frac{\partial v_z}{\partial r}+\frac{v_\theta}{r}\frac{\partial v_z}{\partial \theta}+v_z\frac{\partial v_z}{\partial z}\right)
\end{aligned}\tag{A.9}$$

SPHERICAL COORDINATES

$$\begin{aligned}
&\frac{\partial T_{RR}}{\partial R}+\frac{1}{R}\frac{\partial T_{R\varphi}}{\partial \varphi}+\frac{1}{R\sin\varphi}\frac{\partial T_{R\theta}}{\partial \theta}+\frac{1}{R}(2T_{RR}-T_{\phi\phi}-T_{\theta\theta}+T_{R\phi}\cot\phi)+F_R\\
&=\rho\left(\frac{\partial v_R}{\partial t}+v_R\frac{\partial v_R}{\partial R}+\frac{v_\phi}{R}\frac{\partial v_R}{\partial \phi}+\frac{v_\theta}{R\sin\phi}\frac{\partial v_R}{\partial \theta}-\frac{v_\phi^2+v_\theta^2}{R}\right)\\
&\frac{\partial T_{R\varphi}}{\partial R}+\frac{1}{R}\frac{\partial T_{\varphi\varphi}}{\partial \varphi}+\frac{1}{R\sin\varphi}\frac{\partial T_{\varphi\theta}}{\partial \theta}+\frac{1}{R}[(T_{\phi\phi}-T_{\theta\theta})\cot\phi+3T_{R\phi}]+F_\phi\\
&=\rho\left(\frac{\partial v_\phi}{\partial t}+v_R\frac{\partial v_\phi}{\partial R}+\frac{v_\phi}{R}\frac{\partial v_\phi}{\partial \phi}+\frac{v_\theta}{R\sin\phi}\frac{\partial v_\phi}{\partial \theta}+\frac{v_R v_\phi}{R}-\frac{v_\theta^2\cot\phi}{R}\right)\\
&\frac{\partial T_{R\theta}}{\partial R}+\frac{1}{R}\frac{\partial T_{\varphi\theta}}{\partial \varphi}+\frac{1}{R\sin\varphi}\frac{\partial T_{\theta\theta}}{\partial \theta}+\frac{1}{R}(2T_{\varphi\theta}\cot\phi+3T_{R\theta})+F_\theta\\
&=\rho\left(\frac{\partial v_\theta}{\partial t}+v_R\frac{\partial v_\theta}{\partial R}+\frac{v_\phi}{R}\frac{\partial v_\theta}{\partial \phi}+\frac{v_\theta}{R\sin\phi}\frac{\partial v_\theta}{\partial \theta}+\frac{v_\theta v_R}{R}+\frac{v_\phi v_\theta\cot\phi}{R}\right)
\end{aligned}\tag{A.10}$$

A.4 HOOKE'S LAW

CARTESIAN COORDINATES

$$\begin{aligned}
T_{xx}&=\lambda(\varepsilon_x+\varepsilon_y+\varepsilon_z)+2\mu\varepsilon_x & \varepsilon_x&=\frac{1}{E}\left[T_{xx}-\nu(T_{yy}+T_{zz})\right]\\
T_{yy}&=\lambda(\varepsilon_x+\varepsilon_y+\varepsilon_z)+2\mu\varepsilon_y & \varepsilon_y&=\frac{1}{E}\left[T_{yy}-\nu(T_{zz}+T_{xx})\right]\\
T_{zz}&=\lambda(\varepsilon_x+\varepsilon_y+\varepsilon_z)+2\mu\varepsilon_z & \varepsilon_z&=\frac{1}{E}\left[T_{zz}-\nu(T_{xx}+T_{yy})\right]\\
T_{xy}&=2\mu\varepsilon_{xy} & \varepsilon_{xy}&=\frac{1+\nu}{E}T_{xy}\\
T_{yz}&=2\mu\varepsilon_{yz} & \varepsilon_{yz}&=\frac{1+\nu}{E}T_{yz}\\
T_{zx}&=2\mu\varepsilon_{zx} & \varepsilon_{zx}&=\frac{1+\nu}{E}T_{zx}
\end{aligned}\tag{A.11}$$

CYLINDRICAL COORDINATES

$$
\begin{aligned}
T_{rr} &= \lambda(\varepsilon_r + \varepsilon_\theta + \varepsilon_z) + 2\mu\varepsilon_r \\
T_{\theta\theta} &= \lambda(\varepsilon_r + \varepsilon_\theta + \varepsilon_z) + 2\mu\varepsilon_\theta \\
T_{zz} &= \lambda(\varepsilon_r + \varepsilon_\theta + \varepsilon_z) + 2\mu\varepsilon_z \\
T_{r\theta} &= 2\mu\varepsilon_{r\theta} \\
T_{\theta z} &= 2\mu\varepsilon_{\theta z} \\
T_{zr} &= 2\mu\varepsilon_{zr}
\end{aligned}
\qquad
\begin{aligned}
\varepsilon_r &= \frac{1}{E}\left[T_r - \nu(T_\theta + T_z)\right] \\
\varepsilon_\theta &= \frac{1}{E}\left[T_\theta - \nu(T_z + T_r)\right] \\
\varepsilon_z &= \frac{1}{E}\left[T_z - \nu(T_r + T_\theta)\right] \\
\varepsilon_{r\theta} &= \frac{1+\nu}{E}T_{r\theta} \\
\nu_{\theta z} &= \frac{1+\nu}{E}T_{\theta z} \\
\varepsilon_{zr} &= \frac{1+\nu}{E}T_{zr}
\end{aligned}
\tag{A.12}
$$

SPHERICAL COORDINATES

$$
\begin{aligned}
T_{RR} &= \lambda(\varepsilon_R + \varepsilon_\phi + \varepsilon_\theta) + 2\mu\varepsilon_R \\
T_{\phi\phi} &= \lambda(\varepsilon_R + \varepsilon_\phi + \varepsilon_\theta) + 2\mu\varepsilon_\phi \\
T_{\theta\theta} &= \lambda(\varepsilon_R + \varepsilon_\phi + \varepsilon_\theta) + 2\mu\varepsilon_\theta \\
T_{R\phi} &= 2\mu\varepsilon_{R\phi} \\
T_{\phi\theta} &= 2\mu\varepsilon_{\phi\theta} \\
T_{\theta R} &= 2\mu\varepsilon_{\theta R}
\end{aligned}
\qquad
\begin{aligned}
\varepsilon_R &= \frac{1}{E}\left[T_{RR} - \nu(T_{\phi\phi} + T_{\theta\theta})\right] \\
\varepsilon_\phi &= \frac{1}{E}\left[T_{\phi\phi} - \nu(T_{\theta\theta} + T_{RR})\right] \\
\varepsilon_\theta &= \frac{1}{E}\left[T_{\theta\theta} - \nu(T_{RR} + T_{\phi\phi})\right] \\
\varepsilon_{R\phi} &= \frac{1+\nu}{E}T_{R\phi} \\
\varepsilon_{\phi\theta} &= \frac{1+\nu}{E}T_{\phi\theta} \\
\varepsilon_{\theta R} &= \frac{1+\nu}{E}T_{\theta R}
\end{aligned}
\tag{A.13}
$$

A.5 EQUILIBRIUM EQUATIONS IN TERMS OF DISPLACEMENTS (NAVIER'S EQUATIONS)

CARTESIAN COORDINATES

$$
\begin{aligned}
&\mu\nabla^2 u + (\lambda+\mu)\frac{\partial}{\partial x}\left(\frac{\partial u}{\partial x} + \frac{\partial v}{\partial y} + \frac{\partial w}{\partial z}\right) + F_x = 0 \\
&\mu\nabla^2 v + (\lambda+\mu)\frac{\partial}{\partial y}\left(\frac{\partial u}{\partial x} + \frac{\partial v}{\partial y} + \frac{\partial w}{\partial z}\right) + F_y = 0 \\
&\mu\nabla^2 w + (\lambda+\mu)\frac{\partial}{\partial z}\left(\frac{\partial u}{\partial x} + \frac{\partial v}{\partial y} + \frac{\partial w}{\partial z}\right) + F_z = 0
\end{aligned}
\tag{A.14}
$$

CYLINDRICAL COORDINATES

$$\begin{aligned}
&\mu\left(\nabla^2 u_r - \frac{u_r}{r^2} - \frac{2}{r^2}\frac{\partial u_\theta}{\partial \theta}\right) + (\lambda+\mu)\frac{\partial}{\partial r}\left(\frac{1}{r}\frac{\partial}{\partial r}(ru_r) + \frac{1}{r}\frac{\partial u_\theta}{\partial \theta} + \frac{\partial u_z}{\partial z}\right) + F_r = 0\\
&\mu\left(\nabla^2 u_\theta - \frac{u_\theta}{r^2} + \frac{2}{r^2}\frac{\partial u_r}{\partial \theta}\right) + (\lambda+\mu)\frac{1}{r}\frac{\partial}{\partial \theta}\left(\frac{1}{r}\frac{\partial}{\partial r}(ru_r) + \frac{1}{r}\frac{\partial u_\theta}{\partial \theta} + \frac{\partial u_z}{\partial z}\right) + F_\theta = 0\\
&\mu\nabla^2 u_z + (\lambda+\mu)\frac{\partial}{\partial z}\left(\frac{1}{r}\frac{\partial}{\partial r}(ru_r) + \frac{1}{r}\frac{\partial u_\theta}{\partial \theta} + \frac{\partial u_z}{\partial z}\right) + F_z = 0
\end{aligned} \tag{A.15}$$

SPHERICAL COORDINATES

$$\begin{aligned}
&\mu\left(\nabla^2 u_R - \frac{2u_R}{R^2} - \frac{2}{R^2}\frac{\partial u_\phi}{\partial \phi} - \frac{2u_\phi \cot\phi}{R^2} - \frac{2}{R^2\sin\phi}\frac{\partial u_\theta}{\partial \theta}\right)\\
&+(\lambda+\mu)\frac{\partial}{\partial R}\left(\frac{1}{R^2}\frac{\partial}{\partial R}(R^2u_R) + \frac{1}{R\sin\phi}\frac{\partial}{\partial \phi}(u_\phi\sin\phi) + \frac{1}{R\sin\phi}\frac{\partial u_\theta}{\partial \theta}\right) + F_R = 0\\
&\mu\left(\nabla^2 u_\phi + \frac{2}{R^2}\frac{\partial u_R}{\partial \phi} - \frac{u_\phi}{R^2\sin^2\phi} - \frac{2\cos\phi}{R^2\sin^2\phi}\frac{\partial u_\theta}{\partial \theta}\right)\\
&+(\lambda+\mu)\frac{1}{R}\frac{\partial}{\partial \phi}\left(\frac{1}{R^2}\frac{\partial}{\partial R}(R^2u_R) + \frac{1}{R\sin\phi}\frac{\partial}{\partial \phi}(u_\phi\sin\phi) + \frac{1}{R\sin\phi}\frac{\partial u_\theta}{\partial \theta}\right) + F_\phi = 0\\
&\mu\left(\nabla^2 u_\theta - \frac{u_\theta}{R^2\sin^2\phi} + \frac{2}{R^2\sin^2\phi}\frac{\partial u_R}{\partial \theta} + \frac{2\cos\phi}{R^2\sin^2\phi}\frac{\partial u_\phi}{\partial \theta}\right)\\
&+(\lambda+\mu)\frac{1}{R\sin\phi}\frac{\partial}{\partial \theta}\left(\frac{1}{R^2}\frac{\partial}{\partial R}(R^2u_R) + \frac{1}{R\sin\phi}\frac{\partial}{\partial \phi}(u_\phi\sin\phi) + \frac{1}{R\sin\phi}\frac{\partial u_\theta}{\partial \theta}\right) + F_\theta = 0
\end{aligned} \tag{A.16}$$

APPENDIX

Transformation of Field Variables Between Cartesian, Cylindrical, and Spherical Components

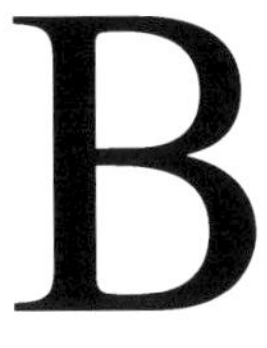

This appendix contains some three-dimensional transformation relations between displacement and stress components in Cartesian, cylindrical and spherical coordinates. The coordinate systems where previously shown in Fig. A.1 and the related stress components are re-illustrated in Fig. B.1. These results follow from the general transformation laws (2.8.1) and (4.3.7). Note that the stress results and can also be applied for any second order tensor including strain transformation.

B.1 CYLINDRICAL COMPONENTS FROM CARTESIAN

The transformation matrix for this case is given by

$$\boldsymbol{Q} = \begin{bmatrix} \cos\theta & \sin\theta & 0 \\ -\sin\theta & \cos\theta & 0 \\ 0 & 0 & 1 \end{bmatrix} \tag{B.1}$$

Displacement Transformation:

$$\begin{aligned} u_r &= u\cos\theta + v\sin\theta \\ u_\theta &= -u\sin\theta + v\cos\theta \\ u_z &= w \end{aligned} \tag{B.2}$$

Stress Transformation:

$$\begin{aligned} T_{rr} &= T_{xx}\cos^2\theta + T_{yy}\sin^2\theta + 2T_{xy}\sin\theta\cos\theta \\ T_{\theta\theta} &= T_{xx}\sin^2\theta + T_{yy}\cos^2\theta - 2T_{xy}\sin\theta\cos\theta \\ T_{zz} &= T_{zz} \\ T_{r\theta} &= -T_{xx}\sin\theta\cos\theta + T_{yy}\sin\theta\cos\theta + T_{xy}(\cos^2\theta - \sin^2\theta) \\ T_{\theta z} &= T_{yz}\cos\theta - T_{zx}\sin\theta \\ T_{zr} &= T_{yz}\sin\theta + T_{zx}\cos\theta \end{aligned} \tag{B.3}$$

Continuum Mechanics Modeling of Material Behavior. http://dx.doi.org/10.1016/B978-0-12-811474-2.00011-3

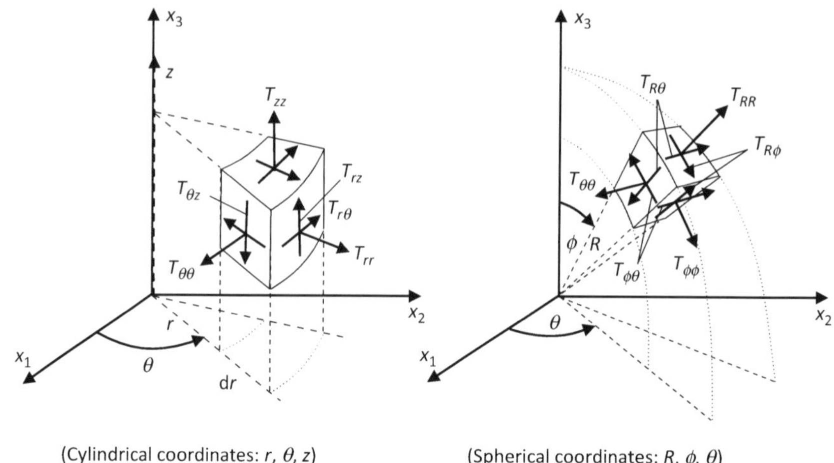

FIGURE B.1

Stress components in cylindrical and spherical coordinates.

B.2 SPHERICAL COMPONENTS FROM CYLINDRICAL

The transformation matrix from cylindrical to spherical coordinates is given by

$$\boldsymbol{Q} = \begin{bmatrix} \sin\phi & 0 & \cos\phi \\ \cos\phi & 0 & -\sin\phi \\ 0 & 1 & 0 \end{bmatrix} \tag{B.4}$$

Displacement Transformation:

$$\begin{aligned} u_R &= u_r \sin\phi + u_z \cos\phi \\ u_\phi &= u_r \cos\phi - u_z \sin\phi \\ u_\theta &= u_\theta \end{aligned} \tag{B.5}$$

Stress Transformation:

$$\begin{aligned} T_{RR} &= T_{rr} \sin^2\phi + T_{zz} \cos^2\phi + 2T_{rz} \sin\phi\cos\phi \\ T_{\phi\phi} &= T_{rr} \cos^2\phi + T_{zz} \sin^2\phi - 2T_{rz} \sin\phi\cos\phi \\ T_{\theta\theta} &= T_{\theta\theta} \\ T_{R\phi} &= (T_{rr} - T_{zz})\sin\phi\cos\phi - T_{rz}(\sin^2\phi - \cos^2\phi) \\ T_{\phi\theta} &= T_{r\theta} \cos\phi - T_{\theta z} \sin\phi \\ T_{\theta R} &= T_{r\theta} \sin\phi + T_{\theta z} \cos\phi \end{aligned} \tag{B.6}$$

B.3 SPHERICAL COMPONENTS FROM CARTESIAN

The transformation matrix from Cartesian to spherical coordinates can be obtained by combining the previous transformations given by (B.1) and (B.4); in the proper order of (B.1) first. Tracing back through tensor transformation theory, this is accomplished by the simple matrix multiplication.

$$
\begin{aligned}
\boldsymbol{Q} &= \begin{bmatrix} \sin\phi & 0 & \cos\phi \\ \cos\phi & 0 & -\sin\phi \\ 0 & 1 & 0 \end{bmatrix} \begin{bmatrix} \cos\theta & \sin\theta & 0 \\ -\sin\theta & \cos\theta & 0 \\ 0 & 0 & 1 \end{bmatrix} \\
&= \begin{bmatrix} \sin\phi\cos\theta & \sin\phi\sin\theta & \cos\phi \\ \cos\phi\cos\theta & \cos\phi\sin\theta & -\sin\phi \\ -\sin\theta & \cos\theta & 0 \end{bmatrix}
\end{aligned} \tag{B.7}
$$

Displacement Transformation:

$$
\begin{aligned}
u_R &= u\sin\phi\cos\theta + v\sin\phi\sin\theta + w\cos\phi \\
u_\phi &= u\cos\phi\cos\theta + v\cos\phi\sin\theta - w\sin\phi \\
u_\theta &= -u\sin\theta + v\cos\theta
\end{aligned} \tag{B.8}
$$

Stress Transformation:

$$
\begin{aligned}
T_{RR} &= T_{xx}\sin^2\phi\cos^2\theta + T_{yy}\sin^2\phi\sin^2\theta + T_{zz}\cos^2\phi \\
&\quad + 2T_{xy}\sin^2\phi\sin\theta\cos\theta + 2T_{yz}\sin\phi\cos\phi\sin\theta + 2T_{zx}\sin\phi\cos\phi\cos\theta \\
T_{\phi\phi} &= T_{xx}\cos^2\phi\cos^2\theta + T_{yy}\cos^2\phi\sin^2\theta + T_{zz}\sin^2\phi \\
&\quad + 2T_{xy}\cos^2\phi\sin\theta\cos\theta - 2T_{yz}\sin\phi\cos\phi\sin\theta - 2T_{zx}\sin\phi\cos\phi\cos\theta \\
T_{\theta\theta} &= T_{xx}\sin^2\theta + T_{yy}\cos^2\theta - 2T_{xy}\sin\theta\cos\theta \\
T_{R\phi} &= T_{xx}\sin\phi\cos\phi\cos^2\theta + T_{yy}\sin\phi\cos\phi\sin^2\theta - T_{zz}\sin\phi\cos\phi \\
&\quad + 2T_{xy}\sin\phi\cos\phi\sin\theta\cos\theta - T_{yz}(\sin^2\phi - \cos^2\phi)\sin\theta \\
&\quad - T_{zx}(\sin^2\phi - \cos^2\phi)\cos\theta \\
T_{\phi\theta} &= -T_{xx}\cos\phi\sin\theta\cos\theta + T_{yy}\cos\phi\sin\theta\cos\theta + T_{xy}\cos\phi(\cos^2\theta - \sin^2\theta) \\
&\quad - T_{yz}\sin\phi\cos\theta + T_{zx}\sin\phi\sin\theta \\
T_{\theta R} &= -T_{xx}\sin\phi\sin\theta\cos\theta + T_{yy}\sin\phi\sin\theta\cos\theta + T_{xy}\sin\phi(\cos^2\theta - \sin^2\theta) \\
&\quad + T_{yz}\cos\phi\cos\theta - T_{zx}\cos\phi\sin\theta
\end{aligned} \tag{B.9}
$$

Inverse transformations of these results can be computed by formally inverting the system equations or redeveloping the results using tensor transformation theory. For example, going from cylindrical to Cartesian would be accomplished by using the inverse of (B.1):

$$
\boldsymbol{Q} = \begin{bmatrix} \cos\theta & \sin\theta & 0 \\ -\sin\theta & \cos\theta & 0 \\ 0 & 0 & 1 \end{bmatrix}^{-1} = \begin{bmatrix} \cos\theta & \sin\theta & 0 \\ -\sin\theta & \cos\theta & 0 \\ 0 & 0 & 1 \end{bmatrix}^{T} = \begin{bmatrix} \cos\theta & -\sin\theta & 0 \\ \sin\theta & \cos\theta & 0 \\ 0 & 0 & 1 \end{bmatrix} \tag{B.10}
$$

APPENDIX

MATLAB Primer and Code Listings

At several locations in the text, numerical methods were used to calculate and plot solutions to a variety of continuum mechanics problems. Although other options are available, the author has found MATLAB software well suited to conduct such numerical work. This particular software has the necessary computational and plotting tools to enable very efficient and simple applications. MATLAB is a professional engineering and scientific software package developed and marketed by MathWorks, Inc. In recent years, it has achieved widespread and enthusiastic acceptance throughout the engineering community. Many engineering schools now require and/or use MATLAB as one of their primary computing tools. Its popularity is due to a long history of well-developed and tested products, to its ease of use by students, and to its compatibility across many different computer platforms. The purpose of this appendix is to present a few MATLAB basics to aid the reader in applying particular software applications and to list several of the codes used in the text. The software package itself contains an excellent *Help* package that provides extensive information on various commands and procedures, and many books are available on the software package. Also much information can simply be found on the web itself through standard keyword search procedures. It is assumed that the reader has some prior computational background and experience and thus has a basic understanding of programming techniques.

C.1 GETTING STARTED

MATLAB is both a computer programming language and a software environment for using the language. Under the MS Windows Operating System, the MATLAB window will appear as shown in Fig. C.1. It is from this window header bar that the *Help* menu can be accessed and this provides extensive information on most topics. In this command window, the user can type instructions after the prompt " > >". However, it is of course much more efficient to create and save application programs within the *Editor* window. This window is activated by going to the File menu in the Command window and selecting either *New Script* to start a new creation or *Open* to open an existing file. MATLAB files are called m-files and have the extension *.m. Within the Editor window, a new application code can be created or an existing one

Continuum Mechanics Modeling of Material Behavior. http://dx.doi.org/10.1016/B978-0-12-811474-2.00012-5

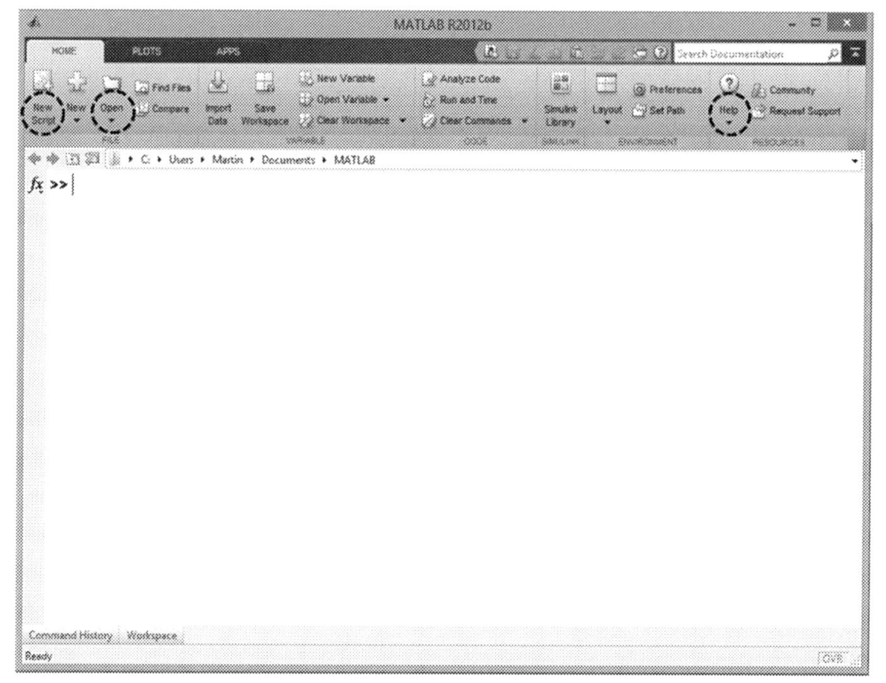

FIGURE C.1

MATLAB command window.

can be modified. In either case, the resulting file can then be saved for later use, and the current file can be run from this window. An example program appearing in the Editor window is shown in Fig. C.2.

C.2 EXAMPLES

We will not attempt a step-by-step explanation of various MATLAB commands, but rather will instead pursue a *learn-by-example* approach. In this fashion, most of the needed procedures will be demonstrated through the presentation of several example codes that have been previously used in the text. Listed codes typically illustrate: use of array I/O; standard calculation steps; various plotting and display schemes (Cartesian, polar, vector, etc.) with labeling methods; "for-end" looping; numerical integration; and a few other more specialized commands. Note that, in general, code lines preceded with a "%" symbol will not be executed and are used for comments to explain the coding. A semicolon ending a code line will suppress screen-printing of that particular calculation. The reader with previous programming experience should

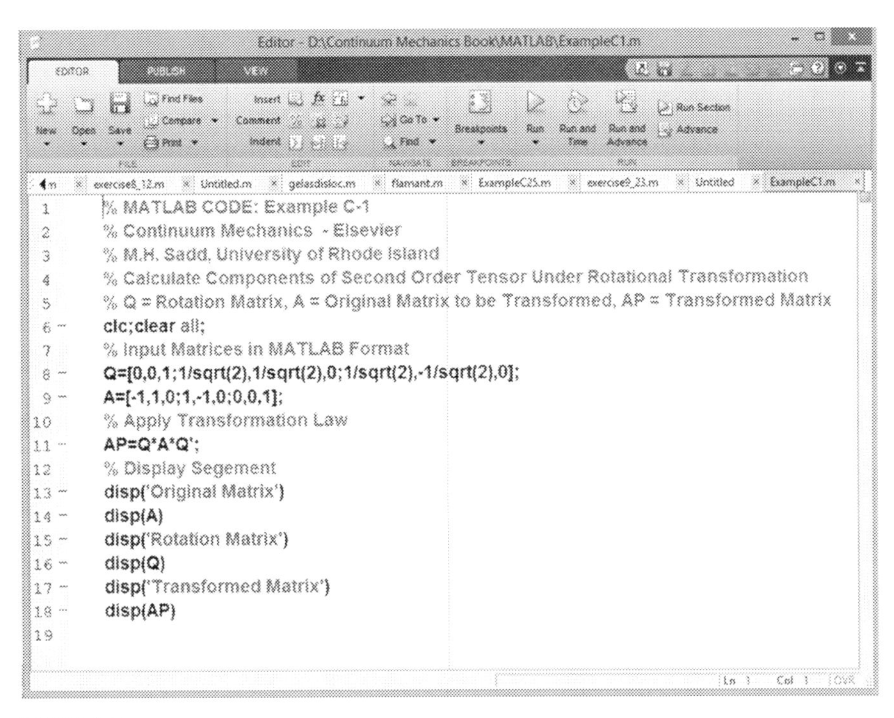

FIGURE C.2

MATLAB editor window.

be able to quickly review these examples and use them as is or make modifications to develop their own codes.

```
% MATLAB CODE: Example C-1
% Continuum Mechanics - Elsevier
% M.H. Sadd, University of Rhode Island
% Calculate Components of Second Order Tensor Under Rotational Transformation
% Q = Rotation Matrix, A = Original Matrix to be Transformed, AP = Transformed Matrix
clc;clear all;
% Input Matrices in MATLAB Format
Q=[0,0,1;1/sqrt(2),1/sqrt(2),0;1/sqrt(2),-1/sqrt(2),0];
A=[-1,1,0;1,-1,0;0,0,1];
% Apply Transformation Law
AP=Q*A*Q';
% Display Segement
disp('Original Matrix')
disp(A)
disp('Rotation Matrix')
disp(Q)
disp('Transformed Matrix')
disp(AP)
```

```
% MATLAB CODE: Example C-2
% Continuum Mechanics  - Elsevier
% M.H. Sadd, University of Rhode Island
% Numerically Calculate Invariants, Principal Values
% and Directions of a Matrix
% Program Uses Matrix from Example 2.11.1
clc;clear all;
% Input Matrix
A=[-1,1,0;1,-1,0;0,0,1]
% Calculate Invariants
invariants=[trace(A),(trace(A)^2-trace(A*A))/2,det(A)]
[V,L]=eig(A);
% Principal Values are the Diagonal Elements of the L Matix
principal_values=[L(1,1),L(2,2),L(3,3)]
% Principal Directions are the Columns of the V Matrix
principal_directions=[V(:,1),V(:,2),V(:,3)]
```

```
% MATLAB CODE: Example C-3
% Continuum Mechanics  - Elsevier
% M.H. Sadd, University of Rhode Island
% 2-D Deformation Examples: Unit Square and Circle
clc;clear all;clf
% Insert Unit Reference Square
figure(1)
X=[0,1,1,0,0];Y=[0,0,1,1,0];
fill(X,Y,[.95 .95 .95],'linestyle','--','linewidth',2)
axis equal;
axis([-0.5 3 -0.5 3])
hold on, grid on
% Add Deformed Element
xx=1.2*X+0.2*Y+1.2;yy=1.2*Y+0.2*X+1.2;
fill(xx,yy,[.8 .8 .8],'linewidth',2)
% Deformation of Unit Circular Area
ang=0:0.01:2*pi; r=0.5;
X=r.*cos(ang)+0.5;
Y=r.*sin(ang)+0.5;
figure(2)
fill(X,Y,[.95 .95 .95],'linestyle','--','linewidth',2)
axis equal;
axis([-0.5 3 -0.5 3])
hold on, grid on
% Add Deformed Circular Element
xx=1.2*X+0.2*Y+1.2;yy=1.2*Y+0.2*X+1.2;
fill(xx,yy,[.8 .8 .8],'linewidth',2)
```

```
% MATLAB CODE: C-4
% Continuum Mechanics  - Elsevier
% M.H. Sadd, University of Rhode Island
% Polar & Cartesian Plot Example 6.2.4 - Figure 6.13
clc;clear all;
% Input (r/a)- Variable and Generate Angular Coordinate Space
r=1;
t=[0:0.01:2*pi];
% Calculation Loop
st=0.5*(1+(1/r)^2)-0.5*(1+3*(1/r)^4)*cos(2*t);
% Plotting Call
figure(1)
polar(t,st,'k')
title('Non-Dimensional Hoop-stress Around Hole')
% Distance Decay Plot
r=[1:0.1:8];
st=0.5*(2+r.^(-2)+3*r.^(-4));
figure(2)
plot(r,st,'k','linewidth',2)
xlabel('Dimensionless Distance, \itr/\ita ')
ylabel('Dimensionless Stress, \itT_\theta_\theta/\itT')
grid on
```

```
% MATLAB CODE: C-5
% Continuum Mechanics  - Elsevier
% M.H. Sadd, University of Rhode Island
% Flamant Problem Example 6.2.5 - Figure 6.15 Plots
clc;clear all;
%Plot Cartesian Stresses
x=[-5:0.1:5];
sy=-2./(pi*(x.^2+1).^2);
txy=-(2.*x)./(pi*(x.^2+1).^2);
figure(1);grid on;hold on
plot(x,sy,'k','linewidth',2)
plot(x,txy,'k','linewidth',2)
xlabel('Dimensionless Distance, \itr/\ita ')
ylabel('Dimensionless Stress')
%Plot Radial Stress Contours
[x,y]=meshgrid(-2:0.1:2,0:0.1:4);
r=sqrt(x.^2+y.^2);
t=asin(y./r);
sr=-(2/pi).*sin(t)./r;
figure(2)
y=-y;
contour(x,y,sr,35,'k','linewidth',1.5)
axis equal
title('\itT_r_r Contours - Flamant Problem')
```

```
% MATLAB CODE: C-6
% Continuum Mechanics - Elsevier
% M.H. Sadd, University of Rhode Island
% Uniform Potential Flow Problem Example 6.3.1 - Figure 6.16 Plot
clc;clf;clear all
[x,y]=meshgrid(0:0.1:1,0:0.1:1);
% Plot velocity vector field (V=1)
vx=1.*x./x;vy=0.*x./x;
figure(1)
nscale=0.08;
quiver(x,y,vx*nscale,vy,'k','linewidth',1.5,'AutoScale','off')
xlabel('x');ylabel('y')
axis([0,1.2,-0.2,1.2])
title('Velocity Field for Uniform Flow (0<x,y<1)')
% Plot Stream & Potential function contours
s=y;p=x;
figure(2)
contour(x,y,s,10,'k','linewidth',1.5)
xlabel('x');ylabel('y')
hold on; axis equal
contour(x,y,p,10,'k--','linewidth',1.5)
title('Stream & Potential Function Contours for Uniform Flow (0<x,y<1)')
```

```
% MATLAB CODE: C-7
% Continuum Mechanics - Elsevier
% M.H. Sadd, University of Rhode Island
% Potential Flow in a Corner Example 6.3.2 - Figure 6.18 Plots
clc;clf;clear all
[x,y]=meshgrid(0:0.1:1,0:0.1:1);
% Plot velocity vector field
vx=x;vy=-y;
figure(1)
quiver(x,y,vx,vy,'k','linewidth',1.5)
xlabel('x');ylabel('y')
title('Velocity Field for Flow in a Corner (0<x,y<1)')
% Plot Stream & Potential function contours
[x,y]=meshgrid(0:0.05:1,0:0.05:1);
s=x.*y;p=0.5*(x.^2-y.^2);
figure(2)
contour(x,y,s,20,'k','linewidth',1.5)
xlabel('x');ylabel('y')
hold on; axis equal
contour(x,y,p,20,'k--','linewidth',1.5)
title('Stream & Potential Function Contours for Corner Flow (0<x,y<1)')
```

```
% MATLAB CODE: C-8
% Continuum Mechanics - Elsevier
% M.H. Sadd, University of Rhode Island
% Potential Flow Around Cylinder Example 6.3.3 - Figure 6.20 Plots
% Must Add Cylinder by Hand
clc;clf;clear all
[x,y]=meshgrid(-4:1:4,-4:1:4);
[t,r]=cart2pol(x,y);
% Plot velocity vector field
```

```
vr=(1-(1./r).^2).*cos(t);
vt=-(1+(1./r).^2).*sin(t);
vx=vr.*cos(t)-vt.*sin(t);
vy=vr.*sin(t)+vt.*cos(t);
figure(1)
nscale=0.4
quiver(x,y,vx*nscale,vy*nscale,'k','linewidth',2,'AutoScale','off')
xlabel('x');ylabel('y');axis equal
title('Velocity Field for Flow Around a Cylinder')
% Plot Stream function contours
[x,y]=meshgrid(-4:0.2:4,-4:0.2:4);
[t,r]=cart2pol(x,y);
s=r.*(1-(1./r).^2).*sin(t);
figure(2)
contour(x,y,s,20,'k','linewidth',1.5)
xlabel('x');ylabel('y');axis equal
title('Stream Function Contours for Flow Around a Cylinder')
```

```
% MATLAB CODE: C-9
% Continuum Mechanics - Elsevier
% M.H. Sadd, University of Rhode Island
% Viscous Flow Examples 6.4.1, 6.4.2, 6.4.3 Figure Plots
clc;clf;clear all
% Example 6.4.1 Velocity Distribution Plot
[x,y]=meshgrid(0,-1:0.1:1);
vx=1-y.^2;vy=zeros(size(y),1);
figure(1)
quiver(x,y,vx,vy,'k','linewidth',2,'AutoScale','off')
title('Velocity Field for Plane Poiseuille Flow')
hold on
plot(vx,y,'k--','linewidth',2)
axis([-0.5,1.5,-1,1])
% Example 6.4.2 Velocity Distribution Plots
y=[0:0.05:1];
for P=[-8,-4,0,4,8]
vx=y+P*(1-y).*y;
figure(2)
plot(vx,y,'k','linewidth',2)
hold on
end
title('Velocity Field for Plane Couette Flow')
% Example 6.4.3 Velocity Distribution Plots
[t,r,z]=meshgrid(0:pi/8:pi,0:0.1:1,0);
[x,y,z]=pol2cart(t,r,z);
u=0*(x.^2+y.^2+z.^2);
v=0*(x.^2+y.^2+z.^2);
w=1-r.^2;
figure(3)
scale=1;
quiver3(x,y,z,u,v,w*scale,'k','linewidth',2,'AutoScale','off')
hold on
T=[0:pi/50:2*pi];R=ones(size(T));Z=zeros(size(T))
X=cos(T);Y=sin(T)
plot3(X,Y,Z,'k','linewidth',2)
axis off
title('Velocity Field for Hagen-Poiseuille Flow')
```

```
% MATLAB CODE: C-10
% Continuum Mechanics  - Elsevier
% M.H. Sadd, University of Rhode Island
% Example 6.5.1 Figure Plots
clc;clf;clear all
t=[0:0.1:10];
for tau=[1,4,8]
  G=exp(-t/tau);
  figure(1)
  plot(t,G,'k','linewidth',2)
  hold on;grid on;
  xlabel('Time , t ')
  ylabel('Normalized Relaxation Function G(t) / E')
end
for tau=[1,4,8]
  J=(tau+t)/tau;
  figure(2)
  plot(t,J,'k','linewidth',2)
  hold on;grid on;
  xlabel('Time , t ')
  ylabel('Normalized Creep Function J(t) E')
end
```

```
% MATLAB CODE: C-11
% Continuum Mechanics  - Elsevier
% M.H. Sadd, University of Rhode Island
% Example 6.5.2 Figure Plots
clc;clf;clear all
t=[0:0.1:10];
G=ones(length(t));
figure(1)
plot(t,G,'k','linewidth',2)
hold on;grid on;
xlabel('Time , t ')
ylabel('Normalized Relaxation Function G(t) / E')
for tau=[1,4,8]
  J=1-exp(-t/tau);
  figure(2)
  plot(t,J,'k','linewidth',2)
  hold on;grid on;
  xlabel('Time , t ')
  ylabel('Normalized Creep Function J(t) E')
end
```

```
% MATLAB CODE: C-12
% Continuum Mechanics  - Elsevier
% M.H. Sadd, University of Rhode Island
% Example 6.5.4 Kelvin Model - Constant Loading Rate
% Solutions & Plots Using Numerical ODE Integrator
clc;clear all;clf
% Input Model Parameters
```

```
eta=100;E=50;
for R=[10,20,40]
% Use Anonymous Function
ode1=@(t,e)(R*t-E*e)/eta
% Call Solver With Anonymous Function Name,
% Independent Variable Range and Initial Value of Dependent Variable
[t,e]=ode45(ode1,[0:0.01:4],0);
% Plotting Segment
T=R*[0:0.01:4];
figure(1)
plot(e,T,'k','Linewidth',2)
xlabel('Strain, \epsilon'),ylabel('Stess, T')
title('Stress - Strain Response')
hold on;grid on
axis([0,0.45,0,60])
figure(2)
plot(t,e,'k','Linewidth',2)
xlabel('Time, t'),ylabel('Strain, \epsilon'),title('Strain-Time Response')
grid on; hold on
figure(3)
plot(t,T,'k','Linewidth',2)
grid on,hold on
xlabel('Time, t'),ylabel('Stress, T'),title('Stress-Time Response')
end
```

```
% MATLAB CODE: C-13
% Continuum Mechanics - Elsevier
% M.H. Sadd, University of Rhode Island
% Example 6.5.5 Kelvin Model Under Loading and Unloading
% Solutions & Plots Using Numerical ODE Integrator
clc;clear all;clf
% Input Kelvin Model & Loading Parameters
eta=100;E=200;
ode1=@(t,e)(60*t-E*e)/eta;
% Determine Loading Strain-Time Values
[t,e]=ode45(ode1,[0:0.001:1],0);
ee=e(end)
% Determine Loading Stress-Time Values
T=60*[0:0.001:1];
% Plot Loading Segment
plot(e,T,'k','Linewidth',2)
xlabel('Strain, \epsilon'),ylabel('Stess, T')
title('Stress - Strain Response')
hold on;grid on
axis([0,0.3,0,70])
% Determine Unloading Strain-Time Values
ode1=@(t,e)(120-60*t-E*e)/eta;
[t,e]=ode45(ode1,[1:0.001:2],ee);
% Determine Unloading Stress-Time Values
T=120-60*[1:0.001:2];
% Plot Unloading Segment
plot(e,T,'k','Linewidth',2)
```

```
% MATLAB CODE: C-14
% Continuum Mechanics - Elsevier
% M.H. Sadd, University of Rhode Island
% Plot Relaxation and Creep Functions
% For Three Parameter Viscoelastic Solid & Fluid Models
clc;clear all;clf
% Input Three Parameter Solid Model Parameters
p1=5;q0=0.5;q1=10;
t=[0:0.01:20];
% Three Parameter Solid Plot
G=(q1/p1)*exp(-t/p1)+q0*(1-exp(-t/p1));
J=(p1/q1)*exp(-q0*t/q1)+(1/q0)*(1-exp(-q0*t/q1));
figure(1)
plot(t,G,'k','linewidth',2)
grid on, hold on
plot(t,J,'--k','linewidth',2)
title('Three Parameter Solid Model')
xlabel('Time,\it t');ylabel('Relaxation and Creep Functions')
axis([0,20,0,2])
legend('Relaxation Function','Creep Function', 0)
% Three Parameter Fluid Plot
% Input Three Parameter Fluid Model Parameters
p1=5;q1=10;q2=10
G=(1/p1)*(q1-(q2/p1))*exp(-t/p1);
J=(t/q1)+(((p1*q1)-q2)/q1^2)*(1-exp(-q1*t/q2));
figure(2)
plot(t,G,'k','linewidth',2)
grid on, hold on
plot(t,J,'--k','linewidth',2)
title('Three Parameter Fluid Model')
xlabel('Time,\it t');ylabel('Relaxation and Creep Functions')
axis([0,20,0,2])
legend('Relaxation Function','Creep Function', 0)
```

```
% MATLAB CODE: C-15
% Continuum Mechanics - Elsevier
% M.H. Sadd, University of Rhode Island
% Example 6.6.1 Thick-Walled Cylinder Problem
% Plot ElastoPlastic In-Plane Stresses
clc;clear all;clf
for rp=[0.5,0.65,0.75,1];
  for k=50:100
r=k/100;
if r<rp
  r=[0.5:0.01:rp];
  Trr=-1+rp.^2-log((rp.^2)./r.^2);
  Ttt=1+rp.^2-log((rp.^2)./r.^2);
else
  r=[rp:0.01:1];
  Trr=-(rp.^2)./(r.^2)+rp.^2;
  Ttt=(rp.^2)./(r.^2)+rp.^2;
```

```
end
plot(r,Trr,'k', 'linewidth', 2)
hold on
plot(r,Ttt,'k--', 'linewidth', 2)
xlabel('Dimensionless Distance, r/r_2')
ylabel('Dimensionless Stress')
grid on;
   end
end
```

```
% MATLAB CODE: C-16
% Continuum Mechanics  - Elsevier
% M.H. Sadd, University of Rhode Island
% Example 6.6.2 Plastic Torsion Example Plot
clc;clear all;clf
a=[1.0:0.1:3];
T=1-0.25*a.^(-3);
plot(a,T,'k', 'linewidth', 2)
ylabel('Dimensionless Torque , T/T_U')
xlabel('Dimensionless Angle of Twist , \alpha/\alpha_p')
grid on
```

```
% MATLAB CODE: C-17
% Continuum Mechanics  - Elsevier
% M.H. Sadd, University of Rhode Island
% Example 7.2.3, Thermal Stresses in Annular Plate
clc;clear all;clf
R=3;
r=[1:0.01:R];
Tr=(1/(2*log(R)))*(-log(R./r)-(R^2-1)^(-1)*(1-(R^2./r.^2))*log(R));
Tt=(1/(2*log(R)))*(1-log(R./r)-(R^2-1)^(-1)*(1+(R^2./r.^2))*log(R));
T=log(R./r)./log(R);
figure(1)
plot(r,T,'k','linewidth',2)
grid on
xlabel('Dimensionless Radial Distance, r/r_i')
ylabel('Dimensionless Temperature, \theta/\theta_i')
figure(2)
plot(r,Tr,'k--','linewidth',2)
hold on;grid on
plot(r,Tt,'k','linewidth',2)
xlabel('Dimensionless Radial Distance, r/r_i')
ylabel('Dimensionless Stress')
```

```
% MATLAB CODE: C-18
% Continuum Mechanics  - Elsevier
% M.H. Sadd, University of Rhode Island
% Poroelasticity Example 7.3.1
% Pore Pressure and Surface Displacement
clc;clear all;clf
% Pore Pressure
t=[0:0.001:0.5];
for m=1:3;
   x=m/4;
p=0;
for n=1:2:25
p=p+(4/(n*pi))*sin(n*pi*x/2).*exp(-n^2*pi^2*t);
end
figure(1)
plot(t,p,'k','linewidth',2.2)
xlabel('Dimensionless Time, \tau');
ylabel('Dimensionless Pore Pressure')
%title(['Temperature Distribution Solution , Time = ',num2str(t)])
axis([0,0.5,0,1])
grid on , hold on
end
u=0;
for n=1:2:25
u=u+(8/(n^2*pi^2))*(1-exp(-n^2*pi^2*t));
end
figure(2)
U=1*(1+1*u)
plot(t,U,'k','linewidth',2.2)
xlabel('Dimensionless Time, \tau');
ylabel('Dimensionless Surface Displacement, u(0,t)/K_1')
grid on
```

```
% MATLAB CODE: C-19
% Continuum Mechanics  - Elsevier
% M.H. Sadd, University of Rhode Island
% Poroelasticity Example 7.3.2
% Pore Pressure Calculation & Plot
clc, clear all
% Input Material Properties
n=0.2;nu=0.5;B=1;
M=(1-n)/(nu-n);
x=[0:0.02:20];
f1=tan(x);f2=M*x;
figure(1)
plot(x,f1,x,f2)
grid on
axis([16 24 0 80])
% Collect roots from graph
%xr=[1.45,1.57,4.67,4.71,7.828,7.842]
xr=[1.3,4.63,7.8,10.97,14.107,17.2];
% Pore Pressure Calculation & Plot
```

```
A1=1/(B*(1-nu));
for t=[0.01,0.1,0.5,1];
x=[0:0.01:1];
p=0;
for n=1:6
   Bn=sin(xr(n))/(xr(n)-sin(xr(n))*cos(xr(n)));
   p=p+Bn*(cos(xr(n)*x)-cos(xr(n)))*exp(-xr(n)^2*t);
end
P=2*p/A1;
figure(2)
plot(x,P,'k','linewidth',2)
%axis([0 1 0 0.4])
grid on, hold on
xlabel('Dimensionless Distance, ')
ylabel('Dimensionless Pore Pressue, p/p_o')
end
```

```
% MATLAB CODE: C-20
% Continuum Mechanics  - Elsevier
% M.H. Sadd, University of Rhode Island
% Example 8.3.1 Uniaxial Nonlinear Elastic Response
% Calculation & Plot Cauchy and PK1 Stresses
clc;clear all ;clf
% Piola-Kirchhoff Stress Plots
% Mooney-Rivlin Material
a1=0.5;a2=0.3;
L=1:0.01:4;
To=(L-(1./(L.^2))).*(a1-(a2./L));
figure(1)
plot(L,To,'k','linewidth',2)
xlabel('Stretch Ratio, \lambda')
ylabel('Piola-Kirchhoff Axial Stress')
hold on
% Neo-Hookean Material
To=a1*(L-(1./(L.^2)));
plot(L,To,'k--','linewidth',2)
legend(['Mooney: \alpha_1 = ',num2str(a1),' \alpha_2 = ',num2str(a2)]...
   ,['Neo-Hookean: \alpha_1 = ',num2str(a1),' \alpha_2 = 0'],2)
grid on
% Cauchy Stress Plots
% Mooney-Rivlin Material
T=(L.^2-(1./L)).*(a1-(a2./L));
figure(2)
plot(L,T,'k','linewidth',2)
xlabel('Stretch Ratio, \lambda')
ylabel('Cauchy Axial Stress')
hold on
% Neo-Hookean Material
T=a1*(L.^2-(1./L));
plot(L,T,'k--','linewidth',2)
legend(['Mooney: \alpha_1 = ',num2str(a1),' \alpha_2 = ',num2str(a2)]...
   ,['Neo-Hookean: \alpha_1 = ',num2str(a1),' \alpha_2 = 0'],2)
grid on
```

```
% MATLAB CODE: C-21
% Continuum Mechanics  - Elsevier
% M.H. Sadd, University of Rhode Island
% Uniaxial Nonlinear Elastic Response
% Calculation & Plot Mooney, Neo-Hookean, Material X Stresses
clc;clear all ;clf
L=1:0.1:2.5;
% Cauchy Stress Plots
% Mooney-Rivlin Material
% Specify Mooney Parameters
a1=??;a2=??;
T=(L.^2-(1./L)).*(a1-(a2./L));
plot(L,T,'k:','linewidth',2.5)
xlabel('Stretch Ratio, \lambda')
ylabel('Cauchy Axial Stress, T_1_1')
hold on
% Neo-Hookean Material
% Specify Neo-Hookean Parameter
a1=??;
T=a1*(L.^2-(1./L));
plot(L,T,'k--','linewidth',2.5)
% Material X Linearized Data Fit
% Specify Linear Equation Fit
Lfit=??;
T=(L.^2-(1./L))*2.*Lfit;
plot(L,T,'k','linewidth',2.5)
legend(['Mooney: \alpha_1 = ',num2str(a1),' \alpha_2 = ',num2str(a2)]...
   ,['Neo-Hookean: \alpha_1 = ',num2str(a1),' \alpha_2 = 0']...
   ,['Material X (Linearized Data Fit)'],2)
grid on
```

```
% MATLAB CODE: C-22
% Continuum Mechanics  - Elsevier
% M.H. Sadd, University of Rhode Island
% Example 8.3.1 Uniaxial Nonlinear Elastic Response
% Calculation & Plot Mooney-Rivlin & Rivlin & Saunders Test Data
clc;clear all; clf
% Load and Plot Data
d=load('elasticdata.txt')
figure(1)
plot(d(:,1),d(:,2),'o','markersize',7,'markerfacecolor','k')
axis([0.2,1,1.7,2.5])
xlabel('1/\lambda'); ylabel('(\alpha_1-\alpha_2/\lambda)/2')
title('Finite Elastic Tensile Data, Rivlin & Saunders (1951)')
hold on
grid on
% Re-Plot and Curve Fit Data
L=d(:,1).^-1;
figure(2)
plot(L,d(:,2),'o','markersize',7,'markerfacecolor','k')
axis([1,4,1.7,2.5])
xlabel('\lambda'); ylabel('(\alpha_1-\alpha_2/\lambda)/2')
title('Polynomial Curve Fit To Tensile Data')
hold on
grid on
```

elasticdata.txt

```
0.27 1.90
0.28 1.88
0.30 1.86
0.31 1.84
0.33 1.83
0.35 1.835
0.375 1.84
0.4 1.85
0.43 1.88
0.47 1.91
0.52 1.95
0.56 2.0
0.625 2.08
0.68 2.15
0.74 2.24
0.79 2.28
0.85 2.31
0.91 2.41
```

```
p=polyfit(L,d(:,2),2)
xp=1:0.1:4;
yp=polyval(p,xp)
plot(xp,yp,'k','linewidth',2.5)
% Plot Stress-Strain Response Using Curve Fit Constitutive Function
T=(xp.^2-(1./xp)).*2.*yp;
figure(3)
plot(xp,T,'k','linewidth',2.5)
xlabel('Stretch Ratio, \lambda')
ylabel('Cauchy Axial Stress, T_1_1')
title('Stress - Deformation Behavior for Finite Elastic Material')
grid on; hold on
% Plot Mooney-Rivlin Prediction
L=1:0.1:4;a1=4;a2=0.3
T=(L.^2-(1./L)).*(a1-(a2./L));
plot(L,T,'k--','linewidth',2.5)
```

```
% MATLAB CODE: C-23
% Continuum Mechanics  - Elsevier
% M.H. Sadd, University of Rhode Island
% Calculation & Plot Gradient Elasticity Dislocation Example 9.5.1
clc;clear all;clf
for n=[1,3,10]
c=.01*n;
r=0.001:0.01:3;
ege=abs((1/(4*pi))*(-(1./r)+(1/sqrt(c)).*besselk(1,r/sqrt(c))));
ece=abs((1/(4*pi))*(-(1./r)));
plot(r,ege,'k-','linewidth',2)
hold on
plot(r,ece,'k--','linewidth',2)
axis([0,3,0,0.5]);grid on
xlabel('Distance, r'); ylabel('| \epsilon_x_z / b |')
title('Strain Field Near Screw Dislocation')
% text(0.03,0.15,'Gradient Elasticity, c=0.01')
% text(0.25,0.35,'Classical Elasticity')
legend('Gradient Elasticity','Classical Elasticity')
end
```

```
% MATLAB CODE: C-24
% Continuum Mechanics  - Elsevier
% M.H. Sadd, University of Rhode Island
% Example 9.5.2 Gradient Elasticity
% Vertical Displacement - Flamant Problem
clc;clear all;clf
c=.1;nu=0.3;cs=sqrt(c);
x=-5:0.01:5;X=abs(x);
Vy=log(X*cs)-0.6;
plot(x,Vy,'k-','linewidth',2)
hold on
Vyg=(log(X*cs)+besselk(0,X))-(1-2./(X.^2)+besselk(2,X))/(2*(1-nu));
plot(x,Vyg,'k--','linewidth',2)
axis([-5,5,-3,0]);grid on
legend('Classical Elasticity','Gradient Elasticity')
xlabel('Dimensionless Distance, x/sqrt(c)');
ylabel('Dimensionless Vertical Displacement')
```

```
% MATLAB CODE: C-25
% Continuum Mechanics - Elsevier
% M.H. Sadd, University of Rhode Island
% Example 9.7.1 Uniaxial Damage Mechanics Model Plot
clc;clear all;clf
for m=[1,1.5,2]
for n=1:100
   e(n)=n/50;
   D(n)=1-exp(-m*e(n));
   s(n)=(1-D(n))*e(n);
end
figure(1)
plot(e,s,'k-','linewidth',2)
xlabel('Strain, \epsilon')
ylabel('Nondimensional Stress, \sigma/E')
grid on; hold on
figure(2)
plot(e,D,'k-','linewidth',2)
xlabel('Strain, \epsilon')
ylabel('Damage Parameter, D')
grid on; hold on
end
```

APPENDIX

Poem

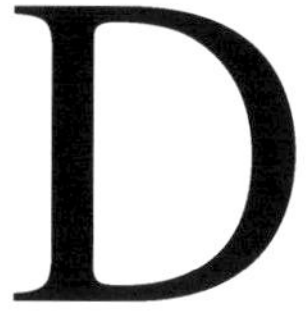

The following poem was written by the author's doctoral advisor at the Illinois Institute of Technology sometime during the 1960s. It is offered as a light-hearted satirical view of continuum mechanics, rheology, and academics. The poem is probably best appreciated after having completed some of the material in the text.

RHEOL-LOGIC

The tensor tramps across the page,
And fields of functions stage by stage
Inform the reader in the know
Of how the stress affects the flow.

No cone or plate need shear a goo
To find the laws of nature true,
But all is known to those who think
And nothing flows but printers' ink.

For by the rules of logic rheol,
To which the erudite appeal,
By word and wit and cogitation
Come true and trusty relevation.

You need not stir a can of paint
To find what is or see what ain't.
Jusy follow through the mathematics
Of stress and flow and creep and statics.

Yet stay before you ope the books,
It's not as easy as it looks.
No spring, no dashpot leads you on
No molecules to lean upon.

Just tensors, fields, and energies,
Just talk and inequalities
In utmost generality.
All else is but banality.

How many heads have spun and reeled
Confronted by a classic field?
How oft have students felt upset
By new, unheard-of alphabet?

The rational assured mechanic
May drive a simple soul to panic,
When concepts new and old are flung
Both here and there in Caesar's tongue.

But hold before you seek a tryst
To meet your psychoanalyst.
Just learn the ways of secret cults
To seek some new and grand results.

New terms you think to add and state
You try your best to complicate,
And should a problem make you weary
You turn instead to newer theory.

And so you think you wield your pen
You write and publish, talk, and then
When all is done and seems perfection,
You print additions and corrections.

... Barry Bernstein

Continuum Mechanics Modeling of Material Behavior. http://dx.doi.org/10.1016/B978-0-12-811474-2.00013-7

Index

D

E

F

S

T

V

W

Y

Printed in the United States
By Bookmasters